Standardization and Tacit Knowledge

WILEY SERIES IN SURVEY METHODOLOGY
Established in Part by WALTER A. SHEWHART AND SAMUEL S. WILKS

Editors: *Robert M. Groves, Graham Kalton, J. N. K. Rao, Norbert Schwarz, Christopher Skinner*

A complete list of the titles in this series appears at the end of this volume.

Standardization and Tacit Knowledge

Interaction and Practice in the Survey Interview

EDITORS

DOUGLAS W. MAYNARD
HANNEKE HOUTKOOP-STEENSTRA
NORA CATE SCHAEFFER
JOHANNES VAN DER ZOUWEN

A Wiley-Interscience Publication
JOHN WILEY & SONS, INC.

This book is printed on acid-free paper. ♾

Published simultaneously in Canada.

For ordering and customer service, call 1-800-CALL-WILEY.

Library of Congress Cataloging-in-Publication Data:

Standardization and tacit knowledge: interaction and practice in the survey interview/ editors, Douglas W. Maynard ... [et al.].
p. cm.—(Wiley series in probability and statistics. Survey methodology section)
Includes bibliographical references and index.
ISBN 0-471-35829-0 (cloth)
1. Social sciences–Research. 2. Social surveys. 3. Interviewing. I. Maynard, Douglas W., 1946—II. Series.

H62.S744 2002
001.4′33–dc21 2001024027

Printed in the United States of America.

10 9 8 7 6 5 4 3 2 1

Contributors

Frederick G. Conrad Bureau of Labor Statistics, 2 Massachusetts Avenue N.E., Room 4915, Washington, DC 20212

Mick P. Couper Joint Program in Survey Methodology, University of Maryland, 1218 Lefrak Hall, College Park, MD 20742

Wil Dijkstra Social Research Methodology, Free University Amsterdam, De Boelelaan 1081C, 1081 HV Amsterdam, The Netherlands

Marek Fuchs Catholic University of Eichstatt, Department of Sociology, Ostenstrasse 26, D-85071 Eischstatt, Germany

Robert M. Groves Joint Program in Survey Methodology, University of Maryland, 1218 Lefrak Hall, College Park, MD 20742

Tony Hak Department of Methodology, Faculteit Bedrijfskunde, Rotterdam School of Management, Erasmus University, P.O. Box 1738, Room F3-54, 300 DR Rotterdam, The Netherlands

John Heritage Department of Sociology, University of California, 405 Hilgard Avenue, Los Angeles, CA 90024-1551

Hanneke Houtkoop-Steenstra Utrecht Institute of Linguistics, Utrecht University, Trans 10, 3512 JK, Utrecht, The Netherlands

Danielle Lavin Department of Sociology, Indiana University, Bloomington, IN 47405

Michael Lynch Science and Technology Studies, Cornell University, 622 Clark Hall, Ithaca, NY 14853-2501

Douglas W. Maynard Department of Sociology, University of Wisconsin, Madison, WI 53706

Robert J. Moore Xerox Palo Alto Research Center, 3333 Coyote Hill Road, Palo Alto, CA 94304

Nora Cate Schaeffer Department of Sociology, University of Wisconsin, Madison, WI 53706

Emanuel A. Schegloff Department of Sociology, University of California, 405 Hilgard Avenue, Los Angeles, CA 90024-1551

Michael F. Schober Department of Psychology, New School for Social Research, 65 Fifth Avenue, New York, NY 10003

Huub van den Bergh Utrecht Institute of Linguistics, Utrecht University, Trans 10, 3512 JK Utrecht, The Netherlands

Johannes van der Zouwen Social Research Methodology, Free University Amsterdam, De Boelelaan 1081C, 1081 HV Amsterdam, The Netherlands

Jocelyn S. Viterna Department of Sociology, Indiana University, Bloomington, IN 47105

Contents

Preface

Survey researchers have had a longstanding interest in interactional facets of the interview. Many subfields within survey methodology — those concerned with response processes, instrument design, and problems of nonresponse — have looked to psychology and anthropology for theories and results to inspire, guide, and improve their work. Survey researchers who are interested in the interviewing process itself and its effects on measurement error have been somewhat slower to locate and collaborate with colleagues in related fields. Interaction-based researchers, on the other hand, have had longstanding concerns with work settings of many kinds. Given the ubiquity of the survey interview, and the number of survey centers that dot the international landscape, it is surprising how infrequently interactionists and ethnomethodologists have visited these centers with their cameras, tape recorders, and procedures of investigation. So from both directions — survey methodology and interaction-based inquiry — this volume and its cross-disciplinary approach to the survey interview represents a significant step.

Collaborations across disciplines are likely to be the most productive when kindred problems are of keen mutual interest, when the same data are mutually workable, and when the relevance and insights of distinct methodologies are mutually appreciated. Interdisciplinary collaborations are risky, however, because one party's jargon may seem overblown and its findings obvious to the other. Or misunderstandings may occur because the two parties have starkly different ways of thinking about, theorizing, and studying phenomena. However, when both parties take the risk, put aside stereotypes of the other's work, and attempt to enter research worlds different from their own, the fruits can be rich and plentiful. We believe this volume represents a unique dialog and presents an array of delightful scholarship regarding the organization of interaction in the survey interview. The chapters are from a variety of traditions, and the researchers have energetically and patiently worked to understand one another as the project made progress. Our readers will now be faced with the same challenge. We think that those who make this effort will, like us, find the different perspectives to be intriguing and the results to be exceptionally useful in promoting knowledge about the interview process.

At the same time, we hope that this volume represents a beginning rather than an end. The project grew from an international workshop on "Interviewer–Respondent Interaction in the Standardized Survey Interview" held in Amsterdam in November 1995. The original idea for this workshop came from Hanneke Houtkoop-Steenstra, and the goal was to bring together people working in the domains of conversation analysis, ethnomethodology, and survey methodology, to see how they could learn from one another. Organized by the editors of this volume, the workshop received financial support from the USF-Research Fund of the Vrije Universiteit in Amsterdam.

The editors represent different perspectives on and interests in studying the survey interview. Two of the editors (van der Zouwen and Schaeffer) are comfortable being called survey methodologists; the other two accept the label ethnomethodologist or conversation analyst. None of us, however, would accept any label as summarizing our full range of interests or training, which may be why we had this unusual convergence of research interest. Each chapter in this volume was reviewed by (at least) one of the "survey methodologist" editors *and* by one of the ethnomethodologist or conversation analyst editors. We wanted to help ensure that our very diverse authors would tread carefully on new territory and not awaken the ire of the native inhabitants. We desire contacts that are stimulating and illuminating rather than simply irritating.

Participants in the workshop were invited to analyze a set of interviews collected at the University of Wisconsin Survey Center (UWSC, which was then the Letters and Sciences Survey Center) during the summer of 1995. Each of 15 interviewers was asked to record two interviews (consecutive interviews if possible) on his or her shift. The instrument the interviewers fielded was the Wisconsin Continuous National Survey, a small, daily RDD (random digit dialing) sample of the continental United States. One recording failed, and a final set of 29 interviews was available. Six of the interviews were transcribed in detail and the balance were roughly transcribed. The interviews in this corpus are referred to with the prefix AW in the chapters in this volume, and one is reproduced in its entirety in the appendix. The survey instrument used in these interviews included questions on a variety of topics written or selected (from previous surveys) by several investigators. Chris Fassnacht helped to oversee and coordinate the process, and Chris Fassnacht, Robert Moore, and Lynn Wimer helped with the transcription. All names and other identifying information in these (and other transcripts of survey interviews in this volume) are fictitious. We make no claims that the practice represented in these interviews (either in the design of the instruments or in the practice of the interviewing) provides either a positive or negative example for the discipline, and we hope that no readers will interpret them in that way. They are, simply, the examples that were available to us, and they are probably neither unique nor completely typical of practice in academic survey research centers.

Coincidentally, after this project was initiated, Doug Maynard became the director of the Sociological Research Practicum (SRP) in the Department of Sociology at Indiana University. Several of the chapters in this volume are an outgrowth of the SRP, which is a device for immersing new graduate students in research. Each year a different director (from within the Indiana University Sociology Department) designs a research effort in line with his or her current interests. When Maynard directed it, the SRP was called "The Ethnography of a Survey Research Center."

Particular gratitude goes to Dr. John Kennedy, the Director of the Indiana University Center for Survey Research, who facilitated the ethnography, including video and audio-recording of interviews. Data collection also was aided by the generous support of the Institute for Research on Learning, in Palo Alto, CA; Jack Whalen, Marilyn Whalen, and Erik Vinkhuyzen (now at Xerox Palo Alto Research Center) were instrumental and indispensable. We thank Jeremy Freese, Bob Moore, and Robert Soto for valuable research assistance, and the participants in the SRP for stimulus, discussion, transcription, and research. These participants included Lisa Aucoin, Kim Darnall, Laura Fingerson, Todd Goodwin, Katy Hadley, Thom Heil, Jocelyn Hopper, Adam King, Danielle Lavin, Jeni Loftus, Chad Menning, Stacy Scherr, Jason Schnittker, Jocelyn Viterna, and Shari Woodbury. The SRP is generously supported financially by the College of Arts and Sciences at Indiana University.

We thank our colleagues and our home institutions for their support of this work in ways small and large. A final word of gratitude to Mike Horgan who worked long, patiently, and meticulously to prepare the manuscript for submission.

DOUGLAS W. MAYNARD
HANNEKE HOUTKOOP-STEENSTRA
NORA CATE SCHAEFFER
JOHANNES VAN DER ZOUWEN

Standardization and Tacit Knowledge

PART 1

Theoretical Orientations

CHAPTER 1

Standardization and Its Discontents

Douglas W. Maynard and Nora Cate Schaeffer

Surveys use questions to collect measurements from a sample in order to estimate characteristics of a population. The "objective" data obtained by survey interviews are achieved in the interaction between respondents and interviewers who have been trained to behave in a standardized manner. Over the years, both critics (e.g., Briggs, 1986; Cicourel, 1974; Mishler, 1986; Suchman and Jordan, 1990) and practitioners (e.g., Kahn and Cannell, 1957) of standardized survey interviewing have considered how the fundamentally social nature of the interview affects the data produced there. Both groups consider the interaction between the interviewer and respondent when they evaluate the quality of survey data (e.g., Cicourel, 1982; Schuman, 1982). Thus, critics have presented examples of the awkward, or even bizarre, interactions that sometimes occur in survey interviews as evidence that the resulting data cannot be valid. Practitioners view such individual incidents as resulting from unusual circumstances or "bad" survey questions,[1] but they consider that adherence to rules of standardization improves the overall quality of the data in the aggregate (e.g., Cannell, Lawson, and Hausser, 1975).

Practitioners and critics of surveys both view respondents' answers as constructed during the interaction, but they differ in how they view the effects on the resulting data.[2] Thus, a critic might argue, for example, that the features of the interview itself—the interviewer's characteristics, her behavior, and the survey question—are such strong influences on the answers that the answers respondents provide are produced by the process of the interview and cannot describe a reality outside the interview. Survey practitioners might argue that the way answers are constructed in a survey interview is similar to the way answers are constructed in other social situations—"in surveys as in life" (Schuman

Standardization and Tacit Knowledge: Interaction and Practice in the Survey Interview
Edited by Douglas W. Maynard, Hanneke Houtkoop-Steenstra, Nora Cate Schaeffer, and Johannes van der Zouwen
ISBN 0-471-35829-0

and Ludwig, 1983)—and that the validity of measurement in surveys must be assessed using traditional scientific criteria, such as the ability to predict an outcome.

In the last decade, survey researchers and other social scientists have revisited the debate about standardization (e.g., Beatty, 1995), a debate that has a long history (O'Muircheartaigh, 1997). The latest round of the debate has been accompanied by a new set of studies of interaction in the survey interview. Some of these studies extend traditional methods used by survey researchers for studying the interview by experimenting with styles of interviewing and analyzing detailed coding of the interaction between the interviewer and respondent. Other studies, deriving from research about work and the production of scientific knowledge, have used concepts and methods from ethnomethodology and conversation analysis.

In this introductory chapter, we first review the justification for standardization in the survey interview: We summarize recent debates, discuss different varieties of standardization, and consider the limitations and limits of standardization. We subsequently argue for studying interaction in the survey interview using an approach we call analytic alternation. By considering the limits of standardization, we draw attention to the role of tacit and commonsense knowledge in the conduct of the survey interview. Because of the tension between the procedures for social measurement and the practices of talk and interaction, interviewers *alternate* between following the rules of standardization and using the tacit knowledge available to competent social actors who must solve problems that arise as their work tasks unfold. *Analytic* alternation means following the situated oscillations between formal rule-following and tacit practices that interviewers enact in concert with survey respondents.

In the second section of this chapter, we outline the role of tacit knowledge in survey interviewing and, to illustrate the deployment of such knowledge, we discuss *interviewing sequences* by which the survey instrument is administered. The third section of the chapter reviews recent studies of interaction in the survey interview. Some of these studies involve coding behavior to examine the effects of different techniques for questioning or probing, other studies are experiments that test how the style of interviewing affects the respondent's motivation, understanding, and other outcomes; and yet another group of studies uses ethnomethodological and conversation analytic approaches to examine the detail of situated work methods and the practices of talk and social interaction. We differentiate these streams of research according to whether the interviewer's questioning tactics are seen as unilateral productions or as bilateral and collaborative productions. As an example of the bilateral approach, we examine the phenomenon of *uncertainty* and how interviewer and respondent deal with it.

In the fourth section to the chapter, we provide a synopsis of the sociological subfields of ethnomethodology and conversation analysis. We do this because these subfields are represented in many of the chapters in the volume and because a number of relevant studies have been published elsewhere. In either case, the subfields and studies may be unfamiliar to some of our readers. Finally, we consider the organization of interaction in the recruitment of respondents and the challenges that research on interaction in the survey interview poses for improving measurement in the social sciences.

JUSTIFICATION FOR STANDARDIZATION: WHY AND WHAT

The justification for standardization lies in the logic of scientific measurement applied to social phenomena. Investigators who analyze survey data would like the variation they observe in data to result from variation in the concept being measured. For example, if various respondents report different levels of self-esteem, the levels recorded in the data should reflect true differences among respondents. If interviewers behave in different ways with different respondents, the distribution of answers may reflect variation in the behavior of interviewers rather than in levels of self-esteem. In addition, interviewers can introduce errors — for example, by their expectations that respondents will have consistent attitudes, or that respondents with specific social characteristics will have certain beliefs — that can affect the accuracy of responses. This is sometimes referred to as biasing answers. The model that divides total error into variance and bias has dominated the way statisticians think about error: "interviewer variance represents the error about the 'expected value' for all the interviewers, while net interviewer bias represents the deviation of this expected value from the true population mean" (Hyman, 1975 [1954]:322).

The varied practices that researchers refer to as "standardization" evolved in response to early studies, such as those described by Hyman (1975 [1954]), demonstrating how the behavior of interviewers affects "error" in survey estimates. Ratings made by interviewers and the number of responses recorded to open questions were substantially affected by interviewer variability (Groves, 1989:380–381; Hyman, 1975 [1954]:257). Individual studies also found substantial errors (e.g., introduced through the expectations or reactions of the interviewer, etc.) that could be present in the work of many interviewers but not increase interinterviewer variation (Hyman, 1975 [1954]:252–274). Summarizing the effects of the interviewer on the distribution of answers, Hyman (1975 [1954]:271) says, "the only reasonable answer seems to be that absolutely anything can happen."

Standardization addresses both components of interviewer error. The rules of standardization attempt to hold the behavior of the interviewer constant, thereby reducing variable error. Recent studies in centralized telephone facilities where standardization is practiced, for example, find that the component of variance due to differences in the behavior of the interviewer across respondents (interviewer variance) is usually quite small for most types of survey items (e.g., Mangione, Fowler, and Louis, 1992; Groves and Magilavy, 1986). This is so even though there are many lapses in achieving the ideal of standardization (e.g., Oksenberg, Cannell, and Kalton, 1991). When standardization is successful, it makes the effect of the interviewer in the aggregate very small, in that it reduces the interviewers' contribution to variance. Standardization also reduces bias by reducing the number of opportunities for the interviewer's expectations or opinions to intrude on the process by which the respondent's answer is generated, interpreted, or recorded.

Standardization does not eliminate the effect of the individual interviewer on the individual respondent. An interviewer who follows the rules of standardization might react to a respondent's ambiguous answer by repeating all the response categories, for example. From the point of view of standardization, this would be preferable to alternative behaviors such as using her beliefs about the respondent's likely answer to choose which response categories to repeat. But the standardized repetition of the response categories influences how respondents express their answers, and, over the course of the interview, probably has the desirable effect of training respondents to choose a category from among those offered. This is clearly an effect of the individual interviewer on the individual respondent. Nevertheless, if standardization were comprehensive and perfectly implemented, we could say that the interviewer did not affect responses — in the very specific sense that the effect of any interviewer would be the same as the effect of any other interviewer. That is, the interviewers are interchangeable.

Recent Debate about Standardization

Hyman (1975 [1954]:20–21) himself eloquently expressed the concern that researchers might focus on improving reliability, and neglect the less tractable problem of finding methods for increasing validity:

> The twin goals of reduction of inter-interviewer variation and an increase in the validity of the results must always be kept in mind... Evaluations oriented purely to the reliability problem... run the danger of conservatism because the standard against which any interviewer's performance is appraised is that of another *current* interviewer, or that of all current interviewers.... It is only as we have as a norm a form of interviewing that approximates close to valid results, that we become radical and experimental. It must be the neglect of this latter concept of interviewer error that accounts for the rarity of innovation.

Decades later, Suchman and Jordan (1990) — something like Briggs (1986), Cicourel (1974), and Mishler (1986) — argued that standardization is itself a substantial threat to the validity of the resulting data. Suchman and Jordan (1990) presented illustrations in which the rigidities of standardization led to awkward interactions and answers that might be inaccurate. They proposed letting both the respondent and the interviewer see the instrument and allowing "the interviewer to talk about the questions, to offer clarifications and elaborations, and to engage in a limited form of recipient design and common-sense inference" (Suchman and Jordan, 1990:40).

Subsequent discussions of standardization among survey practitioners have been influenced by studies that examine social–information-processing models of the process of answering a survey question (e.g., Sudman, Bradburn, and Schwarz, 1996; Tourangeau, Rasinski, and Rips, 2000). Much of the research developed within this framework has focused on questions about events and behaviors and how to improve the respondent's comprehension of survey questions and increase the accuracy of the information a respondent retrieves from memory. Some of this research has experimented with relaxing the constraints of standardization to improve retrieval (e.g., Means et al., 1991). More recent studies have modified standardization to help respondents understand survey concepts.

Nevertheless, suggestions for loosening the constraints of standardization in an effort to improve validity raise the anxieties of many survey researchers. The concerns of survey practitioners include, for example, that questions about things that are more *subjective*, like attitudes, require different measurement practices than questions about things that are more *objective*, such as events and behaviors; that a system of standardization must be suitable for the requirements of recruiting, training, and supervising interviewers for very large surveys; and that innovations not increase interviewer variability. Discussions about standardization in the last decade have focused researchers' attention on several topics, which are examined in the studies we review later: how standardized interviewers actually behave, how their behavior affects the answers given by respondents, how conversational practices from other sorts of talk might be used in standardized interviewing, and how standardization might be changed to improve both the quality of the interaction and that of the resulting data.

Standardization: Varieties and Practices

In debates about interviewing, standardization is sometimes assigned the role of "straw object." It is as if there were one method of standardization that all survey researchers agree is best, which interviewers could implement perfectly despite questions that are sometimes poorly written, and which could provide guidelines for all situations that an interviewer might encounter. The reality has always been more complex than this caricature. There are no tablets on which the rules of standardization are inscribed, and the practices of standardization have developed at various sites, each of which has its own traditions.

The rules that Fowler and Mangione (1990) codified in *Standardized Survey Interviewing* were comprehensive, grounded in the rationale for standardization, and refined through many years of practice and observation. Fowler and Mangione (1990:35) describe four principles of standardized interviewing:

1. Read questions as written
2. Probe inadequate answers nondirectively
3. Record answers without discretion
4. Be interpersonally nonjudgmental regarding the substance of answers

Survey researchers would probably universally accept these principles, although specific practices derived from these principles can vary. At various sites, researchers and staff charged with supervising field operations have developed their own specifications for training interviewers derived from such general principles (Viterna and Maynard, this volume, Chapter 16).

The rules of standardization assembled by Brenner (1981:19–22) illustrate this variation. Brenner allows interviewers to "show an interest in the answers given by the respondent," to "volunteer" clarification "when necessary," and to "obtain an adequate answer by means of nondirective probing, repetition of the question or instruction, or nondirective clarification" when the respondent gives an inadequate answer. If a respondent asks for clarification, an interviewer must provide it, but nondirectively, by using "predetermined clarifications." These predetermined clarifications appear to be similar to the information that is sometimes provided to interviewers in "question-by-question specifications" or to the interviewers in the experiments on "flexible" or "conversational" interviewing described below (Conrad and Schober, 2000; Schober and Conrad, 1997). Both practices offer a contrast to Fowler and Mangione's (1990) recommendations for severely restraining the discretion of interviewers.

The variability in methods of standardization can also be seen by considering how Fowler and Mangione's first principle is implemented. Although the general principle that a question must be read exactly as worded is probably the basic technique of standardization, and is even part of Schober and Conrad's (1997) flexible or conversational interviewing, some survey organizations, such as the Census Bureau, also accept a process that has been called verification as a legitimate technique of standardization. *Verification* acknowledges that respondents sometimes provide information before the interviewer asks for it, and that if an interviewer ignores the fact that she has already heard the answer to a question, the interaction may be awkward for both the interviewer and respondent. For example, an interviewer who has already been told the respondent's age in one context, instead of asking a later, scripted question about how old he or she is, might say, "I think you said earlier that you were 68 years old. Is that correct?" (e.g., Kovar and Royston, 1990:246).

The second principle, to probe nondirectively, can also be implemented in different ways. So, for example, Fowler and Mangione (1990:39–40) require interviewers to repeat all the response categories when they probe closed questions. In support of this rule, there is substantial research showing that the meaning of any individual category depends on the entire set of categories (e.g., Schaeffer and Charng, 1991; Smit, Dijkstra, and van der Zouwen, 1997). But van der Zouwen and Dijkstra (this volume, Chapter 18) describe a practice called *tuning* in which the interviewer needs to repeat only the categories that appear to be in the vicinity of the respondent's answer. Tuning is itself similar to a technique called *zeroing in*, which Fowler and Mangione describe for numerical answers (1990:41).

Practitioners recognize that there are situations in which the behavior of the interviewer cannot be standardized completely, such as the handling of "don't know" answers. Such answers require probing; and probing and "not answered" categories such as "don't know" are associated with interviewer variation (Fowler and Mangione, 1990:44; Mangione, Fowler, and Louis, 1992). Furthermore, when investigators train interviewers using *question-by-question specifications* of objectives, they recognize that the interviewer may need this information, whose application cannot be prespecified, to diagnose and correct respondents' confusions and misunderstandings. That interviewers become less standardized when they actually apply the information provided to them during training is part of the tension latent in traditional styles of standardization.

Limits of Standardization: Alternation to the Tacit Realm

Listening to a survey interview, it is easy to conclude both that standardization is essential and that it will never be perfectly and comprehensively achieved. This is because no set of rules is complete or covers every contingency. If such a set were somehow developed, it would be beyond the capacity of human interviewers to remember and apply. And so survey practitioners often must accept data produced in "messy" interactions as good enough for the purposes and budget at hand. Survey practitioners explicitly or implicitly recognize that interviewers inevitably must *alternate* between the rules of standardization and supplementing those rules. By using the term *alternation*, we mean that interviewers, along with respondents, deploy resources of an *interactional substrate* — basic skills for engaging meaningfully — in pursuit of answers. These skills are presumed by and yet largely ignored in the rules of standardized interviewing (Marlaire and Maynard, 1990; Maynard and Marlaire, 1992). The reliance of survey researchers on their interactional competence acknowledges that *tacit knowledge* is required to implement and supplement rules.

By tacit knowledge, we refer to what Polanyi (1958:20) terms "an essential personal participation of the scientist even in the most exact operations of science." Personal knowledge, drawing on participants' commonsense understanding of everyday courses of action (Schutz, 1973), "cooperates" with formal or impersonal judgments, just as tacit understandings combine with explicit, articulate knowledge, and as "subsidiary" comprehension of how to do some task joins with more "focal" and instrumental awareness of the task (Polanyi, 1958:90–95). It may be that survey researchers, understandably emphasizing standardization so as to improve measurement and the quality of data, have neglected the essential pairing that occurs between the formal and the tacit, or between instructions and the "lived work" and courses of action involved in implementing instructions (Garfinkel, 1988, 1996; Lynch, this volume, Chapter 5). Of course, the title of our chapter adverts to Freud's (1961) famous book on *Civilization and Its Discontents*, in which the reality principle is yoked to its opposite, the pleasure principle. Civilization depends upon sublimation and sometimes denial of the drive for pleasure. Our argument is similar and different. Standardization is essentially linked to tacit knowledge, just as civilization, in Freud's view, is intimately related to the erotic principle. However, standardization depends neither upon sublimation nor suppression of tacit knowledge but on the active deployment of it to *achieve* the procedural uniformity that enhances the quality of survey data. Those who are discontented with strict interpretations of standardization may be reflecting an explicit or implicit understanding of the necessity for tacit knowledge, which, as part of survey interviewing, has not received the scholarly and practical attention it deserves.

TACIT KNOWLEDGE AND INTERACTION IN THE SURVEY INTERVIEW

Tracing the history of the survey interview as an instrument for "viewing mass populations in industrial societies," Converse (1987:408–410) argues that the overwhelming presence of survey research in social science research represents a "historical ascendance of the subjective realm," because survey research is concerned with attitudinal phenomena and self-reports in a way that overshadows actual behavior and experience. However, if we emphasize survey research's concern with *measurement*, then a better, if more cumbersome, phrase might be that there has been a historical ascendance of the *objective* subjective realm. Survey research is not a form of *verstehen*, or interpretive sociology; the preoccupation of surveys with subjectivity involves, as we said, the logic of scientific measurement. The purpose is to gauge the amount and distribution of a population's demographic characteristics and the distribution of attitudes and opinions within that population.

The ascendance of the objective subjective realm in social science is not idiosyncratic, according to Porter (1995) in his book about the massive power and prestige of quantitative methods in the modern world. Rather, this ascendance is part of a broader movement toward standardization in Western societies. Time, space, volume, weight, distance, and value have all been subject to this movement over the past three centuries, and now, psychology (as in standardized testing), biology, medicine, and other disciplines including sociology also participate in this movement. As Porter (1995:74) observes, standardized measurement is a powerful means for rendering both nature and society in objective terms, so that communication about matters affecting diverse locales and peoples can overcome personal interests, prejudices, idiosyncracies, and the particularities of context. *Mechanical objectivity* is Porter's phrase for forms of inquiry in which rules for measurement discipline the subjectivity of interests and experience. Survey research, with its rules for the collection of data via standardized interviews, is a paradigmatic case of mechanical objectivity.

However, as Porter (1995) asserts, the ideal of mechanical objectivity is never fully attainable because of the need for tacit, commonsense knowledge. Early on, Garfinkel (1967:97–98) demonstrated this with regard to rules for coding, and other ethnomethodologists have expanded the point (e.g., Lynch, 1991:83; Zimmerman, 1970).[3] Most recently, sociologists have more fully appreciated that tacit and commonsense understandings are situated within "communities of work" embodying commitments of trust and mutual orientation (Collins, 1985; Lave and Wenger, 1991; Shapin, 1994; Suchman, 1987). In addition, studies of technology and social action are shedding light on, as Heath and Luff (2000:4) put it, "the ways in which individuals, both alone and in concert with each other, use tools and technologies in the practical accomplishment of their daily work."[4] In this volume, Fuchs (this volume, Chapter 20) provides a study of technology by comparing computer-assisted interviewing with surveys using paper instruments, and by comparing different screens as used with computer-aided interviewing for their effects on interview duration and other behaviors.

Relation Between Quantitative and Qualitative Methodologies

The use of commonsense or tacit knowledge is not something that is idiosyncratic; it is exhibited in the orderliness of talk, embodied conduct, and social interaction. Therefore, if survey-based sociology relies on the tacit and craft knowledge embedded in communities of work to derive its findings, this suggests an affinity between survey research as a practice and investigations that probe the orderliness in talk and social interaction. That is, the analysis of interaction, as an enterprise in its own right, also offers a knowledge base about the organization of direct social relations to those who design surveys and train interviewers. This affinity forms the cornerstone for the studies represented in this book. More generally, we are arguing for a particular relation between survey research as a form of *quantitative* sociology and investigations of talk and social interaction, such as conversation analysis, that are *qualitative* in orientation. Before we discuss this affinity, however, we need to clear some brush by reviewing a previous approach to the great quantitative–qualitative methodological divide in social science. We refer to this previous approach as critical remediation.[5]

Critical Remediation

Critical remediation dates to the publication of Cicourel's (1964) *Method and Measurement in Sociology* and continues to this day. In a series of publications spanning three decades (the 1960s through the 1980s), Cicourel (1964, 1974, 1982) has pursued issues surrounding the conduct of survey research. In the late eighties and into the nineties, others, notably Briggs (1986), Mishler (1986), and a few others, have essentially echoed some of his main themes with slightly new twists. Hence, we will concentrate on Cicourel's work and fill in certain areas of his appraisal by referencing this more recent work. The appraisal has two main facets to it.

The first facet of critical remediation has to do with the role of commonsense knowledge in the survey interview. The suggestion is that utterances or questions in the survey interview are never self-evident in terms of meaning; they gain their sense only against a background of respondents' unspoken meanings and aspects of settings (Cicourel, 1974:96). For example, in regard to fertility research, "thought about family size is limited by many common-sense features of everyday life which include such things... as, for example, community conditions in which a childless couple feels 'out of place,' or a 'feeling' on the part of the wife that four children 'would be nice' " (Cicourel, 1964:28). In Mishler's (1986:22–23) terms, the problem is that survey research is a "context-stripping procedure" in which investigators "pretend" that sociocultural meanings residing in the "life setting" of the respondent do not affect the meaning of questions and answers.[6] The neglect of respondents' commonsense knowledge also occurs because investigators' "procedures for categorizing social phenomena are buried in implicit common-sense assumptions about the actor, concrete persons, and the observer's own views about everyday life" (Cicourel, 1964:21). Similarly, Mishler (1986:23–24) contends that the investigator must invoke "a variety of presuppositions and assumptions, and this is usually done implicitly and in an ad hoc fashion," in order to form questions and code responses.[7]

A second aspect to critical remediation follows from the first. Qualitative critics have proposed that "structural and transformational linguistics, cognitive psychology, developmental psycholinguistics, and linguistically oriented anthropological works on componential analysis and the ethnography of speaking" need to be taken into account in the design of survey interviews (Cicourel, 1974:195–196). From Mishler's (1986:12) perspective, survey research exhibits an "almost total neglect... of work by students of language on the rules, forms, and functions of questions and responses," even though philosophers, linguists, sociolinguists, anthropologists, and sociologists have generated a sizable body of work on these topics. Accordingly, the ethnography of speaking or cognitive sociology or narrative analysis would enable the development of "more elaborate and precise theory" (Cicourel, 1964:87). In his study of Argentine fertility, Cicourel (1974:14) uses the term "alternative" to describe his concern with the "actor's lived, everyday experiences" in contrast to the concern with explaining shifts in fertility according to economic and other societal structures (rural versus urban living, community values, familial organization, and the like) that he imputes to survey researchers. Researchers can "uncover social meanings that traditional social research either distorts or fails altogether to handle" (Cicourel, 1974:148). In an even stronger vein, Mishler (1986:132) argues that *his* "alternative" (1986:33) to the standardized survey "radically alter[s]" the researcher's and the respondent's roles, and redistributes power between the parties. With an interest in how the survey interview creates a "special context" for narrative, Mishler (1986:137) argues that the model underlying the uses of surveys so completely neglects narrativity in discourse that a model of gathering information that is entirely different from the standardized survey interview is needed.

Survey Researcher's Lifeworld: Analytic Alternation

We do not conceive of the contributions in this volume as offering an *alternative* to the survey interview. Our attempt to capture *alternation*, by appreciating analytically the tacit work of skillful interviewing, is different from aspiring to change the form of the interview. Previous qualitative approaches to the survey interview, such as those of Cicourel and Mishler, have discussed ideas that survey researchers themselves have long recognized. For example, Kahn and Cannell (1957:16) state that interviews are a "specialized pattern of verbal interaction," and Kidder points out the reliance on "verbal reports" (Kidder, 1981:146).

So appreciating the use of language and the need for its understanding in the measurement process is well established. In a sense, the emphasis on language use and commonsense knowledge with regard to *respondents*' lifeworlds, and the concomitant suggestion among qualitative investigators that survey research disregards or distorts everyday meanings, misses an opportunity. That is, the preoccupation with developing alternatives to the survey interview—the stance of critical remediation—leads qualitative social scientific investigators to do with interviewers precisely what they accuse survey researchers of doing with respect to respondents. They have ignored and misrepresented the *interviewers*' lifeworld—or everyday work world—and failed to comprehend what the survey interview enterprise is as an actual complex set of actions and activities.[8] Or put differently, the qualitative critique tells us more about what survey researchers do *not* do than what they actually do, whereas we are arguing that careful studies of interviewers' workaday world are in order.[9]

To enter the world of the interviewer, we draw on the ethnomethodological studies of work program in the arena of science studies and other professional endeavors (Heritage, 1984b; Lynch, 1992).[10] The thrust of this research is to comprehend systematically the concreteness, specificity, and detail involved in actual work, including how practitioners use technologies and how they talk and act *in concert* with respondents.[11] If we become concerned with the *survey interviewer's* work, and consider the survey enterprise as a kind of "formal analytic" inquiry, then procedures or protocols for standardization, as used in specific instances, necessarily and repeatedly raise an ethnomethodological question. To paraphrase Garfinkel (1996:19), "what more is there to instructions and instructed actions" than a survey center, with its manuals, procedures, instruments, and other kinds of instructions, can provide? The "what more" question points to the tacit knowledge that interviewers use to handle the problems that emerge during the interview.

Accordingly, *analytic alternation*, in the context of the survey interview, means investigating practitioners' work. The work of interviewers, we have argued, involves both adherence to formal inquiry—the use of rules, procedures, and instruments for conducting the interview—and the exercise of taken-for-granted, tacit skills. These skills are exhibited in the produced, orderly detail of everyday talk and embodied action that reflexively supports and helps achieve the accountability of the formal inquiry. *Accountability* means that the interview gets done in ways that are acceptably—for all practical purposes—standardized and scientific. A straightforward example of alternation is to be found in a glance at any survey research center: We can observe interviewers engaged in computer-aided telephone interviewing (CATI) to be gazing at and reading verbatim to their respondents from a scripted interview on the screen in front of them, with hands and fingers tethered to the keyboard on the desk, and entering codes for heard responses that correspond to the proper categories as displayed on the screen. The interviewers, for some duration, appear to be formal or standardized homunculi, enacting what the survey instrument on their computer screen tells them to do as they talk, listen, and type.

At singular and unpredictable moments, however, interviewers will glance away from the screen, move their hands from the keyboard, and gesture more or less expansively, producing talk that is neither scripted nor otherwise predesigned, in order to handle some departure from the routine, with the aim of being able to return to that routine. For example, the interviewer may find it necessary to gesture while explaining something learned during training about the survey to a respondent. Although the gesturing is not something that the respondent sees, it might help the interviewer in articulating a point, or in emphasizing a significant piece of her talk, or some other aspect of the verbal presentation (e.g., this volume, Chapter 10). When she is done with the point, the interviewer returns gaze and hands back to the computer and keyboard. At these moments of departure, interviewers alternate from the formal to the tacit realm and engage in practices designed to provide for alternation again back to the formal.

This is only a more visible manifestation of what we mean by alternation, for even when they read a script more or less verbatim, interviewers, in interaction with respondents, are engaged in performances that improvise on the script. For example, the script does not tell how to manage turn-taking with the respondent; instead interviewers and respondents use a familiar interviewing sequence (which we discuss below). *Analytic* alternation refers to studying departures from, improvisations on, and means for implementing scripts, and capturing these departures, improvisations, and means as orderly practices. That is, interactional organization, as manifestations of tacit and commonsense knowledge, are sites of pattern and organization and need analytic appreciation in their own terms, an appreciation that recent and current (this volume) studies of interaction in the interview are beginning to provide.

The Interviewing Sequence

Some time ago, Schuman (1982:22) observed that "surveys start from two of our most natural intellectual inclinations." Those two inclinations are to ask questions in order to obtain information and to sample in such a way as to have this information generally reflect a larger universe. As an example, Schuman (1982:22) mentions how at professional meetings, one participant might ask another whether there is a good nearby restaurant. Extending this example, we can imagine how the restaurant seeker might decide that information from just one person would not be good enough, and that it would be better, therefore, to ask a few haphazardly if not entirely randomly chosen people. Therein lies the basis for the more systematic and analytic way of doing social measurement. In everyday life, participants have tacit knowledge enabling them to obtain information about worldly matters and to do so in relatively methodical ways that enhance the quality of this information.

One form in which tacit knowledge about how to gather information is exhibited in interaction is through the use of an *interviewing sequence*, which unfolds in three turns: (1) question, (2) answer, and (3) acknowledgment. For example, turns might be: (1) "Do you know of a good restaurant." (2) "Yes try the Grille just two blocks down on the right side." (3) "Great! Thanks!" We can observe this generic interviewing sequence form at the very start of a survey interview. The parts of two interviewing sequences are indicated with arrows in this extract (see Appendix 2 for the entire interview as well as definitions of symbols):

(1) AW01; Appendix, line 76

```
FI: 1→  Oh kay: (.) a::nd >now we'll get started?< .hhh ↑ first
        (.) how many persons live in your househo:ld counting
        all adults and children and including yourself. {q1}
FR: 2→  four
FI: 3→  .hhhh okay(gh): a:: [::nd? now we have some questions=
                            [# #
FI:     =about government agencies. .hhh as you know:? every ten
        years there is a census of the population of the
    1→  United States. .hhh how confident are you: (.) that the
        census bureau protects the privacy of personal information
        about individuals and does not share it with other
        government agencies. .hhh very confident (0.4) somewhat
        ↑confident (0.5) not ↑too confident? (0.2) or not at all
        ↑confident. {q5}
        (1.0)
FR:     Share it with what other governments?[12]
FI:     (tch) .hh well the question doesn't specify: but (0.3) it
        just says other government agen[cie]s
FR:                                    [oh ]
FR: 2→  Probably very confident
        (0.5)
FI: 3→  °Oh kay.° People have different ideas about what the census
        is used for ...
```

Not all survey questions appear in this three part format, with an "okay" or other object in the third position. Often, a *paradigmatic*, two-part question–answer sequence — which omits the third position — appears. For example, in AW01, when the interviewer has a block of questions to ask about a single topic, she tends to omit the acknowledgment and produces paradigmatic sequences:

(2) AW01, Appendix line 300

```
FI: 1→  Okay our next questions have to do with government?
        .hhh on a scale of one to ten (.) where one means very
        ↑poo:r and ten means excellent °.hhh° >how would you
        rate the< job the president is doing. {q49}
        (1.4)
FR: 2→  Uh:::m::: (0.9) six
FI: 1→  .hhh and on the same one to ten scale? where one means
        very ↑poo:r and ten means excellent °.hh° >how would
        you rate< the job the congress is doing. {q49b}
        (1.0)
FR: 2→  Uh::: (0.5) four
FI: 1→  .hhh on the same one to ten scale where one means very
        ↑poor and ten means excellent (.) >how would you rate
        the job< the supreme court is doing? {q49d}
```

(0.8)
FR: 2→ Uh:m five
FI: 1→ .hhh on the same one to ten scale where one means very
poor and ten means excellent (.) >how would you rate the<
job the military is doing? {q49f}
(0.9)
FR: 2→ Uh::m:: (1.3) eight(h)
FI: 3→ Eight ↑okay please tell me whether you agree: or disagree:
with each of the following statements? .hhh the first
statement is (0.4) people like me don't have any say about
what the government does. {q50a}

Notice that at line 22, when the interviewer is done with the questions that ask the respondent to rate the president, Congress, and the Supreme Court, she repeats the respondent's answer ("eight"), and then produces the third-turn acknowledgment, an "okay" that also projects the forthcoming next question (Beach, 1993).

Analyzing the interviewing sequence as a generic form can remind us that the survey is only one type of "interview." Dictionary definitions, for example, suggest that the interview is a "formal consultation" used to evaluate qualifications or "a meeting at which information is obtained (as by a reporter, television commentator, or pollster) from a person" (*Webster's Ninth New Collegiate Dictionary*). Accordingly, the survey has a relationship to other interviews, all of which use the interviewing sequence but which vary in discrete ways (including how the third-turn acknowledgment is configured, as we discuss below). For example, physicians and nurses often interview their patients to obtain a medical history; teachers and psychologists interview students as part of teaching and testing activities, employers interview prospective workers in making hiring decisions; and newscasters interview government officials and others as part of their programming. The survey can be compared to these other interviews to specify some of its distinctive characteristics.

The Interviewing Sequence in the Survey and Other Environments

A prominent feature of the medical interview is how questioning can contain certain biases that surveys attempt to remove. Physicians and other health professionals do not exhibit the disinterest and objectivity required of survey interviewers and instead during history taking "build their questions so as to convey a form of relatedness and concern for the welfare of the patient" (Boyd and Heritage, in press). Questions in medical interviews are fashioned with presuppositions and preferences that suggest optimistic, "no-problem" answers. Medical personnel do not ask questions, such as "Is your father dead?" Rather, they ask, "Is your father alive?" Nor do they ask, "Are your bowel movements problematic?" Rather, "Are your bowel movements normal?" By the same token, British health visitors in interaction with mothers who have recently delivered babies construct their questions about the process of childbirth in ways that favor no-problem responses. The health visitors will ask mothers if they had a "normal pregnancy" a "normal delivery," or if the baby is "bottle feeding," and the like, in a way that expects positive answers. Heritage (this volume, Chapter 14) calls this the principal of optimization, and it can be suggested that if the questions have biases built into them, they may also retrieve biased answers from their recipients.[13]

Interview questions in medical contexts also embody what Sacks, Schegloff, and Jefferson (1974:727) call recipient design, which refers to a "multitude of respects in which the talk by one party in a conversation is constructed or designed in ways which display an orientation and sensitivity to the particular other(s) who are the coparticipants." For example, such recipient design is exhibited in the questions of a physician to a 50-year-old divorced woman, whom the doctor knows to have had a tubal ligation procedure. The question format—"Are you using any contraception? Is that necessary for you?"—exhibits the physician's knowledge of the woman's circumstances (Boyd and Heritage, in press). Heritage (this volume, Chapter 14) demonstrates that health visitors' questions to new parents also exhibit such recipient design. By contrast with earlier "social surveys" (of which the health visitation is a contemporary example) and medical interviews, the standardized survey interview is designed to obviate practices of optimization and recipient design that would increase the likelihood of obtaining biased answers from respondents. Standardized survey interviews are, in Heritage's (this volume, Chapter 14:314) phrase, built to be "essentially anonymous."[14]

Acknowledgment Turns

One contribution of comparing surveys with other forms of interviewing is to make our understanding of what standardization overcomes—such as optimization and recipient design—more precise and specific. A second contribution is to describe issues surrounding third, *acknowledgment*, turns of the interviewing sequence, which survey researchers refer to as *feedback*. Acknowledgments include subtle forms of interviewer behavior, which may be influential, as well as the more formal tokens that interviewers use after a respondent has produced an answer to a survey question. In conversation, such third turns follow information-seeking question–answer sequences (turns 1 and 2) and register that the information is informative to the questioner (Heritage, 1984b:285–286). Turns in the excerpt below are labeled:

```
S:  1→   .hh When do you get out. Christmas week or the week before
         Christmas.
         (0.3)
G:  2→   Uh::m two or three days before Ch[ristmas.]
S:  3→                                    [Oh:,    ]
```

The "Oh" in turn 3 indicates a "change of state" in the answer recipient's knowledge (Heritage, 1984a). In educational settings, *instructional sequences* involve teachers asking "known-information" questions, as in this excerpt from Mehan (1979:52–53):

```
T:  1→  ((Holding up card)) This is the long word. Who knows what it
        says?
S:  2→  Cafeteria.
T:  3→  Cafeteria, Audrey, good for you!
```

In these instructional sequences, the third-turn response evaluates a student's answer.

In contrast with conversational and educational settings, there are other institutional settings in which questioners *withhold* third-turn responses of either variety (indicating a change of knowledge state or evaluation of an answer). Prominent are courtrooms (Atkinson and Drew, 1979), job interviews (Button, 1987), and news interviews (Clayman, 1988; Heritage and Greatbatch, 1991). These venues have in common that interviewers (attorneys or employers or newscasters, as the case may be) are soliciting answers for an *overhearing audience* of some kind who may do their evaluations of interviewees and their answers at some distance from the interview itself. That is, by refraining from postanswer commentary, interviewers exhibit themselves as conduits for those answers to flow to those audiences — who are then in the position of supplying that commentary in another social context, such as a jury room, a recruitment committee, or a political meeting. For example, it is during their later deliberations that jurors, having observed attorneys interview witnesses in the courtroom, may discuss and evaluate witnesses' answers to the attorneys' questions.

An informal or formal rule that forbids third-turn responses, as in news or courtroom environments, may be relatively easy to follow. More difficult are protocols under which the interviewers are permitted or even encouraged to produce third-turn acknowledgments but are to refrain from evaluating the content of the answers they acknowledge. This creates a dilemma of being permitted to respond but in restricted ways, and it characterizes not only the survey interview, but also educational testing interviews. Indeed, the protocols for tests such as the Woodcock–Johnson Psychoeducational Battery may warn administrators, "Be careful that your pattern of comments does not indicate whether answers are correct or incorrect" (Mehan, Hertweck, and Meihls, 1986:96–97). Administrators may produce "neutral" acknowledgments ("okay," "thank you," and the like) after a child has answered a test item. Despite these protocols, research demonstrates that administrators sometimes alter their third-turn responses *systematically* — using "good" when an answer is correct, and "okay" when it is incorrect; they may also give encouraging nonvocal signals by smiling or nodding when a child's answer is right and appearing more taciturn when it is wrong (Maynard and Marlaire, 1992). Such acknowledgments may not affect an individual answer; rather they may have a cumulative influence on the child's performance and at some point alter subsequent performance.

Extrapolating from research on educational testing, the concern that most survey centers have over third-turn feedback is no doubt well-placed. Survey researchers have long recognized that fine gradations in these responses potentially influence the answers of a respondent (Marquis, Cannell, and Laurent, 1972). When survey practitioners first documented how interviewers follow respondents' answers with feedback, the impulse of survey researchers was to standardize that feedback. Controlled feedback (e.g., "thank you" or "that information is helpful") is intended to positively reinforce thoughtful respondent behavior without evaluating the content of the answer. Some experiments with controlled feedback suggest that it may improve the accuracy of reports (Cannell, Miller, and Oksenberg, 1981), although this is not a uniform finding (Miller and Cannell, 1982), possibly because it teaches respondents what kinds of effort and answers the interviewer wants. Although most survey centers have protocols that attempt to standardize the use of feedback, such standards are by no means uniform *across* survey centers (Viterna and Maynard, this volume, Chapter 16).[15] That is, practitioners agree on the principle that feedback should not evaluate the content of the answer, but differ in specifying the frequency, content, and purpose of feedback (whether it should be used to provide reassurance, sustain motivation, praise the respondent's level of effort, or notice that the respondent's answer meets task requirements, and so on).

Conversation with a Purpose

The survey interview has long been described as a "conversation with a purpose" (Bingham and Moore, 1924).[16] Accordingly, administration of the survey is like interviewing behavior in other settings including ordinary conversation (where one participant may casually solicit facts or opinions from another) and institutional settings such as schools, courtrooms, and news organizations. Each party to an interview uses tacit knowledge and taken-for-granted procedures — the interviewer in producing questions, the recipient in answering them, and the interviewer yet again in acknowledging or not in particular ways. Although the survey shares generic practices with other interview forms for asking, answering, and acknowledging answers, it is relatively distinctive in its practices for maintaining the anonymity of questions and acknowledging answers without evaluating their content. Studies of interaction in the survey interview, such as those we review next, recognize the three-part interviewing sequence or two-part, paradigmatic question–answer sequence, as base forms for the interview. There are, however, different approaches to studying this interaction.

APPROACHES TO AND RECENT STUDIES OF INTERACTION IN THE SURVEY INTERVIEW

In the 1990s,[17] the methods used to study what happens during the survey interview and the goals of those studies have become more diverse. Explicitly or implicitly, research efforts recognize how tacit knowledge is used in doing a survey interview, but investigators take different approaches to mapping how this knowledge is used and the effects of tacit knowledge on data quality. For sake of simplicity, we categorize these approaches according to whether the research considers turns of talk in the interviewing sequence as the interviewer's or respondent's *unilateral* productions, or regards these turns as cooperatively, collaboratively assembled and therefore as *bilateral* productions. Both approaches, it needs to be emphasized, deal with the interactive nature of the interview and its constituent question–answer or question–answer–acknowledgment sequences. In the one approach, nevertheless, questions as speech productions "belong" to the interviewer and answers as speech productions "belong" to the respondent. In the other approach, researchers attempt to capture how both questions and answers may be collaboratively accomplished between interviewer and respondent. The research we review includes studies that describe and code the behavior of interviewers and respondents in standardized interviews, experiments that compare the effects of different styles of interviewing, and investigations that are ethnomethodological and conversation analytic.

Interviewer's and Respondent's Turns as Unilateral Productions

Early research about interviewing invoked concepts such as "rapport," acquiescence, and social desirability to speculate about mechanisms that might affect the respondent's answer (e.g., Goudy and Potter, 1975; Weiss, 1968, 1970). But that research did not explicate what these mechanisms might actually look like in the interaction. When Marquis and Cannell (1969), Cannell, Fowler, and Marquis (1968), and Cannell, Lawson, and Hausser (1975) pioneered the effort to examine interactions through coding, their scheme was designed to evaluate interviewers and, as a consequence, neglected respondent behavior.[18] Researchers could code interviewer's turns as "reads question exactly as printed in questionnaire," or "reads question making minor modifications," for example. Morton-Williams (1979), and others, extended the coding scheme to include respondent behavior, such as "asked for repetition of question," or "not answered adequately." More recently, Oksenberg, Cannell, and Kalton (1991) used behavior coding to identify and generate information about problematic questions so that instrument design could be improved, and Brenner (1982, 1985) coded recorded interviews in terms of "rule-breaching" and "rule-following" actions. Although researchers examined how interviewer behavior influences a respondent's answer, and how this answer may affect the interviewer, they regard each of these turns as the speaker's own unilateral production, and the main task has been to identify errors on the interviewer's part or in the design of questions.

The coding scheme that Dijkstra developed departs from the emphasis on interviewer or respondent error and examines a variety of behaviors, using a "multivariate" coding system (Dijkstra, this volume, Chapter 17; van der Zouwen and Dijkstra, this volume Chapter 18). Researchers, using a sophisticated hypercard program (Sequence) that allows online checking of digitized sound, code each turn by an interviewer and respondent on several variables, including who the actor is, what sort of information is exchanged, how close or distant an utterance is from a script, and so on. Once the data are coded, investigators use Sequence Viewer to find utterance frequencies, how given utterances affect what happens in the next turn, recurrent sequences of utterances, and so on. Using these programs, then, investigators analyze very large numbers of recorded interviews and examine particular questions with their words, phrases, pauses, and so on, and their effects on answers and answer patterns. For example, the researcher can discover which questions frequently result in respondents asking for explanation or which questions obtain adequate answers in the next turn. This kind of behavior coding, although it shares an orientation to turns of talk as unilateral productions, is a large step beyond earlier systems and represents a real advance on studying interaction in the survey interview because of its attention to components and strings of utterances.

Reconsidering Standardization

Fuller descriptions of how interaction in the interview actually proceeds, and models of how this interaction affects responses (such as that of van der Zouwen, Dijkstra, and Smit, 1991) could help researchers consider alternatives to strict standardization more systematically. An initial list of activities that we see interviewers doing includes reading questions and response options more or less verbatim, repeating information the respondent has given earlier as a verification of an answer to a current question, responding to requests for elucidation from respondents, proposing solutions to problems that might be comprehension problems, providing acknowledgment or acceptance, and so forth (e.g., Dijkstra, this volume, Chapter 17). Observed behaviors of respondents might include requesting repetition, giving answers that may or may not be formatted in the way that the question "requests," and so on. Behaviors of both interviewers and respondents can be analyzed as part of tasks that might be the subject of interviewing rules and training.

Describing the Behavior of Standardized Interviewers

Recent studies that examine interviewer effects (e.g., Mangione, Fowler, and Louis, 1992) document the effectiveness of standardization in reducing variability due to the behavior of the interviewer. Research also shows how departures from standardization can undermine the validity of answers. For example, Smit, Dijkstra, and van der Zouwen (1997) used a question that evoked a response before the interviewer read the response categories to examine the impact of the way the interviewer presented the response categories after the interruption. After the respondent "interrupted" the reading of the question, the interviewer implemented one of three experimental treatments: (a) she presented three contiguous categories that included the category she thought was suggested by the respondent's answer; (b) she presented a single category in the direction suggested by the respondent; or (c) she suggested the category that she thought was closest to the answer given by the respondent. The results suggest that the categories mentioned by the interviewer influence which category the respondent selects; in addition, the relationship of the answers with the respondent's age varied depending on which technique the interviewer used. An additional experimental manipulation comparing neutral and suggestive probes found that, in one of two cases the researchers examined, respondents answered differently depending on which probe the interviewer used. This experiment complements an earlier one showing that when the interviewer made a brief comment about a topic related to the survey questions, the respondent's answer sometimes shifted in the direction of the interviewer's "opinion" (Dijkstra, 1987). Because some of the interviewing practices examined in these experiments are more "conversational" than those used in strict systems of standardization, the experiments suggest ways in which conversational alternatives to rigid methods of standardization might influence answers.

Other research, however, indicates that some departures from standardization are not associated with response accuracy (Belli, Lepowski, and Kapeto, in press). And some departures might possibly increase accuracy of answers. Dykema, Lepkowski, and Blixt (1997) analyzed a study of health-related events and experiences that included a record check of respondents' answers for several questions. For only one of 10 items was there a relationship between reading the question exactly as worded and the accuracy of the answer. And for that case the relationship was not in the predicted direction. The investigators note that there may be something about the specific items involved that accounts for this finding. Furthermore, the deviations from standardization that account for this relationship when it occurs are unlikely to be random or haphazard alterations of question wording. Rather, the effect may be due either to wording changes that are tailored to the situation in some way or to other interviewer behaviors that happen to be correlated with question wording changes. (For example, experienced interviewers may be both most likely to alter the wording of questions and most adept at identifying when respondents misunderstand questions.) The sole other study of survey interviews that codes behavior and includes a record check reported a similar finding, but those results remain preliminary (Dykema and Schaeffer, 1998). Some of the mechanisms that produce this result may be those examined in recent experiments that let interviewers depart from standardization to try to improve respondents' comprehension of questions (see discussion below).

Interviewing Style

Studies of interviewing "style" and how it can affect the validity or reliability of data are in the tradition that code turns as unilateral speech productions. Experimental variations of interviewing styles can focus on any of the goals of standardized measurement, including reducing interviewer variability and increasing the accuracy of individual answers. Borrowing from social–information-processing models of the question–answer process, one might propose modifying practices of standardization to improve the respondent's comprehension or recall of survey concepts, to clarify the format in which the response is to be provided, or to motivate the respondent to be accurate.

The goal of improving recall has long been a focus of experiments with interviewing techniques, which include having the interviewer provide controlled feedback, asking the respondent to remember events around the boundaries of the reference period to anchor the time frame for the questions, beginning the interview with a free-flowing discussion of the topic to stimulate recall, and using various techniques to stimulate recall of events. Some of the techniques in these experiments could be used in standardized interviews (e.g., controlled feedback), but others involve substantially loosening the practices of standardization (e.g., varying the order of topics, free recall).[19] Although techniques for improving recall have been the focus of many experiments, many—perhaps most—of these techniques have not found their way into large-scale production interviewing. Possible reasons for this are several. The techniques pose challenges (and additional expense) for training and supervising interviewers, for instrument design, and for recording and processing data. In addition, some techniques can consume a substantial portion of the time of the actual interview, and some techniques may be most cost-effective when targeted to groups of respondents who may need help to report more accurately.[20]

The importance of motivating respondents has received some support from experiments. Respondents who are asked to sign a form in which they committed themselves to working hard appear to give more accurate reports (e.g., Cannell, Miller, and Oksenberg, 1981). In Dijkstra's (1987) comparison of formal and personal, elsewhere labeled "socioemotional" interviewing (Dijkstra and van der Zouwen, 1988), both styles are essentially standardized, but the personal style attempts to increase the respondent's motivation by telling the interviewer during training that a good relationship with the respondent is important for getting accurate information and by allowing the interviewer to express sympathy toward the respondent (e.g., "How nice for you!"). Respondents interviewed in the personal style performed better on a map-drawing task and had lower social desirability scores than did respondents interviewed in the formal style, but provided more irrelevant information. Respondents in the two styles were similar in the amount of relevant information they provided and in their tendency to be influenced by an opinion the interviewer provided (as part of an experimental treatment) during the course of the interview.

Subsequent analyses examined how participants behaved in the two styles of interviewing. Interviewers in the formal and personal styles appeared similar in how often they deviated from the wording of the question in the questionnaire and in how often they "chose" an answer based on information provided by the respondent, but interviewers in the personal style were more likely to emit "irrelevant behaviors," and more likely to "hint" at an answer (asking leading questions or suggesting an answer) (Dijkstra and van der Zouwen, 1988). Analyses of the interaction sequences indicate that interviewers trained in the personal style are more likely than interviewers trained in the formal style to use leading probes and that these probes are usually in line with information already provided by the respondent. Respondents are likely to accept suggestions made by the interviewer, particularly when the interviewer uses the personal style (van der Zouwen, Dijkstra, and Smit, 1991).

Interviewer's and Respondent's Turns as Collaborative, Bilateral Productions

Ethnomethodology and conversation analysis (to be discussed later) are among the earliest theories of interaction to argue for an appreciation of the moment-by-moment or temporal and situated construction of meaningful utterances (Garfinkel, 1967:39–41). To say that "talk-in-interaction" is a temporal construction is to emphasize how speakers build turns of talk that exhibit an alertness to cues from recipients (C. Goodwin, 1979, 1981; M.H. Goodwin, 1980; Schegloff, 1997). For the survey interview, the implication is not only that turns of talk are co-constructions and collaborative but that the outputs or results of the interview may be as well. Viewing interview outputs as collaborative productions goes against stimulus–response or other bipartite models of the interviewer–respondent relationship (discussed in van der Zouwen, this volume, Chapter 2), which presume that interviewers are neutral *conduits*[21] of questions that respondents answer according to their own abilities, willingness to disclose, and so on. Research in other interview settings has shown how the interviewer enters into the answering behavior and products of interviewees. In the educational testing situation, children often learn from the examination experience itself, and from such patterns as the pacing of questions, the parsing of stimulus items, and the reactions of administrators, whether their answers are on the right track. On the way to producing an answer, using these cues, they can modify what they end up saying and fashion answers that more or less reflect the local social environment as much as their own solitary efforts (Marlaire and Maynard, 1990; Maynard and Marlaire, 1992; Mehan, 1973; Mehan, Hertweck, Meihls, 1986). Collaborative production of "answers" also occurs in teaching (as opposed to testing) environments, and it is a feature that Schaeffer and Maynard (1996) and Schober and Conrad (this volume, Chapter 3) argue is present in the survey interview.

Example of Collaborative or Bilateral Turn Construction: The Problem of Uncertainty

A recurring challenge to interviewers is the display of uncertainty by a respondent, as she or he struggles to answer a survey question. Expressions of uncertainty can take many forms. By examining situations in which respondents express uncertainty, we can view how interviewers deploy tacit knowledge in the face of routine challenges to standardization. The way interviewers respond to these challenges may increase interviewer variance for questions that elicit expressions of uncertainty. For example, interviewers sometimes *code* respondents' answers (Hak, this volume, Chapter 19) only by ignoring their expressions of uncertainty.

Often, of course, respondents assert their answers in a relatively confident manner. We consider two examples from pretests conducted as part of the redesign of the Current Population Survey[22] in which the respondent displays little, if any, uncertainty.

(3) 006C0204:36

```
I:  Taking into account any extra hours worked or time taken
    off last week (.) how many hours did Nancy Gill actually
    work at her job
R:  Uh (0.4) forty
I:  And now I have...
```

In this exchange, the respondent produces an "Uh" and a brief pause (line 4), which may indicate processing the question cognitively and engaging in limited recall, before giving an unambiguous numerical answer. The interviewer treats the answer as unproblematic by proceeding with the next question (line 5). The next example is similar:

(4) 004C0202:114

```
1   I:  .h Taking into account any extra hours worked or time taken
2       off last week .hh how many hours did Peter King Junior
3       actually work at his job
4   R:  Round fifty
5   I:  .hh Now I have a few questions...
```

In this example, the respondent's answer "Round fifty" (line 4), which is produced without hesitation, has two elements. The first, "Round," presents the answer as an estimate; and the second element, "fifty," provides the numeric answer the question is designed to obtain. In contrast to excerpt 3, in terms of standardized interviewing, the answer here provides the interviewer grounds to question the answer as an estimate. However, by proceeding to the next set of questions, the interviewer treats the respondent's answer as adequate and acceptable.

These two episodes have few indications that the respondent has any uncertainty about answering. When such indicators are minimal, and the interviewer accepts an answer, it appears as the respondent's own answer. Such paradigmatic sequences may camouflage the frequency with which respondents might be unsure about their memory or knowledge. When a respondent does express uncertainty, the interviewer often collaborates in producing an acceptable answer:

(5) 008C0301:60

```
1   I:  And taking into account any extra hours worked or time
2       taken off last week how many hours did you actually work
3       at your job?
4   R:  .hhhhh Uh::: hhhhh let me think here (.) I dunno
5       f::::ifty hours maybe?
6   I:  Kay then (1.0) it would (.) be correct to put in fifty?
7   R:  Nyeah I sup[pose so]
8   I:             [Uh huh.] °okay° (3.5) Now I have...
```

The respondent hesitates at line 4 with a long inbreath–outbreath and elongated "Uh:::" between them, and then precedes ("I dunno") and follows ("maybe") the number of hours with expressions that clearly mark it as uncertain (lines 4–5). The number itself is produced with a stretched "f" sound ("f::::ifty") that further displays hesitancy and uncertainty. The interviewer then uses a probe (line 6) that proposes the correctness of the number. In conversation analytic terms, the turn at line 6 is an "other-initiation" of repair[23] that also offers a *candidate* repair for confirmation by the respondent. Furthermore, in proposing an expected answer, the probe "prefers" a confirmation (Sacks, 1987). In the language of survey interviewing, the probe is directive, although this kind of repair is a regular practice in conversation. The rules of standardized interviewing provide alternative methods of repair: The interviewer could suggest the level of accuracy she actually wants (e.g., "Please be as accurate as you can") or ways the respondent might improve the answer ("Be sure to consider any time you may have taken off or any overtime you might have worked," etc.), but the interviewer does not say these things.

The respondent's answer, "Nyeah" (line 7), conveys continued uncertainty and is achieved as ambiguous (Jefferson, 1978). It presents the hearer (in this case, the interviewer) with the opportunity to make a choice and to help actualize its valence as positive or negative. The respondent adds a suppositional marker to this ambiguous item ("I suppose so," line 7), in a way that further displays uncertainty. However, the interviewer overlaps this display with the start of her receipt and acknowledgment ("Uh huh °okay°," line 8). The acknowledgment proposes closure of the sequence (Beach, 1993), and silence ensues, as the interviewer enters the answer and waits for the next screen (in the 3.5-second silence at line 8). By staying silent after his answer is acknowledged, the respondent appears to accept the interviewer's proposed termination of the sequence. Overall, the interviewer has clearly played an active part in obtaining a "correct" and codeable answer in an environment of palpable uncertainty on the part of the respondent. The answer is very much a collaborative achievement. Yet the collaboration has succeeded in making it possible to see a definite answer as having been produced by the respondent—it is "accountably" the respondent's own answer (cf. Marlaire and Maynard, 1990) in part because of the chain of question–answer sequences by which it is elicited. That is, in formal turn-taking terms, it is the interviewer who officially has asked questions, and the respondent has officially only answered them, even if the interviewer has used an unscripted question–answer sequence to suggest an answer for the respondent to confirm.

When respondents display uncertainty, they may be signaling something about the source of their knowledge or about the basis for a claim that lies behind an answer. The display of uncertainty makes that uncertainty available for pursuit by the interviewer. The fact that standardized interviewers sometimes ignore these expressions (although they do not always do so), is probably due both to the constraints of standardization and to the everyday practices of talk. That interviewers also may more or less actively collaborate in a respondent's production of an answer has several practical implications. For example, conversation analytic (CA) investigations provide a means for identifying interviewing sequences in which expressions of uncertainty frequently occur. This suggests that the recording of interviews and CA investigations are especially relevant during the pretesting of instruments, in the way that Gerber (1999) sees a role for ethnographic studies of respondents' cognitive systems. In addition, because CA provides for the analysis of *interaction*, studies of sequences with apparent uncertainties may be able to identify which instances of hesitation are doing interactional work, such as delaying a "dispreferred" answer,[24] which instances should be probed, which ones may be exhibiting something about the respondent's state of knowledge or strength of opinion,[25] which questions tend to occasion interviewer interventions that are too directive (Schaeffer and Maynard, this volume, Chapter 12), and therefore which questions need redesign and in what particular ways.

Experimental Studies of Collaborative Turn-taking in the Interview

Like the ethnomethodological and conversation analytical approaches just discussed, some psychological and experimental approaches to the survey interview also regard turns of talk as collaboratively produced. Drawing especially on the work of Clark (1992, 1996), Schober and Conrad (this volume, Chapter 3) critique the *message model* or *conduit metaphor* as presupposed in many approaches to communication, which suggest that words are containers of meaning that can be transferred from one head to another through speech.[26] Instead, Schober and Conrad (this volume, Chapter 3) argue that:

> as people speak they carefully monitor their addressees for evidence of understanding or misunderstanding, and they adjust their utterances, moment by moment, to ensure that their addressees are understanding them well enough for current purposes. Addressees, by providing such evidence, help mold the utterances speakers produce.

If turns of talk are collaborative productions, the implications for the survey interview as a structured form of interaction are not exactly clear. Schober and Conrad (this volume, Chapter 3) attempt to derive these implications, in part from previous work in which experimental manipulations test whether flexible forms of interviewing are pitted against more standardized forms.

Under the strongest version of the flexible style of standardization examined by Schober and Conrad (1997; Conrad and Schober, this volume, Chapter 3), the interviewer would be allowed to take initiative in providing definitions or taking other actions to diagnose and remedy the source of the difficulty in a respondent's report. Schober and Conrad tested the effects of allowing interviewers to clarify survey concepts or define survey concepts in their own words, sometimes in combination with instructions to respondents that explained that survey concepts could differ from everyday concepts and that encouraged respondents to ask for help. Respondents are more likely to ask for clarification in the flexible style of interviewing that included extensive instruction about the importance of asking for clarification at the beginning of the interview (Bloom and Schober, 1999; Schober, Conrad, and Fricker, 1999). Either the respondents' requests or the initiative taken by the interviewer or the combination appears to be effective: Respondents with complicated situations were significantly more likely to give accurate answers when they were interviewed in the flexible style. The increased accuracy comes at an increased cost (which is sometimes substantial) in interview time for both straightforward and complex situations (Schober, Conrad, and Fricker, 1999).

Implications for Training

An interviewing technique that authorizes interviewers to diagnose problems in understanding and applying survey concepts and that permits interviewers to intervene to help correct these problems faces a number of challenges. It is true that instruments cannot always anticipate or handle all situations respondents might present, but instruments should be designed to handle complex situations that affect significant proportions of the sample. However, because situations that instruments cannot handle will always arise, training interviewers how to intervene properly could increase validity, even if the procedure took more interviewing time for the affected cases. By observing how interviewers deploy tacit knowledge to solve the problems that arise when instruments and standardization are insufficient, researchers can begin to design interventions that use the skills of the interactants and also maintain the quality of the data as measured by traditional criteria.

The operational problems experimental interviewing techniques face in gaining acceptance include the cost of developing more comprehensive survey definitions, training interviewers in these definitions, and in when and how to use them. There are also potential costs in having even a well-trained interviewer take the initiative. When interviewers are authorized to take the initiative in addressing the problems of respondents, they sometimes do so in situations where it is probably not needed. When this happens, the interview becomes longer without increasing the validity of the data, and break-offs may increase. There are other unknowns: whether respondents in actual surveys will request clarification and continue to ask questions or seek help during long interviews, whether interviewers can remember concepts over long field periods in which any given definition may be required only rarely, how the techniques might be applied to questions about subjective things (or how interviewers might manage alternating between strict standardization for subjective questions and flexible interviewing for other questions), how interviewers might be monitored and supervised in their use of these techniques, and what the impact on interviewer variance in a full-scale study might be. Finally, when interviews are not standardized, and respondents produce their answers in relation to different stimuli, their answers may not be strictly comparable to each other; this can compromise the goal of using survey data to describe a population.[27]

ETHNOMETHODOLOGY AND CONVERSATION ANALYSIS: PREVIOUS RESEARCH ON THE SURVEY INTERVIEW

Many of the chapters in this book draw on ethnomethodology (EM) and conversation analysis (CA), two interrelated subfields in sociology that examine concrete and detailed activities of participants in everyday life as a site of intelligible organization meriting close inspection and scientific analysis (Maynard and Clayman, 1991).[28] These subfields are ripe for application to the survey interview because survey researchers many times in the past have argued that a theoretical and empirical understanding of the survey interview as an *interaction* and not just a stimulus–response mode of inquiry is needed. We have already mentioned van der Zouwen's (this volume, Chapter 2) statement, and similar discussions can be found in Turner (1982:210, fn.37), Groves, Cialdini, and Couper (1992) and, most recently, in O'Muircheartaigh (1999). EM and CA provide both theory and a collection of studies that is now extensive on the organization of direct social relations in ordinary conversation and in work settings.

Some readers of this volume may be unfamiliar with these subfields, so we describe them very briefly before reviewing EM and CA studies of interaction in the survey interview. Ethnomethodology began with a concern for members' *accounting practices*, or the devices by which intelligibility is sustained in and through concrete concerted actions; it has more recently made large contributions to the study of practices in the workplace (Suchman, 1987) and in scientific laboratories (Lynch, 1993). Consequently, it has implications for the study of survey interviewing as a work practice (Lynch, this volume, Chapter 5; Maynard and Schaeffer, 2000). The overriding concern in EM is with the local orderliness of everyday activities. That is, EM suggests that this orderliness does not arise because external forces or internalized values somehow cause participants in daily life to act in patterned ways. Rather, participants endogenously enact procedures, methods, or practices through which they produce and recognize the circumstances and features of settings in which they are embedded.

Conversation analysis (which includes a wide-ranging set of studies over three and a half decades) describes and analyzes the orderly properties of conversational interaction as the achievements of participants. Processes of conversational interaction have their own internal constraints or are under *local control* (Zimmerman, 1988:408). For instance, in taking turns at talk, Sacks, Schegloff, and Jefferson, (1974) find that participants have three ordered options: (1) a current speaker may select the next speaker, (2) a next speaker may self-select, or (3) a current speaker may continue. With this system of turn-taking in conversation, participants have an *intrinsic* motivation for listening to one another, independent of other motivations including interest, deference, or politeness. Potential next speakers must listen to a current speaker to find out whether they have been selected to take the next turn, to discover when the current turn ends and it is appropriate to start speaking, and to know what they may be constrained to say next, given what has been said in the turn of talk underway.

This intrinsic motivation for listening means that conversation is to a large extent self-governing with respect to the orderly achievement of mutual understanding. A "next" speaker, in a following turn, is required to display understanding of the previous turn (and any other talk to which it may be directed), and this provides a methodological resource — what Sacks, Schegloff, and Jefferson, (1974:728) call a "proof procedure" — for the analyst. Because understandings are displayed for co-participants, they are also available to the professional who is inspecting the talk.[29] This feature of talk and social interaction holds for the survey interview to a more limited degree than in conversation, because turns are preallocated (interviewers ask questions, respondents answer) and because both the questions and the answer options are severely constrained. Nevertheless, the methodological orientation to how the interactants themselves — interviewer and respondent — treat one another's talk (particularly when they are off the script) is evident in our chapters that employ CA. Finally, if the recipient of a turn of talk has difficulty in hearing or understanding or otherwise knowing how to deal with it, there are mechanisms for initiating repair on and repairing that talk. By and large, *repair* of all kinds of conversational trouble exhibits sequentially systematic properties (Schegloff, Jefferson, and Sacks, 1977). Moore and Maynard (this volume, Chapter 13) discuss how methods for repair of talk intertwines with respondents' attempts to understand survey questions and with interviewers' *probing* of improperly formatted or otherwise inadequate answers to survey questions.

Transcription and Interactional Detail

As the founder of CA, Sacks (1984) turned to talk and social interaction and to the preservation of detail within this domain because of the sense that social organization and orderliness were to be found in such detail. As Zimmerman (1988:415) puts it, "No scale of detail, however fine, is exempt from interactional organization, and hence must be presumed to be orderly." This implies an interest not just in what participants say but also in silences, in overlapping talk, in sound stretches, breathing, prosody (intonation, stress, volume), and so on. Hence, conversation analysts transcribe tape recordings with much technical specificity to show as many of these features as possible. We have shown this already in transcript extracts in this chapter and the conventions for transcription, we noted earlier, are to be found in Appendix 1.

EM and CA Research on Interaction in the Survey Interview

One of the earliest studies of the survey interview from a conversation analytic perspective was Suchman and Jordan's (1990) publication in the *Journal of the American Statistical Association*. It received a great deal of attention, in part because of the reservations expressed about validity. Suchman and Jordan (1990) observe that survey design governs both what participants in the interview talk about and how they talk (through question–answer sequences). This means that ordinary resources for establishing shared understanding in conversation are prohibited in the interview. For instance, whereas participants in conversation are free to elaborate on what their presumptive worldview is, what their utterances mean, and what the source of any misunderstanding is, interviewers and respondents are severely constrained from such elaboration. In a manner that is similar to that of Cicourel (1974) and Mishler (1986) before them, Suchman and Jordan (1990) argue that participation in the survey interview needs to be more collaborative, and respondents need more freedom to answer questions meaningfully.

Suchman and Jordan (1990) reiterated important questions about the survey interview, but in our view missed an opportunity to positively respecify what the interview is. That is, their critique tells us mostly about how the interview differs from and is inferior to conversation in the establishment of mutual understanding and meaning, rather than what constituent properties characterize the performance of the survey *as a standardized interview* for purposes of social measurement. Schegloff (1990; this volume, Chapter 6), in a response to Suchman and Jordan, (1990), suggests that inquiry is possible into features of the survey interview as, in a sense, having their own integrity. Such inquiry would recognize that "all talk-in-interaction faces certain generic organizational problems, and will perforce adopt some organized solution to them" (Schegloff, this volume, Chapter 6: 152). Generic organizational problems include procedures for allocating turns of talk (assigning questioning and postanswer acknowledging turns to interviewers and answering turns to respondents, for instance), interactional constraints on coherence (such as "preference structure"[30]), repair of troubles in speaking, hearing, understanding, and so forth an overall structural organization, and a variety of other matters depending on the nature of the survey and its substantive topic. This text addresses at least some of these organizational problems.

Various topics adumbrated by Schegloff have already been taken up in journal and book publications. Because the fundamental framework for doing conversation analysis is sequential organization, paradigmatic question–answer sequences and the three-part interviewing sequence, as predominating in the survey interview and as highly regular phenomena, are of natural interest. Houtkoop-Steenstra's (2000) book is the most extensive conversation analytic study of question and answer sequences during the survey interview. However, the book's empirical investigation starts with a consideration of participant roles in the interview, suggesting that the interview implicates more than just an interviewer and respondent.[31] Using Goffman's (1981) concept of *footing*, which refers to the stances that participants take in relation to their talk, Houtkoop-Steenstra (2000) describes how interviewers exhibit themselves as "animators" (Goffman, 1981) or "relayers" (Levinson, 1988) of text that is designed by others, the "author" (Goffman, 1981) or "formulator" (Clark, 1996). Accordingly, when respondents ask for clarification or have other difficulties, interviewers comment in ways that indicate such a participation status: For example, "I just ask questions here." Or, in other circumstances, interviewers may refer to distant participants who design the instrument items, with phrases such as,"They're asking. . . ." From time to time, interviewers may step out of their animating stance altogether, saying (for example) "I'm sorry" when a respondent reports a sad incident. There are complexities to all of this, and Houtkoop-Steenstra (2000) provides a compelling discussion of the interview as an interactive frame, in which two interactants skillfully exhibit a variety of participation statuses relative to the talk and task at hand.

Other chapters in Houtkoop-Steenstra (2000) address the topics of redundant questions, the structure of question turns, and the field coding of questions. *Redundant questions* are those for which a respondent has already provided relevant information in response to a previous question. Although the U.S. Bureau of the Census has a procedure for dealing with redundant information called verification, there are circumstances in which redundant questions still pose problems. In such circumstances, it appears that Dutch survey interviewers become more "conversational" while U.S. interviewers may be stricter in following the rules of standardization (Houtkoop-Steenstra is a Dutch researcher who compares data from the Netherlands with that from the United States). The organization of turns during question-and-answer sequences in interviews is one that Houtkoop-

Steenstra (2000) addresses in a chapter in her monograph and also in her contribution to this book (Chapter 11). Survey researchers had observed that when a question is followed by a definition, respondents sometimes interrupt before the definition is read (e.g., Cannell et al., 1989), and Schaeffer (1991:386–387) notes how work by conversation analysts could be used to understand why such a question might be frequently interrupted. Using specifications provided by Sacks, Schegloff, and Jefferson's (1974) foundational work on turn-taking, Houtkoop-Steenstra (2000; this volume, Chapter 11) shows that many survey questions are vulnerable to interruption because they can be heard to implicate a response from the respondent before the scripted question is complete. Realizing this, interviewers often modify such questions in the course of producing them, although this impedes standardization. In her monograph's final chapter, Houtkoop-Steenstra (2000) deals with field coding of answers to open questions and argues that, in a variety of ways, interviewers are prompted to modify such questions, sometimes, for example, producing candidate answers on behalf of their respondents, who can confirm such candidates. While this may compromise standardization, features of interaction may "naturally" occasion such interviewer behavior.

The other major conversation analytic work[32] on the survey is an unpublished dissertation by Moore (1999) on repair and its relation to probing in the survey interview (cf. Houtkoop-Steenstra, 1996). Moore (1999) discusses, in depth, respondent requests for repeats and clarifications of survey questions and interviewer queries about inadequately formulated answers. A phenomenon related to repair, and to Houtkoop-Steenstra's consideration of question structure (this volume, Chapter 11), is how interviewers handle premature answers or those that occur before a question has been completely read because respondents believe they have enough information. Importantly, repair in conversation is related to the achievement of understanding (Schegloff, Jefferson, and Sacks, 1977), but in the survey interview, standardization may impede practices of repair and therefore interfere with such understanding. Several of these matters are discussed in Moore and Maynard (this volume, Chapter 13).

RECRUITING SURVEY PARTICIPANTS

The importance of the organization of interaction in the survey probably cannot be overstated. To underline this point, we end our introductory chapter by considering the beginning or "front end" of the interview, which is where potential respondents get recruited to participate. Indeed, Groves, Cialdini, and Couper (1992) propose that interaction is one of the critical elements in a theory of survey participation. Conversation analytic investigations add to other studies of survey interview openings that attempt to identify the tacit means by which scripted introductions are made to work on behalf of recruiting respondents to the interview. In one study, Maynard and Schaeffer (1997) examine the practices by which interviewers open their encounters with potential respondents, make requests for participation, and the complementary means by which recipients of calls accept or decline such requests. In this volume (Chapter 8), Maynard and Schaeffer further explicate the opening or closing of the interactional gate separating interviewers from the subjective data they so strongly want to obtain, and find that interviewer optimism as enacted through various conversational practices sometimes works and other times appears counterproductive in the recruitment process.

A concept that Groves, Cialdini, and Couper (1992) introduce for understanding the work of soliciting participation is *tailoring*. Successful interviewers, they argue, tailor the scripted instructions for recruiting participants to specific concerns and objections that potential survey respondents may raise. They engage in a continuous search for cues about a householder who answers the door or telephone, and should be able to quickly size up a situation to apply the appropriate persuasive messages. Couper and Groves (this volume, Chapter 7) report on suggestive evidence that the more interviewers tailor their introductions, the more likely they are to get cooperation. Similarly, Houtkoop-Steenstra and van den Bergh (2000; this volume, Chapter 9), in an experimental study of recruitment to a marketing interview, find that agenda-based conversational introductions are more likely to evoke cooperative behavior from the recipients than any of three scripted, standardized introductions. And Maynard and Schaeffer (this volume, Chapter 10) examine an effort at refusal conversion and the devices of talk and embodied behavior by which an interviewer tailors a callback to a selected respondent who had previously refused participation in a survey. The interviewer successfully "converts" the respondent, through a variety of tacit and persuasive methods, eventually obtaining participation of the household (the respondent's wife).

CONCLUSION

Recent research has provided a dual view of standardization that mirrors the analytic alternation engaged in by standardized interviewers: Studies have examined both how well interviewers implement the rules of standardization and the way interviewers tacitly deploy the resources of the interactional substrate in their pursuit of data. Together, these studies contribute to a better understanding of the fundamentally interactive nature of data collected by interview. Because the rules of standardization can never be complete, interviewers will inevitably alternate between the rules for the standardization of conduct and the interactive resources they have mastered in other, uncontrolled, social contexts for handling emergent, situational contingencies that form their interactions with respondents. By observing what interviewers actually do, we can begin to understand how their various activities might affect the reliability and validity of data, and how they might do so differently for specific kinds of questions — those about events and behaviors and those about evaluations or opinions, for example. This understanding in turn can enhance our vision of how standardization might be augmented to improve the quality of data.

Some tacit knowledge surely can be made explicit, but we first need to know what that knowledge is and how it is deployed in actual circumstances. Such is the overall contribution of the chapters in this volume of studies, which lay bare, as never before, the everyday world of interviewing work and the interactional practices lending to the collection of survey data. Some of these practices are skills that can be taught during interviewer training. Other practices, such as those that respondents deploy to understand and answer questions, are ones to which interviewers can be sensitized, and which survey designers can learn to anticipate in writing instruments for measurement. Of course, the fate of the innovative methods of interviewing explored in recent research suggests that to improve survey measurement, researchers must grapple with the operational issues involved in putting a prototype style of interviewing into production. The challenges include devising methods that large field staffs can be trained to use reliably at an acceptable cost. New interviewing methods also require guidelines for monitoring the quality of interviewers' behavior and appropriate instruments and data processing procedures. Accordingly, we recognize that the improvement of data quality is no easy achievement. However, when we consider the ubiquity of the survey and how it functions in the academy, in government, and in business, and for public as well as private organizations (Maynard and Schaeffer, 2000), then we see that incremental gains might have large benefits.

Other sociological interests are at stake here as well. Survey centers represent a site for inquiry into the orderliness in the concrete human activities that are situated in them, and the study of such centers can contribute to the sociology of (social) scientific knowledge (Maynard and Schaeffer, 2000; also see Lavin and Maynard, 2001, and this volume, Chapter 15), research on human–computer interaction, and the understanding of communities of practice that interviewers occupy. In particular, as a place where standardization of work activities is meant to occur, survey centers are like other places in which an occupation or profession, an agency or bureau, or a business or academic discipline has embraced some from of mechanical objectivity to bring the disparate measures of time, distance, volume, economic performance, program compliance, and the like into uniform alignment. To engage in standardized measurement is a social task and involves lived, interactive, organized practices such as those with which our chapters are concerned. Standardization necessarily and inescapably involves tacit knowledge, and tacit knowledge necessarily and inescapably shapes social measurement.

ACKNOWLEDGMENTS

We would like to thank Frederick Conrad, Jennifer Dykema, John Heritage, Colm O'Muircheartaigh, and Michael Schober for helpful comments on this chapter, which also benefited from earlier discussions with Harold Garfinkel. We also thank the University of Wisconsin Survey Center for collecting interviews for analysis and the U.S. Bureau of the Census for permission to analyze the Current Population Survey pretest interviews. In the excerpts from the interviews, all names of persons, businesses, and locations are fictitious. Some of the analyses reported here were partially supported by a Joint Statistical Agreement (90–45) with the U.S. Bureau of the Census. Supplementary support was provided by the Research Committee of the Graduate School of the University of Wisconsin, Madison, and by the Ameritech Foundation Fellowship Program. Computational support was provided by the Center for Demography and Ecology, which receives core support from the National Institute for Child Health and Human Development (HD05876).

NOTES

1. See, for example, Tourangeau's response (1990) to Suchman and Jordan (1990).

2. For example, compare Holstein and Staples (1992) and Holstein and Gubrium (1995) with Tourangeau, Rasinski, and Rips (2000).

3. In this volume, see Lynch's (Chapter 5: 129) discussion of the relation between instructions and the course of actions that embody it. This discussion draws on Garfinkel's notion of the"Lebenswelt pair," captured in the diagram [Instructions] → {Lived course of action}, with the arrow denoting "the situated work of using the instructions, making out what they say, finding fault with them, re-reading them in light of what is happening *just now*, and other contingent uses and readings."

4. See also Engström and Middleton (1996) and Whalen (1995). For a study of how computer-aided telephone surveys involve not just a relationship between interviewer and respondent but the interviewer's embeddedness in an organized social environment; see Goodwin (1998). The interviewer draws on the "hearable" activities in the survey center and contributes to them in ways that affect the calling and interviewing tasks being performed by other individual interviewers, in their carrels, talking over the phone to respondents.

5. See also the discussion of a third approach to the quantitative–qualitative methodological division, which we term "conditional complementarity" (Maynard and Schaeffer, 2000:332–333).

6. See also Cicourel (1964:111; Briggs, 1986:45).

7. Our own discussion of tacit knowledge (above) makes reference to commonsense knowledge, but we are using this term in a sense that differs from Cicourel's (1974) and Mishler's (1986) usage. Where they point to community and cultural understandings of the terms and categories employed in survey questions, we are referring to the set of taken-for-granted skills (the interactional substrate) whereby interviewers and respondents are able to do the work of completing the survey, whatever its substantive content and however this content is interpreted.

8. The two existing attempts to describe something of the work world of the survey interviewer that we have been able to locate adopt very different perspectives. Converse and Schuman (1974) collect anecdotes that attempt to capture the flavor of the work interviewers do. Peneff (1988) describes the work done by French interviewers as that of "hired hands" (Roth, 1966), who bend the "rules" in ways of which their superiors are ignorant or acquiesce for purposes of getting the work done. The first of these is not intended to be a study of the practice of

science, and the second proceeds from the stance of critical remediation that we have characterized above.

9. In his review of Cicourel's (1974) study of Argentine fertility, Bumpass (1975) charged that the book was a polemic that ignored the survey tradition's self-criticism and scrutiny on the very topics that Cicourel addressed. "There is," Bumpass (1975:476) noted, "an extensive and growing body of research on these problems that is completely ignored in Cicourel's work." Perhaps such a charge could be seen as defensiveness from a prominent demographer whose own mode of inquiry was at stake. On the other hand, it may index a real inadequacy in the qualitative critique of survey research, which is the tendency to avoid immersion in the world of the researcher or the interviewer in the way it is recommended that the researcher and interviewer enter the world of the respondent.

10. See also Button (1993). For more recent discussion of ethnomethodological studies of work, see discussion in Maynard and Schaeffer (2000), and for a recent collection of studies in this tradition, see Luff, Hindmarsh, and Heath (2000).

11. Ethnomethodological studies of work interrelate with the fields of computer supported cooperative work (CSCW) and human-computer interaction (HCI). As Luff, Hindmarsh, and Heath (2000:17) write in an excellent review of the literature in this area, ethnomethodological and conversation analytic approaches to CSCW and HCI are concerned with

> the contingent and situated character of practical organisational conduct, and in explicating the resources on which participants themselves rely in producing and coordinating their actions and activities with each other. They also share an interest in technology... In using studies of work to reconsider the ways in which we understand the relationship between human action and objects, whether highly complex or truly simple.

Surprisingly, not much research has been done on the survey interview as an instance of CSCW and HCI, even though survey centers would seem to be a natural site. See discussion, including an outline of how these fields could inform our understanding of cognition and surveys, in Couper (1999).

12. Here, the respondent initiates a "repair sequence"—a request for clarification. On this kind of repair in the survey interview, see Moore and Maynard (this volume, Chapter 13).

13. See Schaeffer (1991) and this volume, Chapter 4.

14. See also Maynard and Schaeffer (1997) on the anonymous ways in which survey participation is solicited.

15. Another matter for which there are starkly different orientations across survey centers is whether interviewers should be allowed to laugh during an interview. See Lavin and Maynard (this volume, Chapter 15).

16. Cited by Cannell and Kahn (1968); see Schaeffer (1991) and this volume, Chapter 4.

17. For earlier studies, see Schaeffer (1991), and this volume, Chapter 4.

18. For the studies we cite and summarize in this paragraph, we draw on the useful review in Dijkstra (1999).

19. For examples, see discussions in Cannell, Marquis, and Laurent (1977), Means et al. (1989), and Oksenberg et al. (1992).

20. See Dykema and Schaeffer (2000), and for a relevant recent development see Belli, Shay, and Stafford (2000).

21. Below, we discuss the conduit metaphor and message model in relation to the interview.

22. See details in Schaeffer, Maynard, and Cradock (1993).

23. On repair in the context of the survey interview, see Moore (1999) and Moore and Maynard (this volume, Chapter 13).

24. On preference structure in conversation, see note 26 below.

25. We cannot address the complex issues related to response latency here. However, the study of durations between an interviewer's completion of question reading and a respondent's start of an answer may be enhanced both technically and substantively using conversation analysis. For interesting discussions of a technology for measuring response latency, and patterns associated with response latencies, see Bassili (1996). See also the discussion of expression of uncertainty in Mathiowetz (1998).

26. See also Maynard and Whalen's (1995) critique of the conduit metaphor and how it stresses the notion that participants mainly use language to convey information. The view developed from Wittgenstein (1953) and others is that language is used to perform *actions*, of which conveying information is only one.

27. See related discussion in Schaeffer and Royston (1999).

28. For statements on ethnomethodology's program, see Garfinkel (1967) and (1996). For a thorough treatment of both ethnomethodology and conversation analysis, see Heritage (1984b).

29. The methodology of conversation analysis is a topic in its own right and beyond the purview of this chapter. For a brief description of CA methodology, see Psathas (1995:Chapter 4). For a full-length and very thorough monograph on CA methodology, see ten Have (1999), who covers such topics as the relation between ideas and evidence, collecting data through recordings, transcribing talk and interaction, analyzing data systematically, and others.

30. The concepts of preference and preference structure in conversation refer to ways in which agreement (preferred) and disagreement (dispreferred) and other paired conversational actions (acceptance/rejection, offering/requesting, etc.), rather than being equivalent actions, are asymmetric in terms of their conversational enactment. These concepts can be misleading if they are taken to suggest something about psychological attitudes or motivations of participants. Instead, in the literature on conversation analysis, even when participants may "want" to disagree with someone, or "desire" not to accept an invitation to a party, they often display their disagreements or declinations (respectively) in a *dispreferred* fashion, by prefacing and delaying them with various things

including silences, tokens ("Well"), contrast markers ("but"), and the like. Thus, the term *preference* refers to structural properties of and systematic practices in conversation. Primary research in Sacks (1987) and Pomerantz (1984) captures these properties and practices, and secondary discussions, as in Heritage (1984b:264–280) are helpful.

31. Compare Houtkoop-Steenstra's explication of participant roles with Schober's (1999) discussion of audience design. Also see Clark and Schober (1992).

32. A related approach to studying the survey interview can be seen in Coits' (1996) dissertation.

REFERENCES

Atkinson, J. M., and P. Drew. 1979. *Order in Court: The Organisation of Verbal Interaction in Judicial Settings*. London: Macmillan.

Bassili, J. N. 1996. "The How and Why of Response Latency Measurement in Telephone Surveys." Pp. 319–346 in N. Schwarz and S. Sudman (Eds.), *Answering Questions: Methodology for Determining Cognitive and Communicating Processes in Survey Research*. San Francisco: Jossey-Bass.

Beach, W. A. 1993. "Transitional Regularities for 'Casual' 'Okay' Usages." *Journal of Pragmatics* 19: 325–352.

Beatty, P. 1995. "Understanding the Standardized/Non-standardized Interviewing Controversy." *Journal of Official Statistics* 11: 147–160.

Belli, R. F., J. M. Lepkowski, and M. U. Kabeto. In press. "The Respective Roles of Cognitive Processing Difficulty and Conversational Rapport on the Accuracy of Retrospective Reports of Doctors' Office Visits." In *Seventh Health Survey Research Methods Conference Proceedings* (Williamsburg, VA). Chicago: University of Illinois press.

Belli, R. F., W. L. Shay, and F. P. Stafford. 2000. "Event History Calendars and Question List Surveys: A Direct Comparison of Interviewing Methods." Survey Methodology Program Working Paper Series, Survey Research Center, University of Michigan.

Bingham, W., and B. Moore. 1924. *How to Interview*. New York: Harper and Row.

Bloom, J. E., and M. F. Schober 1999. "Respondent Cues That Survey Questions Are in Danger of Being Misunderstood." Pp. 991–996 in *Proceedings of the American Statistical Association Section on Survey Research Methods*. Alexandria, VA: American Statistical Association.

Boyd, E. A., and J. Heritage. In press. "Taking the Patient's Personal History: Questioning During Verbal Examination." In J. Heritage and D. W. Maynard (Eds.), *Practicing Medicine: Talk and Action in Primary Care Encounters*. Cambridge: Cambridge University Press.

Brenner, M. 1981. "Aspects of Conversational Structure in the Research Interview." Pp. 19–40 in P. Werth (Ed.), *Conversation and Discourse*. London: Croom Helm.

Brenner, M. 1982. "Response-Effects of 'Role-Restricted' Characteristics of the Interviewer." Pp. 131–165 in W. Dijkstra and J. van der Zouwen (Eds.), *Response Behaviour in the Survey-Interview*. London: Academic Press.

Brenner, M. 1985. Survey Interviewing. Pp. 9–35 in M. Brenner, J. Brown, and D. Canter (Eds.), *The Research Interview; Uses and Approaches*. London: Academic.

Briggs, C. L. 1986. *Learning How to Ask: A Sociolinguistic Appraisal of the Role of the Interview in Social Science Research*. Cambridge, MA: Cambridge University Press.

Bumpass, L. L. 1975. "Review Symposium." *Contemporary Sociology* 4: 474–76.

Button, G. 1987. "Answers as Interactional Products: Two Sequential Practices Used in Interviews." *Social Psychology Quarterly* 50: 160–171.

Button, G. 1993. *Technology in Working Order: Studies of Work, Interaction, and Technology*. Cambridge: Cambridge University Press.

Cannell, C. F., F. J. Fowler, Jr., and K. H. Marquis. 1968. "The Influence of Interviewer and Respondent Psychological and Behavioral Variables on the Reporting in Household Interviews." *Vital and Health Statistics, Series 2, Data Evaluation and Methods Research, No. 69*. Washington, DC: U.S. Department of Health and Human Services.

Cannell, C. F., and R. L. Kahn. 1968. "Interviewing." Pp. 526–95 in G. Lindzey and E. Aronson (Eds.), *The Handbook of Social Psychology*, Vol. II. Reading, MA: Addison-Wesley.

Cannell, C. F., S. A. Lawson and D. L. Hausser. 1975. *A Technique for Evaluating Interviewer Performance*. Ann Arbor, MI: Survey Research Center of the Institute for Social Research, University of Michigan.

Cannell, C. F., K. H. Marquis, and A. Laurent. 1977. "A Summary of Studies of Interviewing Methodology." *Vital and Health Statistics, Series 2, Data Evaluation and Methods Research, No. 26*. Rockville, MD: National Center for Health Statistics.

Cannell, C. F., P. V. Miller, and L. Oksenberg. 1981. "Research on Interviewing Techniques." Pp. 389–437 in S. Leinhardt (Ed.), *Sociological Methodology*, Vol. 11. San Francisco: Jossey-Bass.

Cannell, C., L. Oksenberg, G. Kalton, K. Bischoping, and F. Fowler. 1989. *New Techniques for Pretesting Survey Questions: Final Report*. Report supported by Grant Number HS 05616. Washington, DC: National Center for Health Services Research and Health Care Technology Assessment.

Cicourel, A. V. 1964. *Method and Measurement in Sociology*. New York: Free Press.

Cicourel, A. V. 1974. *Theory and Method in a Study of Argentine Fertility*. New York: Wiley.

Cicourel, A. V. 1982. "Interviews, Surveys, and the Problem of Ecological Validity." *The American Sociologist* 17: 11–20.

Clark, H. H. 1992. *Arenas of Language Use*. Chicago: University of Chicago Press.

Clark, H. H. 1996. *Using Language*. Cambridge: Cambridge University Press.

Clark, H. H., and M. F. Schober. 1992. "Asking Questions and Influencing answers." Pp. 15–48 in J. M. Tanur (Ed.), *Questions About Questions: Inquiries into the Cognitive Bases of Surveys*. New York: Russell Sage.

Clayman, S. E. 1988. "Displaying Neutrality in Television News Interviews." *Social Problems* 35: 474–492.

Coit, M. V. 1996. "An Interactional Sociolinguistic Approach to the Telephone Survey Interview As a Speech Event." Unpublished Ph.D. dissertation, Department of Linguistics, Georgetown University, Washington, DC.

Collins, H. M. 1985. *Changing Order: Replication and Induction in Scientific Practice*. Chicago: University of Chicago Press.

Collins, M. 1980. "Interviewer Variability, a Review of the Problem," *Journal of the Market Reserch Society* 22: 77–95.

Conrad, F. G., and M. F. Schober. 2000. "Clarifying Question Meaning in a Household Telephone Survey." *Public Opinion Quarterly* 64: 1–28.

Converse, J. M. 1987. *Survey Research in the United States: Root and Emergence 1890–1960*. Berkeley, CA: University of California Press.

Converse, J. M., and H. Schuman. 1974. *Conversations at Random: Survey Research as Interviewers See It*. New York: Wiley.

Couper, M. 1999. "The Application of Cognitive Science to Computer Assisted Interviewing." Pp. 277–300 in M. G. Sirken, D. J. Herrmann, S. Schechter, N. Schwarz, J. M. Tanur, and R. Tourangeau (Eds.), *Cognition and Survey Research*. New York: Wiley.

Dijkstra, W. 1987. "Interviewing Style and Respondent Behavior: An Experimental Study of the Survey Interview." *Sociological Methods and Research* 16: 309–343.

Dijkstra, W. 1999. "A New Method for Studying Verbal Interactions in Survey Interviews." *Journal of Official Statistics* 15: 67–85.

Dijkstra, W., and J. van der Zouwen. 1988. "Types of Inadequate Interviewer Behavior in Survey-Interviews." Pp. 24–35 in W. E. Saris and I. N. Gallhofer (Eds.), *Data Collection and Scaling*. New York: St. Martins.

Dykema, J., M. J. Lepkowksi, and S. Blixt. 1997. "The Effect of Interviewer and Respondent Behavior on Data Quality: Analysis of Interaction Coding in a Validation Study." Pp. 287–310 in L. Lyberg, P. Biemer, M. Collins, E. de Leeuw, C. Dippo, N. Schwarz, and D. Trewin (Eds.), *Survey Measurement and Process Quality*. New York: Wiley-Interscience.

Dykema, J., and N. C. Schaeffer. 1998. "An Analysis of Interviewer-Respondent Interaction on Data-Quality in a Validation Study." Paper presented at the annual meetings of the American Association for Public Opinion Research, May 14–17, St. Louis, MO.

Dykema, J., and N. C. Schaeffer. 2000. "Events, Instruments, and Reporting Errors." *American Sociological Review* 65: 619–629.

Engström, Y., and D. Middleton. 1996. *Cognition and Communication at Work*. Cambridge: Cambridge University Press.

Fowler, Jr., F. J., and T. W. Mangione. 1990. *Standardized Survey Interviewing: Minimizing Interviewer-Related Error*. Newbury Park: Sage.

Freud, S. 1961. *Civilization and Its Discontents*. New York: W. W. Norton & Co.

Garfinkel, H. 1967. *Studies in Ethnomethodology*. Englewood Cliffs, NJ: Prentice-Hall.

Garfinkel, H. 1988. "Evidence for Locally Produced, Naturally Accountable Phenomena of Order, Logic, Reason, Meaning, Method, etc. in and as of the Essential Quiddity of Immortal Ordinary Society (I of IV): An Announcement of Studies," *Sociological Theory* 6: 103–109.

Garfinkel, H. 1996. "Ethnomethodology's Program." *Social Psychology Quarterly* 59: 5–21.

Gerber, E. R. 1999. "The View from Anthropology: Ethnography and the Cognitive Interview." Pp. 217–234 in M. G. Sirken, D. J. Herrmann, S. Schechter, N. Schwarz, J. M. Tanur, and R. Tourangeau (Eds.), *Cognition and Survey Research*. New York: Wiley.

Goffman, E. 1981. "Footing." Pp. 124–59 in E. Goffman (Ed.), *Forms of Talk*. Philadelphia: University of Pennsylvania Press.

Goodwin, C. 1979. "The Interactive Construction of a Sentence in Natural Conversation." Pp. 97–121 in G. Psathas (Ed.), *Everyday Language: Studies in Ethnomethodology*, New York: Irvington.

Goodwin, C. 1981. *Conversational Organization: Interaction between Speakers and Hearers*. New York: Academic.

Goodwin, M. H. 1980. "Processes of Mutual Monitoring Implicated in the Production of Description Sequences." *Sociological Inquiry* 50: 303–317.

Goodwin, T. 1998. "Managing Participation in a Local Work Collectivity: Finding Space, Eavesdropping, and Standardization in the Survey Research Center." Unpublished master's paper, Department of Sociology, Indiana University, Bloomington, IN.

Goudy, W. J., and H. R. Potter. 1975. "Interview Rapport: Demise of a Concept." *Public Opinion Quarterly* 39: 529–543.

Groves, R. M. 1989. *Survey Errors and Survey Costs*. New York: Wiley.

Groves, R. M., R. B. Cialdini, and M. P. Couper. 1992. "Understanding the Decision to Participate in a Survey." *Public Opinion Quarterly* 56: 475–495.

Groves, R. M., and L. J. Magilavy. 1986. "Measuring and Explaining Interviewer Effects in Centralized Telephone Surveys." *Public Opinion Quarterly* 50: 251–266.

Heath, C., and P. Luff. 2000. *Technology in Action*. Cambridge: Cambridge University Press.

Heritage, J. 1984a. "A Change-of-State Token and Aspects of Its Sequential Placement." Pp. 299–345 in J. M. Atkinson and J. Heritage (Eds.), *Structures of Social Action*. Cambridge: Cambridge University Press.

Heritage, J. 1984b. *Garfinkel and Ethnomethodology*. Cambridge: Polity.

Heritage, J., and D. Greatbatch. 1991. "On the Institutional Character of Institutional Talk: The Case of News Interviews." Pp. 93–137 in D. Boden and D. H. Zimmerman (Eds.), *Talk and Social Structure*, Berkeley: University of California Press.

Holstein, J. A., and J. F. Gubrium. 1995. *The Active Interview*. Thousand Oaks, CA: Sage.

Holstein, J. A., and W. G. Staples. 1992. "Producing Evaluative Knowledge: The Interactional Bases of Social Science Findings." *Sociological Inquiry* 62: 11–35.

Houtkoop-Steenstra, H. 1996. "Probing Behavior of Interviewers in the Standardised Semi-Open Research Interview." *Quality and Quantity* 30: 205–230.

Houtkoop-Steenstra, H. 2000. *Interaction and the Standardized Survey Interview: The Living Questionnaire*. Cambridge: Cambridge University Press.

Houtkoop-Steenstra, H., and H. van den Bergh. 2000. "Effects of Introductions in Large-Scale Telephone Survey Interviews." *Sociological Methods and Research* 28: 281–300.

Hyman, H. H. 1975 [1954]. *Interviewing in Social Research*. Chicago, IL: University of Chicago Press.

Jefferson, G. 1978. "What's In a 'Nyem'?" *Sociology* 12: 135–139.

Kahn, R. L., and C. F. Cannell. 1957. *The Dynamics of Interviewing: Theory, Technique, and Cases*. New York: Wiley.

Kidder, L. H. 1981. *Research Methods in Social Relations*. New York: Holt, Rinehart and Winston.

Kovar, M. G., and P. Royston. (1990), "Comment on 'Interactional Troubles in Face-to-Face Survey Interviews.'" *Journal of the American Statistical Association* 85: 246–247.

Lave, J., and E. Wenger. 1991. *Situated Learning: Legitimate Peripheral Participation*. Cambridge: Cambridge University Press.

Lavin, D., and D. W. Maynard. 2001. "Standardization vs. Rapport: Respondent Laughter and Interviewer Reaction During Telephone Surveys." *American Sociological Review* 66: 453–479.

Levinson, S. 1988. "Putting Linguistics on a Proper Footing: Explorations in Goffman's Concepts of Participation." Pp. 161–227 in P. Drew and A. Wootton (Ed.), *Erving Goffman: Exploring the Interaction Order*. Boston: Northeastern University Press.

Luff, P., J. Hindmarsh, and C. Heath. 2000. *Workplace Studies: Recovering Work Practice and Informing System Design*. Cambridge: Cambridge University Press.

Lynch, M. 1991. "Method: Measurement — Ordinary and Scientific Measurement as Ethnomethodological Phenomena." Pp. 77–108 in G. Button (Ed.), *Ethnomethodology and the Human Sciences*. Cambridge: Cambridge University Press.

Lynch, M. 1992. "Extending Wittgenstein: The Pivotal Move from Epistemology to the Sociology of Science." Pp. 77–108 in A. Pickering (Ed.), *Science as Practice and Culture*. Chicago: University of Chicago Press.

Lynch, M. 1993. *Scientific Practice and Ordinary Action: Ethnomethodology and Social Studies of Science*. Cambridge: Cambridge University Press.

Mangione, T. W., F. J. Fowler, Jr., and T. A. Louis. 1992. "Question Characteristics and Interviewer Effects." *Journal of Official Statistics* 8: 293–307.

Marlaire, C. L., and D. W. Maynard. 1990. "Standardized Testing as an Interactional Phenomenon." *Sociology of Education* 63: 83–101.

Marquis, K. H., and C. F. Cannell. 1969. A Study of Interviewer-Respondent Interaction in the Urban Employment Survey. Survey Research Center, Institute for Social Research, University of Michigan, Ann Arbor, MI.

Marquis, K. H., C. F. Cannell, and A. Laurent. 1972. "Reporting Health Events in Household Interviews: Effects of Reinforcement, Question Length, and Reinterviews." *Vital and Health Statistics: Data Evaluation and Methods Research Series 2, No. 45*. Rockville, MD: U.S. Department of Health and Human Services.

Mathiowetz, N. A. (1998), "Respondent Expressions of Uncertainty: Data Source for Imputation." *Public Opinion Quarterly* 62: 47–56.

Maynard, D. W., and S. E. Clayman. 1991. "The Diversity of Ethnomethodology." *Annual Review of Sociology* 17: 385–418.

Maynard, D. W., and C. L. Marlaire. 1992. "Good Reasons for Bad Testing Performance: The Interactional Substrate of Educational Exams." *Qualitative Sociology* 15: 177–202.

Maynard, D. W., and N. C. Schaeffer. 1997. "Keeping the Gate: Declinations of the Request to Participate in a Telephone Survey Interview." *Sociological Methods and Research* 26: 34–79.

Maynard, D. W., and N. C. Schaeffer. 2000. "Toward a Sociology of Social Scientific Knowledge: Survey Research and Ethnomethodology's Asymmetric Alternates." *Social Studies of Science*, 30: 323–370.

Maynard, D. W., and M. R. Whalen. 1995. "Language, Action, and Social Interaction." Pp. 149–175 in K. Cook, G. Fine, and J. House (Eds.), *Sociological Perspectives in Social Psychology*. Boston: Allyn & Bacon.

Means, B., G. E. Swan, J. B. Jobe, and J. L. Esposito. 1991. "An Alternative Approach to Obtaining Personal History Data." Pp. 167–84 in P. P. Biemer, R. M. Groves, L. E. Lyberg, N. A. Mathiowetz, and S. Sudman (Eds.), *Measurement Errors in Surveys*. New York: Wiley.

Means, B., A. Nigam, M. Zarrow, E. F. Loftus, and M. S. Donaldson. 1989. "Autobiographical Memory for Health-Related Events." *Vital and Health Statistics, Series 6, Cognition and Survey Measurement, No. 2*. Washington, DC: National Center for Health Statistics.

Mehan, H. 1973. "Assessing Children's Language Using Abilities: Methodological and Cross Cultural Implications." Pp. 309–343 in M. Armer and A. D. Grimshaw, (Eds.), *Comparative Social Research: Methodological Problems and Strategies*. New York: Wiley.

Mehan, H. 1979. *Learning Lessons*. Cambridge, MA: Harvard University Press.

Mehan, H., A. Hertweck, and J. L. Meihls. 1986. *Handicapping the Handicapped: Decision-Making in Students' Educational Careers*. Stanford: Stanford University Press.

Miller, P. V., and C. F. Cannell. 1982. "A Study of Experimental Techniques for Telephone Interviewing." *Public Opinion Quarterly* 46: 250–269.

Mishler, E. G. 1986. *Research Interviewing: Context and Narrative*. Cambridge, MA: Harvard University Press.

Moore, R. J. 1999. "Achieving Understanding in the Standardized Interview: A Conversation Analytic Study of Talk in Telephone Surveys." Unpublished Ph.D. dissertation. Department of Sociology, Indiana University.

Morton-Williams, J. (1979). "The Use of 'Verbal Interaction Coding' for Evaluating Questionnaire." *Quality and Quantity* 13: 59–75.

Oksenberg, L., T. Beebe, S. Blixt, and C. F. Cannell. 1992. *Research on the Design and Conduct of the National Medical Expenditure Survey Interviews*. Ann Arbor, MI: Institute for Social Research.

Oksenberg, L., C. F. Cannell, and G. Kalton. 1991. "New Strategies for Pretesting Survey Questions." *Journal of Official Statistics* 7: 349–365.

O'Muircheartaigh, C. A. 1997. "Measurement Error in Surveys: A Historical Perspective." Pp. 1–28 in L. Lyberg, P. Biemer, M. Collins, E. de Leeuw, C. Dippo, N. Schwarz, and D. Trewin (Eds.), *Survey Measurement and Process Quality*. New York: Wiley-Interscience.

O'Muircheartaigh, C. A. 1999. "CASM: Successes, Failures, and Potential." Pp. 39–62 in M. G. Sirken, D. J. Herrmann, S. Schechter, N. Schwarz, J. M. Tanur, and R. Tourangeau (Eds.), *Cognition and Survey Research*. New York: Wiley.

Peneff, J. 1988. "The Observers Observed—French Survey Researchers at Work." *Social Problems* 35: 520–535.

Polanyi, M. 1958. *Personal Knowledge: Towards a Post-Critical Philosophy*. Chicago: University of Chicago Press.

Pomerantz, A. 1984. "Agreeing and Disagreeing with Assessments: Some Features of Preferred/Dispreferred Turn Shapes." Pp. 57–101 in J. M. Atkinson and J. Heritage (Eds.), *Structures of Social Action: Studies in Conversation Analysis*. Cambridge: Cambridge University Press.

Porter, T. M. 1995. *Trust in Numbers*. Princeton, NJ: Princeton University Press.

Psathas, G. 1995. *Conversation Analysis: The Study of Talk-in-Interaction.* Thousand Oaks, CA: Sage.

Roth, J. A. 1966. "Hired Hand Research." *American Sociologist* 1:190–196.

Sacks, H. 1984. "Notes on Methodology." Pp. 21–27 in J. M. Atkinson and J. Heritage (Eds.), *Structures of Social Action: Studies in Conversation Analysis.* Cambridge: Cambridge University Press.

Sacks, H. 1987. "On The Preferences for Agreement and Contiguity in Sequences in Conversation." Pp. 54–69 in G. Button and J. R. E. Lee (Eds.), *Talk and Social Organisation.* Clevedon, England: Multilingual Matters.

Sacks, H., E. A. Schegloff, and G. Jefferson. 1974. "A Simplest Systematics for the Organization of Turn-Taking for Conversation." *Language* 50: 696–735.

Schaeffer, N. C. 1991. "Conversation with a Purpose—Or Conversation? Interaction in the Standardized Interview." Pp. 367–92 in P. P. Biemer, R. M. Groves, L. E. Lyberg, N. A. Mathiowetz, and S. Sudman (Eds.), *Measurement Errors in Surveys.* New York: Wiley.

Schaeffer, N. C., and H.-W. Charng. 1991. "Two Experiments in Simplifying Response Categories: Intensity Ratings and Behavioral Frequencies." *Sociological Perspectives* 34: 165–182.

Schaeffer, N. C., and D. W. Maynard. 1996. "From Paradigm to Prototype and Back Again: Interactive Aspects of Cognitive Processing in Standardized Survey Interviews." Pp. 65–88 in N. Schwarz and S. Sudman (Eds.), *Answering Questions: Methodology for Determining Cognitive and Communicative Processes in Survey Research.* San Francisco, CA: Jossey-Bass.

Schaeffer, N. C., D. W. Maynard, and R. Cradock. 1993. "Negotiating Certainty: Uncertainty Proposals and Their Disposal in Standardized Survey Interviews." University of Wisconsin-Madison, Center for Demography and Ecology, Working Paper 93-25.

Schaeffer, N. C., and P. Royston. 1999. "Exploring the Interview Process." Pp. 51–55 in M. Sirkin, T. Jabine, G. Willis, M. Elizabeth, and C. Tucker (Eds.), *A New Agenda for Interdisciplinary Survey Research Methods: Proceedings of the CASM II Conference.* Hyattsville, MD: Centers for Disease Control and Prevention, National Center for Health Statistics.

Schegloff, E. A. 1990. "Comment." *Journal of the American Statistical Association* 85: 248–250.

Schegloff, E. A. 1997. "Turn Organization as a Direction for Inquiry into Grammar and Interaction." in E. Ochs, E. A. Schegloff, and S. A. Thompson (Eds.), *Interaction and Grammar.* Cambridge: Cambridge University Press.

Schegloff, E. A., G. Jefferson, and H. Sacks. 1977. "The Preference for Self-Correction in the Organization of Repair in Conversation." *Language* 53: 361–382.

Schober, M. F. 1999. "Making Sense of Questions: An Interactional Approach." Pp. 77–93 in M. G. Sirken, D. J. Herrmann, S. Schechter, N. Schwarz, J. M. Tanur, and R. Tourangeau (Eds.), *Cognition and Survey Research.* New York: Wiley.

Schober, M. F., and F. G. Conrad. 1997. "Does Conversational Interviewing Reduce Survey Measurement Error?" *Public Opinion Quarterly* 61: 576–602.

Schober, M. F., F. G. Conrad, and S. S. Fricker. 1999. "When and How Should Survey Interviewers Clarify Question Meaning?" Pp. 985–990 in *Proceedings of the American Statistical Association Section on Survey Research Methods.* Alexandria, VA: American Statistical Association.

Schuman, H. 1982. "Artifacts Are in the Eyes of the Beholder." *American Sociologist* 17: 21–28.

Schuman, H., and J. Ludwig. 1983. "The Norm of Even Handedness in Surveys As in Life." *American Sociological Review* 48: 112–120.

Schutz, A. 1973. *Collected Papers I: The Problem of Social Reality*. The Hague: Martinus Nijhoff.

Shapin, S. 1994. *A Social History of Truth: Civility and Science in Seventeenth-Century England*. Chicago: University of Chicago Press.

Smit, J. H., W. Dijkstra, and J. van der Zouwen. 1997. "Suggestive Interviewer Behaviour in Surveys: An Experimental Study." *Journal of Official Statistics* 13: 19–28.

Suchman, L. 1987. *Plans and Situated Actions*. Cambridge: Cambridge University Press.

Suchman, L., and B. Jordan. 1990. "Interactional Troubles in Face-to-Face Survey Interviews." *Journal of the American Statistical Association* 85: 232–253.

Sudman, S., N. M. Bradburn, and N. Schwarz. 1996. *Thinking about Answers: The Application of Cognitive Processes to Survey Methodology*. San Francisco: Jossey-Bass.

ten Have, P. 1999. *Doing Conversation Analysis*. London: Sage.

Tourangeau, R. 1990. "Comment on 'Interactional Troubles in Face-to-Face Survey Interviews.'" *Journal of the American Statistical Association* 85: 250–251.

Tourangeau, R., K. Rasinski, and L. Rips. 2000. *The Psychology of Survey Response*. Cambridge: Cambridge University Press.

Turner, C. F. 1982. "Why Do Surveys Disagree? Some Preliminary Hypotheses and Some Disagreeable Examples." Pp. 159–214 in C. Turner and E. Martin (Eds.), *Surveying Subjective Phenomena*. New York: Russell Sage.

van der Zouwen, J., W. Dijkstra, and J. H. Smit. 1991. "Studying Respondent-Interviewer Interaction: The Relationship Between Interviewing Style, Interviewer Behavior, and Response Behavior." Pp. 419–438 in P. P. Biemer, R. M. Groves, L. E. Lyberg, N. A. Mathiowetz, and S. Sudman (Eds.), *Measurement Errors in Surveys*. New York: Wiley.

Whalen, J. 1995. "A Technology of Order Production: Computer-Aided Dispatch in Public Safety Communications." Pp. 187–230 in P. ten Have and G. Psathas (Eds.), *Situated Order: Studies in the Social Organization of Talk and Embodied Activities*, Washington DC: University Press of America.

Weiss, C. H. 1968. "Validity of Welfare Mothers' Interview Responses." *Public Opinion Quarterly* 32: 622–633.

Weiss, C. H. 1970. "Interaction in the Research Interview: The Effects of Rapport on Response." Pp. 17–20 in *Proceedings of the Social Statistics Section of the American Statistical Association* (Aug.). Washington, DC: The American Statistical Association.

Wittgenstein, L. 1953. *Philosophical Investigations*. Translated by G. E. M. Anscombe. New York: Macmillian.

Zimmerman, D. H. 1970. "The Practicalities of Rule Use." Pp. 221–238 in J. Douglas (Ed.), *Understanding Everyday Life*. Chicago: Aldine.

Zimmerman, D. H. 1988. "On Conversation: The Conversation Analytic Perspective." Pp. 406–432 in J. Anderson (Ed.), *Communication Yearbook II*. Newbury Park, CA: Sage.

CHAPTER 2

Why Study Interaction in the Survey Interview? Response from a Survey Researcher

Johannes van der Zouwen

INTRODUCTION

This volume is about interviewer–respondent interaction in the standardized survey interview. Before we start discussing this topic, we must first address the question of why one should study these interactions at all. What is for survey researchers the "surplus value" of detailed information about the interactional processes that eventually have led to the answers given and the scores obtained? And why is it so important that survey interviews are as standardized as possible?

This volume provides a platform for discussion among conversation analysts, ethnomethodologists, those who study language and interactions, and survey researchers. As the title of this chapter already indicates, my answer to the question mentioned above is that of a survey researcher. People from the area of conversation analysis quite likely will come up with different answers. For a survey researcher an interview is only a means for collecting data about the behavior of people and for measuring opinions, attitudes, and other "subjective phenomena" (Turner and Martin, 1984) of interest to the researcher. For a conversation analyst a survey interview is a, sometimes peculiar, type of talk in interaction, an object of study in its own right.

In this chapter I argue, in line with Fowler and Mangione (1990), that the standardization of interviews is a necessary requirement for a proper quantitative analysis of survey data. But it is also a requirement that is difficult to comply with: Studies of response effects show that characteristics of the cognitive processes

Standardization and Tacit Knowledge: Interaction and Practice in the Survey Interview
Edited by Douglas W. Maynard, Hanneke Houtkoop-Steenstra, Nora Cate Schaeffer, and Johannes van der Zouwen
ISBN 0-471-35829-0

involved in answering a particular question and characteristics of the interactions taking place during the interview may have a considerable, often unintended, impact on the responses given. A detailed analysis of this interaction not only points to difficulties in cognitive processes related to responding, but also shows that the interaction between interviewer and respondent by itself may have a positive or a negative effect on the quality of the data obtained.

In this chapter the concept of *data quality* is treated as one component of survey quality, the one related to nonsampling errors (Lyberg et al., 1997). Data quality as used here is roughly the degree of absence of measurement errors (Groves, 1989). Which particular criterion of data quality has to be used in evaluating the answer to a question in a survey interview depends on the variable that question is assumed to measure and on the format of that question. If the question is related to factual data, answers can be judged to be (in)accurate or (in)valid. For questions about subjective phenomena such as attitudes, where no validating information exists, data quality is assessed by the absence or presence of response effects stemming from the wording of the questions, the behavior of the interviewer, and so forth (Sudman and Bradburn, 1974). The answer to an open question can be judged to be (in)complete, (not) informative, or (im)precise. For closed questions answers are only considered "adequate" if they equal one of the response alternatives offered.

THE NEED FOR STANDARDIZING INTERVIEWS

The variety of issues addressed in survey interviews is enormous: Questions refer to demographic and socioeconomic background variables, they are about actual, past, and even future events and behavior of the respondent; and they relate to opinions and attitudes. For some of these questions there exists external information that can be used to check the accuracy of the answers given by the respondent. For example: What is the highest grade you completed in school? or In what state (country) were you born? However, for the core variables of opinion and attitude research "there is no objective external evidence against which to verify the response" (Sudman and Bradburn, 1974:6–7). Moreover, the mental state of a subject is unobservable for other people: One never knows for sure what another person is really thinking, feeling, or believing. Researchers of such *subjective phenomena* (Turner and Martin, 1984) have to rely on a procedure for indirect measurement. In systems methodology this procedure is known as *black-box analysis*: One cannot observe the inside of the black box, but by comparing its inputs to its outputs, one may make inferences about its contents.

By asking questions (input) and listening to the answers (output) one may, under certain assumptions, make valid inferences about the respondents' opinions and attitudes (the contents of the black box). One of these assumptions is that when the input to different black boxes is similar, but their output is different, this can only be due to differences regarding their contents. And if the outputs are similar, their contents must be equal. Applied to survey methodology, this assumption implies that when a group of respondents answer the same question, those who give the same answer have the same value on the unobservable variable, and those who give different answers have different values. A less trivial, and more precise, formulation of this auxiliary hypothesis states that the set of different values of the unobserved variable is mapped onto the set of response alternatives by a so-called *response function*. This response function is for every respondent the same as the one intended by the designer of the instrument. By using the *inverse* of the assumed response function, that is, by going back from the response to the opinion, one may infer from the response alternative chosen the value on the unobserved variable (Dijkstra and van der Zouwen, 1977, 1982).

The logic of reasoning, borrowed from black-box analysis, requires that the researcher offer the same stimulus (question) to all respondents: For if the stimulus varies, variation in responses may be due to variation in the stimulus rather than to variation in the unobserved variable. In that case, the output, the response, cannot be used for making inferences about the unobservable variable that we try to measure. This not only implies that the wording of the question has to be identical for all respondents, but also that the meaning of the question for the respondents has to be approximately equal to the meaning intended by the designer of the question. And the way interviewers present the questions has to be about equal for every respondent and every interviewer. For that reason Smit (1995) has proposed the term stimulus–response model for this highly standardized method of data collection.

DEVIATIONS FROM THE STIMULUS–RESPONSE MODEL

Deviations Related to the Respondent's Task

If the requirements of the *stimulus–response model* are completely fulfilled, the answers given depend only on the opinions the respondents have or, more generally, on the values on the variable one tries to measure indirectly. The model thus assumes that answers are not directly correlated with other variables, including those that summarize features of the interview. However, methodological research has shown time and again the existence of so-called response effects, influences on the response to a question, of variables different from the variable one wants to measure with that question.

By the early 1970s the number of studies of response effects had become so large that Sudman and Bradburn (1974) could perform a kind of meta-analysis of the outcomes of these studies. For this meta-analysis they used a conceptual framework of the interview, with three important components: the interviewer, the respondent, and the task. Within this third component they distinguished between variables related to "(1) task structure, (2) the degree to which the task engages problems of self-presentation for the respondent, and (3) the salience of the required information" (p. 8). Variables related to task structure included the use of open or closed questions, the use of supplementary devices such as aided-recall techniques and showcards, and the mode of administration of the questionnaire (e.g., in a face-to-face interview, a telephone interview, or using a self-administered questionnaire). It turned out that the variables related to the (degree of) task structure, such as question format and mode of administration of the questionnaire, may have a significant influence on responses given by the respondent.

Four strategies are employed to investigate task-related response effects. The first strategy is an experimental one: the so-called *split-ballot experiment*. Different versions of the question, different orderings of the questions, or different modes of administration of the questionnaire, for example, are randomly assigned to subsamples of the population. Statistically significant differences between the response distributions of these subsamples are then attributed to differences between the two versions of the questionnaire or modes of administration applied. For example, Schuman and Presser (1981) administered opinion questions in which one random subsample was offered a "don't know" category and another random subsample was not. By comparing the response distributions obtained with the two subsamples, Schuman and Presser (1981) could examine the impact of offering the response category and could explore the implications of their observations for theories about attitudes.

The second strategy is a statistical procedure by which not the separate responses are analyzed but combinations or *patterns of responses* of the same respondent to different, but related, stimuli. If an instrument contains questions or statements that are assumed to measure the same unobservable variable, the responses to these items have to form a "plausible" pattern. For example, if a respondent says "yes" to a rather extreme item, he or she will also be expected to say "yes" to a somewhat less extreme item. This strategy is closely related to the procedures used by psychometricians assessing the reliability and scalability of multi-item scales. Thus, if a series of statements with the same response alternatives express conflicting views about an attitude object, the pattern of responses should be plausible. If they are not, researchers suspect that responses are affected by response tendencies such as "yeah-saying" or "nay-saying."

A third strategy is that interviewers are interviewed by the researcher about their experiences during the fieldwork (Cannell et al., 1989; Converse and Schuman, 1974; Presser and Blair, 1994). In this kind of *debriefing* session the interviewers may point at questions that are difficult to pose or to answer; questions that may raise misunderstandings or spoil the rapport between interviewer and respondent.

A fourth procedure to investigate task-related response effects in a sense tries "to open up" the black box; that is, to reconstruct the process inside the head of the respondent after the stimulus is given and before the answer is received (Schwarz and Sudman, 1996). Respondents are asked to "think aloud" while they are looking for an answer, leading to a *verbal report* of their cognitive activities (Bolton and Bronkhorst, 1996). After respondents have answered the question, or part of the questionnaire, they are asked what they thought and did when they tried to find an answer, how they understood the different terms used in the question (Belson, 1981), how they selected one option from a list of response alternatives, and so forth. Such an "interview about an interview" is called a *cognitive interview*. It is based on introspection, the ability of respondents to observe and report about their own cognitive processes.

The study of task-related response effects was strongly stimulated by cognitive psychologists who became interested in survey interviewing (Jabine et al., 1984). They conducted wording experiments and cognitive interviews about the processes respondents are dealing with when interpreting a question, looking for information in long-term memory, selecting the most appropriate response alternative, finding compromises between true and socially desirable answers, and the like. And they found tentative explanations for response effects that are related to topic, format, wording, and ordering of the questions (Strack, 1994; Sudman, Bradburn, and Schwarz, 1996; Tanur, 1992).

Deviations Related to the Interviewer

Some task-related response effects can be explained by close reading of the instrument and by a reconstruction of the cognitive processes involved in answering the questions. However, this procedure cannot explain why the same question asked in a face-to-face interview, or in a telephone interview, or in a self-administered (mail) instrument, may receive different responses from random subsets of respondents. As mentioned above, split-ballot experiments have shown the effects of the mode of administration (mode effects) on the (quality of the) data collected. For example, both face-to-face and telephone interviewing are better regarding the completeness of the collected data (i.e., less item nonresponse). Self-administered instruments perform better with respect to the accuracy of the data; "accuracy" here means that the information contained in the answers equals some kind of external, validating information (e.g., medical records) (de Leeuw, 1992).

If the question is the same and the subsamples are similar, but the distributions of the responses are different, the interviewer will come into focus: What is her or his impact on the responses given? One may tackle this problem by randomly assigning respondents to interviewers and computing the correlations between the scores obtained by the respondents and characteristics of the interviewers. The correlation coefficients provide the answers to questions like: Does it make a difference whether we use male or female interviewers, younger or older ones, experienced or inexperienced ones?

We now have many studies of so-called interviewer effects. An interesting outcome is that the interviewer's general social background variables, such as gender, age, race, and social status, make a difference only under very specific conditions. For example, the interviewer's race has an effect, but only if racial policies form the questionnaire topic [see reviews in Dijkstra and van der Zouwen (1982)].

However, stronger effects come from opinions the interviewers themselves have about the topics included in the instrument and from expectations they have about the answers the respondent will give (Sudman et al., 1977). These latter effects (i.e., the effects of information stored in the head of the interviewer) on the responses given by the respondent remain mysterious, a type of "mind reading," unless one assumes that interviewers display these ideas to their respondents in one way or another. To further analyze these effects, one can no longer restrict oneself to the responses as recorded on the questionnaire, but one has to observe and record the interview process itself.

Much traditional research on interviewer effects is relatively straightforward as compared to the difficulties of studying the interview process itself. It is only with reluctance that methodological researchers spend the time and effort needed to investigate the details of interactions in interviews.

One convincing argument for studying interactions in interviews came from the late Michael Brenner (1982). Expanding the work started by Cannell, Marquis and their colleagues (e.g., Cannell, Fowler, and Marguis, 1968; Cannell, Lawson, and Hausser, 1975), he conducted a detailed action-by-action analysis of audiotapes of survey interviews. Brenner showed that a normal standardized survey interview is far less standardized than the survey researcher likes to admit: Interviewers alter the wording of questions, respondents in fact respond to questions other than those intended by the researcher, interviewers suggest response options, and so forth. And Brenner's conclusion is corroborated by studies by Billiet and Loosveldt (1988), Dijkstra and van der Zouwen (1988), Houtkoop-Steenstra (1996), and by Schaeffer and Maynard (1996).

The observation that survey interviews sometimes seriously deviate from the standardization as required by the stimulus–response model has led to two opposite reactions. The first one is to restore the stimulus–response model to the extent possible. The most radical way is to completely eliminate the interviewer, by replacing the face-to-face interview or telephone interview by a self-administered instrument. And survey researchers may have good reasons for choosing that radical solution: The quality of the data collected by a competently designed self-administered instrument is often nearly as good as the quality of interview data, and the data are often cheaper to obtain (cf. de Leeuw, 1992). Moreover, eliminating the interviewer by using a self-administered instrument is an effective method of eliminating interviewer variance (see, e.g., O'Muircheartaigh, 1977). The same applies to the recently developed procedures for computerized self-interviews, or teleinterviewing (Saris, 1991, 1998).

Another radical reaction is to give up the strong pretensions and assumptions of the stimulus–response model, especially the idea that we need to strive for standardization in order to improve data quality (Silverman, 1985). In this approach, an interview is conceived as a type of conversation; and conversations have their own goals and rules. The interaction in an interview is not a component of the interview; the interaction *is* the interview, or rather, the interview *is* the interaction. And good interviews are those that best meet the rules and maxims that hold for good conversations (Suchman and Jordan, 1990, 1992).

These two radical proposals, eliminating the interviewer or giving up the requirement of standardization, each have their drawbacks. Careful study of interview protocols teaches us that the presence of an interviewer is sometimes vital to motivate the respondent, to explain the intended meaning of the question, and to prevent and correct misunderstandings and errors by respondents. On the other hand, giving up the standardization required by the stimulus–response model makes answers of different respondents difficult to compare, thereby making designs such as the survey and the panel study hardly usable and theories containing unobservable variables untestable (Beatty, 1995).

A less radical reaction to the observed violations of the stimulus–response model is constraining the variety of the interviewer's behavior (Fowler and Mangione, 1990). This may be effected by intensively training interviewers, by computer-assisted presentation of the questions, and by close supervision of the interviewers. Billiet and Loosveldt (1988) have shown that these measures indeed have an effect, in the sense that the variety in behavior decreases, and that unwanted interviewer behavior is reduced.

From the mode experiments mentioned above, it is clear that interviewers may have a positive effect not only on the decisions of (potential) respondents to participate in the survey, but also to answer all relevant questions. This explains the better performance of interviewing methods in comparison to the self-administered instrument, regarding the criteria of representativeness and completeness of the data. The presence of the interviewer "makes a difference" indeed.

But what is it that makes the difference? In his field experiments on the effect of different styles of standardized interviewing on responses in surveys, Dijkstra (1987) found that of the two styles compared, the formal style and the socioemotional one, the latter one leads to more frequent unwanted behavior of the interviewer. But the socioemotional style also led to better data quality: Respondents gave more accurate information about their neighborhood and fewer socially desirable answers. In his explanation of this unexpected result, Dijkstra pointed to the better motivation of respondents interviewed in the warm, caring socioemotional style for giving answers to difficult questions that were as good as possible.

In another study (van der Zouwen and Dijkstra, 1995) it appeared that interviewers played an important role by explaining to respondents the intended meaning of questions, by correcting inadequate or incomplete answers through all kinds of repair activities, and by further probing when the provided answer initially needs clarification. In her study "meeting both ends" Houtkoop-Steenstra (1995) has shown that interviewers sometimes skillfully adapt the wording of the questions to the particular situation of a respondent, by making use of information already provided by that respondent. These adjustments not only make the questions sound more appropriate but may also increase the respondent's motivation: "Look, this interviewer is really listening to me." So there is evidence for the beneficial influence the interviewer might have with respect to the adequacy, completeness, and accuracy of the data obtained.

On the other hand, also considerable empirical evidence exists that interviewers' deviations from the questionnaire may worsen data quality: Answers from different respondents become incomparable, and suggestively posed questions may have disturbed the information collected. To disentangle these opposite effects of the interviewer on data quality, we will have to look at the process of interaction between interviewer and respondent.

INTERACTION ANALYSIS

Introduction

In the last section we began answering the central question: Why study interactions in survey interviews? The provisional answer is twofold. First, the interaction by itself always affects the responses obtained, and the effect may be positive or negative. This also means that the interaction may improve, or hamper, the comparability of the responses obtained. And, second, a detailed analysis of interactions deviating from the one "outlined" by the designer of the instrument may inform the survey researcher about difficulties respondents and interviewers have with the tasks they are supposed to perform. This means, in part, that the interaction may function as an indicator, or "inspection door," of the cognitive and interactional process that eventually have led to a score on the interview form.

Types of Interaction Analysis

The detailed study of interactions in interviews requires tape-recorded interviews. Although Brenner (1982) has proved that an interaction analysis directly from tape (i.e., coding actions by listening to tapes) is possible, the audiotape is usually first transcribed into a verbatim protocol.

The procedures for analyzing these protocols differ with respect to the foci of interest (e.g., cognitive aspects of respondent's answering process, task-related behavior of interviewers, or interactional problems of both participants). And they differ with respect to the degree of standardization of the procedure and, related to that, the degree to which coders and computers are used in the analysis. The combination of these two differences leads to at least three different types of interaction analysis.

In the first type of analysis, the interaction itself is the focus of interest. Schaeffer and Maynard (1996) have shown how conversation analysis of only a few parts of detailed interview transcripts may reveal "the collaborative process of producing, understanding, and answering a question" (p. 82) as it takes place in survey interviews. Neither coders nor computers are used in this type of analysis.

In the second type of interaction analysis, the focus is the frequency with which certain task-related behaviors of interviewers and respondents occur. These behaviors are used as indicators of problems that particular survey questions may pose to respondents or interviewers. Fowler and Cannell (1996), for example, developed a protocol for behavior coding of audiotaped standardized telephone interviews. If a question leads to a relatively high proportion of problem indicators this question is classified as *problematic* and needs to be improved. Problem indicators are codes indicating changes in the reading of the question by the interviewer, or interruptions and requests for clarification by the respondent, or refusals to answer, or qualified and inadequate answers. In this second type of analysis coders are used.

The third type of interaction analysis is the computer-assisted coding of all speech acts included in the selected question–answer sequences (Dijkstra and Draisma, 1995; Dijkstra, 1999). The analysis of these coded speech acts may focus on the occurrence of *speech acts* with particular codes (e.g., for a question that is suggestively posed). Or it focuses on *patterns* of a few succeeding speech acts (e.g., a certain action of the respondent directly followed by a certain action of the interviewer). Or the analysis may focus on question–answer *sequences* as a whole, that is, the string of all speech acts starting with the reading out of a question from the questionnaire by the interviewer and ending just before the next question from the questionnaire will be read out.

This volume presents examples of all three procedures. The remainder of this chapter will be restricted to the last procedure mentioned, computer-assisted coding and analysis of complete question–answer sequences.

The Paradigmatic Sequence

In standardized interviews, many question–answer sequences have the same pattern. The question from the instrument, including its response alternatives (if applicable), is correctly read out by the interviewer, the respondent mentions his or her choice from the set of alternatives, the interviewer acknowledges this response, notes it on the instrument, and moves on to the next question. This type of standard interaction sequence has different labels: It is called "straightforward" (Sykes and Collins, 1992), "trivial" (van der Zouwen and Dijkstra, 1995), or "paradigmatic" (Schaeffer and Maynard, 1996). These labels indicate that this string of coded speech acts is precisely the one intended by the designer of the instrument (in that sense it is paradigmatic), that the process of communication is as efficient as possible, without any detours or redundancy (it is straightforward), and that the sequence does not contain additional information about the communication process (the things you might say about this sequence sound quite trivial). To avoid further confusion around these labels, we will use the term *paradigmatic* in the rest of this chapter to indicate such sequences. Paradigmatic sequences cannot explain differences regarding the responses obtained because they do not show any variety.

The interview included in this volume (Wisconsin AW01; see Appendix 2) contains a fair number of paradigmatic sequences. For example, all 12 questions concerning "the ways you might have felt or behaved during the past weeks; please tell me how many days you've felt this way?" (questions {q150} through {q205}) are correctly read out by the interviewer and, very briefly, answered with "none." There is no evidence here of the interviewer inappropriately influencing the respondent, nor do these interactions contain any information about the cognitive processes that have brought the respondent to choose just this response alternative. The same applies to the first three questions about her marriage, following this series ({q265} through {q267a}), and the four questions where the respondent is asked to evaluate government agencies on a scale from 1 to 10 ({q49} through {q49f}).

The probability that a sequence belongs to the category of paradigmatic sequences varies with all three components of the interview: the respondent, the interviewer, and especially with the task they have to fulfill, that is, with the topic, wording, and format of the question. Van der Zouwen and Dijkstra (1995) found that for 17 interviewers, involved in a survey of young adults, the percentage of paradigmatic sequences ranged from 27 to 72 percent. For nine questions, used in these face-to-face interviews, paradigmatic sequences ranged from 30 to 80 percent. Sykes and Collins (1992) report that especially single-response questions, where the respondent chooses one answer from a showcard, lead to high percentages of paradigmatic sequences: 87 percent of all such questions were executed in a straightforward manner. Van der Zouwen and Dijkstra (1996a; Chapter 18, this volume) also found that use of a showcard, and of response alternatives that fit well with the wording of the question and that are evenly placed within the total range of the variable that the question is intended to measure, strongly increases the probability that the question–answer sequence becomes paradigmatic.

Minor Deviations from the Paradigmatic Sequence

It sometimes occurs that a question–answer sequence deviates slightly from the paradigmatic one. For example, the interviewer does not read the question exactly as it is printed but does not change the meaning of the question. Another example of such a minor deviation is that the respondent starts answering by repeating the just mentioned response alternative, without waiting till the whole series of alternatives is presented. In the Wisconsin interview AW01, the respondent reacts to the question about the federal income taxes ({q30}: "Do you think federal income taxes are excessively high, about right, or low?") with immediately repeating "excessively high," after which the interviewer still mentions the other two alternatives. This also happens with the question about political party affiliation ({q56}: "Generally speaking, do you usually think of yourself as a republican, democrat, independent, or something else?"): The alternative "republican" is immediately repeated, but the interviewer still proceeds with the other alternatives. It seems as if the respondent tries to speed up the interview by interrupting the interviewer: "There is no need for summing up the other alternatives, because this one already is the right one."

These minor deviations from the paradigmatic sequence do not contain much interesting information for the survey researcher about the task to be fulfilled, nor is it likely that these small deviations will affect data quality.

Sequences Indicating Problems

The interaction in sequences that deviate significantly from paradigmatic sequences, the nonparadigmatic sequences for short, may reveal something about the cognitive and interactional process that eventually led to a score on the interview form. In this section we will give five examples, mostly from the Wisconsin AW01 interview, of (parts of) nonparadigmatic sequences that suggest problems the respondent, the interviewer, or both have with fulfilling their task.

1. *Unclear Purpose or Meaning of the Question or the Response Alternatives.* The respondent does not understand the question and asks for a repetition or an explanation. For example, in reaction to question {q5} about (sharing information collected by the Census Bureau with other government agencies), the respondent asks: "Share it with what other governments?" In reaction to question {q7} she asks "but see now, who's the different agencies of government?." Question {q76} reads: "Now turning to business conditions in the country as a whole, do you think that during the next twelve months we'll have good times financially, bad times or what?" The respondent does not understand what is meant here by "or what," and after a pause the interviewer explains "whatever you wanna say," which eventually leads to "not sure." Question {q359f} "What were your most important activities or duties on this job?" sounds pretty clear. But respondent and interviewer have difficulty with it. The respondent first asks "Well you mean of teaching?" After the interviewer has rephrased the question into "What do you do teaching?" the respondent gives a warning "that could take a long time," and it ends up with the "deal" between interviewer and respondent that the respondent will only mention the top three topics she teaches during health education. Eventually, after this negotiation, the respondent mentions five topics she teaches, topics that have to be typed letter by letter by the interviewer, leaving long pauses when only typing is heard.

2. *Incorrect Background Assumption.* Sometimes the respondent has difficulty with answering a question because the assumption "behind" the question does not hold, or at least seems not to hold, in his or her case. Although audience designed (Houtkoop-Steenstra, 1995), that is, worded in view of this particular survey population, some questions seem to be not applicable to certain categories of respondents. For example, question {q359}, "Did you do any work for pay last week?" is meant to assess the employment status of the respondent. The question gets the respondent into trouble: Because of the school summer holidays she does not *work* for her school, yet she is still *paid* by her school. So what is the correct answer? "No" because she did not work at school last week or "yes" because she was paid by the school? The next question "How many hours did you work last week?" {q359a} causes the same problem again. After a long discussion between interviewer and respondent, the interviewer cries "okay, I'm just gonna put a note in my computer to explain this."

3. *Too Difficult-to-Answer Question.* Answering a question sometimes requires complex "computing" by the respondent. The Wisconsin interview contains a lot of questions that are difficult to answer. But the questions {q461k to q461n} asking for an estimate of the percentage chance that respondent's total income before taxes will be under a certain number of dollars in the upcoming 12 months beat them all. No wonder that even the clever female teacher in AW01 makes a computing error and gives an answer opposite to the answer she seems to intend.

As we will see in the next section, this error is observed and corrected by the interviewer.

4. *Inadequate or Missing Response Alternatives.* The response alternatives sometimes do not fit well with the underlying dimension of the question. In a study by van der Zouwen and Dijkstra (1996a) this design failure appeared to be "responsible" for a great many nonparadigmatic question–answer sequences. For example, the question "Have you ever thought about how it would be if you became involuntary unemployed?" was followed by the official response alternatives: hardly ever, little, now and then, often, and very often. However, the most frequently spontaneously given responses to this question (i.e., no and never) were not included in the list, which led to much negotiating between interviewer and respondent, and thus to many nonparadigmatic sequences.

5. *Unclear Turn-Taking Indications.* The proper moment to start answering may not be clear to the respondent. This might happen when the question itself is introduced or followed by an explanation that confuses the respondent. After hearing the first sentence of the long introduction to question {q7d} the respondent asks: "I forgot, what was your question?" In her study of telephone surveys Houtkoop-Steenstra (1995) mentioned examples of respondents already starting answering a question before the mandatory additional explanation of some terms used in the question was presented. Problems with transition points for turns have been noted earlier by Cannell et al. (1989) and by Schaeffer (1991:386). A similar confusion, regarding the appropriate moment to start answering the question, occurs when the question is asked and then, in a next sentence is followed by a series of response alternatives. A quick respondent, like the female teacher in the interview AW01, will already have given the answer to the question before the alternatives are read out. This respondent leaves the interviewer with the decision, either to mention all the "official" alternatives or just those that fit best with the provisional answer; see the interactions following the questions {q9e}, {q9f}, {q30}, and {q848} in Appendix 2. These examples suggest that when the interaction following a question does not conform to the paradigmatic sequence of standardized interviewing, the design of the question may be at fault.

Positive Effects of the Interaction on Data Quality

Deviations from the paradigmatic sequence may not only point to problems the participants in the interview may have with the fulfillment of their task; deviations may also have a positive or negative effect on data quality. In this section we give four examples of positive effects, that is, of actions that may lead to data of better quality than if these actions had not taken place.

1. *Motivating and Reassuring the Respondent.* The interviewer may try to motivate the respondent to give the best possible answer or at least a substantive one. In reaction to two questions about the purposes for which the census is used, {q5b} and {q5d}, the respondent says "no idea." In an effort to reduce the number of nonsubstantive answers, the interviewer says "we don't expect you to know

the solid hard fast truth, it's just whatever your perception may be." This effort, however, is not successful: The answer to the next question in this series {q5h} is again "no idea." With question {q50d} the interviewer is more successful. Asked whether she agrees or not with the statement "the government knows more about me than it needs to," the respondent says "Well I don't know. I don't know what they know about me. I don't really care what they know." The interviewer probes further ("so would you agree, disagree?"), and eventually, after a short pause, the respondent reacts with "agree."

2. *Tuning the Provisional Response.* It quite often occurs that a respondent does not answer a closed question by choosing one of the response alternatives offered but by giving an answer of "his own making." If the interviewer transforms this provisional answer into one of the official response alternatives, we call this *tuning* of the response (van der Zouwen and Dijkstra, 1996a). Tuning is a frequently occurring activity of the interviewer; it is often followed by a question of the interviewer whether the chosen alternative is the right one. For example, the interviewer checks whether the official "nothing at all" in response to question {q6} equals the provisional answer "none" and whether the official "favor it strongly" in question {q9e} equals "good idea" uttered by the respondent.

3. *Correction of the Response.* The interviewer may hear an answer that is probably incorrect, or at least not plausible, given other information the respondent has already provided. In these cases the interviewer may confront the respondent with her doubts and see whether something is wrong indeed. A good example of such a correction activity appears in the interaction following question {q461n}, the very complex question about the chance "that the own total income before taxes will be under forty thousand dollars in the upcoming twelve months." The interviewer realizes that the respondent makes a mistake when "computing" her response: She says "zero percent," while the answer "hundred percent" is much more likely, given her response to question {q460c}, that is, that she expects to earn no more than $38,000 in the next 12 months. And indeed, the respondent acknowledges that she had made a mistake. Thus the nonparadigmatic action by the interviewer in this case had a positive effect on data quality, that is, on the accuracy of the response.

A less spectacular instance of a response correction, initiated by the interviewer, is the interaction following the question about the evaluation of the sales tax {q36}, with response alternatives "excessively high," "about right," and "low." The respondent answers with "the second one," and the interviewer, already informed about the negative opinion of the respondent regarding U.S. taxation, checks "about right," which eventually leads to "Oh no, excessively high."

4. *Requesting and Offering Additional Explanation about the Task.* The respondent who has missed part of the text or the meaning of the question may ask for a repetition or an explanation of the question or the response alternatives. An example is given by the interaction following question {q28}: "Do you think the present federal income tax system is fair to most people or unfair to most people?" The respondent asks whether the first alternative is "very fair," and the interviewer reacts with "just fair or unfair," followed by the question of the

respondent "nothing in between?" whereupon the interviewer says "No middle of the road on this one." Eventually the response becomes "unfair."

The examples above make clear that the same sequence may point at a problem with the task to be fulfilled and at an effort of one or both participants to solve that problem to the extent possible.

Negative Effects of the Interaction on Data Quality

In the Wisconsin interview, used earlier to illustrate different types of nonparadigmatic sequences, both interviewer and respondent do a very good job. The interviewer rarely makes mistakes, and the only criticism of the respondent might be that she answers too quickly, often forcing the interviewer into tuning activities. That so many nonparadigmatic sequences occur is due mainly to the complexity of the questions asked.

This means that for examples of nonparadigmatic sequences that might have negative effects on data quality we have to look for another data set. The data of this section form part of the survey of young adults conducted in the Netherlands in 1987 that was earlier mentioned. The 1227 question–answer sequences analyzed stem from 185 interviews, conducted by 17 interviewers. Nineteen percent of these sequences contain at least one action that is inadequate, as seen from the viewpoint of standardization. Here the six most often observed inadequacies, which probably threaten data quality, follow (van der Zouwen and Dijkstra, 1996a; Chapter 18, this volume).

1. *Changing the Meaning of the Question.* The interviewer makes a mistake when reading out the question and does not correct this mistake. The respondent thus answers a question other than that meant by the researcher. For example, the question "How much trouble do you take to find living accommodations?" was meant to relate to the *present* situation. For many of the young adults interviewed this means: "living in the parental home." But this question is preceded by a question that relates to a *future* situation, that is, if the respondent leaves the parental home. No wonder that some respondents were uncertain about the time referent of this question. A couple of interviewers prevented this uncertainty on the part of the respondent by adding words like "then," ("How much trouble will you take then to find living accommodations?"), thereby not only changing the meaning of the question but also distorting the responses, resulting in a higher score: By the time you leave the parental home, you naturally have to find another place to live. The meaning of the question is changed in 3 percent of all (1227) sequences.

2. *Changing the Response Categories.* Some alternatives are omitted, misread, or presented in an incorrect order, and the error is not corrected. This mistake occurs in 6 percent of all sequences.

3. *Offering a Risky or Wrong Interpretation of the Initial Response.* Here the tuning of the provisional answer of the respondent to one of the official alternatives has gone wrong, or there is at least not enough evidence in the interaction

that it went right, because the interviewer did not check with the respondent the result of the tuning. This happened in 2 percent of the sequences.

4. *Suggesting an Answer.* The interviewer offers a "candidate answer": "It is X, isn't?" This type of inadequate interviewer behavior has to be distinguished from an appropriate "verification" by the interviewer of information the respondent has already provided. In the type of mistake meant here the answer offered to the respondent is not completely implied by earlier responses. This inadequacy occurred in 3 percent of all sequences.

5. *Insufficient Probing.* In 2 percent of the sequences the respondent did not provide an adequate answer, but the interviewer records some inferred score on the instrument and proceeds to the next question.

6. *Other.* For example, a third party actively engages in a discussion about a proper response (1 percent). In another study—an interactional analysis of telephone interviews about opinions concerning television advertisement (van der Zouwen and Dijkstra, 1996b; this volume, Chapter 18)—we came across a risky type of feedback given by the interviewers to the answers of the respondents. Quite often the interviewers reacted positively to critical opinions regarding this type of advertisement: "I can understand that very well"; "I think just the same way," and so forth. This type of improper feedback might have a response effect on subsequent questions about the same topic because this interviewer reaction enforces the utterance of negative opinions concerning commercials.

SUMMARY AND CONCLUSION

We have seen that for an adequate measurement of unobservable variables in surveys, especially for those for which there is no "validating information" available, like for the measurement of opinions and attitudes, it is vital that all the respondents react to the same input. This means that the question is presented to, and understood by, all respondents in the way as meant by the designer of the instrument. The easiest way to make inputs equal is to standardize all actions of the participants. However, methodological research has made clear that the goal of standardization is not always reached in practice. This observation has urged further investigation into the areas where, and the conditions under which, the basic requirement of survey research, that is, comparability of answers, is at risk.

The study of response effects may point to ways in which standardization has not been successful and identify determinants of response bias. Split-ballot experiments may indicate where differences with respect to question wording, question ordering, and question format or regarding the selection, instruction, and training of interviewers become crucial. Analyses of response patterns, debriefings of interviewers, think-aloud protocols, and cognitive interviewing of respondents may inform the researcher about the difficulties respondents have in fulfilling their tasks, about misunderstanding of questions, and about the reasoning behind the choice of a particular answer. All these methods together may provide the survey researcher with a large amount of information that can be used for improving the questionnaire and interviewer training and for producing a more sophisticated interpretation of the outcomes of a survey.

The analysis of interactions, described briefly in the preceding section, may provide the survey researcher with additional information about the problems the participants in the interview have with adequately fulfilling their tasks. Moreover, a detailed analysis of question–answer sequences may tell us how participants, especially the interviewer, try to overcome these problems by motivating and reassuring the respondents, and by performing various kinds of checks and by doing repair activities. But we can also observe in detail how some interviewers make serious mistakes, mistakes that are hardly perceptible by inspecting and analyzing completed instruments, debriefing interviewers, or reinterviewing respondents. Summarized, analysis of interaction can provide necessary and otherwise unobtainable information for the survey researcher.

However, interaction analysis has three drawbacks. The first is that it is time consuming. If the interviews are recorded on audiotapes, the fieldwork as such does not take time, but the transcription and coding of the recorded sequences certainly does. That is the reason some researchers are developing computer-assisted coding procedures that will speed up the coding process (Dijkstra, this volume, Chapter 17; Dijkstra and Draisma, 1995).

The second is that only a minority of the coded interactions, the nonparadigmatic sequences, contains information about the processes investigated. However, a sequence proceeding in a paradigmatic way is neither a guarantee that participants do not have problems nor that the quality of the data collected via this sequence is optimal. For example, a respondent's misunderstanding of the question is only recognizable if the respondent does something other than selecting one of the response alternatives offered.

The last drawback is paradoxical: The method of interaction analysis is threatened by its own success. Like other procedures aimed at improving methods of data collection, interaction analysis suffers from a feedback loop between the "controllability of the process" and the "observability of the process" (van der Zouwen, 1990). The more we learn from interaction analyses about how to properly design instruments and instruct interviewers, the more we decrease the proportion of nonparadigmatic sequences and, consequently, the less information we will gather from these sequences. Survey researchers will fight nonparadigmatic interactions, but by doing so they also dry out the source of information the nonparadigmatic sequence provides. However, at present there is still plenty of room and need for using this elegant procedure as the remainder of this book will demonstrate.

REFERENCES

Beatty, P. 1995. "Understanding the Standardized/Non-Standardized Interviewing Controversy." *Journal of Official Statistics* 11: 147–160.

Belson, W. A. 1981. *The Design and Understanding of Survey Questions*. Aldershot: Gower.

Billiet, J., and G. Loosveldt. 1988. "Improvement of the Quality of Responses to Factual Survey Questions by Interviewer Training." *Public Opinion Quarterly* 52: 190–211.

Bolton, R. N., and T. M. Bronkhorst. 1996. "Questionnaire Pretesting; Computer-Asssisted Coding of Concurrent Protocols." Pp. 37–64 in N. Schwarz and S. Sudman (Eds.), *Answering Questions: Methodology for Determining Cognitive and Communicative Processes in Survey Research.* San Francisco: Jossey-Bass.

Brenner, M. 1982. "Response-Effects of 'Role-Restricted' Characteristics of the Interviewer." Pp. 131–165 in W. Dijkstra and J. van der Zouwen (Eds.), *Response Behaviour in the Survey-Interview*. London: Academic Press.

Cannell, C. F., F. J. Fowler, Jr., and K. H. Marquis. 1968. "The Influence of Interviewer and Respondent Psychological and Behavioral Variables on the Reporting in Household Interviews." In *Vital and Health Statistics: Data Evaluation and Methods Research Series 2*, No. 26. Washington DC: U.S. Department of Health and Human Services.

Cannell, C. F., G. Kalton, L. Oksenberg, K. Bischoping, and F. J. Fowler. 1989. "New Techniques for Pretesting Survey Questions." Final report, Grant Number HS 05616, National Center for Health Services Research and Health Care Technology Assessment.

Cannell, C. F., S. A. Lawson, and D. L. Hausser. 1975. *A Technique for Evaluating Interviewer Performance; A Manual for Coding and Analyzing Interviewer Behavior from Tape Recordings of Household Interviews.* Ann Arbor, MI: Survey Research Center of the Institute for Social Research, University of Michigan.

Converse, J. M., and H. Schuman. 1974. *Conversations at Random; Survey Research as Interviewers See it.* New York: Wiley.

Dijkstra, W. 1987. "Interviewing Style and Respondent Behavior: An Experimental Study of the Survey-Interview." *Sociological Methods and Research* 16: 309–334.

Dijkstra, W. 1999. "A New Method for Studying Verbal Interactions in Survey-Interviews." *Journal of Official Statistics* 15: 67–85.

Dijkstra, W., and S. Draisma. 1995. "A New Method for Studying Verbal Interactions in Survey Interviews." Pp. 296–301 in *Proceedings of the International Conference on Survey Measurement and Process Quality in Bristol* (UK). Alexandria, VA: American Statistical Association.

Dijkstra, W., and J. van der Zouwen. 1977. "Testing Auxiliary Hypotheses Behind the Interview." *Annals of Systems Research* 6: 49–63.

Dijkstra, W., and J. van der Zouwen (Eds.). 1982. *Response Behaviour in the Survey-Interview*. London: Academic Press.

Dijkstra, W., and J. van der Zouwen. 1988. "Types of Inadequate Interviewer Behaviour in Survey-Interviews." Pp. 24–35 in W. E. Saris and I. N. Gallhofer (Eds.), *Sociometric Research*, Vol. 1: *Data Collection and Scaling*. London: MacMillan.

Fowler, F. J., Jr., and C. F. Cannell. 1996. "Using Behavioral Coding to Identify Cognitive Problems with Survey Questions." Pp. 15–36 in N. Schwarz and S. Sudman (Eds.), *Answering Questions: Methodology for Determining Cognitive and Communicative Processes in Survey Research*. San Francisco: Jossey-Bass.

Fowler, F. J., Jr., and T. W. Mangione. 1990. *Standardized Survey Interviewing: Minimizing Interviewer-Related Error*. Newbury Park: Sage.

Groves, R. M. 1989. *Survey Errors and Survey Costs*. New York: Wiley.

Houtkoop-Steenstra, H. 1995. "Meeting Both Ends. Between Standardization and Recipient Design in Telephone Survey Interviews." Pp. 91–106 in P. ten Have and

G. Psathas (Eds.), *Situated Order: Studies in the Social Organization of Talk and Embodied Activities*. Washington, DC: University Press of America.

Houtkoop-Steenstra, H. 1996. "Probing Behaviour of Interviewers in the Standardized Semi-open Research Interview." *Quality and Quantity* 30: 205–230.

Jabine, T. B., M. L. Straf, J. M. Tanur, and R. Tourangeau (Eds.). 1984. *Cognitive Aspects of Survey Methodology: Building a Bridge between Disciplines*. Washington, DC: National Academy Press.

de Leeuw, E. D. 1992. *Data Quality in Mail, Telephone and Face to Face Surveys*. Amsterdam: TT-Publikaties.

Lyberg, L., P. Biemer, M. Collins, E. de Leeuw, C. Dippo, N. Schwarz, and D. Trewin (Eds.). 1997. *Survey Measurement and Process Quality*. New York: Wiley.

O'Muircheartaigh, C. A. 1977. "Response Errors." Pp. 193–239 in C. A. O'Muircheartaigh and C. A. Payne (Eds.), *Analysis of Survey Data*, Vol. 2, London: Wiley.

Presser, S., and J. Blair. 1994. "Survey Pretesting: Do Different Methods Produce Different Results?" Pp. 73–104 in P. V. Marsden (Ed.), *Sociological Methodology*, Washington, DC: Jossey-Bass.

Saris, W. E. 1991. *Computer-Assisted Interviewing*. Newbury Park, CA: Sage.

Saris, W. E. 1998. "Ten Years of Interviewing without Interviewers: The Telepanel." Pp. 409–429 in M. P. Couper, R. P. Baker, J. Bethlehem, C. Z. F. Clark, J. Martin, W. L. Nicholls II, and J. M. O'Reilly (Eds.), *Computer Assisted Survey Information Collection*. New York: Wiley.

Schaeffer, N. C. 1991. "Conversation with a Purpose—or Conversation? Interaction in the Standardized Interview." Pp. 367–391 in P. P. Biemer, R. M. Groves, L. E. Lyberg, N. A. Mathiowetz, and S. Sudman (Eds.), *Measurement Errors in Surveys*. New York: Wiley.

Schaeffer, N. C., and D. W. Maynard. 1996. "From Paradigm to Prototype and Back Again: Interactive Aspects of Cognitive Processing in Standardized Survey Interviews." Pp. 65–88 in N. Schwarz and S. Sudman (Eds.), *Answering Questions: Methodology for Determining Cognitive and Communicative Processes in Survey Research*. San Francisco: Jossey-Bass.

Schuman, H., and S. Presser. 1981. *Questions and Answers in Attitude Surveys; Experiments on Question Form, Wording, and Context*. New York: Academic.

Schwarz, N., and S. Sudman (Eds.). 1996. *Answering Questions. Methodology for Determining Cognitive and Communicative Processes in Survey Research*. San Francisco: Jossey-Bass.

Silverman, D. 1985. *Qualitative Methodology and Sociology*. Aldershot: Gower.

Smit, J. H. 1995. "Suggestieve vragen in survey interviews; vóórkomen, oorzaken en gevolgen" "[Leading Questions in Survey Interviews, Occurrence, Determinants and Effects]." Unpublished Ph.D. dissertation, Vrije Universiteit, Amsterdam.

Strack, F. 1994. *Zur Psychologie der standardisierten Befragung; Kognitive und kommunikative Prozesse*. Berlin: Springer.

Suchman, L., and B. Jordan. 1990. "Interactional Troubles in Face-to-Face Survey Interviews." *Journal of the American Statistical Association* 85: 232–241.

Suchman, L., and B. Jordan. 1992. "Validity and the Collaborative Construction of Meaning in Face-to-Face Surveys." Pp. 241–267 in J. Tanur (Ed.), *Questions about Questions*. New York: Russell Sage.

Sudman, S., and N. M. Bradburn. 1974. *Response Effects in Surveys; A Review and Synthesis*. Chicago: Aldine.

Sudman, S., N. M. Bradburn, E. Blair, and C. Stocking. 1977. "Modest Expectations. The Effects of Interviewers' Prior Expectations on Responses." *Sociological Methods and Research* 6: 171–182.

Sudman, S., N. M. Bradburn, and N. Schwarz. 1996. *Thinking about Answers: The Application of Cognitive Processes to Survey Methodology*. San Francisco: Jossey-Bass.

Sykes, W., and M. Collins. 1992. "Anatomy of the Survey Interview." *Journal of Official Statistics* 8: 277–291.

Tanur, J. M. (Ed.). 1992. *Questions about Questions*. New York: Russell Sage.

Turner, C. F., and E. Martin (Eds.). 1984. *Surveying Subjective Phenomena*. New York: Russell Sage.

van der Zouwen, J. 1990. "The Impact of Self-Referentiality of Social Systems on Research Methodology." Pp. 59–68 in F. Geyer and J. van der Zouwen (Eds.), *Self-Referencing in Social Systems*. Salinas, CA: Intersystems.

van der Zouwen, J., and W. Dijkstra. 1995. "Trivial and Non-trivial Question-Answer Sequences; Types, Determinants and Effects on Data Quality." Pp. 81–86 in *Proceedings of the International Conference on Survey Measurement and Process Quality in Bristol* (UK). Alexandria, VA: American Statistical Association.

van der Zouwen, J., and W. Dijkstra. 1996a. "The Impact of the Question on the Interactions in Survey-Interviews." Paper presented at the Fourth International ISA Conference on Social Science Methodology (Essex '96), July 1–5, University of Essex, Colchester, United Kingdom.

van der Zouwen, J., and W. Dijkstra. 1996b. "Testing (CATI-) Questionnaires Using Computer Assisted Interaction Coding." Paper presented at the International Conference on Computer-Assisted Survey Information Collection (InterCASIC '96), December 11–14, San Antonio, TX.

CHAPTER 3

A Collaborative View of Standardized Survey Interviews

Michael F. Schober and Frederick G. Conrad

What role should interviewers play in standardized surveys? Under the prevailing view, interviewers should be little more than purveyors of scripted questions; survey data are most valid when interviewers are least likely to have biased the answers, or at least to have biased them in different ways. This view, we propose, rests on problematic underlying assumptions about communication: that words contain meaning and that sameness of words entails sameness of meaning. Here we examine these assumptions more closely and demonstrate how interviewers can influence responses even when they follow the most strictly standardized prescriptions to the letter. We present a more collaborative view of interviewer–respondent interaction and explore how and when it might be beneficial to adopt such a view.

STANDARDIZING WHAT INTERVIEWERS SAY

Most large-scale surveys — including the one on which this volume focuses — try to standardize what interviewers say. As described by Fowler and Mangione (1990) in an influential work on the theory and practice of standardized survey interviewing, interviewers must read questions exactly as worded, they must probe neutrally, and they must never allow their own ideas to influence the respondents' answers. According to Fowler and Mangione (1990), this practice reduces or eliminates what they call *interviewer-related error* — any systematic effect of particular interviewers on survey responses.

Standardization and Tacit Knowledge: Interaction and Practice in the Survey Interview
Edited by Douglas W. Maynard, Hanneke Houtkoop-Steenstra, Nora Cate Schaeffer, and Johannes van der Zouwen
ISBN 0-471-35829-0

These prescriptions reflect how most researchers have conceived of the sources of measurement error in survey interviews. Under the usual view, various kinds of error contribute to overall measurement error: interviewer error, respondent error, sampling error, problems of question design (including wording problems), and errors in coding or classifying responses (see Groves, 1989, 1991). Some of these kinds of errors are more under researchers' control than others. Sampling, question design, and interviewer training are up to the researchers, but other kinds of error are less clearly controllable, for example, those due to the accuracy of respondents' memories or their willingness to respond honestly.

Under this view, the different kinds of error are seen as independent (although they can affect one another), and they are reduced by different methods. Sampling error can be reduced by improved statistical methods. Question design problems can be reduced through laboratory pretesting. Interviewer error can be reduced through standardized interviewing practices. Classification errors can be reduced by more strictly monitoring or automating the process of classifying the data.

Error due to the respondent, while not directly controllable, can be minimized by reducing the other sorts of error. So if questions have been well pretested, respondents should be less likely to misunderstand them. And if interviewers behave in a standardized way, that is, expose respondents to the same question stimulus, researchers can be confident that differences in the answers stem only from the respondent, and not from anything in the interviewer's behavior (Fowler and Mangione, 1990:14–15).

THE MESSAGE MODEL OF COMMUNICATION

By completely separating interviewer behavior, respondent behavior, and question wording, this prevailing approach relies on a view of communication that has been discredited, at least for ordinary spontaneous conversations. This view, which dates back to John Locke or even earlier, has been called, variously, the *message model* (Akmajian et al., 1990), the *conduit metaphor* (Reddy, 1979/1993), and the *meaning-in-words assumption* (Schober, 1998). On this view, speakers encode their thoughts into linguistic messages and send these messages to recipients, who decode them into their own thoughts. Thoughts and conceptual material are thus transferred from one head to the other via words.

Although the message model captures most people's intuitions about how communication works, it cannot account for all of what goes on in ordinary conversations (see also Clark, 1992, 1996; Gibbs, 1994; Maynard and Whalen, 1995). The problem is that the message model assumes that the meaning of speakers' words is in the words themselves. But it isn't. Rather, the same words can have vastly different meanings depending on what speakers intend them to mean on particular occasions (see Akmajian et al., 1990). In fact, meaning is even more complicated than this: Utterance meaning depends not only on what speakers mean on particular occasions but also on how addressees take those meanings — what speakers and addressees together determine the words mean. If this is true, the meanings that a speaker and addressee jointly determine may differ from the meanings inferred by someone who isn't participating in their conversation.

An empirical demonstration of this is found in a laboratory experiment by Schober and Clark (1989). In the study, pairs of strangers performed a matching task. One person, the *director*, saw an array of unfamiliar abstract geometric figures. The other person, the *matcher*, sat behind a screen with a scrambled set of the same abstract figures. The task was for the director to get the matcher to arrange his figures in the same order as the director's figures, saying anything it took for the two to perform the task. Unbeknownst to the director and matcher, a third person, the *overhearer*, listened to every word they said to each other from the moment they met. As the director and matcher conversed to arrange the figures, the overhearer tried to arrange the same set of figures based on what she heard.

The logic of the study was this: If a speaker's meaning is contained in the words themselves, then overhearers should match the directors' (speakers') array of figures at least as well as matchers do, because the overhearers hear every word the matchers (the addressees) hear. In fact, overhearers may perform better than matchers, because the overhearers don't have to expend any attentional resources on letting directors know that they have understood. In contrast, if speakers and addressees jointly determine the meaning of what is said (i.e., if comprehension requires participation in the interactive process), then overhearers should not match the figures as well as the participating matchers, even though they have heard every word uttered from the moment the director and matcher have met.

The results were plain: Conversational participants (the matchers) matched figures with nearly perfect accuracy. Nonparticipants (overhearers) matched figures reliably less accurately. And this continued to be the case over time. After directors and matchers completed the task, the directors' figures were reshuffled and the director and matcher began again. Overhearers' performance was poorer than the matchers for even the sixth reshuffling, even though they heard everything the conversational participants said. And another set of overhearers listening to tape-recordings of the conversations, who could press the pause button and stop and think as long as they liked, performed no better.

A message model can't account for findings like these. If meaning is in the words themselves, overhearers should understand at least as well as addressees. Because on some occasions they don't, the conclusion must be that meaning is not only in the words; the words themselves do not always bring speakers' and listeners' conceptualizations into sufficient alignment [for further discussion see Schober (1998)]. We need an alternative model to explicitly include what both parties in a conversation do together.

One proposal is Clark and Wilkes-Gibbs' (1986) *collaborative* model (see also Clark, 1992, 1996; Clark and Brennan, 1991; Clark and Schaefer, 1987, 1989; Schober and Clark, 1989). This model generalizes observations by Grice, Schegloff, and others, bringing them into the psychological realm in ways that can be modeled and tested precisely (e.g., Cahn and Brennan, 1999; Traum, 1994). Under this approach, as people speak they carefully monitor their addressees for evidence of understanding or misunderstanding, and they adjust their utterances, moment by moment, to ensure that their addressees are understanding them well enough for current purposes. Addressees, by providing such evidence, help mold the utterances speakers produce.

Speakers and addressees thus coordinate in speaking much as dance partners coordinate in dancing or acquaintances coordinate in shaking hands (see Clark, 1992, 1996; Clark and Brennan, 1991). Consider shaking hands. When you extend your hand to initiate a handshake, you can't begin the handshake until your partner has extended her hand in just the right way. You can only continue shaking hands as long as your partner is also engaged in the activity in the right manner, with appropriate hand pressure and movement. You adjust the force and character of your handshake to coordinate with what your partner is doing. And you can only stop shaking hands once your partner implicitly agrees to stop. The argument is that language use is much like a handshake or any other kind of joint action: Both parties together continually adjust their behavior on the basis of what their partners do.

On this view, no utterance is complete until it has been *grounded*—until both participants have accepted that it has been understood. Understanding a reference in any particular utterance requires active participation by both speaker and addressee, and this can take several turns. Note that the point isn't that words don't have conventional meanings; the conventional meanings of words provide important constraints on speakers' meaning. But speakers regularly use words in idiosyncratic ways that go far beyond dictionary definitions (see Clark, 1991; Clark and Gerrig, 1983). Speakers produce utterances based on their *common ground* with their conversational partners—that is, what they presume that they and their conversational partners mutually know, believe, and assume.

THE MESSAGE MODEL AND STANDARDIZED INTERVIEWS

So the message model doesn't sufficiently account for ordinary conversations. [See Akmajian et al. (1990) for other reasons that a message model is both theoretically and empirically untenable for describing language use. For further arguments about how meaning is socially or interactionally constructed see Button (1987), Cicourel (1973), Goffman (1981), Goodwin (1981), Gumperz (1982), Heritage (1984), Krauss and Fussell (1996), Maynard and Whalen (1995), Rogoff (1990), Sacks, Schegloff, and Jefferson (1974), Schiffrin (1994), among many others.] But could the message model hold for survey interviews?

On the face of it, it could: Standardized survey interviews differ from ordinary conversation in several ways that have been documented elsewhere (see, e.g., Clark and Schober, 1991; Schaeffer, 1991; Strack and Schwarz, 1992; Suchman and Jordan, 1990, 1991). The message model might hold in standardized surveys because survey interviewers (according to the theory of standardization), unlike ordinary speakers, use words carefully scripted to be interpretable by the general population. Enough respondents may understand questions as intended, given that the questions have been carefully pretested, that a message model really can describe communication in survey interactions. In a large sample survey, misunderstandings may be sufficiently rare that noncollaborative, standardized exchanges lead to adequate responses overall.

Evaluating the message model's applicability to standardized surveys requires closer examination of standardized interviewing techniques. In strictly standardized interviews, interviewers are warned that interpreting questions for respondents often can change the meaning of the question, and so they are required to follow standardized *neutral probing* techniques. Consider these sample probes from the interviewer training materials from the University of Wisconsin Survey Center (October 1994), which are consistent with Fowler and Mangione's prescriptions. Interviewers are advised to say:

"Is there anything else?"

"Yes, I see" or "Uh-huh" in an expectant manner and followed by a pause.

"Could I read back what I have taken down to be sure I have exactly what you wanted to say?"

"What do you mean by that?"

"Could you be a little bit more specific?"

"Which comes closest to the way you feel?"

If a respondent asks what a question means, the interviewer can say:

"Whatever it means to you."

"I'm sorry, but I don't have that information."

"It's important that the question be answered as best you can, in terms of the way it's stated. Perhaps I could read it to you again."

These probing techniques are designed to prevent the interviewer's point of view from intruding on respondents' interpretations of questions (and thus their answers). But a little reflection shows that they nonetheless can. Each of these techniques can provide some information to the respondent about how the question should be interpreted—they aren't entirely neutral (see also Marlaire and Maynard, 1990; Maynard and Marlaire, 1992; Schaeffer and Maynard, 1996, this volume, Chapter 12; Suchman and Jordan, 1990, 1991). (Whether the information potentially provided is *harmful* or not we will take up later).

If an interviewer says "Is there anything else?" she provides the respondent with some important information, namely that it is entirely possible that the answer the respondent has thus far provided is insufficient, and that the respondent should work to elaborate the answer if possible. In contrast, if the interviewer doesn't use this neutral probe, she may have signaled that what the respondent provided thus far is sufficient. The same logic holds for saying "I see" or "uh-huh" in an expectant manner, which can signal that more is expected of the respondent. While this may seem to be an innocuous intervention, it can actually substantively mold the response; if the interviewer didn't say "uh-huh" expectantly, the respondent might infer that the initial response was sufficient (see Cannell, Miller, and Oksenberg, 1981, for a discussion of how other types of interviewer feedback may be able to improve response accuracy).

Other neutral probes don't simply request more information; they suggest that the respondent ought to provide a *different* answer than he has already given. If an interviewer, following the suggestion in the training materials, says "Could I read back what I have taken down to be sure I have exactly what you wanted to say?" this can suggest that there is something inadequate about the answer. Saying "What do you mean by that?" more directly implies that the answer is unclear or unacceptable and requires the respondent to present more.

Even those probes that seem as if they *must* be neutral—the "whatever it means to you" sort that inform the respondent that the respondent's own interpretation of the question is required—provide real information about question interpretation. By using such probes, interviewers indicate that the respondent's own idiosyncratic interpretation of the question is the desired one. This can lead to respondents interpreting questions quite differently than survey designers intend (see Belson, 1981, 1986; Schober and Conrad, 1997). Take this example:

(1) AW01:620

```
IV:   ... and >now we'd would like to ask about< y↑our
      ↑employment ↑status .hhh did you do any work (.) for pay
      (.) last week {q359}
      (1.1)
FR:   °.hh° Well
      (0.6)
FR:   I'm still getting paid but school's out (.) So
      (0.7)
IV:   Okay s:o (.) would you say (0.4) I mean (0.3) [it's=
FR:                                                 [well
IV:   it's your c[all
FR:              [I got paid (.)
FR:   [for work  (.)  But  I  was]n't at work
IV:   [Ok(h)ay(h) huh huh huh huh]
IV:   .hhh o↑kay (.)
```

Here the respondent doesn't answer the question with a "yes" or a "no," but rather describes her circumstances ("reports," in Drew's [1984] terms; see also Schaeffer and Maynard [1996]), implicitly leaving the interpretation of "work for pay" up to the interviewer. The interviewer refuses to interpret "work for pay" for the respondent, telling her that "it's your call" (line 11). This indicates that the respondent's interpretation of "work for pay" is adequate. The respondent answers that she was "paid for work," which the interviewer takes as a "yes" answer (we can tell this from the subsequent questioning). The interviewer's neutral probe molds the respondent's answer such that she picks one interpretation. (Later in the interview, it turns out that the respondent's personal interpretation of "work for pay" here was not appropriate, and the response must be changed.)

So interviewers, following standard neutral probing techniques, provide evidence for respondents of whether they have interpreted questions appropriately, and thus whether they have provided sufficient responses.

The issue here is an even larger one. It isn't just that neutral probing techniques aren't really neutral: Interviewers can influence answers even when they do *not* probe (see also Marlaire and Maynard, 1990; Maynard and Marlaire, 1992, for discussion of this phenomenon in the domain of standardized testing). Consider this example, where all the interviewer does is accept the respondent's answer by saying "okay":

(2) AW02:691

```
IV:   [°Thee(gh) U S .hhh ar-my° .hhh what wr:- what were]=
      [                  ((typing))                      ]
      =your [most important activities? or duties on this=
      [#
      =jo:b {q359f}
      (1.8)
MR:   uh:: This past week I had to lead uh: my
      [platoon out to uh (0.2) field exercise]
      [                  ((typing))           ]
IV:   [.hhh Oka(gh):::y]
      [((typing))      ]
      (3.7)  ((typing))
IV:   Field exerci(gh)::se:
      (5.1)  ((typing))
      [((typing))]
IV:   [.hhh      ] -Oh kay:
```

By saying "okay" (line 16) the interviewer seems to indicate that the respondent's answer was sufficient. In conversational analysis terms, *response tokens* such as "okay" and "uh-huh" suggest that a speaker was understood and doesn't need to initiate a repair. In the second example, if the interviewer hadn't said "okay" and had asked for further or different information, the respondent would almost certainly have produced a substantively different answer.

Even when interviewers say absolutely *nothing*, they are informative about how questions should be interpreted. In the next example, after the respondent provides an answer in line 8 ("repair and maintaing payphones") the interviewer types the answer and goes on to the next question:

(3) AW21:643

```
IV:   ((typing)).hh And what were your most important
      activities or duties on this job? {q359f}
      (3.7)
MR:   Mos' importan'
      (0.7)
MR:   .hhh duties er activities=
IV:   =mhmm
MR:   repair and maintaing (.) -payphones
      (8.5)
IV:   ((typing)).hh Do you work for yourself in a family
      business or for someone else. {q359h}
```

Merely by going on to the next question without comment the interviewer implicitly indicates that the answer was appropriate for current purposes (see also Button, 1987; Heritage and Sorjonen, 1994; Schaeffer, 1991, among others). So even when interviewers don't probe, they indicate whether respondents have interpreted questions appropriately. The interviewer's behavior is informative in light of what *else* the interviewer might have done — the evidence the interviewer did *not* give that the answer was inappropriate.

All this shows that even the most seemingly neutral interviewer behaviors may not be neutral (see also Schaeffer and Maynard, 1996; Schober and Conrad, 1997:592–593). As conversational analysts have pointed out in nonsurvey contexts (e.g., Button, 1987; Marlaire and Maynard, 1990), in standardized interactions, everything a questioner says — and doesn't say — can affect the respondent's interpretations. This leads us to conclude that the message model of communication — a model that assumes that interviewer and respondent behaviors are separable — does not sufficiently characterize survey interactions to be a theoretically appropriate account.

CONSEQUENCES OF ADOPTING A COLLABORATIVE MODEL

What are the practical consequences of changing our underlying assumptions, of adopting a collaborative model and partitioning the measurement error differently? Does changing our assumptions about meaning require us to change what we train interviewers to do?

The answer to this question isn't straightforward; in fact, there is a range of possible answers. To take one extreme position, one could argue that absolutely strict standardization should be the norm. Even if the message model is inappropriate for survey interviews, the prescriptions of standardization are nonetheless practically useful, or even collaboratively appropriate. Perhaps the standardized prescriptions lead to desirable interactions and deviations from standardized practice lead to undesirable interactions. This, of course, requires elaboration of what counts as a desirable interaction.

To take an alternative extreme position, one could argue that the prescriptions of standardized interviewing should be entirely rethought. Since standardization's underlying model of communication is problematic, interviewers should be trained to use the full range of conversational resources that ordinary speakers use to make sure they are understood, even if this requires deviating from an official script. If there is no longer such a clear dividing line between those interviewer behaviors that affect a response (unlicenced probes) and those that supposedly don't (neutral probes) — because every interviewer behavior can influence responses — then perhaps no probes should be out of bounds.

To evaluate these two extreme positions, and the range of possibilities in between, we first examine standardized interviewer behaviors, sanctioned and unsanctioned, from a collaborative viewpoint. To do this, we outline further details of a collaborative view of spontaneous (nonsurvey) conversations. Then we examine interviewer behaviors in standardized surveys from this viewpoint. This analysis shows that the answer to our question — are the prescriptions for standardized interviewing appropriate? — depends on how informative interviewers' probes are about question meaning.

A Model of Grounding

Conversational partners continuously provide each other with evidence of whether the conversation is on track — whether they are understanding each other well enough for current purposes (see Brennan, 1990; Clark and Brennan, 1991). The conversation can go off track at several levels (Clark and Schaefer, 1987). A speaker can produce an utterance that the addressee doesn't even notice. An addressee can notice that there was an utterance but fail to have heard it. Or an addressee can hear an utterance (understand all the words) but fail to have understood what the speaker meant.

How do people provide each other with evidence of understanding or misunderstanding? They do it by accepting (or failing to accept) each presentation their partner makes. The idea is this: Every utterance a speaker produces constitutes the first part — the *presentation* phase — of what Clark and Schaefer call a *contribution.* For a presentation to become part of what both parties believe has been understood in the conversation, it must be *accepted* by the other party. And then *that* acceptance must be accepted as well.

The contribution structure of a conversation (the details of which are only evident in retrospect) is hierarchical. Consider the following episode, which in Schegloff's (1972) terminology would involve an *insertion sequence* within an adjacency pair. Speaker A asks partner B a question. Rather than answering the question, partner B asks A another question that requires an answer. Only once A has answered B's question does B proceed to answer A's original question. In Clark and Schaefer's terms, this final answer constitutes B's acceptance of A's original question; the question and answer that formed the insertion sequence form their own presentation and acceptance at a lower level of the hierarchy. So acceptances occur at the same level of the conversational hierarchy that their presentations occurred in (for details, see Clark and Schaefer, 1989).

It is only once a speaker and addressee have given each other evidence that a particular utterance has been understood that the content of that utterance officially becomes part of the conversational record, or part of the pair's common ground — the set of beliefs, assumptions and knowledge they can presume they mutually know. When presentations are attempted but aren't accepted, they can't be considered to be a part of common ground; if a speaker assumes that an unaccepted presentation has succeeded, this can lead to misunderstanding.

Speakers and addressees use several techniques to show each other whether they have understood each other well enough for their current purposes — that is, whether they have *grounded* their utterances. By using any of the following five techniques (from Clark and Schaefer, 1989:267) an addressee B shows he has understood what speaker A means; by not presenting any of these kinds of evidence, he shows that the conversation has gone off track.

1. *Continued Attention.* B shows he is continuing to attend and therefore remains satisfied with A's presentation.

2. *Initiation of the Relevant Next Contribution.* B starts in on the next contribution, which would be relevant at a level as high as the current one.
3. *Acknowledgment.* B nods or says "uh huh," "yeah," or the like.
4. *Demonstration.* B demonstrates all or part of what he has understood A to mean.
5. *Display.* B displays verbatim all or part of A's presentation.

As an illustration, consider Clark and Schaefer's example from the London-Lund corpus (p. 270):

```
A. How far is it from Huddersfield to Coventry .
B. um . About um a hundred miles -
A. So, in fact, if you were . living in London during that
   period, . you would be closer - .
```

According to Clark and Schaefer, in saying "um . about um a hundred miles" B presents evidence that she has understood A's question. By initiating the next relevant contribution (by answering the question), B shows that she has understood that A presented a question (as opposed to some other sort of speech act), and the content of her answer shows how she interpreted the question. A must also accept B's acceptance of A's question; here she does this by going on with the next relevant contribution, rather than correcting A's interpretation of the question (e.g., "No, I mean how long does it take to get there?").

Note that with these ordinary grounding techniques *both* parties have input into what is understood. Each party's actions ultimately affect the other's, and we would be missing out on a crucial part of the process if we focused only on what A does or what B does. B's conversational move of responding to the question as she does affects A's actions; if B had displayed any evidence of a problem, A would have repeated the question (if B hadn't heard it) or rephrased it (if B had misunderstood it). A's subsequent acceptance of B's answer similarly affects B's actions; if A hadn't gone on to a relevant next utterance but had demonstrated that he hadn't understood B's utterance, B would have repeated or reframed her utterance.

A final point about the contribution model: Speakers and addressees ground utterances "to a criterion sufficient for current purposes" (Clark and Wilkes-Gibbs, 1986). When both parties desire extremely precise and accurate understanding—say, when a telephone caller wants a telephone number from an operator (Clark and Schaefer, 1987)—people present explicit displays of understanding, like overt "uh-huh"s. When people are engaged in small talk or banter, demonstrating precise understanding may be less important, and less explicit grounding techniques like continued attention may be more prevalent.

The most important point, for our purposes, is that the use of any grounding technique by one party *always* affects what the other party does. No grounding technique is neutral: A partner's response presents evidence about his or her states of understanding and thus shapes what the other party says.

Grounding in Standardized Surveys

Licensed Probes

If we examine Fowler and Mangione's (1990) prescriptions for interviewer behavior from a collaborative viewpoint, one feature is striking. All the neutral probing techniques—with one major exception—are part of speakers' ordinary repertoire of ways to provide evidence of how well they have understood each other. Ironically, standardized interviewers are licensed to provide (or at least not prevented from providing) exactly the kinds of evidence of understanding that Clark and Schaefer (1989) have outlined, and that have been shown to affect addressees' interpretations of speakers' utterances: continued attention, initiation of the relevant next contribution, acknowledgment, demonstration, and display. And, just like ordinary conversationalists, interviewers are licensed to probe when they suspect that respondents have not understood questions as intended or when respondents fail to provide appropriate answers. Consider again those neutral probes listed in the Wisconsin training materials. Each of them implements one or more of the kinds of grounding techniques delineated by Clark and Schaefer (1989). Some correspond to techniques for presenting positive evidence of understanding. For example, when interviewers say "Yes, I see" or "Uh-huh," in an expectant manner and followed by a pause, they are presenting an acknowledgment (technique 3) of what the respondent has said thus far, but by not going on to the next question they are failing to fully accept the respondent's answer and indicating that more is required. When interviewers say "Could I read back what I have taken down to be sure I have exactly what you wanted to say?" they are proposing to display (technique 5) the respondent's answer verbatim, allowing the respondent to correct the display if it doesn't accurately reflect what the respondent intended.

Other neutral probes facilitate grounding because they allow interviewers to give negative evidence, that is, evidence that they are not willing to accept their partner's presentation. When interviewers say "What do you mean by that?" they demonstrate that they are not yet willing to accept the respondent's answer. The same is true when interviewers say "Is there anything else?" or "Could you be a little bit more specific?" The probe "Which comes closest to the way you feel?" can be used when a respondent presents an answer that doesn't conform to a question's response categories, for example, an "I'm not sure" in response to a question about "Do you think the economy is improving, staying about the same, or getting worse?" This probe has the same character as the other negative probes: It shows that the interviewer is unwilling to accept the respondent's current formulation of the answer.

The exceptions to this—the neutral probing techniques that *don't* correspond to the kinds of grounding techniques used in spontaneous conversation—are the scripted responses to requests for clarification: "Whatever it means to you" or "I'm sorry, I don't have that information." In spontaneous conversations, speakers generally take some responsibility for what they mean (Clark and Schober's [1991] *principle of speaker's meaning*). If a speaker's question (presentation) isn't immediately understood (accepted) by the addressee, it is most often up to the speaker to help clarify what she means. When a standardized interviewer apologizes that she doesn't have the information the respondent needs, she behaves in a way that ordinary conversationalists do not: She is refusing to help clarify the terms of her own question.

The "Whatever it means to you" probe is even stranger—it doesn't even attempt to excuse the interviewer's failure to help. Imagine how odd you might find this invented interchange with a friend:

```
Friend: How many hours per week do you usually work?
You:    Well, that depends...What do you count as work?
Friend: Whatever it means to you.
```

Your friend's answer would be highly uncooperative; in fact, Grice (1975) would characterize your friend as opting out of the cooperative principle that underlies conversation altogether. You would be justified in retorting "What are you talking about? *You* asked the question!" Whether or not survey respondents perceive these "whatever it means to you" probes as odd, they certainly differ from the other types of neutral probes. They differ because they place the burden of interpreting the interviewer's meaning on the respondent (or rather, to be more accurate, they place the burden of interpreting the *survey designer's* meaning on the respondent—see Clark and Schober, 1991; Schober, 1999; Suchman and Jordan, 1990, 1991) and they prevent the grounding of meaning. The other probes are consistent with the ordinary conversational assumption that speakers are responsible for what they mean. Unlike the "whatever it means to you" probes, they help the respondent to infer whether they have interpreted the questions as the survey designers intended.

Deviations from Standardization

What do deviations from standardization look like from a collaborative viewpoint? Many deviations from standardization seem to serve collaborative functions (see also Houtkoop-Steenstra, 2000; Schaeffer and Maynard, this volume, Chapter 12; Schober, 1999; Schober and Conrad, 1998, Suchman and Jordan, 1990, 1991). That is, interviewers' deviations from the script can be seen as implementations of grounding techniques from spontaneous conversations, often in the form of demonstrations. Take this example:

(4) AW21:370

```
IV:  mhmm # .hh And how about a year from now, do you expect
     business conditions will be better or worse than they
     are at present, or just about the same {q0080}
     (0.6)
MR:  tch I don't believe they'll improve a whole lot
IV:  mhmm s¯ (.) So do you think
     (1.0)
IV:  is that worse or about the same
     (1.8)
MR:  I:t (0.4) probably (would) stay about the same [ as ] it
     is now
```

The interviewer's probe "So do you think is that worse or about the same" (lines 6–8) is nonstandardized because the interviewer fails to re-present *all* the response alternatives (better, worse, or just about the same). Standardization theory considers probes like this potentially biasing because the meaning of the response alternatives depends on the entire set; by limiting the response alternatives the interviewer is providing an interpretation of what the respondent has said thus far (of "I don't think they'll improve a whole lot"). Such probes can lead respondents to pick the mentioned alternatives more often than they otherwise would (e.g., Schaeffer and Charng, 1991; Smit, Dijkstra, and van der Zouwen, 1997).

From a collaborative viewpoint, on the other hand, the interviewer's probe is simply an example of ordinary grounding behaviors. The interviewer's probe does indeed indicate an interpretation of what the respondent said: It is precisely through this interpretation that an interviewer demonstrates her current state of understanding (technique 4) and helps ground the response. By presenting this demonstration, the interviewer also provides the respondent with the opportunity to reject the interviewer's interpretation.

Here is another example of an unlicenced probe that can be seen as helping ground understanding:

(5) AW21:488

IV: ... .h um On how many days during the past week did you
not feel like eating-your appetite was poor {q0155}
(0.7)
MR: I don't have a problem with that
IV: hhh .h So zero?
(1.0)
MR: Zero=
IV: =mhmm # .

According to norms of strict standardization, the interviewer has potentially biased the response by proposing the answer ("So zero?" line 5) for the respondent.[1] On a collaborative view, the interviewer has simply demonstrated understanding of the respondent's utterance "I don't have a problem with that" (line 4). The respondent, by repeating "zero" (line 7), then confirms (or at least does not challenge) that the interviewer's interpretation was appropriate; for Clark and Schaefer (1989), the respondent has accepted the interviewer's interpretation through a display (technique 5).

Interviewers' deviations from standardization don't take the form only of demonstrations. In this example, the interviewer uses acknowledgment (technique 3) and display (technique 5) to accept the respondent's interpretation of the question:

(6) AW02:133

IV: ... A:nd how much have you heard or read .hhh about the
issue: of federal agencies sharing information about
individuals .hhh a great deal?
(0.5)

IV: some (0.4) not very much or nothing at all? {q6}
(0.9)
MR: uh:: Lately I haven't heard anything at [all
[# #
[#
IV: [.hhh A:nd how concerned would you say you are about
this issue? (0.3) very concerned somewhat -concerned
not very concerned or not at all concerned? {q7}
(1.0)
MR: Oh you're talkin' about individuals you're (talk-)
like- you would u:se- like my n- my name in and to say
.hhh this person said so and so (.) a:nd the:n (.)
they'd give it to another agency to like follow up or
(0.8)
MR: use for their purposes?
(0.6)
IV: Y[eah
MR: [Is that what you're- (0.3) (l[ike)
IV: [That-
IV: [That's what's
MR: [the actual individual
(0.5)
IV: Thee individual right
(3.7)
MR: I:'m not very concerned about it 'cause [I don't-
IV: [huh
MR: the census really can't
(0.8)
MR: .hhh
(0.8)
MR: (it really doesn't come up with)
(2.5)
MR: anything that would be too damaging I guess to a person
IV: Okay? So (0.4) you said not very concerned about it?
(0.5) ((#))
MR: No
IV: .hhh -Okay

When the respondent proposes an interpretation of the question (the utterance that ends "they'd give it to another agency to like follow up or use for their purposes," lines 16–19), the interviewer accepts this interpretation in line 21 with "yeah" (an acknowledgment) and then accepts a further refinement (the respondent's "the actual individual," line 25) with a display ("thee individual right," line 27).

Interviewers also use ordinary grounding techniques when they answer a respondent's request for clarification substantively rather than saying "whatever it means to you":

(7) AW02:104

IV: .hhh -A:nd is the census used to decide how many
representatives each state will have in congress {q5d}
(1.1)
MR: (n- that-) You said this is a business census?
IV: .hhh No [this is-

```
MR:          [or just population census?=
IV:   =Th- this is a population census the U S
      (0.8)
IV:   population census done by: (.) the census bureau
      (0.6)
MR:   Yes I think it-
      (0.5)
MR:   it's used for uh:
      (0.7)
MR:   alocatin' seats in the house
IV:   Okay:?
```

By norms of standardization, this interviewer should have repeated the question or left the interpretation up to the respondent; certainly the interviewer should not have defined "the census" (lines 7–9). On a collaborative view, of course, the interviewer has simply taken some responsibility for explaining her own utterance (Clark and Schober, 1991).

So to sum up, a collaborative analysis of interviewer behaviors in standardized surveys shows that most licensed neutral probes, and many prohibited nonstandardized probes, fulfill ordinary grounding functions. The one kind of probing that does not is the "Whatever it means to you" kind of probe.

Interviewer Training from a Collaborative Perspective

What does this analysis suggest for training interviewers? As we noted earlier, accepting a collaborative viewpoint still allows for a range of possible approaches to training interviewers. On one extreme (let us call it approach 1), one might argue that standardized prescriptions lead to desirable interactions and deviations from standardized practice lead to undesirable interactions. On the other extreme (approach 2), one might argue that a collaborative viewpoint requires us to abandon the notion that standardized meaning can be achieved through standardized wording and neutral probes. Since all interviewers' actions influence respondents' interpretations of questions, and thus their responses, no particular ways of clarifying meaning should be forbidden.

The real question for approach 1 is: What constitutes a desirable interaction? The logic underlying standardization is that desirable interactions are those in which interviewers don't bias responses—or at least one interviewer doesn't bias responses differently than another. By this logic, the "whatever it means to you" probes lead to desirable interactions because interpretations are left up to respondents. Paradoxically, our analysis has shown that the other neutral probes can lead to interactions that are undesirable because they are not standardized; interviewers can affect responses idiosyncratically by probing or not probing, by choosing one probe instead of another, or because one probe affects one respondent differently than another. A collaborative viewpoint suggests a very different notion of what constitutes a desirable interaction: Desirable interactions are those in which respondents interpret questions accurately—that is, as intended by the researchers (see Suchman and Jordan, 1990, 1991). By this logic, any interviewer probes that help respondents interpret questions accurately are desirable. Those probes that don't lead to accurate interpretations are undesirable. So "whatever it means to you" probes are highly *un*desirable because they leave interpretations up to the respondent; any probes that facilitate grounding are desirable.

As for approach 2, if interviewers are licensed to say anything at all to clarify question meanings, we have no idea what the ultimate effects might be. Standardization theorists like Fowler and Mangione claim that the effect would be a disastrous return to all the ills that led to standardized practice in the first place (see Beatty, 1995, for a review). Interviewers, for example, might be more likely to influence respondents' opinions, as Dijkstra's (1987) results (see also Smit, Dijkstra, and van der Zouwen, 1997) suggest can happen when interviewers allow their own opinions to become evident. Perhaps standardization's goal of "holding potentially important influences on answers constant across respondents" (Schaeffer, 1991:377, note) is worth upholding at all costs. But an alternative view (e.g., Mishler, 1986; Suchman and Jordan, 1990, 1991) is that empowering interviewers would have beneficial effects. Who is right?

Corpus-based evidence doesn't resolve the issue. From the Wisconsin transcripts, for example, we can find interactions that deviate from strictly standardized prescriptions and that seem to help ground understanding, presumably resulting in more accurate responses. Take this example:

(8) AW01:907

```
IV:  .hh And what is the percent chance that your own total
     income before taxes will be under forty thousand
     dollars in the next twelve months {q461n}
     (2.6)
FR:  Say that one again
IV:  .hhh What [is the percent chance that your own total=
FR:            [huh
IV:  =income before taxes will be under forty thousand
     (0.7)
FR:  before taxes? {q461n}
IV:  mm hmm
     (2.0)
FR:  Uh::
     (1.3)
FR:  So what you're asking is forty- (0.4) A:m I gonna make
     forty thousand dollars
IV:  .hhh Well they're asking- (0.6) I mean just what the
     question says what is the percent chance that your
     income before taxes is gunna be under forty thousand
     (0.6)
IV:  in the upcoming twelve months {q461n}
FR:  Zero percent
     (0.6)
IV:  eh- [(0.4) Okay (.) S:o
         [#
     (0.6)
IV:  there's- (0.4) Do you think it will be under forty
     thousand
```

```
        (0.6)
FR:     Yes
IV:     Then you would want a high percent [<cause there's=
FR:                                        [Oh
IV:     =a good chance [(it=ll be-)
FR:                    [hundred percent
IV:     oh(gh) kay(gh) [.hhh it's a confusing section I know=
                       [## #
IV:     =a lot of people get real messed up (h)on th(h)at .hhh
        okay what was your age at your last birth day? {q800}
```

In this example it seems that the interviewer's extremely unstandardized interventions (e.g., "do you think it *will* be under forty thousand?" lines 27–28, and "then you would want a *high* percent," line 31) helped the respondent to provide an accurate answer; if the respondent had not been guided by the interviewer, she would have given a response exactly wrong for what her circumstances warranted.

But for every positive example like this, one can find other examples where nonstandardized interventions may well be having ill effects, obscuring the intended meaning of questions and leading respondents to answer based on inappropriate interpretations. Consider this example:

(9) AW01:698

```
IV:     .hh A:nd what were your most important activities or
        duties on this jo(gh)::b hh {q359f}
        (2.8)
FR:     Well you mean of teaching?
IV:     .hhh ye- Well I mean I know I just need you to tell me
        (.) so I can put it in the computer
        (0.5)
IV:     What do you do
        (0.7)
IV:     in your jo(gh):b what do you do teaching
        (1.1)
FR:     (h)Oh:: [well] Tha- that could take a long time
IV:             [hhh ]
IV:     Okay [so-
FR:          [You mean you wanna know all the things that I
        teach?
IV:     No: let's say top three
        (1.0)
FR:     [uh:::m:
IV:     [jst-
        (0.6)
IV:     What are your: activitie[s
FR:                             [In health? I- eh:: my top
        [three things I probably teach are: .hhh drugs (h)]
        [                 ((typing))                      ]
        (1.2)  ((typing))
FR:     [sex education]
        [ ((typing))  ]
        (1.8)  ((typing))
```

```
30   FR:  [uh::::m    ]
31        [((typing))]
32        (0.5)   ((typing))
33   FR:  C P R?
34        (1.0)   ((typing))
35   IV:  Okay?
36        (1.4)   ((typing))
37   FR:  [a::nd in physical education:]
38        [          ((typing))         ]
39        (0.6) ((typing))
40   FR:  [.hhh Oh gosh I teach every sport]
41        [           ((typing))           ]
42        (0.5)   ((typing))
43   IV:  Okay? [So teaching sports that's: good] enough
44              [          ((typing))           ]
45   FR:  [Okay
46        [# #
47        (4.0)   ((typing))
48   IV:  .hhh A::nd do you work for yourself in a family business
49        or for someone else:? {q359h}
```

This interviewer explicitly defines "most important activities or duties" as meaning the top three (line 17); she only goes on to the next question once all three activities are reported and recorded. But was this intervention desirable? Might the respondent have reported more or other activities if the interviewer had not intervened, or had intervened differently?

As we see it, response accuracy should be the ultimate arbiter of which approach we should take; the whole point of doing a survey is to find out about what is really going on in respondents' lives. By our view, accuracy involves the way respondents map the intended question meaning to their own circumstances. For example, if a respondent is asked whether he has been "out of work" in the last 12 months and answers affirmatively, the accuracy of his response depends on whether his interpretation of his circumstances actually corresponds to the researcher's definition of "out of work." If cooperative respondents understand the question as intended, there is a high probability their responses will be accurate. In this sense, accuracy reflects comprehension. (Of course, respondents can misunderstand a question and answer accurately just by chance or they can understand a question as intended and respond erroneously, including lying about their circumstances.)

Response accuracy can be quite difficult to measure (van der Zouwen, Dijkstra, and Smit, 1991; Wentland, 1993). It is especially difficult to measure by relying only on conversational examples. For our "out-of-work" respondent, unless he happens to describe the details of his employment situation in the course of answering the question, the transcript will not provide sufficient evidence for us to evaluate this correspondence and, therefore, the accuracy of the response. This is one reason we adopt the experimental approach described in the next section. From a practical standpoint, approach 1 may be easier to adopt than approach 2, since it may be closer to standard practice in many organizations (although practice can vary substantially—see Schober and Conrad, 1998). But under either approach—or under some intermediate approach—taking a collaborative viewpoint may require redefining what we mean by the right and wrong kinds of interviewer influence.

EXPERIMENTAL EVIDENCE

In this section we describe a series of experiments in which we have evaluated response accuracy and interaction patterns for strictly standardized and more collaborative *conversational* interviews, and for various interviewing styles in between. In these studies, professional interviewers telephoned naïve respondents either in the laboratory (Schober and Conrad, 1997; Schober, Conrad, and Fricker, 1999) or at home (Conrad and Schober, 2000) and asked questions from large US government surveys. In the laboratory studies, the "respondents" answered the questions on the basis of scenarios that described the work, housing, and purchases of fictional people. Because we created the scenarios, we knew the correct answer, according to official definitions, for each question–scenario combination and so could determine response accuracy. In the household study, the respondents answered about their own lives; we evaluated the response accuracy by less direct methods.

In each study, respondents participated in either strictly standardized interviews, following Fowler and Mangione's prescriptions (1990), or in various sorts of less standardized, more collaborative interviews. In all interviews, questions were first posed exactly as worded. In the more collaborative interviews, interviewers were then encouraged to ground understanding of question meaning, for example, by providing scripted definitions when respondents explicitly asked for them (Schober, Conrad, and Fricker, 1999) or by using whatever words interviewers liked to make sure respondents understood the questions as intended (Schober and Conrad, 1997). We believe such studies are necessary because of the variability in interviewer training across (and within) different organizations that consider themselves to practice standardization (Conrad and Schober, 2000) and because of the variability in interviewer behavior within organizations (Schober and Conrad, 1998). Therefore in these studies we have implemented "pure" versions of various techniques in order to see the effects of each technique clearly.

In all studies our basic question is which interviewing technique leads to more accurate responses. Of course, even in purely standardized surveys interviewers can affect responses. What we have tested here is how the kinds of influence that occur in strictly standardized interviews (approach 1) affect response accuracy, as compared to the kinds of influence that occur in more collaborative interviews (approach 2).

Mappings Between Scenarios and Questions

In all studies we used pretested questions from ongoing surveys whose words and grammar have been shown to be understandable, but whose interpretation for some respondents on some occasions might be unclear. Consider a question like "During the past year, have you purchased or had expenses for household furniture?" A respondent who has bought an end table shouldn't have much trouble answering "yes," but a respondent who has bought a floor lamp may be less sure. Or consider a question like "Last week, how many hours did you work?" This should be clear for a respondent who has a 9-to-5 job that includes overtime, but less clear for a respondent who does business over lunch or solves work-related problems while jogging. We presume that the second respondent would be more likely than the first to request clarification such as "What do you mean by work?" or "Should I consider business lunches to be work?" Interviewers following strictly standardized procedures cannot answer these questions in a meaningful way; they would be obliged to use the "whatever it means to you" neutral probing techniques. More conversationally flexible interviewers could provide information to the respondent that would help clarify what the author of the question had in mind.

In the laboratory studies, we designed the fictional scenarios so that half corresponded to the questions in a straightforward way and, for the other half, the mapping was more complicated. In the household study, we had no control over the frequency of complicated mappings. Our prediction was that response accuracy would be high for both standardized and more conversational interviewing when the mappings were straightforward; when the mappings were complicated, accuracy should suffer for strictly standardized interviewing but not as much — maybe not at all — for conversational interviewing. This pattern of results in the household study, where mappings were not under our control, would indicate that complicated mappings are frequent enough in the real world to warrant further exploration of collaborative techniques.

Concept Training

To assure that conversational interviewers could answer respondents' substantive questions, we needed to teach them the official definitions of key concepts in the questions. Providing standardized interviewers with this knowledge might seem to violate the principles of standardization: The only role for definitions in the standardized interview is for them to be read in their entirety to all respondents on all occasions. But the logic of our experiment required us to train all interviewers together on the concepts so that any accuracy differences could not somehow be attributed to different levels of knowledge between standardized and conversational interviewers. (Standardized interviewers were told that the concepts training was necessary so that interviewers would be able to judge when respondents had answered a question completely; see Beatty, 1995.) In all studies the training lasted about 90 minutes; interviewers first studied the official definitions and then actively carried out exercises to ensure that they had grasped the concepts in detail.

Interviewing Techniques

After the concepts training, interviewers were then trained in their respective interviewing techniques. The standardized instructions were based on guidelines that appeared in an interviewing manual for a survey on which many of these interviewers regularly worked, and were consistent with Fowler and Mangione's (1990) approach. Using this material, we reviewed standardized question-asking and neutral probing techniques and supplemented this with role-playing exercises.

The interviewers who were trained to use more conversational techniques were instructed to initially read the question as worded. Then (depending on the study) they could substantively answer respondents' requests for clarification, either following a script or in their own words; or they could also provide unsolicited clarification (scripted or in their own words) when respondents seemed to need it, even if respondents hadn't asked for help. In another study (Schober and Conrad, 1998), interviewers were not trained in a particular technique but were told to do whatever they ordinarily do.

Procedure

In the laboratory studies, 130 respondents were given a packet of scenarios to study, and then they were questioned over the telephone about the scenarios. The 87 respondents in various sorts of conversational interviews were instructed to work with the interviewers to make sure they had interpreted the questions as the survey designers intended; they were encouraged to ask for clarification if they needed it. Response accuracy was measured as the percentage of questions for which responses matched what the official definitions required.

Although the interviewers knew that respondents were answering on the basis of fictional situations, they were not familiar with the content of the individual scenarios, and so knowledge was allocated much as it is in real surveys: Interviewers knew the questions, and respondents knew about their own circumstances. We counterbalanced the items so that the respondents who were assigned to a particular interviewer always received different versions of the scenarios. This way the interviewers could not become familiar with the scenarios based on anything the previous respondents might have said.

In the household study 227 respondents were telephoned at home and asked about their own lives; no scenarios were involved. Because we could not directly determine the accuracy of the respondents' answers, we designed the experiment to provide two indirect measures. One measure was response change between interviews. All respondents participated in two interviews: The first was strictly standardized for all respondents; the second was strictly standardized for half of the respondents and conversational for the other half. If respondents' circumstances mapped in a complicated way to the question concepts, they should be more likely to change their answers between an initial standardized interview and a subsequent conversational interview than between two standardized interviews. This is because in the conversational interviews the interviewers should clarify question meaning, correcting respondent misconceptions, which in turn should lead to different answers than in the initial interview. In a second standardized interview, in contrast, the interviewers should not clarify meaning, and initial respondent misconceptions should remain uncorrected and responses unchanged.

The other measure in the household survey was the "legality" of respondents' explanations for their responses. If respondents answered "yes" when asked if they had made certain types of purchases, they were asked to briefly describe the purchase(s). These were then coded for their consistency with official definitions—their legality. When interviewers could clarify question meaning (in a second conversational interview), respondents should be more likely to base their responses on legal purchases than when interviewers could not clarify question meaning (in any of the standardized interviews).

Results

In each study we first verified that interviewers had followed our instructions and implemented the appropriate technique, based on transcripts of the interviews. One way we showed this was to code the interviewers' various deviations from strict standardization: Rephrasing all or part of the question, providing all or part of a definition (either verbatim or paraphrased), classifying the respondents' descriptions of their circumstances (the fictional scenario in the laboratory studies), offering to provide clarification, confirming or disconfirming the respondent's interpretation of the question, and requesting particular information about the respondent's circumstances. For example, in the following exchange (from Schober and Conrad, 1997), the conversational interviewer paraphrased the long government definition of "household furniture" to answer the respondent's question:

```
I:  Has Kelly purchased or had expenses for household
    furniture.
R:  Um...is a lamp furniture?
I:  No sir, we do not include lamps and lighting fixtures.
R:  Okay, no.
(I goes on to next question)
```

In strictly standardized interviewing, the interviewer should not have answered the respondent's request for clarification because by doing so she interpreted the survey question for the respondent. Across the various studies our coding leads us to be confident that our interviewer training led to fundamentally different types of interaction. In the Schober and Conrad (1997) study, for example, 85 percent of the question–answer sequences in conversational interviews contained deviations from standardization, compared with only 2 percent in strictly standardized interviews.

We can now turn to response accuracy. Again, an accurate response in our experiments is one that is consistent with the official definition of the relevant concept. In our earlier lamp example, the correct answer is "no" because a lamp purchase does not qualify as a furniture purchase, and so the respondent's answer was accurate.

Across all our lab studies, a general pattern emerges. When mappings between question concepts and people's circumstances (scenarios) are straightforward, all interviewing techniques lead to nearly perfect accuracy; virtually all respondents interpret question concepts in the ways that survey designers intended. But when mappings between question concepts and people's circumstances are complicated, strictly standardized interviewing leads to quite poor response accuracy (28 percent in Schober and Conrad's 1997 study). Response accuracy improves when interviewers provide clarification on request (54 and 59 percent in Schober, Conrad, and Fricker, 2000), and it improves substantially more when interviewers are also empowered to use the full resources of ordinary conversational interaction (87 percent in Schober and Conrad, 1997). It does not matter much whether interviewers' clarification is scripted or unscripted (Schober, Conrad, and Fricker, 2000). Even when interviewers do what they ordinarily do (Schober and Conrad, 1998) response accuracy is much higher when they provide clarification (79 percent) than when they leave the interpretation of questions entirely up to respondents (23 percent).

These results are mirrored in the household study (Conrad and Schober, 2000) where we used indirect measures of response accuracy. Respondents changed their answers more when their second interview was conversational (22 percent) than when it was strictly standardized (11 percent). In addition, more responses were based on legal purchases when the second interview was conversational (95 percent) than when it was standardized (57 percent). The improved response accuracy in conversational interviews suggests that respondents' actual circumstances (as opposed to the fictional scenarios presented in the lab studies) are complicated often enough—at least for these questions—to justify exploring the technique further.

Fowler and Mangione (1990) have raised the concern that interviewers empowered to collaborate will mislead respondents by providing inaccurate information. In our experiments this hasn't seemed to be the case. For example, in the Schober and Conrad (1997) study, the information about the question that conversational interviewers provided was accurate (conformed to the official definitions) in 93 percent of the cases where it was given. For 87 percent of all cases respondents received accurate information from interviewers and provided accurate answers; for only 6 percent of all cases did respondents receive accurate information from interviewers but provide inaccurate answers. For the 7 percent of cases where interviewers provided inaccurate information, respondents were still accurate about half the time. In 4 percent of all cases, respondents received inaccurate information from interviewers yet provided accurate answers; in only 3 percent of all cases did respondents receive inaccurate information and answer the questions inaccurately. So overall, conversational interviewers provided highly accurate information. When they provided inaccurate information, this did not necessarily lead respondents to produce incorrect answers; in fact, respondents produced incorrect answers resulting from inaccurate information only 3 percent of the time.

Closer analysis of the interviewer–respondent interaction (see Conrad and Schober, 2000; Schober and Conrad, 1997) shows that it really was interviewers' deviations from standardization that led to the increases in response accuracy. And it seems that interviewers' interventions improved response accuracy whether respondents had requested clarification or not (i.e., even when interviewers provided the information without the respondents having asked for it). So, for example, in the Schober and Conrad (1997) study, for the 64 complicated-mapping cases where interviewers provided unsolicited help, respondents produced accurate answers for 55, a rate of 86 percent accuracy. In contrast, for the 11 complicated-mapping cases where interviewers did not provide any help, respondents only produced four accurate answers, a rate of 34 percent. This figure is close to the 28 percent accuracy for complicated mappings in standardized interviews, suggesting that when conversational interviewers behave like standardized interviewers and don't provide clarification, response accuracy will suffer.

Across our studies, this improvement in response accuracy came at a significant cost. Conversational interviews took much longer than standardized interviews (from 80 to 300 percent longer, in the various studies), and this was true for both straightforward and complicated mappings. It remains to be seen whether this is a necessary by-product of conversational interviewing or whether conversationally flexible interviewers could be trained to achieve the same improvements in response accuracy in shorter interviews.

The results of our studies should not be taken as the final word on the issue, nor should they be taken as showing definitively that conversational interviewing is always a good idea. Our studies have examined nonsensitive fact-based questions, and the results may not generalize to questions about sensitive topics or opinions in a straightforward way. Just like fact-based questions, opinion questions contain phrases with alternative possible interpretations — consider "abortion" or "approve" — and thus opinion surveys might benefit from more collaborative approaches to interviewing. But whether this can be done without influencing the opinions is unclear, especially because response accuracy for opinions can't be validated as directly as it is for the fact-based questions in our studies.

Our results also don't take into account the potential real-world costs of implementing more collaborative interviewing techniques. Beyond the potential expenses associated with increased interview length, interviewer training might have to be more intensive than it often is now. Interviewer behavior would have to be monitored even more closely to ensure that question meanings were being clarified appropriately and uniformly, without increasing interviewer variance. Far more effort would have to go into developing clear definitions for question concepts. To the extent that respondents found increased collaboration a burden (which could depend on several factors — see Schober, Conrad, and Bloom, 2000), response rates could be affected.

Ultimately the generalizability of our experimental findings depends on the frequency of complicated mappings between questions and respondents' circumstances in real surveys, and this may vary from survey to survey. If complicated mappings are known to be rare, then strictly standardized techniques should be optimal, leading to accurate responses at reasonable costs. If the complicated mappings are known to be frequent, or if (more realistically) their frequency is unknown, more collaborative techniques might be worth the increased costs their use would, no doubt, entail.

CONCLUSIONS

We have seen that the theory of standardization assumes a view of communication, the message model, that does not hold in spontaneous conversations. We have argued on conceptual grounds that it doesn't hold in standardized surveys either. Interviewers can't help but influence responses even if they use only the neutral probing techniques that are supposed to prevent such influence. Because different probes have different collaborative effects, even allowing interviewers to choose which probes to provide could lead one interviewer to bias responses differently than another.

Since interviewers always influence responses, this raises the question of which kinds of influence are benign and which are not. We argue that the criterion should be how interviewer behaviors affect response accuracy — that is, how well responses correspond with the definitions the survey authors had in mind. Using this criterion, we have demonstrated experimentally that standardized interviewing techniques lead to poor response accuracy when respondents are unclear about how questions map onto life circumstances. Interviewers empowered to use the full range of techniques for grounding understanding in natural conversation — providing substantive clarification, both solicited and unsolicited — help improve response accuracy.

We are aware that in some surveys such clarifications are considered part of standardized practice. But under the purest definition of standardization, interviewer clarifications are not standardized. Survey researchers who allow clarification should be aware that they are advocating a certain degree of nonstandardization. While we believe our experimental data and our conceptual arguments support some version of conversational interviewing, it is unclear whether the mix of standardized and conversational interviewing in current practice is optimal.

Adopting a collaborative model of interviewer–respondent interaction forces us to redraw various theoretical boundaries. We need to reconsider the assumption that respondent error, interviewer error, and error due to question wording are independent. We need to dispense with the notion that interviewers can behave in ways that don't influence responses. We need to examine further whether the deviations from strict standardization that occur in current interviewing practice are harmful or helpful. If we abandon the underlying message model assumption that meaning resides in words, we are forced to take seriously the proposal that interviews are only standardized when respondents interpret questions the same way (see Suchman and Jordan, 1990, 1991). Ultimately we may need to redefine what standardization ought to be.

NOTES

1. Some organizations that consider themselves to promote standardized interviewing would not consider this a deviation from standardization. But technically, because not every respondent is provided with this feedback, this intervention isn't standardized, in the strictest sense (Fowler and Mangione, 1990).

REFERENCES

Akmajian, A., R. A. Demers, A. K. Farmer, and R. M. Harnish. 1990. *Linguistics: An Introduction to Language and Communication*, 3rd ed. Cambridge, MA: MIT Press.

Beatty, P. 1995. "Understanding the Standardized/Non-Standardized Interviewing Controversy." *Journal of Official Statistics* 11: 147–160.

Belson, W. A. 1981. *The Design and Understanding of Survey Questions*. Aldershot: Gower.

Belson, W. A. 1986. *Validity in Survey Research*. Aldershot: Gower.

Brennan, S. E. 1990. "Seeking and Providing Evidence for Mutual Understanding." Unpublished Ph.D. dissertation, Department of Psychology, Stanford University.

Button, G. 1987. "Answers as Interactional Products: Two Sequential Practices Used in Interviews." *Social Psychology Quarterly* 50: 160–171.

Cahn, J. E., and S. E. Brennan. 1999. "A Psychological Model of Grounding and Repair in Dialog." Pp. 25–33 in S. E. Brennan, A. Giboin, and D. Traum (Eds.), *Proceedings of the American Association for Artificial Intelligence Fall Symposium, Psychological Models of Communication in Collaborative Systems*. Menlo Park, CA: AAAI Press.

Cannell, C. F., P. V. Miller, and L. Oksenberg. 1981. "Research on Interviewing Techniques." Pp. 389–437 in S. Leinhardt (Ed.), *Sociological Methodology,* Vol. 11, San Francisco: Jossey-Bass.

Cicourel, A. V. 1973. *Cognitive Sociology: Language and Meaning in Social Interaction.* New York: Free Press.

Clark, H. H. 1991. "Words, the World, and Their Possibilities." Pp. 263–277 in G. Lockhead and J. Pomerantz (Eds.), *The Perception of Structure*. Washington, DC: American Psychological Association.

Clark, H. H. 1992. *Arenas of Language Use*. Chicago: University of Chicago Press.

Clark, H. H. 1996. *Using Language.* Cambridge: Cambridge University Press.

Clark, H. H., and S. E. Brennan. 1991. "Grounding in Communication." Pp. 127–149 in L. B. Resnick, J. M. Levine, and S. D. Teasley (Eds.), *Perspectives on Socially Shared Cognition*. Washington, DC: American Psychological Association.

Clark, H. H., and R. J. Gerrig. 1983. "Understanding Old Words with New Meanings." *Journal of Verbal Learning and Verbal Behavior* 22: 591–608.

Clark, H. H., and E. F. Schaefer. 1987. "Collaborating on Contributions to Conversations." *Language and Cognitive Processes* 2: 19–41.

Clark, H. H., and E. F. Schaefer. 1989. "Contributing to Discourse." *Cognitive Science* 13: 259–294.

Clark, H. H., and M. F. Schober. 1991. "Asking Questions and Influencing Answers." Pp. 15–48 in J. M. Tanur (Ed.), *Questions About Questions: Inquiries into the Cognitive Bases of Surveys*. New York: Russell Sage.

Clark, H. H., and D. Wilkes-Gibbs. 1986. "Referring as a Collaborative Process." *Cognition* 22: 1–39.

Conrad, F. G., and M. F. Schober. 2000. "Clarifying Question Meaning in a Household Telephone Survey." *Public Opinion Quarterly* 64: 1–28.

Dijkstra, W. 1987. "Interviewing Style and Respondent Behavior: An Experimental Study of the Survey-Interview." *Sociological Methods and Research* 16: 309–334.

Drew, P. 1984. "Speakers' 'Reportings' in Invitation Sequences." Pp. 129–151 in J. M. Atkinson and J. Heritage (Eds.), *Structures of Social Action: Studies in Conversation Analysis*. Cambridge: Cambridge University Press.

Fowler, F. J., and T. W. Mangione. 1990. *Standardized Survey Interviewing: Minimizing Interviewer-Related Error.* Newbury Park, CA: Sage.

Gibbs, R. W. 1994. *The Poetics of Mind*. Cambridge: Cambridge University Press.

Goffman, E. 1981. *Forms of Talk*. Philadelphia: University of Pennsylvania Press.

Goodwin, C. 1981. *Conversational Organization: Interaction between Speakers and Hearers*. New York: Academic.

Grice, H. P. 1975. "Logic and Conversation." Pp. 225–242 in P. Cole and J. L. Morgan (Eds.), *Syntax and Semantics*, Vol. 3: *Speech Acts*. New York: Academic.

Groves, R. M. 1989. *Survey Errors and Survey Costs*. New York: Wiley.

Groves, R. M. 1991. "Measurement Error across the Disciplines." Pp. 1–25 in P. P. Biemer, R. M. Groves, L. E. Lyberg, N. A. Mathiowetz, and S. Sudman (Eds.), *Measurement Errors in Surveys*. New York: Wiley

Gumperz, J. J. 1982. *Discourse Strategies*. Cambridge: Cambridge University Press.

Heritage, J. 1984. *Garfinkel and Ethnomethodology*. New York: Basil Blackwell.

Heritage, J., and M. L. Sorjonen. 1994. "Constituting and Maintaining Activities across Sequences: *And*—Prefacing as a Feature of Question Design." *Language in Society* 23: 1–29.

Houtkoop-Steenstra, H. 2000. *Interaction and the Standardized Survey Interview: The Living Questionnaire*. Cambridge: Cambridge University Press.

Krauss, R. M., and S. R. Fussell. 1996. "Social Psychological Models of Interpersonal Communication." Pp. 655–701 in E. T. Higgins and A. Kruglanski (Eds.), *Social Psychology: Handbook of Basic Principles*. New York: Guilford.

Marlaire, C. L., and D. W. Maynard. 1990. "Standardized Testing as an Interactional Phenomenon." *Sociology of Education* 63: 83–101.

Maynard, D. W., and C. L. Marlaire. 1992. "Good Reasons for Bad Testing Performance: The Interactional Substrate of Educational Testing." *Qualitative Sociology* 15: 177–202.

Maynard, D. W., and M. R. Whalen. 1995. "Language, Action, and Social Interaction." Pp.149–175 in K. Cook, G. Fine, and J. House (Eds.), *Sociological Perspectives in Social Psychology*. Boston: Allyn & Bacon.

Mishler, E. G. 1986. *Research Interviewing*. Cambridge, MA: Harvard University Press.

Reddy, M. J. 1979/1993. "The Conduit Metaphor: A Case of Frame Conflict in Our Language about Language." Pp. 164–201 in A. Ortony (Ed.), *Metaphor and Thought*, 2nd ed. Cambridge: Cambridge University Press.

Rogoff, B. 1990. *Apprenticeship in Thinking: Cognitive Development in Social Context*. New York: Oxford University Press.

Sacks, H., E. A. Schegloff, and G. Jefferson. 1974. "A Simplest Systematics for the Organization of Turn-Taking in Conversation." *Language* 50: 696–735.

Schaeffer, N. C. 1991. "Conversation with a Purpose—or Conversation? Interaction in the Standardized Interview." Pp. 367–391 in P. P. Biemer, R. M. Groves, L. E. Lyberg, N. A. Mathiowetz, and S. Sudman (Eds.), *Measurement Errors in Surveys*. New York: Wiley.

Schaeffer, N. C., and H.-W. Charng. 1991. "Two Experiments in Simplifying Response Categories: Intensity Ratings and Behavioral Frequencies." *Sociological Perspectives* 34: 165–182.

Schaeffer, N. C., and D. W. Maynard. 1996. "From Paradigm to Prototype and Back Again: Interactive Aspects of Cognitive Processing in Standardized Survey Interviews." Pp. 65–88 in N. Schwarz and S. Sudman (Eds.), *Answering Questions:*

Methodology for Determining Cognitive and Communicative Processes in Survey Research. San Francisco, CA: Jossey-Bass.

Schegloff, E. A. 1972. "Notes on a Conversational Practice: Formulating Place." Pp. 75–119 in D. Sudnow (Ed.), *Studies in Social Interaction.* New York: Free Press.

Schiffrin, D. 1994. *Approaches to Discourse.* Cambridge, MA: Basil Blackwell.

Schober, M. F. 1998. "Conversational Evidence for Rethinking Meaning." *Social Research* 65: 511–534.

Schober, M. F. 1999. "Making Sense of Questions: An Interactional Approach." Pp. 77–93 in M. G. Sirken, D. J. Herrmann, S. Schechter, N. Schwarz, J. M. Tanur, and R. Tourangeau (Eds.), *Cognition and Survey Research.* New York: Wiley.

Schober, M. F., and H. H. Clark. 1989. "Understanding by Addressees and Overhearers." *Cognitive Psychology* 21: 211–232.

Schober, M. F., and F. G. Conrad. 1997. "Does Conversational Interviewing Reduce Survey Measurement Error?" *Public Opinion Quarterly* 61: 576–602.

Schober, M. F., and F. G. Conrad. 1998. "Response Accuracy When Interviewers Stray from Standardization." Pp. 940–945 in *Proceedings of the American Statistical Association, Section on Survey Methods Research.* Alexandria, VA: American Statistical Association.

Schober, M. F., F. G. Conrad, and J. E. Bloom, 2000. "Clarifying Word Meanings in Computer-Administered Survey Interviews." Pp. 447–452 in L. R. Gleitman and A. K. Joshi (Eds.), *Proceedings of the 22nd Annual Conference of the Cognitive Science Society.* Mahwah, NJ: Erlbaum.

Schober, M. F., F. G. Conrad, and S. S. Fricker. 1999. "When and How Should Survey Interviewers Clarify Question Meaning?" In *Proceedings of the American Statistical Association, Section on Survey Methods Research.* Alexandria, VA: American Statistical Association.

Smit, J. H., W. Dijkstra, and J. van der Zouwen. 1997. "Suggestive Interviewer Behaviour in Surveys: An Experimental Study." *Journal of Official Statistics* 13: 19–28.

Strack, F., and N. Schwarz. 1992. "Communicative Influences in Standardized Question Situations: The Case of Implicit Collaboration." Pp. 173–193 in G. R. Semin and K. Fiedler (Eds.), *Language, Interaction and Social Cognition.* London: Sage.

Suchman, L., and B. Jordan. 1990. "Interactional Troubles in Face-to-Face Survey Interviews." *Journal of the American Statistical Association* 85: 232–253.

Suchman, L., and B. Jordan. 1991. "Validity and the Collaborative Construction of Meaning in Face-to-Face Surveys." Pp. 241–267 in J. M. Tanur (Ed.), *Questions about Questions: Inquiries into the Cognitive Bases of Surveys.* New York: Russell Sage.

Traum, D. R. 1994. "A Computational Theory of Grounding in Natural Language Conversation." Unpublished Ph.D. dissertation, Department of Computer Science, University of Rochester.

van der Zouwen, J., W. Dijkstra, and J. H. Smit. 1991. "Studying Respondent-Interviewer Interaction: The Relationship Between Interviewing Style, Interviewer Behavior, and Response Behavior." Pp. 419–437 in P. P. Biemer, R. M. Groves, L. E. Lyberg, N. A. Mathiowetz, and S. Sudman (Eds.), *Measurement Errors in Surveys.* New York: Wiley.

Wentland, E. J. 1993. *Survey Responses: An Evaluation of Their Validity.* San Diego, CA: Academic.

CHAPTER 4

Conversation with a Purpose—or Conversation? Interaction in the Standardized Interview

Nora Cate Schaeffer

INTRODUCTION

The classic characterization of the interview as a "conversation with a purpose" (Bingham and Moore, 1924; cited by Cannell and Kahn, 1968) has been used by both survey practitioners and critics. One recent critical examination concludes that ways in which the standardized interview is *not* a conversation may undermine its purpose (Suchman and Jordan, 1990). Survey researchers might respond that although the standardized interview resembles a conversation and uses participants' conversational skills, it is not a conversation, nor is it meant to be. The interaction between the interviewer and respondent is, however, one potential source of observational errors. Moreover, measurement-by-talk imposes constraints that require complicating theories of measurement. Understanding how talk in interviews differs from talk in other situations is needed to improve measurement theory and practice. Critics are correct that locating where standardization fails and suggesting solutions require a closer look at interaction in the interview. Overlooking the problems of standardization poses the risk of canonizing current failures. But reforms that ignore the justification for standardization run the risk of repeating old mistakes.

Standardization and Tacit Knowledge: Interaction and Practice in the Survey Interview
Edited by Douglas W. Maynard, Hanneke Houtkoop-Steenstra, Nora Cate Schaeffer, and Johannes van der Zouwen
ISBN 0-471-35829-0

THE STANDARDIZED INTERVIEW: THE CASE FOR AND AGAINST

Underlying competing assessments of the survey interview are different views of three interrelated topics: the appropriate goals of research, the role of the interviewer, and validity and error. Issues of validity (reviewed by Groves, 1991) motivate concern with the first two issues, and together they suggest taking a closer look at the site of data collection.

The Case for the Standardized Interview

The standardized survey interview takes measurements on a large number of persons, for a set of topics selected by a researcher, in such a way that measurements can be aggregated to describe a population. Aggregating measurements is only straightforward if the same question is asked of all respondents; each respondent has one of a set of mutually exclusive answers; and no respondent–answer combinations are empty (Galtung, 1967). Strictly speaking, answers are only comparable if all respondents are also asked to accept or reject the same choices (see Schaeffer and Thomson, 1992). Ideally, errors are taken into account by theories of construct validity and models of response errors.

Survey researchers train hundreds of interviewers of varying abilities to collect thousands of interviews. Differences between interviewers in the way they understand the investigator's purpose, administer questions, and probe answers increase interviewers' contributions to variable survey errors and bias. These in turn undermine the goal of obtaining comparable answers from all respondents. Two standardizing procedures designed to reduce between-interviewer variation and meet the demands of comparability are using highly scripted closed questions and restricting the ways interviewers may deviate from this script. By training interviewers to read questions as written and to use neutral probes, and by turning certain requests back to the respondent, survey researchers attempt to hold key interviewer behaviors constant. The relationship between the interviewer and respondent is also constrained so that the demands of the personal relationship will not outweigh those of measurement. Ironically, perhaps, it is because their research tradition recognizes the power of the interviewer that survey researchers have made the standardized interviewer almost powerless over a broad range of decisions, in the judgment of some observers.

That reporting context affects answers is implicit in the notion of a "classical true score" (see Groves, 1991, and Bohrenstedt, 1983). Standardization takes seriously the effects of context on validity by attempting to hold crucial aspects of context constant, although this is overlooked by arguments that standardization typically sacrifices validity for reliability (e.g., Cicourel, 1964:76–81). But critics are correct that the substantive implications of the context implemented by standardization are rarely considered (Mishler, 1979).

And the Case Against It

Some critics argue that the goal of mass measurement may be fine in principle, but impossible in practice. The researcher's agenda may be unrelated to the lives of respondents (Cicourel, 1964, 1974; Mishler, 1986a). Partly as a result, answers in an interview lack ecological validity—the meaning conferred by grounding in the everyday (Cicourel, 1982). Because they do not examine what actually happens in the standardized interview, survey researchers pressure respondents into categories that fail to capture their experience. By sweeping problematic answers into categories such as "don't know" or "refused," survey researchers ignore signals that negotiations about meaning have broken down. These related concerns focus on what survey answers do and do not mean, and together implicate the validity of responses in the standardized interview.

One solution emerging from such criticisms is to "empower" respondents and interviewers. At one extreme are proposals that the respondent should be invited to collaborate in defining the research question.[1] This goal entails revising the power asymmetry of the interview, so that interviewers become "learners" and "advocates" (Mishler, 1986a; see also, Paget, 1983). Validity would be improved if interviewers could pursue important issues and respondents could describe their experiences using culturally "natural" forms of talk. Respondents' stories could then be analyzed using the tools of narrative analysis (Mishler, 1986a).

A less extreme proposal is to reformulate the role of the interviewer to minimize the damage standardization does to validity. Respondents present unpredictable situations, and attempting to script the interview in advance deprives interactants of interactional resources to use in negotiating the unexpected (Suchman and Jordan, 1990). Giving the interviewer flexibility in diagnosing and correcting misunderstandings could increase validity: Question wording might not be standardized, but *meaning*, which is more important, would be (see Briggs, 1986:24; Suchman and Jordan, 1990:240).

Talk—But Not a Conversation

Some of the critics' illustrations include what most survey practitioners would recognize as "bad" questions (e.g., Mishler, 1986a:60) or rare situations (e.g, Suchman and Jordan, 1990:237). But practitioners themselves sometimes produce questions that in retrospect are clearly bad and encounter situations that require reconsidering the assumptions of their questions. Criticisms run the risk of being ignored, however, when they fail to confront the general goals of survey research. Simply rejecting surveys may mask an unwillingness to face similar problems posed by other styles of research (e.g., Briggs, 1986:10; Hyman 1975 [1954]:4–9). Three interrelated difficulties survey researchers encounter in attempting to benefit from these criticisms are the use of "everyday conversation" as a standard (e.g., Mishler, 1986a; Suchman and Jordan, 1990), a focus on the structure rather than the content of the interview (e.g., Mishler, 1986a), and an implicitly deterministic model of the interview (e.g., Mishler, 1986a; Suchman and Jordan, 1990).

Use of Conversation as a Standard

Objections that the standardized interview is not a conversation—at least not an "ordinary" conversation—play on the ambiguity in the word "conversation".

Sometimes conversation is used synonymously with "talk in interaction," which could include interviews, and sometimes it is used more restrictively. It seems useful to recognize "interview" and "conversation" as distinct classes of talk-in-interaction (e.g., Schegloff, 1990), which can be compared with each other. This requires specifying what is distinctive about "conversation" and what is added or subtracted by "ordinary." Almost any interview (see Werner and Schoepfle, 1987:306–310; Button, 1987) is different from conversation — speech exchange in which the "order, lengths, and contents of turns are specified neither by prior social prescriptions nor by persons with socially prescribed responsibility for managing the interaction" (Wilson, Wiemann, and Zimmerman, 1984:160). But the inequality in control over topics and turns at talk that characterizes interviews also occurs in other native speech events (e.g., a parental lecture on good conduct), some of which might be considered conversations.

The contrast between standardized interviews and conversation is most instructive in that respondents may apply rules learned in other speech events and reject attempts to impose interview rules (e.g., Briggs, 1986, and the discussion below). Interviewers may begin to discard interview rules in response (van der Zouwen, Dijkstra, and Smit, 1991). Although norms with which interviewer and respondent are more familiar may never be completely suppressed, it is an open question whether systematically giving participants more access to "normal" conversational resources would improve the quality of the interaction or the resulting data. Any movement to remove constraints on the interviewer must pay attention to effects on interviewer variability.

There is no simple criterion to use in evaluating how the specialized nature of the standardized interview affects validity. The artificiality of the standardized interview does not necessarily invalidate responses. After all, both the "true self" we reveal with friends and the "polite self" we present to strangers are social constructions, as is the sense that the former is a more "real" self (Lofland, 1969). People express important things about themselves in situations even odder than the survey interview, and keep parts of themselves secret in everyday life. The relationship of answers produced in a standardized interview — with the kind of validity they have to offer — to other kinds of data — with the kind of validity they have to offer — requires more investigation than it usually receives. Recognizing the distinctive character of interviews is important to understanding the relationship between respondents' expressions in interviews and in other contexts.

Focus on Structure

In his discussion of the research interview, Mishler (1986a) advocates less structured interviews and the techniques of narrative analysis. Such proposals entail two problems for survey researchers. First, there is the problem of how to combine and compare the answers of different respondents, particularly when interviewer variability may be substantial. Second, the analytic techniques described seem more suitable for understanding the process and structure of interviewing than for analyzing the *content* of answers across respondents. This content is central for survey researchers, who worry that a respondent might produce different stories in different unstandardized interviews. The criticisms do, however, emphasize that the structure of interaction is one source of its meaning and that the line between "structure" and "content" is often fine. "Nonattitudes" (Converse, 1970) illustrate this close connection between structure and content in standardized interviewing.

Deterministic Models of Interaction

Discussions of interaction—for example, ones that describe how a respondent understood a particular question in a particular interaction—appear to use an implicitly deterministic model (for an exception, see Churchill, 1978; see also the discussion of turn-taking models in Wilson, Wiemann, and Zimmerman, 1984). The "validity" of a response is situated—it is simply that respondent's answer in that conversation. Analyzing validity would require seeking criteria that would be constructed in other interactions. Emphasizing a specific interaction as a source of meaning is also compatible with a focus on structure and a lack of concern for rules for combining stories and generalizing about their content. At its most extreme, such determinism neglects rules for making generalizations about content or implies that they are impossible. This determinism is, however, a reminder that responses are always negotiated in specific cases, and there may be systematic causes for what appears as "error" in aggregate models.

SURVEY RESEARCHERS EXAMINE INTERACTION IN THE INTERVIEW

The models that survey researchers use to study issues related to interaction in the interview accept the goals outlined earlier. Sudman and Bradburn's (1974:4–18) model emphasizes the structure of the interview: Variables of interest are associated with the interviewer role (role behavior and extra-role characteristics), the respondent role (role behavior), and the structure of the task (question form, mode of administration, etc.). Cannell and Kahn (1968) propose a "motivational model which treats the interview as a social process and regards the interview product as a social outcome" (1968:538). Although they also characterize the role structure of the interview, Cannell and Kahn underscore the importance of motivational and cognitive factors, such as the respondent's perceptions of the purpose of the interview, and the participants' actual behavior. A third class of model combines these elements somewhat differently and focuses on the process by which the respondent arrives at and reports an answer (Cannell, Marquis, and Laurent, 1977; Cannell, Miller, and Oksenberg, 1981:393; Tourangeau, 1984).

These models can be seen as nested and complementary: The structural model identifies central elements of the interview; the social–psychological model suggests that they must be subjectively transformed to be relevant to participants; the cognitive model shows how they enter into the response process. Similarly, much of the research relevant to understanding interaction in the interview can be roughly classified according to whether it measures structural features of the interview (e.g., interviewer race, question form, etc.) or participants' perceptions of these features and their behavior in the interview. Here I focus on the latter.[2]

Participants' Orientations and Definitions

The importance of a *phenomenology* of the interview, which would capture the participants' shifting definitions of the interview, has long been recognized (Hyman, 1975 [1954]). Studying participants' perceptions of the interview, however, requires either constantly interrupting the interview to ask about current perceptions or undertaking a second interview (see Forsyth and Lessler, 1991). In the first case, the intervention is likely to affect the perceptions asked about; the second approach adds problems of memory, dissonance reduction,[3] and social desirability. Consequently, there are few studies of participants' perceptions of the interview.

An exception is interviewers' expectations, some of which can be measured before interviewing begins. Hyman (1975 [1954]) distinguished among expectations based on interviewers' theories about attitude structure, about respondents' social roles, and about the probability distribution of responses in the population. He concluded that interviewers' expectations about respondents have more influence on what interviewers record than the interviewer's own beliefs, partly because interviewers are trained not to communicate their beliefs. Later studies have distinguished among general expectations about a study and those for specific questions and among effects on cooperation, item nonresponse, and responses (e.g., Bailar, Bailey, and Stevens, 1977; Singer, Frankel, and Glassman, 1983; Singer and Kohnke-Aguirre, 1979; Sudman, et al., 1977).

Respondents' global perceptions of the interview have received little attention recently. One reason is probably the conclusion of early investigations that participants' behavior had a greater impact on the course of the interview than did their orientations to the interview (e.g., Cannell, Fowler, and Marquis, 1968). An important exception is that the respondent's definition of the task appears to affect reporting (Cannell, Fowler, and Marquis, 1968:31–32; see also Forsyth and Lessler, 1991).

Behavior in the Interview

A series of systematic studies has provided an increasingly complex view of interaction in the interview. In one of the earliest studies, interviewers' and respondents' different orientations to the interview were suggested by the much greater frequency with which respondents initiated talk unrelated to the interview (Cannell, Fowler, and Marquis, 1968). The finding that feedback from interviewers affected the amount of information respondents provided (Cannell, Marquis, and Laurent, 1977; Marquis and Cannell, 1969, cited in Cannell, Miller, and Oksenberg, 1981) led to devising feedback that reinforces appropriate role behavior by respondents (Cannell, Miller, and Oksenberg, 1981) and to recommendations for more scripting of the standardized interview (e.g., Fowler and Mangione, 1990).

These studies have been important in developing ways to identify and record features of interaction. Without such methods, the components of traditional concepts such as rapport—which captures a range of qualities—remain unanalyzed (see Goudy and Potter, 1975–1976; Hyman, 1975 [1954]). Rapport has recently been replaced with the more neutral concept of "interviewing style". Behavior coding systems are essential for comparing professional or formal styles, on the one hand, and interpersonal, socioemotional, or personal styles, on the other.

Coding Behavior in the Interview

Studying interaction requires classifying a continuous stream of behavior. Three related systems for coding interaction have evolved. The earliest of these systems (Marquis, 1971b) associates a series of codes for interviewer and respondent behaviors with each questionnaire item. Codes indicate, for example, whether the interviewer modifies the question wording or the respondent requests clarification. The coding system has been used in evaluating interviewer performance (Cannell, Lawson, Hausser, 1975; Cannell and Oksenberg, 1988; Mathiowetz and Cannell, 1980) and in pretesting instruments (Cannell and Robison, 1971; Morton-Williams, 1979; Morton-Williams and Sykes, 1984; Oksenberg, Cannell, and Kalton, 1989).

A different approach begins with a detailed analysis of the rules of standardized interviewing and the role requirements and actions they imply (Brenner, 1981c). The resulting coding system classifies each action by whether it meets the actor's role requirements or how it fails to do so (Brenner, 1981b, 1982). The behaviors can then be compared across interviewers (Brenner, 1982) or types of questions (Brenner, 1981b), using tree diagrams that show the sequence of actions that begins with asking a questionnaire item. Dijkstra, van der Veen, and van der Zouwen (1985) extended previous coding systems for an experiment in which interviewers were trained to behave in either a formal or a personal style (see also van der Zouwen, Dijkstra, and Smit, 1991). Each speech act is coded for four variables that indicate the actor (interviewer or respondent), speech category (e.g., question or answer), orientation (task or personal), and adequacy with respect to the prescriptions of standardized interviewing [see also the elaboration in Dijkstra and van der Zouwen (1988) and van der Zouwen, Dijkstra, and Smit (1991)].

Issues in Studying Interaction in the Interview

Studies of interaction in the interview raise several issues. First, all the interviewers in these comparisons are trained to be standardized (though this is less true in Brenner's studies), and interviewers' "personal" behaviors are highly restricted. Thus, the comparison is not between standardized and unstructured interviewing. The dangers of not controlling interviewer behavior are suggested by the results of an experiment in which the interviewer expressed an opinion about whether transportation or recreational facilities would be important to her in deciding where to move. To a later question about what would be important in a move, respondents in both styles were equally likely to echo the interviewer's opinion (Dijkstra, 1987:327). A second issue is that interviewers are usually charged with persuading respondents to participate. If studies do not control how interviewers recruit respondents and success in recruitment is correlated with interviewing style, results could reflect differences among the respondents that each type of interviewer persuaded.

As these coding systems illustrate, decisions about what to record determine which features of the interview are visible and how they can be analyzed. Coding interaction requires identifying meaningful actions and their boundaries. Cannell and Kahn (1968:572) identify a basic cycle within the interview: The interviewer reads a question, the respondent answers, and the interviewer improvises an indication of acceptance or rejection. This basic structure is prescribed in standardized interviewing, but it can be augmented by participants. Brenner briefly discusses the difficulties encountered in segmenting interaction (1981a:27–28, 1981b:138–139), such as overlapping speech and incomplete utterances (see also van der Zouwen, Dijkstra, and Smit, 1991).

These coding schemes are evaluative because survey researchers are concerned with measurement issues, and they entail compromise because the material in any interview is overwhelming. But continuing experimentation with ways of segmenting and coding interaction is needed. Otherwise we risk concluding that features of interaction are inconsequential when we have only failed to identify what is important. For example, only when the coding system records respondents' interruptions of the initial question reading can this be revealed as a problem in question design.[4] Similarly, respondents often provide more material than a question specifically requests (Cannell, Fowler, and Marquis, 1968:19). The implicit request for additional material characterizes some questions in other social situations (e.g., Clark, 1985; Francik and Clark, 1985), and respondents may assume that survey questions are intended similarly. Extended answers may occur more frequently in the first of a series of related survey questions, and the spontaneously provided information may affect the interaction for an entire series of questions (Cannell, personal communication). Although available coding systems could be used to analyze some features of such phenomena, they cannot yet describe the impact of the *content* of respondents' spontaneous offerings through time or how interviewers respond to that content.

The data produced by coding interaction pose difficult statistical and methodological problems. Analyses must consider that multiple question-and-answer sequences are provided by each respondent, questions-and-answers are developed through time, and respondents are nested within a small number of interviewers (see Hox, de Leeuw, and Kreft, 1991). Statistical tests should consider these features. Furthermore, patterns of error correlations under different interviewing styles need attention.

Finally, *how* the style of interviewing affects respondents' answers must also be examined. Interviewing style might affect whether respondents attempt to be accurate or to ingratiate themselves with the interviewer (Dijkstra, 1987). Mediating processes could also include effects on the respondent's role expectations (e.g., attending to interviewer cues), the interviewer's task behavior (e.g., providing inadequate cues), reinforcement patterns in the interview, and the respondent's motivation or mood (Dijkstra and van der Zouwen, 1987:201; see also van der Zouwen, Dijkstra, and Smit, 1991). This list should also include effects on how the respondent understands the *point* of a question and uses this understanding to formulate an answer. The sequential structure of the tasks presented by different questions must be analyzed and explicitly considered in analyses of interviewing style [van der Zouwen, Dijkstra, and Smit, (1991) begin this task]. The apparent contradiction that a professional interviewing style yielded more accurate reports about accidents as the time since the accident increased (Henson, Cannell, and Lawson, 1976) but a personal interviewing style resulted in more accurate maps of the neighborhood (Dijkstra, 1987) suggests that the difficulty of the task must also be considered. A personal style may better motivate respondents, but make it difficult for interviewers to impose the demands of difficult recall tasks. Lastly, because the actions of respondents vary with style of interviewing, the effect of the respondent on the interviewer may also so vary.

STRATEGIES FOR STUDYING INTERACTION

Studies by survey researchers primarily focus on whether behavior meets the criteria of standardization. In this section I draw on research outside the survey tradition to consider (1) conceptualizations of the context of interaction, (2) interaction as a medium of comprehension, and (3) detailed studies of interaction that can be used in studying the survey interview. The first two motivate the third. The need to broaden measurement theory to consider the context and practice of measurement motivates all three.

The measurement models that survey researchers typically work with are more refined versions of one we use every day: If I ask someone how she is, I recognize that social factors as well as her true state influence her self-report. Similarly, measurement in surveys views self-reports as functions of underlying unobservables. And at their best, measurement models include other influences on self-reports, and studies measure these influences (e.g., interviewer race) or hold them constant (e.g., question wording).[5] Improving the "measurement" conducted in interviews requires understanding how social and psychological processes are used in constructing self-reports. Just as the findings of psychologists require incorporating heuristics into measurement models (e.g., Nisbett and Ross, 1980; Tversky and Kahneman, 1973), studies of behavior in the interview can identify interactional influences on responses. An understanding of the range and power of these influences is needed to appreciate the kind of knowledge self-reports have to offer. Social context both constrains self-reports and establishes what they mean. Specifying when a relationship between variables holds (Mishler, 1979) requires refining our conceptualization of the social context in which self-reports are produced.

"Context" is one way of referring to, among other things, the global structure of the interview as a speech event, the social situation surrounding the interview, or the role relationships (both inside and outside the interview) among the participants. Context can be defined at a very local level, for example, as the interpretive framework resulting from a series of turns at talk or as the immediately preceding sentence that provides the antecedent for a pronoun. Context can also be described from the point of view of goal-directed participants or from that of an observer. Finally, discussions of context can focus on the content or structure of talk or show how they are interrelated.

Global Context: The Interview as Speech Event

A speech event, such as an interview, is an activity "directly governed by rules or norms for the use of speech" (Hymes, 1972:56) and is made up of speech acts. Briggs (1986:40–60) presents a model of the interview that directs attention to (a) the frame supplied by the social situation and type of communicative event constituted by the interview; (b) the interviewer and respondent, their respective roles, and interactional goals; and (c) the referents, message forms, channels, and codes of communication. Explicitly viewing the interview as a speech event stresses the role of communicative norms in guiding talk and interaction. The distinction between the interview as a speech event and its constitutive speech acts (e.g., questions and answers, stories, side sequences) emphasizes that context is both global and local—participants orient both to their perceptions of the purpose and structure of the interview as a whole and to their immediate talk, and both figure in participants' understandings. In addition, Briggs' model directs attention to speech events that may frame or interrupt the interview, such as the conversation in which the interviewer obtains the respondent's participation.

Global Definitions of the Speech Event

As evidence that respondents recognize the interview as a speech event, Wolfson (1976) cites the existence of the word *interview* and respondents' exasperation when the interviewer, in an effort to impose less structure, fails to ask questions. The prescriptions respondents ascribe to the cognitive "world" of the interview (Agar and Hobbs, 1982) may vary. A respondent's global definition of the interview may be an intimate conversation or therapy session, an exercise in filling out a form, a referendum, a test, or a game (Turner and Martin, 1984:262–270), a "talk show" (Suchman and Jordan, 1990:236), or an opportunity to construct a coherent positive identity (Mishler, 1986b).

People telling narratives select those details that are relevant to the point they are trying to make (Moerman, 1973), so that defining what the point of the story is to be becomes important. Similarly, a respondent's global definition of the interview's purpose influence his contribution to the interaction, and hence the resulting data. Global perceptions may motivate the respondent's participation but subvert the interviewer's control of the situation or undermine the goal of candid and accurate responses. Or a respondent can use global perceptions of the interview's purpose to compensate for an interviewer's failures: Agar and Hobbs (1983) describe an interview with an experienced informant, Jack, to learn why he had moved from Chicago to New York. When that story proved shorter than their typical interview, Jack demonstrated orientation to a "norm" that an interview should produce at least one side of tape, which he apparently inferred from their past interviews. The interview rambled after its initial purpose was completed because the interviewer was not prepared for another topic. But Jack attempted to produce material relevant to the research topic—he talked about his first arrest and sexual experiences in prison—illustrating his orientation to what he perceived as the interviewer's goals, at a level that subsumed the specific interview.

Global Rules Governing Talk

Talk in interviews is subject to formality constraints that have yet to be described. Formality may help ensure that the interactional tasks of the interview are accomplished (Atkinson, 1982) and may affect the use of language. Compared to language in other social situations, formal talk may be more highly structured and more consistent. Speakers may invoke their positional identities or roles more readily and use language to maintain focus on the situation (Irvine, 1979). The formal relationship between interview participants also undoubtedly constrains what can be discussed and how. For example, Jefferson (1985) compares a mother's conversations about her son with two different neighbors. Statements the mother makes in one conversation "gloss," or refer to, something that is "unpackaged" in a later conversation with a more receptive conversational partner. Only when a gloss is unpackaged, either spontaneously or in response to direct questioning, is its status as a gloss revealed. Jefferson's discussion suggests examining more closely how the interviewer collaborates in providing a social context in which formal glosses can be recognized and unpackaged. (One could also argue, of course, that it is just this variability by social context that justifies standardization.)

In addition to being guided by their global definitions of the interview and rules about formal interaction, respondents' talk uses rules learned in other talk. For example, respondents may use conversational maxims not to be redundant (Grice, 1975), as shown by answers to a series of closed questions (Schwarz and Strack, 1988; Schwarz, Strack, and Mai, 1991). Similarly, respondents asked a series of open questions refer to previous comments and rely on the sequence of answers to clarify what they mean (Bobo, personal communication).

Studying the Context Supplied by the Interview

The work of survey researchers suggests that respondents' definitions of the interview have important consequences, but that their understanding of the researcher's goals may remain vague even after the interview (Cannell, Fowler, and Marquis, 1968). Briggs (1986:95, 106–107) suggests analyzing the structure of actual interviews and paying particular attention to comments participants make about the interview itself—their metacommunications—to uncover global perceptions of the interview.

Our understanding of the social constraints on talk in interviews would be helped by comparative studies of the types of formality that characterize talk between strangers in a variety of roles on a variety of topics. Identifying appropriate comparisons requires an analysis of the dimensions of the social context of talk (see above) and may prove difficult. There are many studies of talk between doctors and patients (see references in Maynard and Clayman, 1991), for example, but the actors are often not strangers, and the patient typically seeks the contact and wants something from the doctor. Similarly, we need comparative studies of how statements respondents make in the standardized interview are related to statements they make in other social situations and the contextual sources of similarities and differences (see Cicourel, 1982; Dean and Whyte, 1958; Turner and Martin, 1984:86–89). Such studies would take seriously that there is no "context-free" measurement and aid the theoretical task of understanding the expression of "unobservables" in different social contexts.

Local Comprehension: Questions and Responses

Because surveys prescribe the questions interviewers ask and how they may react to respondents' answers, studies of interaction in the interview must attend to characteristics of questions, both as scripted and as enacted, and the rules guiding interviewers' behavior. Because questions and answers are enacted in talk, the participants' mutual comprehension is fundamentally interactive. The coordination between interviewer and respondent uses context as it appears in their common ground (Clark and Marshall, 1981). *Common ground* includes everything said in an encounter (linguistic evidence), the common experience of the participants (perceptual evidence), and things that participants believe known to others in their communities (community membership). The common ground thus changes over the course of talk. Questions and answers "become" problematic when they introduce interactional "troubles" into the common ground.

Improving Common Ground: Questions

Simply by being asked, almost any question conveys that it is intelligible, salient, answerable, and that the person asked is capable of providing an answer. Some questions also suggest what is feasible, viable, or appropriate, or what does or does not exist (May, 1989). The suggestions contained in questions are problematic for all interviews. In addition, poor question design—particularly problems of question relevance and meaning—can impair comprehension of questions and cause interactional problems. The "correct" answer to a question that makes a false assumption, for example, is rejecting the assumption (Kaplan, 1981). Such rejections require the interviewer's intervention, probably increasing interviewer variability. Researchers repeatedly face the problem that what a question turns out to have been—for example, a simple request or an admonition—depends on what the respondent does with it (e.g., McHoul, 1987).

Careful question design can address some problems in comprehension and relevance that become problems in interaction. Studies of conventional questions and spontaneous answers suggest that question writers could better exploit respondents' tendency to answer the question behind the question. For example, when restaurant owners were asked if they accept credit cards, half spontaneously listed the credit cards they accept (Clark, 1979). Werner and Schoepfle (1987:319–321) illustrate the use of ethnographic techniques to "negotiate the question." An informant is presented with an "answer" and asked to provide the question that would elicit it. The informant successively formulates a series of questions and answers to his own questions. The process shows both how questions restrict possible answers and how a desired answer restricts possible questions. Ethnographic and focused interviews used with individuals or groups (e.g., Merton, Fiske, and Kendall, 1990) may illuminate matters as simple as the appropriate vocabulary (e.g., Bauman and Adair, 1989) and as complex as the kinds of uncertainty respondents experience (e.g., Schaeffer and Thomson, 1992). Such material can improve the relevance and intelligibility of questions by providing investigators a description of the "cognitive worlds" of potential respondents.

Improving Common Ground: Responses

To infer what a question asks for, respondents may apply rules of deletion or selection, generalization, and construction (van Dijk, 1980:46–50) to the question as it appears in the common ground. Some comprehension problems in interviews arise when participants apply these rules using community-membership-based assumptions about common ground that differ from the researcher's assumptions. A respondent may assume, for example, that no one would ask about "unimportant" visits to the doctor. When read a list of health practitioners to consider when calculating the number of medical visits, the respondent may then consider some practitioners on the list "irrelevant details" (deletion/selection). From the practitioners remaining on the list, the respondent may generalize that the question asks about visits to specialists. If the interviewer previously accepted a rough estimate as an answer, the respondent may also draw on this common ground to conclude that the interviewer wants a rough estimate of visits to specialists.

Responses are a standard mechanism by which speakers display both understanding and acceptance of prior talk. In some social situations, for example, a listener to a story may indicate how she interpreted the story by commenting or by telling a similar story (Moerman, 1973). This response allows the original story teller to determine how he has been understood. By providing fixed response categories, standardized interviewing attempts to minimize the need for interviewers to check their understanding of a respondent's answer. But standardization provides interviewers few resources to use in diagnosing and correcting problems of comprehension when this is needed (see example in Suchman and Jordan, 1990:239–240). Some methods of displaying understanding must be used cautiously in any interviewing and are inappropriate for large-scale surveys. But interviewers need neutral methods of verifying their understanding with respondents. One approach is to distinguish between scripted questions and side sequences (see below) in which the interviewer may use prescribed techniques for offering clarifications and pursuing ambiguities.

Enacting Local Context: Analysis of Speech Acts

The survey data we analyze are products of talk: Interviewers enact questions and negotiate answers with respondents. Studies of talk in other contexts can contribute to understanding interview data in two ways—by illuminating persistent problems in surveys and by providing concepts and methods for basic research on interaction in the interview. Because all talk occurs in *some* social context, both strategies are inherently comparative. Comparative studies are needed to identify interactional influences on survey answers for inclusion in measurement models—as other "method" factors currently are—where appropriate. To capture the close connection between structure and meaning in talk, such studies must examine details of interaction.

Conversation analytic (CA) studies offer results, concepts, and methods useful in analyzing the sequential organization of talk and the *interactional substrate* of the interview (Marlaire and Maynard, 1990; Maynard and Marlaire, 1987).[6] The interactional substrate consists of the interactional skills that interviewers and respondents use to arrive at "accountable" survey answers (see Button, 1987; Garfinkel and Sacks, 1970; Groves et al., 1990). An accountable answer is one an interviewer can justify as having been properly arrived at. For example, an interviewer using a neutral probe to help a respondent select an answer is applying interviewing rules to cooperatively produce an answer that researchers using the data can then regard as "objective." CA studies also contribute to a comparative foundation for studies of interaction in the interview, but they must be used carefully. Many CA studies do not describe the role relationship between the speakers whose talk is being analyzed or the situation in which the speech fragment occurred (see Cicourel, 1987). Thus, it is not always clear how the social context might be similar to or different from the interview (see Goffman, 1981:29–34).

Applying Results from Conversation Analytic Studies

Some results of conversation analysis can show how common conversational practices might affect the quality of data collected by surveys. I present two examples, one dealing with a nagging theoretical issue, the other with a practical problem that can increase interviewer variability.

Acquiescence

Acquiescence, the tendency of respondents to agree with questions or statements using an agree/disagree format, has been described as a "quagmire" (Schuman and Presser, 1981). This swamp may be everywhere because one way people talking together structure minimal affiliation is to design questions or assessments for agreement (Sacks, 1987; Pomerantz, 1984a). The relative preponderance of "yes" over "no" answers in talk suggests speakers often proceed in this way (Sacks, 1987:57). But speakers know that listeners will not always agree with them and need a mechanism to engineer agreement. The following pattern is one mechanism: question_1, pause, question_2, answer. The question recipient supplies a pause that lets the speaker redesign the question for agreement. The pause serves as a signal, because disagreeing answers tend to be preceded by delay (Table 4.1, sections A and D). When disagreement does occur, the question recipient may first

Table 4.1 Agreement and Disagreement in Talk

A. Delayed Disagreement

A: Yuh comin down early?
B: Well, I got a lot of things to do before gettin cleared up tomorrow. I don't know. I w-probably won't be too early.

B. Agreement Followed by Disagreement

A: How about friends. Have you friends?
B: I have friends. So called friends. I had friends. Let me put it that way.

C. Strong Agreement Followed by Disagreement

A: 'N they haven't heard a word huh?
B: Not a word. uh-uh. Not- Not a word. Not at all. Except Neville's mother got a call . . .

D. Question Reformulated for Agreement

A: They have a good cook there?
((pause))
A: Nothing special?
B: No, everybody takes their turns.

E. Agreement to Both Parts of Forced-Choice Question

```
A:   ((clears throat)) Was yer lenience in yer opinion because
     of the fact thatchu uh thought this would be a better way of
     controlling uh the members
                              [
B:                            Yes because-
A:   or was it because of the fact thatchu wanted them to like you.
B:   Well partly it was because I wanted them to like me and they-
     I mean they were all my friends. Everybody in that dorm had
     known me fer years. An-
```

Source: Simplified from Sacks, 1987:58, 62–64. Section E conforms to Sacks' comments in the text. Reprinted with permission from Multilingual Matters, Ltd., Bank House, 8a Hill Rd., Clevedon, Avon BS21 7HH, England.

agree and then disagree, or co-participants may negotiate by proposing candidate formulations until agreement is reached (Table 4.1, sections B, C, and D).

Drawing on their experience in other social contexts, some respondents may assume that many survey items have been formulated for agreement. Respondents who disagree may initially signal their response as they might in other situations; they then find that the interview does not allow them to negotiate agreement by modifying the question or statement. Because "neutral" probes do not recognize signals of disagreement, respondents may also perceive probes as requesting "agreement." Forced-choice questions may offer less escape than is

Table 4.2 Downgraded Conventional Response Followed by Trouble

Bob:	[How are you] feeling now.
Jayne:	Oh? pretty good I guess [
Bob:	Not so hot?
	(0.8 second pause)
Jayne:	I'm just sort of waking up,
Bob:	Hmm,
	(3.6 second pause)
Jayne:	Muh- ((hiccup)) (0.9 second pause) My (ear),
Bob:	Huh?
Jayne:	My (ear) doesn't hurt (0.4 second pause) my head feels ((brief pause)) better,

Source: Simplified from Jefferson, 1980:153–154.

sometimes hoped (Converse and Presser, 1986:38). Both parts of the question may evoke agreement (Table 4.1, section E), and respondents may then appear ambivalent (see Schaeffer and Thomson, 1992). A similar problem may affect other questions. For example, a *downgraded conventional response* such as "I'm pretty good" may announce that I am *not* so good and invite further inquires (Table 4.2).

Standardized interviewing often implicitly assumes that a respondent's first answer indicates salience or attitude strength (Kane and Schuman, 1991). Accepting the respondent's first answer without pursuing it minimizes negotiations between respondent and interviewer (and opportunities for the interviewer to influence the answer), and is thus also part of constituting the answer as produced by the respondent (see Button, 1987). Some practices depart from this model [e.g., warning interviewers to probe initial "don't know" answers, Survey Research Center (1976:17)]. By generally preferring the respondent's first answer, however, practice does not consider that respondents may first agree and then disagree (see also Suchman and Jordan, 1990:236). Interviewer variability probably increases when respondents spontaneously develop a second answer, because interviewers' responses may vary.

In surveys as in life (Schuman and Ludwig, 1983), quagmires offer no easy escapes, only the opportunity to study the spot where one is sinking. CA studies suggest that acquiescence sometimes inheres in self-reports, but also that some statements, such as self-deprecations, are formulated for disagreement. [See Pomerantz (1984a). This may contribute to race-of-interviewer effects.] Background expectations about survey questions probably vary over respondents and questions and may dominate when the respondent does not feel strongly about an issue (Schuman and Presser, 1981). Or such expectations may be more important in talk between strangers, where affiliation depends on the immediate interaction.

This discussion suggests that we study closely how respondents agree and disagree, as well as "don't know" answers, other possible disagreements, and subsequent probes and resolutions. CA studies of talk in other contexts provide comparisons for research on acquiescence, particularly if the participants' relationship and the interaction's level of formality can be considered. Moreover, studying how people talk in other situations may help investigators face the difficult problem of deciding which answer they want, when respondents have more than one to offer, and the implications of that choice for measurement. Interviewing techniques — particularly for subjective items — that let respondents talk about a topic before presenting them with fully standardized questions may allow respondents who have several answers to resolve them.

Turn-Taking and a Simple Issue in Question Design

Part of displaying competence in everyday talk is appropriately projecting and using possible transition points for turns (Sacks, Schegloff, and Jefferson, 1984). A question projects a transition point, and speakers appear to prefer placing questions at the end of turns (Sacks, 1987). Hearers project the beginning of their turn by recognizing the upcoming completion of the question. Writers of standardized questions frequently violate the rule of placing the actual question at the end of the interviewer's turn. This may happen because the topic of a questionnaire item is indicated in a question, which is then followed by restrictions, definitions, or response categories (Table 4.3). Although some interviewers succeed in holding their turn with a "rush-through" (Schegloff, 1982:76), the resulting confusion for respondents is evident in the frequency with which they "interrupt" the initial question once they recognize its point [see examples in Cannell et al. (1989) and Jefferson (1973; 1986) for discussions of overlapping talk at transitions]. The interviewer may have to cut off the respondent's premature answer or repeat definitions or categories until the respondent learns that response categories *may* follow a question.[7] Replacing a long item with several questions or placing the actual question at the end of an item greatly reduces the frequency with which respondents interrupt (Table 4.3).

Basic Research Using Conversational Analytic Techniques

Some details of interaction, such as feedback by interviewers, are known to affect respondents' answers. But further studies of interaction may reveal previously unnoticed features as important. CA studies have identified important components of the interactional substrate of talk generally. Basic research on the interview can compare how these components appear in interviews and other interactions, in different styles of interviewing, and in interviews that work and those that do not. Such basic research could improve questionnaire design and interviewer training, identify interactional sources of meaning, and specify how the effects of race of interviewer or different styles of interviewing are accomplished (see Marquis, 1971a; van der Zouwen, Dijkstra, and Smit, 1991).

Side Sequences and Interview Structure

The flow of the standardized interview may be interrupted by respondents' requests for clarifications or comments. These *side sequences* may be structured as insertion sequences in which a preliminary question must be resolved before

Table 4.3 A Frequently Interrupted Question and Its Revision

First Pretest	Second Pretest
During the past 12 months, since January 1, 1987, how many times have you seen or talked with a doctor or assistant about your health? Do not count any times you might have seen a doctor while you were a patient in a hospital, but count all the other times you actually saw or talked to a medical doctor of any kind about your health.	Have you been a patient in a hospital overnight in the past 12 months, since July 1, 1987? (Not counting times when you were in a hospital overnight,) During the past 12 months, since July 1, 1987, how many times did you actually *see* any medical doctor about your own health?
	During the past 12 months since July 1, 1987, were there any times when you did not actually *see* the doctor, but saw a nurse or other medical assistant working for the doctor?
	(IF YES) How many times?
	During the past 12 months since July 1, 1987, did you get any medical advice, prescriptions, or results of tests over the telephone from a medical doctor, nurse, or medical assistant?
	(IF YES) How many times?

Source: Adapted from Cannell et al., 1989, Appendix A.

the initial question can be answered (Jefferson, 1972; Psathas, 1986; Schegloff, 1972). Less often, such a side sequence may be initiated by an interviewer who infers that a respondent has misunderstood a question. As in other kinds of talk, something only "becomes" a misapprehension if a co-participant makes it so (see Jefferson, 1972:305). The principal types of side sequences that occur in standardized interviews are probably personal business or "irrelevant" talk, metatalk, respondents' requests for clarification, and repairs initiated by the respondent or interviewer (primarily probing). We know little about the structure or content of these side sequences, but they become part of the common ground of the interview and may have effects beyond the question–answer sequence in which they occur.

Repair

Repair occurs when one party corrects her own contribution to the talk or attempts to correct another party's (see Schegloff, Jefferson, and Sacks, 1977). A common repair side sequence in interviews is interviewer-initiated probing. Probing is frequent [85 percent of interviewers' task behavior in Cannell, Fowler, and Marquis (1968:18)], and it may be the most important indication that a question is susceptible to substantial interviewer variability (Fowler and Mangione, 1985:136).

Probing often begins after an answer that is perfectly adequate in other talk, but not acceptable in the standardized interview. For example, a respondent may answer, but not select a response category. When the respondent finishes answering, the interviewer can use this transition point to display that the respondent's answer was adequate by continuing to the next question (Jefferson, 1985:458) or initiate repair, usually by probing. In some kinds of talk, a speaker seeking to confirm or disconfirm an assertion that initially receives no response may clarify the assertion, review knowledge shared with the hearer, or modify his position (Pomerantz, 1984b). Speakers asking questions that receive inadequate responses appear to use similar strategies [suggested by transcripts in Sacks (1987)]. Probing in the standardized interview, however, is largely restricted to repeating the question or response categories (Fowler and Mangione, 1990). If producing a remedy for a complaint legitimizes it (Jefferson, 1972:310), a respondent may feel that an interviewer who simply repeats the response categories is delegitimizing his complaint.

What actually happens during probing has received little study. These negotiations, however, have considerable impact on the social atmosphere of the interview and the reliability and validity of the resulting data. In addition, interviewers self-repair question delivery, and respondents self-repair their own answers (see, e.g., Jefferson, 1974). Studying respondents' self-repairs, in particular, could tell us how respondents find their way into standardized categories and what information is lost in the process.

Response Tokens

Response tokens, such as "mm hmm" or "uh huh," are highly flexible. They can be used to invite a speaker to continue talking or to indicate that an alternative response, such as laughter, is being withheld. As an invitation to continue, a response token may be seen as passing an opportunity to initiate a repair, such as asking for clarification, and thereby communicate that the speaker was understood (Schegloff, 1982:86, 88). Speakers use the token "yeah" to change or curtail a topic, and "oh" to signal receipt of new information (see Heritage, 1984, 1985:5). Given the limited range of responses available to interviewers, they may use tokens in distinctive ways (e.g., Jefferson, 1984b), particularly in telephone interviews. The way interviewers use tokens may be difficult for respondents to interpret, may inadvertently reinforce some behaviors of the respondent, or may be perceived as nonresponsive. Survey researchers have recognized the importance of response tokens [e.g., research to that date reviewed in Cannell and Kahn (1968:581–582)]. But distinctions among tokens are needed to identify the interactional problems they are used to solve [as in the coding categories developed by Johannes Smit (van der Zouwen, personal communication)].

Laughter
Proffering, joining, and withholding laughter are all patterned uses of this conversational resource. A speaker who laughs is regularly joined by her co-participant (Jefferson, 1979). Laughter is also used to affiliate with a speaker or is withheld to disaffiliate or control escalation of a topic [Jefferson, Sacks, and Schegloff (1987); see also the discussion of laugh tokens in Jefferson, (1972:299–301, 449)]. In talk about troubles, the troubles teller may laugh as a display of his ability to manage the trouble, while the recipient responds to the content of the statement, rather than to the laughter (Jefferson, 1984a:348, 358). When both parties laugh together in talk about troubles, it may indicate a *time-out* (Jefferson, 1984a:356). The little evidence available suggests that a large proportion of respondents' talk unrelated to the interview is laughter, and that interviewers laugh much less frequently than respondents (Cannell, Fowler, Marquis, 1968:19–20; see also Bradburn and Sudman, 1979). Although questions distinctively susceptible to interviewer effects are not distinctive for eliciting laughter, interviewer laughter is significantly correlated with some question characteristics: difficulty (but not sensitivity), opinion questions, and open questions (Fowler and Mangione, 1985:137–139).

The work laughter does for either participant in an interview has not been explored. Given the uneven distribution of laughter in the interview, we can expect that work to differ for respondents and interviewers. Interviewers may use laughter to manage problems in their roles (such as problems caused by inappropriate questions). Laughter may also help establish and maintain affiliation (or disaffiliation) with the respondent, and hence the character of rapport and the reciprocity that underpins the interview. When a respondent greets a question with laughter, it may be the occasion for the interviewer's collaborating in or avoiding a side sequence (Jefferson, 1972) or time-out (Jefferson, Sacks, and Schegloff, 1987) in which the interviewer steps out of role and then must negotiate a return to the interview (e.g., Goffman, 1981:45–46, footnote).

CONCLUSION

All interviewing — standardized or not — relies on implicit or explicit models of the meaning of self-reports, just as observational studies rely on models of the meaning of behavior. The meanings interactionally created and recorded during data collection require the attention of those who rely on the spoken words of others to describe the world. The research reviewed and proposed here suggests closer study of how conversational practices are enacted within the prescribed structure of the interview to create data, and comparisons between talk in interviews and talk in other situations. Such studies potentially have both practical and theoretical results: improvements in question design and interviewing technique, as well as increased understanding of the substantive implications of measurement based on talk.

The problem we face in improving measurement in surveys was summarized by Herbert Hyman (1975 [1954]:20–21) in words that presaged those of critics of the standardized interview:

> The twin goals of reduction of inter-interviewer variation and an increase in the validity of the results must always be kept in mind. ... Evaluations oriented purely to the reliability problem ... run the danger of conservatism because the standard against which any interviewer's performance is appraised is that of another *current* interviewer, or that of all *current* interviewers. ... It is only as we have as a norm a form of interviewing that approximates close to valid results, that we become radical and experimental. It must be the neglect of this latter concept of interviewer error that accounts for the rarity of innovation.

Hyman's reflection on survey interviewing 20 years later puts the burden where it belongs: "The fault sometimes is not in our interviewers, but in ourselves" (Hyman, 1975 [1954]:xviii).

ACKNOWLEDGMENTS

This chapter omits material that was specific to the volume in which the text originally appeared. Citations and references to works in press were also updated. This chapter benefited from a 1987 workshop on the role of the interviewer sponsored by the Social Science Research Council (SSRC) and a similar workshop in 1989 sponsored by SSRC and the Russell Sage Foundation. Discussions with Robert M. Groves, Douglas W. Maynard, and Harold Garfinkel also contributed to the ideas developed here in ways for which they bear no responsibility. Robert M. Groves, Robert Hauser, Douglas W. Maynard, Howard Schuman, and Johannes van der Zouwen provided very helpful comments on earlier drafts. My thanks to Corey Keyes for his help with the tables.

NOTES

1. This recommendation attempts to resolve some of the political issues inherent in research (e.g., Briggs, 1986:120–125; Mishler, 1986a:122–132) but raises problems of its own. Community residents may be unaware of issues that the researcher judges important to their lives; or communities may comprise factions with competing goals, among which the researcher must choose.

2. Research that examines structural features of the interview is discussed elsewhere (see, e.g., Fowler, 1991; Hill, 1991; Hox, de Leeuw, and Kreft, 1991; Lyberg and Kasprzyk, 1991). Studies of interviewer effects and variability describe the impact of the respondent–interviewer interaction without studying it directly (see Hyman, 1975 [1954]; and reviews in Groves, 1989; Hagenaars and Heinen, 1982; Lessler, 1984). I also do not discuss studies of interviewers'

vocal characteristics (Barath and Cannell, 1976; Blair, 1977–1978; Oksenberg, Coleman, and Cannell, 1986; Oksenberg and Cannell, 1988).

3. For example, respondents overwhelmingly prefer that the interviewer behave on the second interview just as she did on the first (Cannell, Fowler, and Marquis, 1968:12) and also prefer the mode of administration to which they were first exposed (Groves, 1979).

4. Marquis (1971b) codes a behavior called *blocking* (p. 235). It is unclear whether this code would be used if the respondent interrupted the reading of a question with an answer. The accompanying analysis of questions does not appear to use the code (Cannell and Robison, 1971). Oksenberg, Cannell, and Kalton (1989) use a separate code if the respondent interrupts the initial question reading with an answer.

5. Although investigators attempt to eliminate some sources of bias (e.g., leading questions), this is not the principal technique, as critics sometimes suggest (e.g., Briggs, 1986:24). Even using neutral probes implements the more general strategy of holding potentially important influences on answers constant across respondents.

6. For this review, I treat conversational analysis as a subfield of ethnomethodology. Although I focus on CA studies here, other types of ethnomethodological studies may also help us understand survey interviewing (see Groves et al., 1990; Maynard and Clayman, 1991).

7. Once trained, respondents may patiently wait for response categories to be presented for open questions. This suggests that questionnaire design does not sufficiently signal different kinds of questions.

REFERENCES

Agar, M., and J. Hobbs. 1983. "Natural Plans: Using AI Planning in the Analysis of Ethnographic Interviews." *Ethos* 1: 33–48.

Agar, M., and J. Hobbs. 1982. "Interpreting Discourse: Coherence and the Analysis of Ethnographic Interviews." *Discourse Processes* 5: 1–32.

Atkinson, J. 1982. "Understanding Formality: The Categorization and Production of 'Formal' Interaction." *British Journal of Sociology* 33: 86–117.

Bailar, B., L. Bailey, and J. Stevens. 1977. "Measures of Interviewer Bias and Variance." *Journal of Marketing Research* 14: 337–343.

Barath, A., and C. Cannell. 1976. "Effect of Interviewer's Voice Intonation." *Public Opinion Quarterly* 40: 370–373.

Bauman, L., and E. Adair. 1989. "The Use of Ethnographic Interviewing to Inform Questionnaire Construction." Paper presented at the annual meeting of the American Association for Public Opinion Research, St. Petersburg, FL, May.

Bingham, W., and B. Moore. 1924. *How to Interview*. New York: Harper and Row.

Blair, E. 1977–1978. "More on the Effects of Interviewer's Voice Intonation." *Public Opinion Quarterly* 41: 544–548.

Bohrnstedt, G. 1983. "Measurement." Pp. 70–122 in P. Rossi, J. Wright, and A. Anderson (Eds.), *Handbook of Survey Research*. New York: Academic.

Bradburn, N., S. Sudman, and Associates. 1979. *Improving Interview Method and Questionnaire Design*. San Francisco: Jossey–Bass.

Brenner, M. 1981a. "Aspects of Conversational Structure in the Research Interview." Pp. 19–40 in P. Werth (Ed.), *Conversation and Discourse: Structure and Interpretation*. London: Croom Helm.

Brenner, M. 1981b. "Patterns of Social Structure in the Research Interview." Pp. 115–158 in M. Brenner (Ed.), *Social Method and Social Life*. New York: Academic.

Brenner, M. 1981c. "Skills in the Research Interview." Pp. 28–58 in M. Argyle (Ed.), *Social Skills and Work*. London and New York: Methuen.

Brenner, M. 1982. "Response-Effects of 'Role-Restricted' Characteristics of the Interviewer." Pp. 131–165 in W. Dijkstra and J. van der Zouwen (Eds.), *Response Behaviour in the Survey Interview*. London: Academic.

Briggs, C. 1986. *Learning How to Ask*. Cambridge: Cambridge University Press.

Button, G. 1987. "Answers as Interactional Products: Two Sequential Practices Used in Interviews." *Social Psychology Quarterly* 50: 160–171.

Cannell, C., F. Fowler, and K. Marquis. 1968. "The Influence of Interviewer and Respondent Psychological and Behavioral Variables on the Reporting in Household Interviews." *Vital and Health Statistics*, Series 2, No. 26, Washington, DC: U.S. Government Printing Office.

Cannell, C., and R. Kahn. 1968. "Interviewing." Pp. 526–595 in G. Lindzey and E. Aronson (Eds.), *The Handbook of Social Psychology*, Vol. 2. Reading, MA: Addison-Wesley.

Cannell, C., G. Kalton, L. Oksenberg, K. Bischoping, and F. Fowler. 1989. "New Techniques for Pretesting Survey Questions." Unpublished report, final report for grant HS05616, National Center for Health Services Research, Ann Arbor, MI, University of Michigan.

Cannell, C., S. Lawson, and D. Hausser. 1975. *A Technique for Evaluating Interviewer Performance*. Ann Arbor, MI: Survey Research Center, University of Michigan Press.

Cannell, C., K. Marquis, and A. Laurent. 1977. "A Summary of Studies of Interviewing Methodology." *Vital and Health Statistics*, Series 2, No. 69, Rockville, MD: National Center for Health Statistics.

Cannell, C., P. Miller, and L. Oksenberg. 1981. "Research on Interviewing Techniques." Pp. 389–437 in S. Leinhardt (Ed.), *Sociological Methodology*. San Francisco: Jossey-Bass.

Cannell, C., and L. Oksenberg. 1988. "Observation of Behavior in Telephone Interviews." Pp. 475–495 in R. Groves, P. Biemer, L. Lyberg, J. Massey, W. Nichols II, and J. Waksberg (Eds.), *Telephone Survey Methodology*. New York: Wiley.

Cannell, C., and S. Robison. 1971. "Analysis of Individual Questions." Chapter 11 in J. Lansing, S. Withey, and A. Wolfe (Eds.), *Working Papers on Survey Research in Poverty Areas*. Ann Arbor, MI: Institute for Social Research, University of Michigan.

Churchill, L. 1978. *Questioning Strategies in Sociolinguistics*. Rowley, MA: Newbury House.

Cicourel, A. 1964. *Method and Measurement in Sociology*. New York: Free Press.

Cicourel, A. 1974. *Theory and Method in a Study of Argentine Fertility*. New York: Wiley.

Cicourel, A. 1982. "Interviews, Surveys, and the Problem of Ecological Validity." *American Sociologist* 17: 11–20.

Cicourel, A. 1987. "The Interpenetration of Communicative Contexts: Examples from Medical Encounters." *Social Psychology Quarterly* 50: 217–226.

Clark, H. 1979. "Responding to Indirect Speech Acts." *Cognitive Psychology* 11: 430–477.

Clark, H. 1985. "Language and Language Users." Pp. 179–232 in G. Lindzey and E. Aronson (Eds.), *Handbook of Social Psychology*, Vol. 2. New York: Random House.

Clark, H., and C. Marshall. 1981. "Definite Reference and Mutual Knowledge." Pp. 10–63 in A. Joshi, B. Webber, and I. Webber (Eds.), *Elements of Discourse Understanding*. Cambridge: Cambridge University Press.

Converse, J., and S. Presser. 1986. *Survey Questions: Handcrafting the Standardized Questionnaire*. Sage University Paper series on Quantitative Applications in the Social Sciences, 007–063. Beverly Hills: Sage.

Converse, P. 1970. "Attitudes and Non-attitudes: Continuation of a Dialogue." Pp. 168–189 in E. Tufte (Ed.), *The Quantitative Analysis of Social Problems*. Reading, MA: AddisonWesley.

Dean, J., and W. Whyte. 1958. "How Do You Know If the Informant Is Telling the Truth?" *Human Organization* 17: 34–38.

Dijkstra, W. 1987. "Interviewing Style and Respondent Behavior." *Sociological Methods and Research* 16: 309–334.

Dijkstra, W., L. van der Veen, and J. van der Zouwen. 1985. "A Field Experiment on Interviewer-Respondent Interaction." Pp. 37–63 in M. Brenner, J. Brown, and D. Canter (Eds.), *The Research Interview: Uses and Approaches*. London: Academic.

Dijkstra, W., and J. van der Zouwen. 1987. "Styles of Interviewing and the Social Context of the Survey-Interview." Pp. 200–211 in H. Hippler, N. Schwarz, and S. Sudman (Eds.), *Social Information Processing and Survey Methodology*. New York: Springer-Verlag.

Dijkstra, W., and J. van der Zouwen. 1988. "Types of Inadequate Interviewer Behavior in Survey-Interviews." Pp. 24–35 in W. Saris and I. Gallhofer (Eds.), *Sociometric Research, Vol. 1: Data Collection and Scaling*. New York: St. Martin's Press.

Forsyth, B. H., and J. T. Lessler. 1991. "Cognitive Laboratory Methods: A Taxonomy." Pp. 393–418 in P. B. Biemer, R. M. Groves, L. E. Lyberg, N. A. Mathiowetz, and S. Sudman (Eds.), *Measurement Errors in Surveys*. New York: Wiley.

Fowler, F. J., Jr. 1991. "Reducing Interviewer-Related Error through Interviewer Training, Supervision, and Other Means." Pp. 259–278 in P. P. Biemer, R. M. Groves, L. E. Lyberg, N. A. Mathiowetz, and S. Sudman (Eds.), *Measurement Errors in Surveys*. New York: Wiley.

Fowler, F. J., and T. Mangione. 1985. "The Value of Interviewer Training and Supervision." Final Report to the National Center for Health Services Research, Grant #3-R18-HS04189.

Fowler, F. J., Jr., and T. W. Mangione. 1990. *Standardized Survey Interviewing: Minimizing Interviewer-Related Error*. Newbury Park: Sage.

Francik, E., and H. Clark. 1985. "How to Make Requests That Overcome Obstacles to Compliance." *Journal of Memory and Language* 24: 560–568.

Galtung, J. 1967. *Theory and Methods of Social Research*. London: George Allen and Unwin.

Garfinkel, H., and H. Sacks. 1970. "On Formal Structures of Practical Actions." Pp. 337–366 in J. McKinney and E. Tiryakian (Eds.), *Theoretical Sociology: Perspectives and Development*. New York: Appleton-Century-Crofts.

Goffman, E. 1981. "Replies and Responses." Pp. 5–77 in *Forms of Talk*. Philadelphia: University of Pennsylvania Press.

Goudy, W., and H. Potter. 1975–1976. "Interviewer Rapport: Demise of a Concept." *Public Opinion Quarterly* 39: 529–543.

Grice, H. 1975. "Logic and Conversation." Pp. 41–58 in P. Cole and J. Morgan (Eds.), *Syntax and Semantics*. New York: Academic.

Groves, R. M. 1979, "Actors and Questions in Telephone and Personal Interview Surveys." *Public Opinion Quarterly* 43: 190–205.

Groves, R. M. 1989. *Survey Errors and Survey Costs*. New York: Wiley.

Groves, R. M. 1991. "Measurement Error Across Disciplines." In P. P. Biemer, R. M. Groves, L. E. Lyberg, N. A. Mathiowetz, and S. Sudman (Eds.), *Measurement Errors in Surveys*. New York: Wiley.

Groves, R. M., C. Cannell, D. Maynard, L. Polanyi, M. Salo, and N. C. Schaeffer. 1990. "Improving Data Quality from the Survey Interview Through Studies of Interaction." Unpublished report, University of Michigan. Ann Arbor, MI.

Hagenaars, J., and Heinen, T. 1982. "Effects of Role-Independent Interviewer Characteristics on Responses." Pp. 91–130 in W. Dijkstra and J. van der Zouwen (Eds.), *Response Behaviour in the Survey Interview*. London: Academic.

Henson, R., C. Cannell, and S. Lawson. 1976. "Effects of Interviewer Style on Quality of Reporting in a Survey Interview." *Journal of Psychology* 93: 221–227.

Heritage, J. 1984. "A Change-of-State Token and Aspects of Its Sequential Placement." Pp. 299–345 in J. Atkinson and J. Heritage (Eds.), *Structure of Social Action*. Cambridge: Cambridge University Press.

Heritage, J. 1985. "Recent Developments in Conversation Analysis." *Sociolinguistics* 15: 1–19.

Hill, D. H. 1991. "Interviewer, Respondent, and Regional Office Effects on Response Variance: A Statistical Decomposition." Pp. 463–483 in P. P. Biemer, R. M. Groves, L. E. Lyberg, N. A. Mathiowetz, and S. Sudman (Eds.), *Measurement Errors in Surveys*. New York: Wiley.

Hox, J. J., E. D. de Leeuw, and I. G. G. Kreft. 1991. "The Effect of Interviewer and Respondent Characteristics on the Quality of Survey Data: A Multilevel Model." Pp. 439–462 in P. P. Biemer, R. M. Groves, L. E. Lyberg, N. A. Mathiowetz, and S. Sudman (Eds.), *Measurement Errors in Surveys*. New York: Wiley.

Hyman, H. 1975 [1954]. *Interviewing in Social Research*, Chicago: University of Chicago Press.

Hymes, D. 1972. "Models of the Interaction of Language and Social Life." Pp. 35–71 in J. Gumperz and D. Hymes (Eds.), *Directions in Sociolinguistics: The Ethnography of Communication*. New York: Holt, Rinehart, and Winston.

Irvine, J. 1979. "Formality and Informality in Communicative Events." *American Anthropologist* 81: 773–790.

Jefferson, G. 1972. "Side Sequences." Pp. 294–238 in D. Sudnow (Ed.), *Studies in Social Interaction*. New York: Free Press.

Jefferson, G. 1973. "A Case of Precision Timing in Ordinary Conversation: Over-Lapped Tag-Positioned Address Terms in Closing Sequences." *Semiotica* 9: 47–96.

Jefferson, G. 1974. "Error Correction as an Interactional Resource." *Language in Society* 3: 181–199.

Jefferson, G. 1979. "A Technique for Inviting Laughter and Its Subsequent Acceptance/Declination." Pp. 79–96 in G. Psathas (Ed.), *Everyday Language: Studies in Ethnomethodology*. New York: Irvington.

Jefferson, G. 1980. "On 'Troubles-Premonitory' Response to Inquiry." *Sociological Inquiry* 50: 153–185.

Jefferson, G. 1984a. "On the Organization of Laughter in Talk about Troubles." Pp. 346–369 in J. Atkinson and J. Heritage (Eds.), *Structure of Social Action*. Cambridge: Cambridge University Press.

Jefferson, G. 1984b. "Notes on a Systematic Deployment of the Acknowledgement Tokens 'Yeah' and 'Mm hmm.' " *Papers in Linguistics* 17: 197–216.

Jefferson, G. 1985. "On the Interactional Unpackaging of a 'Gloss.' " *Language in Society* 14: 435–466.

Jefferson, G. 1986. "Notes on 'Latency' in Overlap Onset." *Human Studies* 9: 153–183.

Jefferson, G., H. Sacks, and E. Schegloff. 1987. "Notes on Laughter in the Pursuit of Intimacy." Pp. 152–205 in G. Button and J. Lee (Eds.), *Talk and Social Structure*. Clevedon, Avon, England: Multilingual Matters.

Kane, E. W., and H. Schuman. 1991. "Open Survey Questions as Measures of Personal Concern with Issues: A Reanalysis of Stouffer's Communism, Conformity, and Civil Liberties." Pp. 81–96 in P. Marsden (Ed.), *Sociological Methodology*. Oxford: Blackwell.

Kaplan, S. 1981 "Appropriate Responses to Inappropriate Questions." Pp. 127–144 in A. Joshi, B. Webber, and I. Sag (Eds.), *Elements of Discourse Understanding*. Cambridge: Cambridge University Press.

Lessler, J. 1984. "Measurement Error in Surveys." Pp. 405–440 in C. Turner and E. Martin (Eds.), *Surveying Subjective Phenomena*, Vol. 2. New York: Russell Sage.

Lofland, J. 1969. *Deviance and Identity*. Englewood Cliffs, NJ: PrenticeHall.

Lyberg, L., and D. Kasprzyk. 1991. "Data Collection Methods and Measurement Error: An Overview." In P. P. Biemer, R. M. Groves, L. E. Lyberg, N. A. Mathiowetz, and S. Sudman (Eds.), *Measurement Errors in Surveys*. New York: Wiley.

Marlaire, C., and D. Maynard. 1990. "Standardized Testing as an Interactional Phenomenon." *Sociology of Education* 63: 83–101.

Marquis, K. 1971a. "Effects of Race, Residence, and Selection of Respondent on the Conduct of the Interview." Chapter 12 in J. Lansing, S. Withey, and A. Wolfe (Eds.), *Working Papers on Survey Research in Poverty Areas*. Ann Arbor, MI: Institute for Social Research, University of Michigan.

Marquis, K. 1971b. "Purpose and Procedure of the Tape Recording Analysis," Chapter 10 in J. Lansing, S. Withey, and A. Wolfe (Eds.), *Working Papers on Survey Research in Poverty Areas*. Ann Arbor, MI: Institute for Social Research, University of Michigan.

Marquis, K. H., and C. F. Cannell. 1969. "A Study of Interviewer-Respondent Interaction in the Urban Employment Survey." Ann Arbor, MI: Survey Research Center, Institute for Social Research, University of Michigan.

Mathiowetz, N., and C. Cannell. 1980. "Coding Interviewer Behavior as a Method of Evaluating Performance." Pp. 525–528 in *Proceedings of the Section on Survey Research Methods*, American Statistical Association.

May, J. 1989. "Questions as Suggestions: The Pragmatics of Interrogative Speech." *Language and Communication* 9: 227–243.

Maynard, D. W., and S. E. Clayman. 1991. "The Diversity of Ethnomethodology." *Annual Review of Sociology* 17: 385–418.

Maynard, D., and C. Marlaire. 1987. "The Interactional Substrate of Educational Testing." Paper presented at the Conference on Video Analysis, University of Surrey, Guilford, England.

McHoul, A. 1987. "Why There Are No Guarantees for Interrogators." *Journal of Pragmatics* 11: 455–471.

Merton, R., M. Fiske, and P. Kendall. 1990. *The Focused Interview: A Manual of Problems and Procedures*, 2nd ed. New York: Free Press.

Mishler, E. 1979. "Meaning in Context: Is There Any Other Kind?" *Harvard Educational Review* 49: 1–19.

Mishler, E. 1986a. *Research Interviewing*. Cambridge, MA: Harvard.

Mishler, E. 1986b. "The Analysis of Interview Narratives." Pp. 233–255 in T. Sarbin (Ed.), *Narrative Psychology: The Storied Nature of Human Conduct*. New York: Praeger.

Moerman, M. 1973. "The Use of Precedent in Natural Conversation." *Semiotica* 9: 193–218.

Morton-Williams, J. 1979. "The Use of 'Verbal Interaction Coding' for Evaluating a Questionnaire." *Quality and Quantity* 13: 59–75.

Morton-Williams, J., and W. Sykes. 1984. "The Use of Interaction Coding and Follow-up Interviews to Investigate Comprehension of Survey Questions." *Journal of the Market Research Society* 26: 109–127.

Nisbett, R., and L. Ross. 1980. *Human Inference: Strategies and Shortcomings of Social Judgment*. Englewood Cliffs, NJ: Prentice-Hall.

Oksenberg, L., and C. F. Cannell. 1988. "Effects of Interviewer Vocal Characteristics on Nonresponse." Pp. 257–272 in R. M. Groves, P. P. Biemer, L. E. Lyberg, J. T. Massey, W. L. Nicholls II, and J. Waksberg (Eds.), *Telephone Survey Methodology*. New York: Wiley.

Oksenberg, L., C. F. Cannell, and G. Kalton. 1989. "New Methods for Pretesting Survey Questions." Pp. 30–62 in C. Cannell, L. Oksenberg, G. Kalton, K. Bischoping, and F. Fowler (Eds.), *New Techniques for Pretesting Survey Questions*. Ann Arbor, MI: Survey Research Center, University of Michigan.

Oksenberg, L., L. Coleman. and C. F. Cannell. 1986. "Interviewers' Voices and Refusal Rates in Telephone Surveys." *Public Opinion Quarterly* 50: 97–111.

Paget, M. 1983. "Experience and Knowledge." *Human Studies* 6: 67–90.

Pomerantz, A. 1984a. "Agreeing and Disagreeing with Assessments: Some Features of Preferred/Dispreferred Turn Shapes." Pp. 57–101 in J. Atkinson and J. Heritage (Eds.), *Structure of Social Action*. Cambridge: Cambridge University Press.

Pomerantz, A. 1984b. "Pursuing a Response." Pp. 152–163 in J. Atkinson and J. Heritage (Eds.), *Structure of Social Action*. Cambridge: Cambridge University Press.

Psathas, G. 1986. "Some Sequential Structures in Direction-Giving." *Human Studies* 9: 231–246.

Sacks, H. 1987. "On the Preferences for Agreement and Contiguity in Sequences in Conversation." Pp. 54–69 in G. Button and J. Lee (Eds.), *Talk and Social Organization*. Clevedon, Avon, England: Multilingual Matters.

Sacks, H., E. Schegloff, and G. Jefferson. 1984. "A Simplest Systematics for the Organization of Turn Taking in Conversation." Pp. 7–56 in J. Schenkein (Ed.), *Studies in the Organization of Conversational Interaction*. New York: Academic. (Original version published in 1974 in *Language* 50: 696–735.)

Schaeffer, N. C., and E. Thomson. 1992. "The Discovery of Grounded Uncertainty: Developing Standardized Questions about Strength of Fertility Motivation." Pp. 37–82 in P. V. Marsden (Ed.), *Sociological Methodology 1992*. Oxford: Basil Blackwell.

Schegloff, E. 1972. "Notes on a Conversational Practice: Formulating Place." Pp. 75–119 in D. Sudnow (Ed.), *Studies in Social Interaction*. New York: Free Press.

Schegloff, E. 1982. "Discourse as an Interactional Achievement: Some Uses of 'Uh-Huh' and Other Things That Come between Sentences." Pp. 71–93 in D. Tannen (Ed.), *Analyzing Discourse: Text and Talk*. Washington, DC: Georgetown University Press.

Schegloff, E. 1990. Comment. *Journal of the American Statistical Association* 85: 248–250.

Schegloff, E., G. Jefferson, and H. Sacks. 1977. "The Preference for Self-Correction in the Organization of Repair in Conversation." *Language* 53: 361–382.

Schuman, H., and J. Ludwig. 1983. "The Norm of Even-Handedness in Surveys as in Life." *American Sociological Review* 48: 112–120.

Schuman, H., and S. Presser. 1981. *Questions and Answers in Attitude Surveys: Experiments on Question Form, Wording, and Context*. Orlando, FL: Academic.

Schwarz, N., and F. Strack. 1988. "The Survey Interview and the Logic of Conversation." Paper presented at the annual meeting of the American Association for Public Opinion Research. Working paper from the Zentrum für Umfragen, Methoden, und Analysen, e. V., Mannheim.

Schwarz, N., F. Strack. and H. P. Mai. 1991. "Assimilation and Contrast Effects in Part Whole Question Sequences: A Conversational Logic Analysis." *Public Opinion Quarterly* 55: 3–23.

Singer, E., M. Frankel, and M. Glassman. 1983. "The Effect of Interviewer Characteristics and Expectations on Response." *Public Opinion Quarterly* 47: 68–83.

Singer, E., and L. Kohnke-Aguirre. 1979. "Interviewer Expectation Effects: A Replication and Extension." *Public Opinion Quarterly* 43: 246–260.

Suchman, L., and B. Jordan. 1990. "Interactional Troubles in Face-to-Face Survey Interviews." *Journal of the American Statistical Association* 85: 232–241.

Sudman, S., and N. Bradburn. 1974. *Response Effects in Surveys*. Chicago: Aldine.

Sudman, S., N. Bradburn, E. Blair, and C. Stocking. 1977. "Modest Expectations: The Effect of Interviewers' Prior Expectations on Responses." *Sociological Methods and Research* 6: 171–182.

Survey Research Center. 1976. *Interviewer's Manual*, rev. ed. Ann Arbor, MI: Survey Research Center, Institute for Social Research.

Tourangeau, R. 1984. "Cognitive Sciences and Survey Method." Pp. 73–100 in T. Jabine, M. Straf, J. Tanur, and R. Tourangeau (Eds.), *Cognitive Aspects of Survey Methodology: Building a Bridge between Disciplines*. Washington, DC: National Academy Press.

Turner, C., and E. Martin (Eds.). 1984. *Surveying Subjective Phenomena*, Vol. 1. New York: Russell Sage.

Tversky, A., and D. Kahneman. 1973. "Availability: A Heuristic for Judging Frequency and Probability." *Cognitive Psychology* 5: 207–232.

van Dijk, T. 1980. *Macrostructures: An Interdisciplinary Study of Global Structures in Discourse, Interaction, and Cognition*. Hillsdale, NJ: Lawrence Erlbaum.

van der Zouwen, J., W. Dijkstra, and J. H. Smit. 1991. "Studying Respondent-Interviewer Interaction: The Relationship between Interviewing Style, Interviewer Behavior, and Response Behavior." Pp. 419–438 in P. P. Biemer, R. M. Groves, L. E. Lyberg, N. A. Mathiowetz, and S. Sudman (Eds.), *Measurement Errors in Surveys*. New York: Wiley.

Werner, O., and M. Schoepfle. 1987. *Systematic Fieldwork*, Vol. 1: *Foundations of Ethnography and Interviewing*. Newbury Park, CA: Sage.

Wilson, T., J. Wiemann, and D. Zimmerman. 1984. "Models of Turn Taking in Conversational Interaction." *Journal of Language and Social Psychology* 3: 159–183.

Wolfson, N. 1976. "Speech Events and Natural Speech: Some Implications for Sociolinguistic Methodology." *Language in Society* 5: 189–209.

CHAPTER 5

The Living Text: Written Instructions and Situated Actions in Telephone Surveys

Michael Lynch

Considered as a methodological instrument, the standardized survey interview is a tool for measuring the distribution of attitudes and other social characteristics in a population. Considered as an ethnomethodological phenomenon, the interactional achievement of a standardized survey interview is an instance of how members of a society locally enact prestructured standardized protocols, resulting in data that are treated by survey analysts as measures of that self-same society. These two points of view are not incompatible, although they certainly differ. To say that survey interviews "achieve" or "produce" standardized data does not necessarily imply that, because survey results are constructed by interviewers, respondents, and analysts, their objectivity is doubtful. Instead, it can just as easily imply that the local production of survey interviews is an unremarkable and uncontroversial accomplishment: that during any single interview, interviewers and respondents are able to proceed through the sequence of items that makes up the interview schedule, and that their interaction results in standardized, textually formatted, sociological data.

The survey interview is sometimes called an *instrument*, which suggests that, like other scientific instruments, it mediates relations to an order of things that would remain invisible or unintelligible to unaided observation. This instrument may be more like a laboratory protocol than a telescope, insofar as it consists of a sequence of embodied actions that is repeated and systematically modified from case to case (Jordan and Lynch, 1998). This particular protocol is a partly scripted interactional production performed again and again at different

Standardization and Tacit Knowledge: Interaction and Practice in the Survey Interview
Edited by Douglas W. Maynard, Hanneke Houtkoop-Steenstra, Nora Cate Schaeffer, and Johannes van der Zouwen
ISBN 0-471-35829-0

sites, which can also be likened to a stage play. Singular performances are integrated by means of the centrally designed and globally distributed text, and interviewers are trained and rehearsed in order to use the script artfully and naturally.

The survey interview is a means for collecting sociological data, but it is also a social object in its own right. The sociality of this object is more immediate, proximal, and concrete than the social distributions of attitudes and demographic characteristics that are reconstructed from survey data. This chapter explores the interactional achievement of this standardized social object by focusing on two related themes: the relations between formal instructions and situated actions, and the production of accountable responses.

The first theme is a variant of the *micro–macro* problem in sociology. More specifically, it has to do with the relationships between de-situated textual forms and situated conduct. Textual devices have been construed as crucial facilities for linking extensive actor-networks to *centers of calculation* (Latour 1987, 1990), and written forms of representation and the standardized relations they serve have been said to extend *relations of ruling* into everyday lifeworlds (Smith, 1992). When this variant of the micro–macro problem is ethnomethodologically respecified, it becomes a locally investigatable *topic of order* (Garfinkel, 1991). In the present case, a programmatic equivalence between a written questionnaire and a standardized phone survey will bring into relief how interviewers and respondents interactionally produce distributable textual products.

The second theme has to do with the specificity of survey interview responses. To begin to appreciate this theme, consider a remark by Foucault about Immanuel Kant's essay "What Is Enlightenment?". Foucault (1984:32) observes that Kant's essay was written as a response to a question administered by the editors of the German periodical *Berlinische Monatschrift*:

> Today when a periodical asks its readers a question, it does so in order to collect opinions on some subject about which everyone has an opinion already; there is not much likelihood of learning anything new. In the eighteenth century, editors preferred to question the public on problems that did not yet have solutions. I don't know whether or not that practice was more effective; it was unquestionably more entertaining.

Professional survey analysts are unlikely to agree that nothing new can be learned from collecting opinions.[1] And, as I shall demonstrate, whether or not respondents "already have" opinions can be problematic, and the struggle to solicit such opinions may be more entertaining than Foucault imagined. Nevertheless, Foucault's remark illuminates the peculiarity of the *kind* of questioning featured in a standardized survey. Consider, for example, the absurd case of a passage from Kant (1988:462) presented in the format of a modern survey item:

Interviewer: Please tell me whether you agree or disagree with each of these statements. The first statement is:

"Enlightenment is man's release from his self-incurred tutelage."

```
<1>  AGREE
<2>  DISAGREE
<8>  DON'T KNOW
<9>  REFUSED
```

Whether read in Kant's original text or in the format of the survey item, the statement in question can be counted as an expression of opinion. If a respondent "agrees" with the statement, then it seems reasonable to say that the respondent holds the same opinion as Kant expresses in his essay. But, as Foucault suggests, there are profound differences between the two. In the context of his essay, Kant's "opinion" is part of an argument. When transposed into the survey item, it becomes an isolated, self-contained statement with which to agree or disagree. One may be tempted to locate a great historical shift in the interval between the two political orderings of opinion: In Kant's eighteenth century essay, the opinion is an expression of commitment, whereas in the modern survey it is a standard item for measuring a distribution of agreement and disagreement. In both cases, the opinion is connected with the work of building community: In the essay it is an expression of a heroic attempt to *move* community opinion by inciting its members to "dare to know" and think for themselves; whereas in the survey it is part of an attempt to measure a communal distribution that already exists at large. A specific respondent's espousal of opinion has no special, or leading, significance except as part of the distribution.

The incongruity created by transposing a passage from Kant into the frame of an opinion survey is not intended here to expose the survey as a Philistine operation. Instead, it is intended to elucidate how expressing an opinion, belief, or attitude in a survey requires a distinctive competency. Responding to a survey certainly involves a different order of competency than does writing a learned essay. By comparison to Kant's achievement, the survey respondent's competency may seem trivial indeed, but, as we shall see, it is a necessary, and yet hidden, feature of the orderly production of survey data. In the second part of this chapter, I examine a particular disruption in an interview that illuminates this competency.

FORMAL STRUCTURES AND SITUATED ACTIONS

In literate societies, written rules, plans, recipes, methods, programs, directives, maxims, maps, and models are used to instruct, guide, and record social actions. When considered in a polarized way, such actions involve two interacting elements: (1) broadly distributed, stable, formal, and (relatively) context-free formulations, and (2) actual courses of action that *somehow* (and just how is problematic) are produced in (or out of) accord with the formulations. According to the conceptions of situated practice, contextual language use, and locally organized action, which are put forward in the philosophy of ordinary language, ethnomethodology, interpretative sociology and anthropology, and social studies of science, formal accounts (rules, texts, programs, etc.) do not contain or causally determine actions performed in accord with them, but they are inseparable from the integrity and identity of such actions. Philosophical investigations and empirical studies indicate that there is no single logical or empirical relation between actions and formal accounts of action.[2] Different skills, facilities, degrees of precision, coercions, resistances, and allowances for judgment and improvisation come into play in different cases of building a machine from plans, replicating a classic experiment, learning to drive in accordance with the traffic code, working out the answer to a mathematical problem, preparing a dish in accordance with a recipe, or playing a game with rules. Each case can be said to instantiate a socially structured activity, but the collection of cases does not illustrate a single model of the relations between stable cultural forms and situated practices. No single case study, and no collection of case studies, is likely to resolve long-standing debates about structure versus agency, global versus local determination, and caused behavior versus reasoned action.

Reviews and critiques of ethnomethodological research often associate its program with a one-sided position in an antinomy between agency and structure. According to this view, ethnomethodologists stress the contingent and indeterminate nature of local action, and they ignore and downplay the importance of trans-situational structures (Bourdieu, 1977; Giddens, 1993). Such criticisms tend to confuse ethnomethodology with a skeptical variety of social constructivism and/or a commitment to methodological individualism. As I understand it, the aim of ethnomethodological investigations is not to demonstrate the incompleteness or indeterminacy of formal accounts, but to discover how different modes of practical conduct make use of such accounts. To maintain an orientation to discoverable contextures of practical action, it can be helpful to presume that no *single* relationship obtains between instructions, rules, written texts, or plans, on the one hand, and the actions that cite, read, interpret, follow, disregard, or otherwise use texts and formalisms, on the other. More importantly, as Garfinkel and Sacks (1970) demonstrate, interactional conduct does not require separate *formulations* of meaning or instruction to be intelligible and doable. Unlike many analytic sociologists, structural anthropologists, and cognitive psychologists who endow formal structures with a normative, and even determinate, existence, ethnomethodologists emphasize the singular use, misuse, and disuse of formal structures in practical action. This emphasis neither follows from nor entails a skeptical view of structures, rules, or norms. Ethnomethodologists do not investigate instructed actions to reaffirm the uncanny freedom of human agency. Instead, they pursue an investigative task: to examine how specific formalisms are made relevant, deployed, and (re)configured in here-and-now performances of practical actions.

Garfinkel (1989) occasionally used a formal device he called "the *Lebenswelt* pair" to elucidate the relationship between formulations and local actions.[3] This device teases apart the primordial unity of, for example, actions performed in accordance with instructions, by distinguishing such actions into two parts. The first part of the pair is a formal set of instructions: a list of rules in the traffic code, a musical score, a sketch map showing directions to a destination, a set of instructions on how to play a board game, or a mathematical formula awaiting proof. The second part consists in an actual course of "lived work": driving in accordance (or out of accordance) with rules in the code, playing (or practicing) music according to the score, following directions in the course of a journey, actually playing the game, or working out a proof of the formula. This order can be schematically represented by using the term *instructions* to stand for the various texts, rules, models, and so forth that make up the formal part of the pair:

`[Instructions] → {Lived course of action}`

The arrow denotes the situated work of using the instructions, making out what they say, finding fault with them, re-reading them in light of what is happening *just now*, and other contingent uses and readings. Despite the directionality of the arrow, the route traveled from instructions to lived course of action is not a one-way street, nor does the "lived work" follow a single pattern. Instead, the formal account of the *Lebenswelt* pair itself acts as a gloss that requires the (re)discovery of the local work of *doing* what the pair formulates. The point of juxtaposing the two pair parts is neither to generate ironies nor to celebrate indeterminacy; instead, the point is to encourage investigations of *just how* the lived work of any given pair is achieved.

THE LIVING TEXT

The type of standardized interview discussed in this chapter is an especially strong case of instructed action. It is strong in the degree to which a computer-assisted telephone interview relies upon an explicit written text, together with instructions embedded in a computer program. Compared to more open-ended forms of instruction (e.g., "Okay, you know what to do; just do it!"), the written instructions and computer program furnish the interview with an exhaustive plan, a detailed script, and a network of contingent possibilities. The programmatic instructions furnish the interview with the following resources:

- List of questions and instructions for reciting them
- Exact wording of each item to be recited by the interviewer
- List of possible answers or categories of answer for each question
- Less exact instructions about what to do or say in specific circumstances
- Keyboard instructions for entering encoded responses
- Instructions for the computer program, triggered by response entries, to "go to" a numbered item and skip the intervening items so that the appropriate next item appears on screen.

The printed text of instructions is about a hundred pages long, but an appreciation of its features can be gained by examining a few items. A preliminary item tells the interviewer and computer program what to do when a respondent refuses to begin the interview because of "bad health." The text instructs an interviewer who receives such a response to follow a script (note that the instructions in square brackets (e.g., [goto bar1]) do not appear on the screen, as they are automatically executed by the computer program).

```
I'm sorry to hear that. Have you been sick long? I would be happy
to call back another time.
( IF LENGTHY OR SERIOUS ILLNESS THAT MAKES DOING AN INTERVIEW
IMPOSSIBLE, EXCUSE YOURSELF AND SAY THAT WE WILL NOT CALL AGAIN.)
Older people's opinions are just as important as anyone else's.
In order for the results of this survey to be representative for
all residents of the country, we need to be sure that older people
have as much chance to give their opinion as anyone else does.
We really want YOUR opinion.
( PRESS F7 KEY TO RETURN TO LAST OPEN QUESTION, AND EITHER
CONTINUE OR REFUSE REMAINDER OF INTERVIEW.)
<1>   TOO BUSY [goto bar1]
<2>   FEEL INADEQUATE: DON'T KNOW ENOUGH TO ANSWER [goto bar3]
<4>   NOT INTERESTED [goto bar4]
<5>   NO ONE'S BUSINESS WHAT I THINK / CONFIDENTIALITY [goto bar5]
<6>   OBJECTS TO SURVEYS [goto bar6]
<7>   OBJECTS TO PHONE SURVEYS [goto bar7]
<8>   NONE OF THESE [goto bar4]
```

This preliminary item instructs the interviewer on how to contend with, and make a record of, a respondent's refusal to undergo the interview for reasons of bad health.[4] The item specifies what the interviewer should do or say and provides a list of possible types of response, but it is less precise in its design than many of the items in the body of the interview. For example, consider two items that are part of a series of questions about the U.S. census (I shall return to these questions later):

```
>5b<
People have different ideas about what the census is used for, and
I'm going to read you some of them. As I read each one, please
tell me whether you think the census is used for that purpose.
To find areas of the country that need government help.
Do you think the census is used for that purpose?
<1>  YES
<2>  NO
<8>  DON'T KNOW
<9>  REFUSED

>5d<
Is the census used to decide how many representatives each state
will have in Congress?
<1>  YES
<2>  NO
<8>  DON'T KNOW
<9>  REFUSED
```

These items constitute a phase of an interactionally administered questionnaire form — a living text.[5] The idea of a living text brings into relief a variety of settings in which actors, players, performers, and other participants organize their speech and bodily conduct in relation to a script, score, set of instructions, or other text. Oral reading lessons in elementary classrooms, political speeches, university lectures, poetry readings, sermons, stage plays, and many other occasions of "live" conduct are more or less closely organized around prepared (typically printed) scripts. In the case of the phone survey above, the interviewer reads aloud the preliminary instruction for item 5b. The instruction forecasts an indefinite number of items to follow and indicates the appropriate kind of response. The interviewer then reads a series of questions and waits for the response. After receiving a response, the interviewer enters the appropriate code into the computer database.[6]

The interviewer and respondent have differential access to the instructional text, and some instructions about what to do next (e.g., "goto bar1") are automatically programmed into the computer. For the most part, the interviewer's monitor dictates the exact words she should say. But, unlike an exhaustively scripted dialog, the instructions do not dictate both parts of the dialog. The respondent's access to the written instructions is entirely mediated by the interviewer's voice. A kind of instructional transformation is thereby produced: The instructions and automated movements through a sequence of questions direct the interviewer to read a next item, and while it is read over the phone the item becomes an instruction to the respondent on how to reply. So, for example, the text for 5b (see above), furnishes the interviewer with a script to read. The script includes instructions for the respondent on how to answer the item: "I'm going to read you some of them. As I read each one, please tell me whether you think the Census is used for that purpose." The invitation to "tell me whether..." grammatically keys the relevance of a yes–no answer, but the respondent does not actually view the four possible response categories on the interviewer's monitor. When reading item 5b, the interview does not (and perhaps does not need to) say in so many words that it is possible to respond by saying "don't know" or refusing to answer.[7] In addition to covering logically possible responses to yes–no questions, "don't know" and "refuse" fill out a set of options the interviewer can use to code a range of responses, including those that are not responsive to the question.[8] This design achieves closure, in principle, by covering residual responses in addition to the binary (affirmative and negative) responses.

The interviewer's and respondent's differential relations to the instructional script serve obvious methodological functions. To appreciate this point, imagine the absurd case of an item that dictated both a specific question *and* a specific response:

```
>5d<
Is the census used to decide how many representatives each state
will have in Congress? Say ''Yes.''
<1>  YES
<9>  REFUSED [Absence of <1>]
```

Such an item might be useful as a test of the degree of respondent compliance, but the answer would have no value as a datum about the respondent's beliefs about census data. The actual interview avoids such absurdity by allowing the respondent a degree of freedom. At the same time it provides the interviewer with a closed system of alternatives for coding the response. In this way, the interview design serves an empirical research program in which a controlled intervention takes the measure of a circumscribed, but free, choice. The interview is designed to produce a quantified measure of aggregate tendencies to choose among standard alternatives. This design sets up a standardized frame in which any unique interview produces a distinctive profile of responses selected from a finite array of choices:

- Interview design and computer program guide the interviewer to present the same items in the same way to each respondent
- Design of items allows for contingent variation in the production of responses
- Design of particular items anticipates and instructs a respondent's answers
- Closed set of possible responses for each item enables the interviewer to encode an open-ended range of actual responses

The analytical aspect of the interview — the production and encoding of responses — introduces a second pair device that operates in the reverse of the *Lebenswelt* pair outlined earlier. Garfinkel and Wieder (1992) call this second device the *rendering theorem*[9]: Beginning with the 'lived work' of, say, playing 'live' music on an actual occasion, the rendering theorem describes how that work is analytically encoded as a *signed object* (a recorded, encrypted text):

```
{Lived course of action} → (Analytical object)
```

Putting together the *Lebenswelt* pair and the rendering theorem, we get a double metamorphosis:

```
[Instructions] → {Lived course of action} → (Analytical object)
```

A standard survey interview can thus be construed as a sociotechnical production in which the interviewer and computer act in accordance with a set of instructions, initiate a "lived course of action" in the form of a dialog over the phone, and then analytically encode the response as a signed object. The lived course of action in this case is a conditionally relevant interactional production.[10] Consequently, the {Lived course of action} is a dialogical production of an item–response pair, and not an action produced by one part alone:

```
{Interviewer: Item
Interviewee: Response}
```

We must keep in mind, however, that this interactional order — the response, produced with a locally evident, recognizable, accountable relation to the item — is set up by a standard set of instructions, and then rendered as a singular coded entry (the analytical object):

```
[Instructions] → {Interviewer: Item
                  Interviewee: Response} → (Analytical object)
```

According to this formulation, the *work* of the interview is preanalytical, and yet it is framed by, and presumably oriented to, the analyzability of talk as part of a data-productive instrument. The analyzability of a coded entry as a datum indexed to a particular interviewee depends upon the instrumental transparency of the singular, interactionally analyzable production of the interview response. Although the interviewer is primarily responsible for soliciting and encoding data on the respondent's background, identity, and opinions, both parties must collaborate in the enframing, extracting, and encoding of such data. By "data" I simply mean a series of entries into a cumulative database consisting of the survey results from numerous respondents on a set of prespecified topics.[11]

When considered as a data-productive instrument, the survey interview is both more and less than a dialog between two persons. From the perspective of the interviewer, the sequence of actions includes *more* than talk. The phases of the dialog are literally keyed off by keyboard actions: An entry for each item is produced by typing the relevant code for the response; and each answer entered instructs the computer program to "go to" a next item. In other words, the interview is as much an act of *writing* as it is of talking. Phone surveys are like written questionnaires, except that an interviewer, aided by a computer, reads a sequence of items aloud and records the responses. In a written questionnaire, the respondent both reads the items (along with any preliminary instructions) and marks the responses. Of course, it would be necessary to modify the grammar and formatting of a phone survey to convert it into a questionnaire mailed to potential respondents, and there would be no guarantee that responses to the two instruments would be equivalent. However, there does seem to be a rough functional equivalence between the oral-electronic administration of an interview and the literate self-administration of a questionnaire.

Viewed as a methodological instrument roughly equivalent to a written questionnaire, the standardized phone interview also is *less* than a dialog between two persons. Training, skill, and singular attention to the contingencies of the interaction surely are required for competent phone interviewing, but for methodological purposes the interviewer is an "invisible technician" (Shapin, 1994:361ff.). Survey interviewers are lowly adjuncts of a research enterprise who read questionnaire items designed by researchers, and who act as amanuenses who inscribe what respondents say into a database. The interviewer is part of an instrumental circuit: The interviewer's voice acts together with the computer and phone connection to relay items to the respondent, and the responses are relayed through the phone receiver to the interviewer, who in turn relays each response through the keyboard into the database. The computer is the proximal source of the question, the respondent is the source of response, and the database is the destination for the encoded data. The interviewer, phone, and keyboard act as relays:

Computer	←——→	Interviewer	←——→	Phone	←——→	Respondent
Question	----→	Relay	----→	Relay	----→	Answer
Data entry	←----	Relay	←----	Relay	←----	

While an actual phone interviewer is never actually reduced to a cipher, her improvisations and interpretations are limited. It might be said that the interviewer embodies a kind of machine intelligence. This is not an instantiation of artificial intelligence, but the opposite: a person voicing the dictates of a program and acting as an interface between the respondent and the database.[12]

The Living Text in Action

When we examine tape recordings of actual phone surveys, it becomes apparent that the intermediary technologies (the interviewer and phone relays) are not entirely transparent, and responses do not always respect the closed system of alternatives. Tape recordings of actual survey interviews exhibit interesting contrasts between the textual design and the actual production of standardized interviews. A number of such differences can be illustrated by comparing the instructions for an item with a transcript of its delivery in a specific case.

Instructions for item 7d (this occurs as part of a series of questions about the U.S. census exploring respondents' beliefs about possible interagency uses of census data):

```
>7d<
Sharing information between different agencies of government when
both want the same information saves time and money, but it also
means some loss in privacy for the individual. Do you think the
benefits of saving time and money outweigh the costs in privacy, or
do you think loss of privacy outweighs the benefits of saving
time and money?
<1>  BENEFITS OF SAVING TIME/MONEY OUTWEIGH COSTS IN PRIVACY
<2>  LOSS OF PRIVACY OUTWEIGHS BENEFITS OF SAVING TIME/MONEY
<8>  DON'T KNOW
<9>  REFUSED
```

A particular delivery of item 7d: This occurs during a tape-recorded interview and exhibits a number of departures from the sequential design of the instructions (IV is interviewer; FR is female respondent):

(1) AW01:147

```
IV:  ↑ A:n' sharing information between different agencies
     of government when both want the same information
     (0.4) saves time [( )
FR:                   [But see now who's the different
     agencies of government.
IV:  .hhhh we:ll (0.4) the- you know the government
     (0.2) is made up of different asiancies: one of
```

which: is the census bureau
(0.9)
IV: So (0.2) they're just talking abou:t
(0.5)
IV: whatever: (0.4) the other government agencies are:
FR: Within the United States
IV: >Right<
FR: Oh okay=
IV: =so w- y'know
FR: Now I'm more clear
IV: Ok(h)ay(h) huh huh.hhh AH: Sharing information
between different agencies of government
[when both-]
FR: [I forgot] what was your question?
(0.3)
IV: Oh I'm- this is just the lead in
(0.3)
IV: [We're just starting. .hhh ah:: sharing ...

Several departures from the interview text are noticeable. Many of the differences between a questionnaire and a spoken interview appear superficial, unremarkable, and methodologically uninteresting, but some of them may be more significant for sustaining the interview in the face of local troubles.

In line 4, the respondent interrupts an item before its scripted completion.[13] The interruption begins just prior to the interviewer's completion of a sentence that, according to the script, precedes the question. What the respondent says interruptedly reacts to, and calls for clarification of, the particular reference of "agencies." She gives priority to a question of understanding just "who" the agencies are. The sequential contingencies of item at hand are thus suspended in the face of a *hermeneutical* question about the interview text. The interviewer cuts off the item and in line 6 begins to clarify the unspecified reference to "agencies." At this point, both parties are off the script. They have suspended the order of questions and answers that constitutes the interview's scripted organization. Nevertheless, their talk remains intelligible and is accountably ordered as conversational dialog. Moreover, it is intelligible as a "conversational" interlude that serves the continuation of the interview. Although the interchange does not follow the interview script, it evidently functions in the service of the interview by collaboratively clarifying the sense and sequential implicature of the item at hand. Both parties can be imagined to be working earnestly to achieve a sensible interview. They are not, for example, taking a break and engaging in an exchange of pleasantries.

The interviewer's point of clarification (line 12: "whatever the other government agencies are") is especially interesting for its referentially empty and noncommittal treatment of the respondent's request (or demand) for clarification. The interviewer does not respond to the demand for "who," or which agencies in particular, might be involved in a hypothetical "sharing of information." Instead of, perhaps, prejudicing the case by mentioning particular agencies (e.g., the CIA, FBI, INS, or IRS), she gives nothing away and simply mentions that the government is made up of different agencies. The only agency she names is the, already given, Census Bureau, and she leaves the rest of them up to the respondent's judgment ("whatever the other government agencies are"). Perhaps the interview's noncommittal move respects demands for a standard and neutral delivery of items, but to recall a line from the earlier-quoted passage from Foucault (1984:32), this is a dialog in which, for the participants at least, "there is not much likelihood of learning anything new." Where the respondent deferred compliance to the sequential design of the interview in the face of an ambiguity in *understanding*, the interviewer puts the design back on track by instructing her to gloss "agencies" as "whatever."

After the respondent accepts the clarification (lines 15, 17), the interviewer reintroduces the question with an extended prefatory component [line 18: "ok(h)ay(h) huh huh.hhh AH:"]. Perhaps this holds the turn while she finds her place on the script. The laughter laced into the preface can also be heard to treat the respondent's "Now I'm more clear" as a joke or ironic comment on the interviewer's rather empty clarification.[14] She then begins to recite the scripted question. The respondent interrupts again (line 21), only a bit earlier this time, just after the interviewer completes the first phrase in the preliminary statement. The respondent's question, "I forgot, what was your question?" might seem puzzling, in light of the fact that she interrupts a preface to a question that has not yet been asked. It is not so puzzling when we take into account that the respondent cannot see the scripted question, so that the interviewer's utterance may sound like it resumes somewhere in the middle of an item that had been read prior to the intervening side sequence.

In line 23, the interviewer, after a brief pause, explains that "this is just the lead in," thereby instructing the respondent to wait for the question she has not yet recited. She thus formulates in so many words what otherwise goes without saying when members achieve accountable sequential orders in the collaborative way they time and place their utterances (Garfinkel and Sacks, 1970).[15] After the respondent acknowledges this last point of clarification, the interviewer reiterates the point (line 25, "We're just starting") and then begins the item again.

This extract, like many other recorded interviews, demonstrates that a standardized recitation of interview items over the phone is difficult to sustain in the face of the local contingencies of the specific interview. The difficulties with executing the scripted interview, together with the resources for contending with those difficulties, arise from a surplus order of interactional relevancies. In other words, the actual delivery of the interview opens up local sequential options that are not part of the scripted design.[16] Moreover, some of these sequential departures, such as the respondent's "interruption" of item 7d, have to do with working up a locally adequate understanding of the standard interview text; an understanding sufficient to "go on" with the interview. Interpretive confusions and logical conundrums are thus exhibited, and bypassed, in and through sequential departures from the interview script.

The interview questions are designed to be read in standardized fashion, so that the interviewer does not induce the respondent to give an answer other than the one she would elect if the item had appeared on a questionnaire. The interview is designed to collect and compare respondent beliefs and to hold constant the contingencies of the interview and the peculiarities of the interviewer. However, as Suchman and Jordan (1990) have argued, confusion and misunderstanding can result when the context-free design is contextually produced as part of a singular phone call. Consequently, a degree of flexibility is necessary to sustain the intelligibility of the standardized interview. Departures from the scripted order of the interview, such as the side sequence exhibited above, can serve to clarify the "original" sense of what a question is *asking* (and not asking). If the above interviewer were to adhere rigidly to the script, she would be unable to address the respondent's requests for clarification, and it seems likely that the interview would degenerate into an empty and uncoordinated interchange. At the same time, however, conversational (and hermeneutic) possibilities remain subordinated to the task at hand: getting through the interview.

The parties to the interview produce a sequence of talk that unfolds in real time, but the relevant analytical basis for the talk — a stable, context-free frame of reference — is furnished by the textual program that dictates what the interviewer reads, and encodes how the interviewee responds. The instructions and the code are intertwined because the enactment of the instructions sets up the enactment of a series of relevant (and encodeable) responses.

PRODUCTION OF ACCOUNTABLE RESPONSES

The living text is not simply a device for enunciating questions and responses over the phone. It is a methodological instrument. In the particular case at hand, many of the questions invite the respondent to consider a personal statement that some people are likely to agree with and others not. For example:

```
>50a<
Please tell me whether you agree or disagree with each of these
statements. The first statement is:
People like me don't have any say about what the government does.
<1>  AGREE
<2>  DISAGREE
<8>  DON'T KNOW
<9>  REFUSED
```

In short, this is an item designed to solicit opinions.[17] The survey contains many such items for identifying and assessing the distribution of beliefs, views, perceptions, and attitudes. Survey items exhibit a clear logical design that identifies a relevant opinion, belief, and so forth with a statement. The above item is typical: The interviewer asks the respondent to agree or disagree with a statement: "People like me don't have any say about what the government does." The statement is written in the first-person case. Presumably an affirmative answer takes ownership of the opinion it expresses. The interviewer then encodes the answer in terms of one of four possibilities: agree, disagree, don't know, refuse. The instruction and the statement set up the relevance of agreeing, disagreeing, refusing to do either, or professing ignorance. As we have seen, in actual cases a respondent can ask for clarification, fail to understand the question, or object to the terms of reference, but whatever she says can be coded (made accountable) in terms of the four options included under the item.[18] Consequently, the distribution of responses on this item across the entire survey sample can be used to measure the differential distribution of the relevant opinions.

There is another way to think about opinions. Consider the possibility that expressing (or in this case affirming) an opinion involves a competency, and that aside from indicating the existence and measuring the distribution of opinions, survey interviews instruct respondents on how to have opinions, attitudes, beliefs, and other survey-analytic phenomena. *That* survey interviews may do so is hidden by the formal design of items. To understand what sort of competence is at stake, consider an analogy with the Catholic confessional. The institution of the confessional requires the parishioner to examine his or her conscience and confess any sins committed since the last confession. According to Jonson and Toulmin (1988:96–97), this institution dates back to the sixth century of the Christian era, a time when theological conceptions of the person began to focus more closely on individual spiritual choices and responsibilities and less on the status of the individual under the law. Books called *Penitentials* were distributed to priests to instruct the laity on how to examine their consciences in preparation for confessing their sins. The competence these books instructed—the ability to search one's soul and to recognize which acts counted as sins under church doctrine—eventually became taken for granted. Historical research on the effort to establish the conditions for standard confession and penance alerts us to the fact that confession requires more than a guilty conscience.[19] Having sins on one's soul requires an ability to recognize and codify them appropriately. According to this analogy, we might suppose that having opinions in one's mind also requires a competency, and that opinion surveys do not simply tap into a natural reservoir of public opinion. The smooth and seamless operation of questionnaires gives little or no indication of such a competence; it simply relies upon its effective performance. This is where instances of the "living text" of survey interviewing becomes particularly instructive.

In light of the foregoing, consider the following transcript of a tape-recorded interview. This is an instance in which the interviewer diverges from the formatted items to instruct the respondent on how to answer a series of "opinion" questions. More precisely, these questions solicit what the respondent "thinks" (or believes) is done with U.S. census data. The improvised instruction is offered as a remedy for the respondent's way of handling prior items. The transcript is of a particular enactment of items 5b and 5d:

(2) AW01:97

```
IV:  °Oh kay° people have different ideas about what
     the census is used for [and now] I'm going to=
FR:                         [uh huh ]
IV:  = read some of them?.hhh As I read each one (0.3)
     please tell me whether you think the census is used
     for that purpose:.hhh To find areas of the country?
     that need government help? (0.4) Do you think the
```

```
         census is used for that purpose?  {q5b}
         (1.3)
    FR:  I have no idea.
    IV:  .hhh Okay? a:nd is the census used to decide how many
         representatives each state will have in congress?
         (1.8)
    FR:  No idea
    IV:  .hhh Okay these are?- (0.3) they're just guesses,
         you know, we don't [(        )] expe(h)ct (h)you=
    FR:                     [uh huh::]
    IV:  =t(h)o know the: eh- solid hard fast truth it's just
         (0.3) whatever your perception may be:.hhh So: if
         you just wanna take a gue(gh)ss go ahea(gh)d. .hhh
         Is it used to see what changes have taken place  in
         the United States. {q5f}
    FR:  Yes
```

When reading items 5b and 5d, the interviewer closely follows the wording in the interview schedule. In lines 1 and 4 the interviewer reads 5b, which (as already noted) is scripted as follows in the interview schedule:

```
People have different ideas about what the census is used for, and
I'm going to read you some of them. As I read each one, please tell
me whether you think the census is used for that purpose. To find
areas of the country that need government help. Do you think the
census is used for that purpose?
```

In line 11 of the transcript, the interviewer reads 5d, scripted as follows: "Is the census used to decide how many representatives each state will have in Congress?"

There are some minor discrepancies between script and transcript, like the transitional preface, ".hhh okay? a:nd" at the beginning of line 11 in the transcript, but I assume that these discrepancies did not interfere with the delivery of the items. I also assume that the consecutive responses "I have no idea" and "no idea" were easily coded under the category of "DON'T KNOW." Nevertheless, something interesting happens. Upon receiving the response to 5d ("no idea," line 14), the interviewer breaks out of her subservience to the script and initiates an unprogrammed emergency action (line 15). As I understand this action, it is an unprogrammed instruction designed to repair an emergent problem with the way the respondent is treating the series of items. Conventionally transliterated,[20] the instruction reads like this:

```
These are... just guesses, you know, we don't expect you to know
the... solid hard fast truth, it's just whatever your perception
may be, so if you just want to take a guess, go ahead.
```

This ad hoc instruction is immediately followed by the next question "is it used to see what changes have taken place in the United States" (line 21), to which the respondent says "yes" (line 23).

So what could be the trouble, and what is the repair *for*? In many respects, there is no trouble and no intervention seems called for. The interviewer begins her intervention just after the response to item 5d ("no idea" in line 14 of the transcript). This response is an attenuated version of the response to item 5b ("I have no idea" in line 10). Item 5b is formatted as the first in a series of similar items, and the respondent may have picked up on the word "ideas" in the prefatory instruction to that item: "People have different ideas about what the census is used for" (line 1). It is easy enough to imagine that many U.S. citizens in fact have "no idea" what the government does with census data. Consequently, "I have no idea" is a contextually relevant and reasonable response. The repetition and attenuation of the second response ("no idea," the response to 5d) appears to project more of the same for any further questions on the topic.[21]

Why, then, would the interviewer want to intervene at that point? Any answer to this question is speculative, but exploring the sequential and conceptual environment of the repair may shed light not only on this incident, but also on the survey instrument's interactional design. Unlike "yes" and "no," "no idea" is *conceptually* nonresponsive to the terms of the question. The prefatory instruction ("tell me whether you think...") and the wording of the question "Do you think the census is used for that purpose?" explicitly make an issue of what the respondent "thinks" (in the vernacular sense of "believes"). A "yes" or "no" response retains the relevance of "thinks," but a "don't know" response does not count as part of a distribution of affirmative and negative *beliefs* ('beliefs' being the conceptually reified object indexed by the question). "Don't know" is a perfectly intelligible response, but it does not code as a yes–no answer to the question, "Do you think the census is used for that purpose?"[22] The response is accountable, and survey analysts may have some interest in the aggregate distribution of "don't know" responses to the item, but those who respond "don't know" are not counted as part of the relevant distribution of "belief." They are conceptually off scale.

This chapter is not written from the point of view of a survey analyst, and I am not prepared to say whether the interviewer's move should be admonished as a mistake or praised as an artful repair. On the one hand, I can imagine that survey administrators might want to warn their employees against giving such improvised instructions because they confound the comparability of responses by different respondents to the "same" series of items. The interviewer's vernacular advice "if you just want to take a guess, go ahead" may license a more casual approach to the questions for this respondent, as compared with other respondents who were not given such advice. On the other hand, the advice takes the form of a nondirective "probe," which interviewers are sometimes encouraged to initiate to invite more substantive responses than "Don't know."[23] Although such practical, methodological issues may be pertinent, they are not what interests me about the interviewer's ad hoc instruction. What I find interesting is that the instruction takes notice of a trouble with the *accountability* of a response that is formally permitted by the exam design. Although there should be no difficulty with treating "[I have] no idea" as a relevant, normal, codable response to the item (namely, a "don't know" response), the interviewer's probe indicates a difficulty with the accountability of such a response. The formal codability of the response therefore does not exhaust the matter of its accountability. The code, although it *can* cover the case, fails to resolve *just what* the interviewer suggests is contextually relevant for giving such a response here and now.

The nonscripted ad hoc instruction, "if you just want to take a guess, go ahead," draws attention to a local order of relevancy that is hidden by the apparently adequate formal relations between the prior item–response pairs. This ad hoc instruction is interesting for the following two reasons:

1. It distinguishes between the form of a response, as a selection of one of the four categories making up the set of relevant responses, and the local accountability of that response. Another way of putting this is that the interviewer's intervention implies that producing an appropriate *type* of response in a relevant sequential slot does not necessarily constitute the most appropriate response to *that* item by *this* respondent, now.
2. It associates a competent sequential production with a specific orientation to knowledge and truth. In this case, the interviewer informs the respondent that she can take a relatively relaxed approach to answering the questions: "they're just guesses... we don't expect you to know the solid hard fast truth, it's just whatever your perception may be."

It may be pertinent that these particular questions contain factual statements with which the respondent is asked to agree or disagree. For example, question 5d "Is the census used to decide how many representatives each state will have in Congress?" could easily be used to test a recipient's knowledge of the relevant government operations. Presumably, someone who is well-informed about government uses of census data would know the *correct* answer to this question, and others either would "think" they knew (and be wrong) or know they had "no idea." In this interview, however, the item is designed not to test whether the respondent *knows* or *does not know* the correct answer, but to solicit what she *believes* in terms of two possibilities (indicated by "yes" and "no" responses). The apparent confusion the interviewer addresses in the above sequence thus may be a consequence of poorly designed questions,[24] but I will not pursue that possibility. What I find interesting about the interviewer's instruction is the way it invites the respondent to perform a kind of gestalt switch, to view her responses from the vantage point of a survey of beliefs rather than an examination of what she knows: "These are... just guesses, you know, we don't expect you to know the... solid hard fast truth." Accordingly, she should view her responses as part of an objective distribution of beliefs and not as an examination of her knowledge of objective government operations. The interviewer thus gives her permission to register her belief as part of an aggregate, much in the way she is entitled to register as a member of a political party. In both cases the entitlement to register is independent of any *knowledge about* the political entities in question.

CONCLUSION

The standardized interview is a measuring instrument, but it is also a highly structured social object. The stable form of the instrument facilitates a two-way communicative link between an administrative center and the "capillaries" of the body politic (Foucault, 1980). In classic sociological terms, it is a concrete embodiment of a micro–macro link (Alexander et al., 1987): Instructions are designed by the center to assure standardized administration of discursive objects (interviews) to representative members of a populace, and the concrete responses to interview questions are aggregated into quantitative measures of distribution and variance. When viewed from the center, the integrity of the micro–macro link depends upon a disciplined production of local interactions in accordance with a standard design. But, as this chapter has illustrated, the production of singular instances of a survey interview can involve ad hoc departures from an interview schedule that contend with complaints, apparent confusions, and other emergent problems. In many cases, such departures from the schedule preserve the survey's design in the face of possible rigidity, misunderstanding, and even absurdity.

The standardized survey is a concrete instance of an actor network (Latour, 1987). The survey instrument is a two-way text: It disseminates a standardized, centrally designed script to actors in the provinces, and it records data that are sent back to the center. The "actors" in this instance include survey interviewers, who, like stage actors, are trained to read lines on a script. The interviewer's arts may be less highly valued, but like the more prestigious performer she is charged with bringing a script to life and contending with the exigencies of an actual performance. The interviewer may not be expected to embellish a character in a stylish way — indeed her art involves a systematic effacement of character — and the anonymous interview writers have little in common with Shakespeare. Nevertheless, there is a sense in which the picture of script writing and acting applies rather closely to the production of interview texts and the dissemination of interview performances. There is a key difference, however. A Shakespeare does not write scripts to collect data, though the market researchers working for the theater company may collect data on audience preferences and reactions. The "others" in the survey interview — the respondents — are neither actors controlled by the script nor members of an audience who react *en masse* to the lines delivered by the actors. The respondents are treated as a collectivity, but the interviews are delivered individually, interactively, and line by line. The responses are more precisely formatted and forecasted by the design of the survey instrument, and they are recorded for posterity. Most importantly, the responses are not treated as audience reactions to the elements of a performance, they are encoded and aggregated as measures of naturally distributed population characteristics.

The discussion of "the living text in action" in the first part of this chapter suggested that the effectiveness of the linkage between the center of calculation and a network of interviewers depends upon ad hoc efforts to counteract troubles that arise in the sequential production and interpretation of survey items.

Superficially understood, this means that the local, singular, moment-to-moment production of phone interviews subserves the global operation of the network. It means more than that, however. What is at stake is the very specificity of the interview, as a survey instrument. The specificity of the interview pervades the network; it is not limited to the face-to-face interaction between interviewers and respondents.[25] Troubles in the implementation of surveys can elucidate what the interview *is* and is *about*. It is not a scripted dialog, a conversation, a test, or an interrogation. It is not sufficient to say that the interview is textually mediated, or that texts and inscriptions are intermediary agencies in the network. The question is, *what* are the texts mediating? The "voiced" survey is interesting not only because, in contrast to a written self-reported questionnaire, it introduces further interactional contingencies; it also enables us to appreciate the vicissitudes of writing, reading, and recording that are integral to the survey. We need not suppose that what is said aloud over the phone by interviewers and respondents is equivalent to what a questionnaire respondent silently reads and thinks. All we need to suppose is that the confusions and local instructions that are voiced when surveys are delivered over the phone make an explicit issue of the local competencies associated with a survey's analytical operations. The second half of the chapter pursued one such area of competency, associated with the production of particular items designed to assess beliefs about the census.

The discussion of beliefs explored the linkage between questions and responses. Unlike the linkage between a centrally written item and the locally situated production of the "same" item, this linkage was internal to the relations between an item and response in an opinion survey. In the sequence examined, the interviewer intervened to correct a response that was formally covered by the coded response option "Don't Know." Whether or not the interviewer's intervention was justified on methodological grounds, it attempted to address an apparent misunderstanding on the part of the respondent. The possible misunderstanding arose from factual questions that were used for eliciting beliefs, and the interviewer attempted to emphasize the salience of belief by downplaying the salience of truth and knowledge. This ad hoc intervention compensated for an ambiguity in the formal design of particular survey questions and invited the respondent to treat her particular answers with the indifference of a survey analyst.

Like the humans who serve machine translation programs with grammatical repairs of the machine's inputs and outputs, interviewers and respondents manage to set up the analyzability of their talk for the survey instrument. Harvey Sacks (1972:44) once marvelled at the extraordinary lengths to which members will go to respect the logical analyzability of their talk. It is no less wonderful to find that persons on both ends of a phone connection will often go to great lengths to set up the transparency of their own "conversation" for the purpose of collecting survey data. There is no reason to suppose that members cannot achieve the adequate equivalency, the adequate objectivity, and the adequate representativeness of their talk for survey purposes, although such achievement may require ad hoc work that it not easily justified under the analytic program it serves. If the history of analytic sociology is any indication, members *can* achieve the analyzability of their activities when acting in accord with the design of a survey, but this very achievement introduces elements of surprise and instability that furnish survey analysts with troubles that supply ethnomethodologists with data.

NOTES

1. Foucault uses the term *opinion* in the vernacular sense of a generally held view or belief. I am told that survey analysts operationalize a more restricted concept of *opinion question*. Such a question contains an implicit or explicit "should" with which someone could agree or disagree (e.g., "Representatives should be limited to three consecutive terms.") or favor or oppose (Nora Cate Schaeffer, personal communication, December 1999).

2. Philosophical investigations of actions in accord with rules include Wittgenstein (1958) and Winch (1990). Numerous ethnomethodological studies can be cited in this context, including investigations of situated uses of formulations (Garfinkel and Sacks, 1970; Heritage and Watson, 1980), records (Garfinkel, 1967; Zimmerman, 1974), codes (Wieder, 1974), mathematical proof statements (Livingston, 1986), plans (Suchman, 1987), and instructions (Bjelic and Lynch, 1992; Lynch and Jordan, 1995). In social studies of science and technology a skeptical reading of Wittgenstein's account of actions in accordance with rules has been developed in studies of mathematics (Bloor, 1973), experimental replication (Collins, 1975), and artificial intelligence (Collins, 1990). For a debate on the differences between the ethnomethodological and science studies versions of actions in accord with rules see Lynch (1992) and Bloor (1992). For a systematic argument on how ethnomethodology, conversation analysis, and sociology of scientific knowledge can be used in an investigation of survey interviews, see Maynard and Schaeffer (2000).

3. The theme of *Lebenswelt* pair is discussed in an unpublished manuscript by Garfinkel et al. (1989). For published sources, see Livingston (1987:119ff.) and Lynch (1993:287ff.). The term *Lebenswelt pair* is a somewhat irreverent citation to Husserl's (1970) and Schutz's (1972) phenomenological explications of the life-world. It is irreverent because ethnomethodological investigations of *Lebenswelt* actions do not respect the first-person, reflective standpoint that, for Husserl and Schutz, is essential to the discovery and description of the *Lebenswelt*.

4. See Maynard and Schaeffer (this volume, Chapter 8) for a detailed account of respondent declinations during early phases of the interview. The interviewers' efforts to contend with such declinations and related aspects of recruitment work, while being essential to the success of the survey enterprise, are not extensively scripted.

5. Paul ten Have (1996) uses a similar expression ("the spoken questionnaire") to describe such interviews. Also see Houtkoop-Steenstra (1996) for a conversation analytic study that examines the interplay between standardized and locally improvised (recipient-designed) features of interviewing.

6. This account of what the interview does is inferred from audiotapes and texts of instructions. A close ethnographic study of interviewers at work would very likely

reveal a more detailed, and perhaps different, sequence of readings, recitings, and keyboard actions.

7. Some interview items have the interviewer read aloud the response items to the respondent in order to exemplify the appropriate response type. (I thank Doug Maynard, personal communication, December 1999, for pointing to these features.)

8. Not every sequentially relevant utterance is covered by the four categories. For example, respondents sometimes defer responding, or refusing to respond, by asking for clarification. See discussion of item 7d later in this chapter.

9. Garfinkel and Wieder (1992) do not present the "rendering theorem" as a directional opposite to the *Lebenswelt* pair. It is quite likely that I am simplifying the workings of the several pairs (p. 176) to which they allude.

10. Conditional relevance is a general ethnomethodological theme that has become closely identified with the relationship between sequentially ordered utterances in conversation (Sacks, Schegloff, and Jefferson, 1974). In its most general sense, conditional relevance respecifies the methodological problem of relevance, so that the problem becomes a researchable topic of order [see Garfinkel (1991) for a discussion of ethnomethodological respecification]. The classic problem of relevance arises when a scholar explicates a text or an observer describes a social scene. When putting forward a particular interpretation of the text or scene, the interpreter faces the burden of showing that *that* interpretation, and not some other imaginable interpretation, is the one that is relevant in context. Schutz (1964) observed that the problem of relevance applies in social interaction as well as in professional sociological methodology, and Garfinkel (1967) developed the insight with his brilliant series of exercises on the documentary method of interpretation. So, for example, for someone who asks a question, the context and wording of the question provide conditions for determining the sense and relevance of a one-word answer (e.g., "Yes"). From the point of view of the asker, the answer thereby becomes an intelligible, conditionally relevant *document* of the recipient's singular understanding and treatment of the question. Garfinkel demonstrated that such a document should not be viewed as objective indication of a speaker's intention, or of an unequivocal relationship between, for example, a question and an answer to that question. Instead, the conditional relevancy of questions and answers is a locally discovered property of a dialog that is internal to its production.

11. This information extraction machine is a human-interactional equivalent of the modern technologies Heidegger (1977) subsumes under the theme of enframing. *Enframing* is a matter of challenging (or setting upon) nature to extract use value; for example, damming a river to extract energy. In this case, a restricted discursive program is implemented to extract machine-readable data from a natural conversation. Of course, in this instance, the interview is not a "natural conversation" from which data are extracted, since it is itself fashioned as an instrument of data production.

12. See Collins (1993) and Button et al. (1995) for critical discussions of interactive programs used in connection with the Turing test. The standardized

survey interviewer's actions can be construed as a special case of what Collins and Kusch (1995) call *mimeomorphic actions*, actions that are performed as though to mimic a machine's capacities for exact duplication and timing.

13. Characterizing an 'interruption' can be a delicate matter [see Sacks (1992:295) on priority items, and Schegloff (1987)]. In this case, the interruption is accountable as such by reference to the design of the interview item. However, the interruptive utterance does claim priority because it objects to, and demands clarification of, the terms of reference of the incipient question. As I understand the objection, the respondent is saying that she cannot respond until she has a better idea of just which agencies are covered by the question.

14. I thank Doug Maynard (personal communication, December, 1999) for this observation.

15. The confusion (and local exploitation) of turn-transitional relevancies in prefaced questions produced as constituents of news interviews is discussed by Schegloff (1988/1989) and Clayman and Whalen (1988/1989).

16. See McHoul (1987) and Lynch and Bogen (1996) for discussions of local exploitations of alternative sequential relevancies in interrogations.

17. Here, I am using the word *opinions* in the same, generic, sense as is used in the survey text when it invites respondents to "give YOUR opinion." Opinions in this, nontechnical, sense include the whole range of responses the survey is designed to collect. Survey analysts sometimes employ a more restricted sense of the word (see note 1).

18. It might be reasonable to suppose that the item specifies what counts as the relevant opinion (or absence of opinion). Accordingly, it could be said that the item sets up a systematic, and rather restricted, variant of what Heritage (1984:239) has called the "seamless web of accountability." Heritage exemplifies this notion of seamless web with a discussion of the accountable relationship between the first- and second-pair parts of a greeting sequence, and he also relates it to examples from interpretive sociology and anthropology. The classic case of the latter is the Azande oracle. In that case the seamless web of accountability is set up by the incorrigible premises of the oracle. For the Azande, the events of the oracle, and the consequences of the oracular prophecy, are all accountable in terms of a magical system that sustains the local rationality of the institution.

19. Stories told by novices of their first confessions also demonstrate that competent confession is not the natural by-product of a normally sinful life. See McCourt (1997:126–127) for an amusing example of a boy's disappointment when, after confessing what he thought was a fine example of a dreadful sin, he hears the priest stifle a laugh.

20. The transliteration removes, among other things, numerous indications in the transcript of a nervous, halting delivery. These features can, of course, be relevant for an analysis, but they are not germane to the point I want to make about the instruction.

21. See Lynch (1984) and Maynard and Marlaire (1992) for discussions of repetition and attenuation in a series of test items.

22. Courtroom interrogators often complain that a witness who has responded to a question (often repeatedly) has not "answered" it. "Don't recall" is a prototypical interrogatory response that circumvents the logic of "yes" and "no." In effect, it is an unresponsive response to a question. See Lynch and Bogen (1996:Chapter 4) for a discussion of such unresponsive response. "No idea" may be similar in the respect that it is a response that does not *answer* a yes–no question.

23. I am indebted to Douglas Maynard (personal communication, June 5, 1997) for informing me about the use of probes in this context.

24. The poor design would have to do with asking what the respondent "thinks" about matters of fact (statements that are either correct or not about the census and its uses) that very few respondents are likely to know the facts about.

25. Latour (1996), like too many others who should know better, sells ethnomethodology short by suggesting that it is a study of concrete face-to-face relations that ignores the nonhuman agencies through which stable action and communication at a distance are achieved. I am suggesting that ethnomethodology elucidates pervasive orders of specificity that are missed by constructive analysts and constructivist theorists alike. In Coulter's (1991) terms, this is a matter of logical grammar, which is informed by studies of face-to-face communication without being limited to *micro* configurations of space and body.

REFERENCES

Alexander, J., B. Giesen, R. Münch, and N. Smelser (Eds.). 1987. *The Micro-Macro Link.* Berkeley: University of California Press.

Bjelic, D., and M. Lynch. 1992. "The Work of a (Scientific) Demonstration: Respecifying Newton's and Goethe's Theories of Prismatic Color." Pp. 52–78 in G. Watson and R. Seiler (Eds.), *Text in Context: Contributions to Ethnomethodology*. London: Sage.

Bloor, D. 1973. "Wittgenstein and Mannheim on the Sociology of Mathematics." *Studies in the History and Philosophy of Science* 4: 173–191.

Bloor, D. 1992. "Left and Right Wittgensteinians." Pp. 266–282 in A. Pickering (Ed.), *Science as Practice and Culture*. Chicago: University of Chicago.

Bourdieu, P. 1977. *Outline of a Theory of Practice*. Translated by R. Nice. Cambridge: Cambridge University Press.

Button, G., J. Coulter, J. R. E. Lee, and W. W. Sharrock. 1995. *Computers, Minds and Conduct*. Oxford: Polity.

Clayman, S., and J. Whalen. 1988/1989. "When the Medium Becomes the Message: The Case of the Rather/Bush Encounter." *Research on Language and Social Interaction* 22: 241–272.

Collins, H. M. 1975. "The Seven Sexes: A Study in the Sociology of a Phenomenon, or the Replication of Experiments in Physics." *Sociology* 9: 205–224.

Collins, H. M. 1990. *Artificial Experts: Social Knowledge and Intelligent Machines*. Cambridge, MA: MIT Press.

Collins, H. M. 1993. "The Turing Test and Language Skills." Pp. 231–245 in G. Button (Ed.), *Technology in Working Order: Studies of Work, Interaction, and Technology*. London: Routledge.

Collins, H. M., and M. Kusch. 1995. "Two Kinds of Action: A Phenomenological Study." *Philosophy and Phenomenological Research* 60: 799–819.

Coulter, J. 1991. "Logic: Ethnomethodology and the Logic of Language." Pp. 20–50 in G. Button (Ed.), *Ethnomethodology and the Human Sciences*. Cambridge: Cambridge University Press.

Foucault, M. 1980. "Two Lectures." Pp. 37–54 in C. Gordon (Ed.), *Power/Knowledge: Selected Interviews and Other Writings 1972–1977*. New York: Pantheon Books.

Foucault, M. 1984. "What Is Enlightenment?" Pp. 32–50 in P. Rabinow (Ed.), *The Foucault Reader*. London: Penguin.

Garfinkel, H. 1967. *Studies in Ethnomethodology*. Englewood Cliffs, NJ: Prentice-Hall.

Garfinkel, H. 1991. "Respecification: Evidence for Locally Produced, Naturally Accountable Phenomena of Order, Logic, Reason, Meaning, Method, Etc. in and as of the Essential Haecceity of Immortal Ordinary Society (I) — an Announcement of Studies." Pp. 10–19 in G. Button (Ed.), *Ethnomethodology and the Human Sciences*. Cambridge: Cambridge University Press.

Garfinkel, H., E. Livingston, M. Lynch, D. Macbeth, and A. B. Robillard. 1989. "Respecifying the Natural Sciences as Discovering Sciences of Practical Action (I and II): Doing So Ethnographically by Administering a Schedule of Contingencies in Discussions with Laboratory Scientists and by Hanging Around Their Laboratories." Unpublished paper, Department of Sociology, University of California at Los Angeles.

Garfinkel, H. and H. Sacks. 1970. "On Formal Structures of Practical Actions." Pp. 337–366 in J. C. McKinney and E.A. Tiryakian (Eds.), *Theoretical Sociology: Perspectives and Development*, New York: Appleton-Century-Crofts.

Garfinkel, H. and D. L. Wieder. 1992. "Two Incommensurable, Asymmetrically Alternate Technologies of Social Analysis." Pp. 175–206 in G. Watson and R. M. Seiler (Eds.), *Text in Context: Contributions to Ethnomethodology*. London: Sage.

Giddens, A. 1993. *New Rules of Sociological Method*, 2nd ed. Stanford, CA: Stanford University Press.

Heidegger, M. 1977. *The Question Concerning Technology and Other Essays*. New York: Harper & Row.

Heritage, J. 1984. *Garfinkel and Ethnomethodology*. Oxford: Polity.

Heritage, J. C., and D. R. Watson. 1980. "Aspects of the Properties of Formulations in Natural Conversations: Some Instances Analyzed." *Semiotica* 30: 245–262.

Houtkoop-Steenstra, H. 1996. "Probing Behavior of Interviewers in the Standardized Semi-open Research Interview." *Quality and Quantity* 30: 205–230.

Husserl, E. 1970. *The Crisis of European Sciences and Transcendental Phenomenology: An Introduction to Phenomenological Philosophy*. Translated by D. Carr. Evanston, IL: Northwestern University Press.

Jonson, A. R., and S. Toulmin. 1988. *The Abuse of Casuistry: A History of Moral Reasoning*. Berkeley, CA: University of California Press.

Jordan, K., and M. Lynch. 1998. "The Dissemination, Standardization, and Routinization of a Molecular Biological Technique." *Social Studies of Science* 28: 773–800.

Kant, I. 1988. "What is Enlightenment?" Pp. 462–67 in L. W. Beck (Ed.), *Kant Selections*. London: Macmillan.

Latour, B. 1987. *Science in Action: How to Follow Scientists and Engineers through Society*. Cambridge, MA: Harvard University Press.

Latour, B. 1990. "Drawing Things Together." Pp. 19–68 in M. Lynch and S. Woolgar (Eds.), *Representation in Scientific Practice*. Cambridge, MA: MIT Press.

Latour, B. 1996. "On Interobjectivity." *Mind, Culture, and Activity* 3: 228–245.

Livingston, E. 1986. *The Ethnomethodological Foundations of Mathematics*. London: Routledge and Kegan Paul.

Livingston, E. 1987. *Making Sense of Ethnomethodology*. London: Routledge and Kegan Paul.

Lynch, M. 1984. "Turning Up Signs in Neurobehavioral Diagnosis." *Symbolic Interaction* 7: 67–86.

Lynch, M. 1992. "Extending Wittgenstein: the Pivotal Move from Epistemology to the Sociology of Science." Pp. 215–265 in A. Pickering (Ed.), *Science as Practice and Culture*. Chicago: University of Chicago Press.

Lynch, M. 1993. *Scientific Practice and Ordinary Action: Ethnomethodology and Social Studies of Science*. New York: Cambridge University Press.

Lynch, M., and K. Jordan. 1995. "Instructed Actions in, of and as Molecular Biology." *Human Studies* 18: 227–244.

Lynch, M., and D. Bogen. 1996. *The Spectacle of History: Speech, Text and Memory at the Iran-Contra Hearings*. Durham, NC: Duke University Press.

Maynard, D., and C. Marlaire. 1992. "Good Reasons for Bad Testing Performance: The Interactional Substrate of Educational Exams." *Qualitative Sociology* 15: 177–202.

Maynard, D., and N. C. Schaeffer. 2000. "Toward a Sociology of Social Scientific Knowledge." *Social Studies of Science* 30: 323–370.

McCourt, F. 1997. *Angela's Ashes: A Memoir*. New York: Scribner.

McHoul, A. W. 1987. "Why There Are No Guarantees for Interrogators." *Journal of Pragmatics* 11: 455–471.

Sacks, H. 1972. "An Initial Investigation of the Usability of Conversational Data for Doing Sociology." Pp. 31–74 in D. Sudnow (Ed.), *Studies in Social Interaction*. New York: Free Press.

Sacks, H. 1992. *Lectures on Conversation*, Vol. 1. Oxford: Basil Blackwell.

Sacks, H., E. A. Schegloff, and G. Jefferson. 1974. "A Simplest Systematics for the Organization of Turn-taking in Conversation." *Language* 50: 696–735.

Shapin, S. 1994. *A Social History of Truth: Civility and Science in Seventeenth-Century England*. Chicago: University of Chicago Press.

Schegloff, E. A. 1987. "Between Micro and Macro: Contexts and Other Connections." Pp. 207–234 in J. Alexander, B. Giesen, R. Münch, and N. Smelser (Eds.), *The Micro-Macro Link*. Berkeley: University of California Press.

Schegloff, E. A. 1988/1989. "From Interview to Confrontation: Observations of the Bush/Rather Encounter." *Research on Language and Social Interaction* 22: 215–240.

Schutz, A. 1964. "The Problem of Rationality in the Social World." Pp. 64–90 in *Collected Papers, Vol. 2*. The Hague: Martinus Nijhoff.

Schutz, A. 1972. *The Phenomenology of the Social World*. London: Heinemann.

Smith, D. 1992. "Sociology from Women's Experience: A Reaffirmation." *Sociological Theory* 10: 88–98.

Suchman, L. 1987. *Plans and Situated Actions*. Cambridge: Cambridge University Press.

Suchman, L., and B. Jordan. 1990. "Interactional Troubles in Face-to-Face Survey Interviews." *Journal of the American Statistical Association* 85: 232–241.

ten Have, P. 1996. "The Spoken Questionnaire: Harvesting Answers." Unpublished paper, Department of Sociology, University of Amsterdam.

Wieder, D. L. 1974. *Language and Social Reality: The Case of Telling the Convict Code*. The Hague: Mouton.

Winch, P. 1990. *The Idea of a Social Science and Its Relation to Philosophy*, 2nd Ed. Atlantic Highlands, NJ: Humanities Press.

Wittgenstein, L. 1958. *Philosophical Investigations*. Translated by G. E. M. Anscombe. Oxford: Basil Blackwell.

Zimmerman, D. H. 1974. "Record Keeping and the Intake Process in a Public Welfare Agency." Pp. 128–143 in R. Turner (Ed.), *Ethnomethodology*. Harmondsworth: Penguin.

CHAPTER 6

Survey Interviews as Talk-in-Interaction

Emanuel A. Schegloff

Introducing the import of the conversational turn-taking system before explicating a version of our understanding of it, in Sacks, Schegloff, and Jefferson (1974:701–702) we noted that a turn-taking system operates, as an economy does, to distribute a resource — turns at talk — among participants, and that the shape of that distribution should be supposed to affect what is being distributed, that is, the talk.

> Until we unravel its organization, we shall not know what those effects consist of and where they will turn up. *But, since all sorts of scientific and applied research use conversation now, they all employ an instrument whose effects are not known. This is perhaps unnecessary* [emphasis added].

In a review article in the *Journal of the American Statistical Association*, Suchman and Jordan (1990) have cogently brought several lines of inquiry in recent studies of conversation and other talk-in-interaction to bear on the survey interview as a research instrument. An earlier version of the present chapter was prepared as one of several "Comment" pieces solicited by the *Journal* to respond to the Suchman and Jordan article (Schegloff, 1990) and has been revised only to incorporate summary glosses of points in their article no longer directly accessible to the reader of this volume (and several citations to subsequent literature). In what follows, for the more general reference I use the term *talk-in-interaction* rather than *conversation* so as to include a range of "speech-exchange systems" (Sacks, Schegloff, and Jefferson, 1974:729–731), as well as variations of genre and setting within particular speech-exchange systems. I reserve the term

Standardization and Tacit Knowledge: Interaction and Practice in the Survey Interview
Edited by Douglas W. Maynard, Hanneke Houtkoop-Steenstra, Nora Cate Schaeffer, and Johannes van der Zouwen
ISBN 0-471-35829-0

conversation for that underlying, ordinary, unmarked speech-exchange system of which meetings, ceremonies, debates, talk-show formats, and interviews of diverse sorts appear to be systematic structural or procedural transformations.

Let us leave aside for the moment the ways in which talk-in-interaction is used by inquiry in the physical, biological, and medical sciences — for example, in the conduct of laboratory work or clinical research, and the effects that it has there. In the social sciences, the three most widely employed forms of talk-in-interaction are the psychological experiment, ethnographic participant observation, and the interview. In each of these, interaction between the investigator (or the investigator's agent) and the "subject(s)" is the instrument for data collection. It is high time indeed that we began to examine the actual, detailed course of these events *as interactions*, for it is only by arbitrary stipulation that what happens in the course of these interactions has been ruled irrelevant for the inquiries being conducted through them, and for the assessment of their results. One direction in which such an examination may be pressed with respect to behavioral science experiments is explored in Schegloff (1991:54–57).

Of course, interviews are but one of several forms for organizing talk-in-interaction that contrast with ordinary conversation. And survey interviews are but one of a range of forms of talk to which the term *interview* is applied. So there are at least two lines of inquiry relevant here: One is the examination of survey interviews as one distinctive modality among the range of forms of talk-in-interaction; the other is the bearing of this "interview" way of organizing talk on what is to be made of its products. Here I can offer only brief remarks on each of these lines of inquiry.

Note, then, that several of the features that Suchman and Jordan (1990) noted that distinguish survey interviews from ordinary conversation distinguish other forms of talk as well, and something may be gained for the understanding of survey interviews from studies of these other forms. For example, the fact that talk is being done "on behalf of" an absent third party was treated by Heritage (1985) as critical in understanding certain features of the conduct of news interviews and is relevant as well in understanding aspects of the turn-taking organization prevalent in official courtroom proceedings (Atkinson and Drew, 1979). The conduct of interaction with a focus on filling out a form — a project that is mutually oriented to supplying the relevancies and directing the course of the interaction — is common to a wide variety of bureaucratic encounters and professional interviews, for example, medical history taking. Button (1987) described the consequences of a withholding of repair practices in interviews of a very different sort — employment interviews. And so forth.

The conduct of survey interviews, then, represents one configuration of organizational features and shares some of these features with other forms of talk-in-interaction. Note that this is inescapable. The upshot of Suchman and Jordan's (1990) article *cannot* be efforts to remove or neutralize the interactional features of interviews. All talk-in-interaction faces certain generic organizational problems and will perforce adopt some organized solution to them. Thus we have the following.

1. If (as is virtually always the case) the talk is to be organized to have one participant speaking at a time, there will be some organized procedure for allocating opportunities to talk, and usually for restricting their size; that is, there will be some form of turn-taking organization, and it will have consequences that go well beyond simple "traffic management."
2. If there are to be coherence constraints, that is, if there are to be constraints on what can go in some turn at talk given what occurred in a preceding one or preceding ones, then there will be an organization of sequences (if actions are the relevant units), topics (if discursive contributions are the relevant unit), or both. In any case, interactional considerations are likely to be involved concerning both agreement and disagreement, and alignment and opposition, and aspects of the talk will be formed by reference primarily to these.
3. If there are to be ways of coping with transient or persistent troubles in speaking, hearing, understanding, or remembering, and so on, then some organization of repair will be in operation.
4. Almost certainly an overall structural organization will be in operation to shape the events composing the occasion, and to set the boundaries between it and surrounding events in time and space. Suchman and Jordan (1990:235) refer, for example, to one interview in which members of the family being interviewed opt out of the occasion and leave the room as the repetitive patterning of the questions becomes apparent, and another interview in which the adoption of "don't know" answers by a respondent seems prompted by reference to managing the duration and shape of the interview's occasion—"expedit[ing] the business of getting on with the interview."
5. There will be relevant other sets of practices—regarding the telling of stories and the shaping of accounts, regarding the ways in which persons and places and events should be referred to or formulated (e.g., Pomerantz, 1987), and so forth—that will of necessity enter into the constitution of the event called the interview.

And, almost inescapably, these will all be drawn from (or designed to contrast with) organized solutions to these problems in ordinary conversation, where they are designed to supply the interactional infrastructure for social organization—to allow the society to work in whatever way it works—and not to produce reliable and valid grist for social scientists' computational and analytic mills.

The last observation is especially important. The aspects of talk-in-interaction on which Suchman and Jordan (1990) focused are central properties of that medium through which the major institutions of society are embodied and enacted by its members. These properties thus inform ordinary persons' conduct of their affairs and are not merely social-scientifically motivated analyses of them. This can be seen in various features of the excerpts with which Suchman and Jordan (1990) exemplify their points.

For example, one difference between survey interviews and conversation taken up by Suchman and Jordan (1990) concerns standardization of questions to ensure comparability of responses, on the one hand, and recipient design of questions sensitive to local context, on the other. With the standardization of questions in advance and a prohibition on adaptation and redesign for a local context, some questions can have been rendered redundant by the time they are to be asked, but methodological canons require them to be asked nonetheless, engendering anomalous exchanges by conversational standards. For example, they describe (1990:234) an exchange in which a respondent's prompt response to a question about the family's combined income exceeding $20,000 is followed by a further expansion of the question that could only be redundant with the answer already given.

```
I:      Was the total combined family income during the past
        twelve months, that is, yours, your wife's, Judith's
        and Jerry's more or less than twenty thousand dollars.
Mrs. E: More.
I:      Include money from jobs, social security, retirement
        income, unemployment payments, public assistance, and so
        forth. Also include income from interest, dividends, net
        income from business, farm, or rent, and any other money
        income received.
Mrs. E: More. It was more income.
```

Suchman and Jordan (1990) note that the interviewer's continuation requires that Mrs. E "reiterate an answer she has already provided." Note as well, however, that Mrs. E does more than simply reiterate her earlier answer. She has already answered "More." Now she responds, "More. It was more income." That addition by the respondent displays that *she* [and not just Suchman and Jordan (1990)] took that continuation of the question to be problematic in the following respects (among others): (a) it implicated whether the interviewer had understood her answer before; (b) it raises the issue for her whether her second response would be understood as directed to the question as initially asked [note that her addition to the answer ("more income") echoes elements of the first question, and ties her answer to it]; and (c) she displays an understanding that the reply "more" by itself might not be recognizable as an adequate/sensible answer to the question as expanded. All of this is built into her response and displays her uptake of the interviewer's conduct and a stance toward it. What response that stance elicits in turn from the interviewer I cannot tell from the excerpt.

The point then is that modifications of ordinary practices of talk-in-interaction enacted by interviewers to underwrite the data requirements of social scientific inquiry may continue to be grasped and responded to by interviewees by reference to the practices of ordinary talk-in-interaction. And this same point must be appreciated for each of the elements in Suchman and Jordan's (1990:233–239) telling account of a range of differences between interviews and conversation: local versus external control, requirements on answers (including disallowance of elaboration on answers, on the one hand, and pursuit of such elaboration, on the other), canons of relevance, detection and repair of misunderstanding, and so forth. They are describing orientations that inform the conduct of participants to talk-in-ordinary-interaction and that are not automatically held in abeyance in interviews; indeed, whose relevance to answering interview questions can be seen in the details of the answers produced. I can cite but a single additional example.

This is a striking case of an interview with a doctor, with which Suchman and Jordan (1990:236) exemplify the potential validity problems hidden by a failure to allow stories and elaborations following initial responses.

```
I:  When you think about other doctors in general, how
    would you compare yourself to them. Are you very similar
    or different?
R:  I think I'm pretty similar to most doctors. Except that
    a lot of doctors try to stay right in the mainstream of
    medicine. They don't like to be out, away from the drug-oriented
    type of medical treatment. In other words, you
    have a problem, you have a drug for it, and that'll take
    care of it. Or surgery or something. Cut it off, and
    you'll be fine (laughs). And most doctors have that
    attitude. Then there's a small group that believe in the
    reason you have doctors in the first place. And
    that is that we're more holistic. So we can use a more natural
    approach. The hippocratic approach. So I think I'm
    more like that group.
I:  You think that's a smaller group.
R:  Yes, that's a smaller group now.
```

This exchange, in which an answer is totally reversed after elaboration, illustrates a point made by Sacks (1987:62–63), that there is a systematic place for "exceptions" in conversation, namely, at ends of turns. Indeed, I would guess that in this interview there was a moment's silence following the initial response ("I think I'm pretty similar to most doctors"), a silence that is regularly treated in conversation as prefiguring disagreement, to which the "exception" or backing off is a response. In light of the operation of such a mechanism, one wonders what systematic effects are introduced into surveys by what used to be called "the silent probe" following a respondent's initial response.

An impossible conundrum may seem to be posed here. On the one hand Suchman and Jordan (1990) appear to object to disallowing elaboration, and on the other they appear to object to its pursuit. The interviewer's silence, which might be the vehicle for allowing elaboration, can appear to prefigure rejection or disagreement with a response and prompt backdowns by a respondent in some contexts. Indeed! "All of the above!" What a silence or a pursuit of elaboration conveys and embodies interactionally *does*, of course, depend on context—on what the question is, what type of response has been offered, and so forth. And this is what underlies Suchman and Jordan's (1990) recommendation that global or formulaic devices will not work as solutions to the management of interactional contingencies in interviews, that interviews be designed to exploit the properties of the interactions that they are in any case destined to inhabit, and that interviewers be better armed to do so in informed consonance with the goals of the inquiry.

Although an interview cannot in any case be like a thermometer [to cite Suchman and Jordan's (1990) instructive contrast with another measurement instrument], reliable exploitation even of a thermometer requires knowing the properties of mercury, the glass in which it is encased, and so on, and incorporating these properties in the extraction of the desired information from the measurement device. It is by no means clear that we have such elementary understanding of the constitutive components of the survey interview. In addition to the research that Suchman and Jordan (1990) recommend into the possibilities of a more overtly collaborative design for the interaction between interviewer and respondent, a more general inquiry into the features of the survey interview as an organized occasion of talk-in-interaction may help us think through in a thoroughly informed way how exactly to understand the methodological, epistemological, and theoretical features and status of the interview as a tool of inquiry.

We can then turn to the other settings in which talk-in-interaction is used as an instrument of research, and explore the effects of the forms that it takes there, and their effects on the goals of those lines of inquiry (see, e.g., Goodwin, 1994; Heeschen and Schegloff, in press; Marlaire and Maynard, 1990; Schegloff, 1999).

REFERENCES

Atkinson, J. M., and P. Drew. (Ed.). 1979. *Order in Court: The Organization of Verbal Interaction in Judicial Settings*. London: Macmillan.

Button, G. 1987. "Answers as Interactional Products: Two Sequential Practices Used in Interviews." *Social Psychology Quarterly* 50: 160–171.

Goodwin, C. 1994. "Professional Vision." *American Anthropologist* 96: 606–633.

Heeschen, C., and E. A. Schegloff. (In press). "Aphasic Agrammatism as Interactional Artifact and Achievement." In C. Goodwin (Ed.), *The Situation of Language in Brain Damaged Persons: Conversation and Language Impairment*. New York: Oxford University Press.

Heritage, J. 1985. "Analyzing News Interviews: Aspects of the Production of Talk for an Overhearing Audience." Pp. 95–119 in T. A. van Dijk (Ed.), *Handbook of Discourse Analysis*, Vol. 3. New York: Academic.

Marlaire, C., and D. Maynard. 1990. "Standardized Testing as an Interactional Phenomenon." *Sociology of Education* 63: 83–101.

Pomerantz, A. 1987. "Descriptions in Legal Settings." Pp. 226–243 in G. Button and J. R. E. Lee (Eds.), *Talk and Social Organization*. Clevedon, England: Multilingual Matters.

Sacks, H. 1987 [1973]. "On the Preferences for Agreement and Contiguity in Sequences in Conversation." Pp. 54–69 in G. Button and J. R. E. Lee (Eds.), *Talk and Social Organization*. Clevedon, England: Multilingual Matters.

Sacks, H., E. A. Schegloff, and G. Jefferson. 1974. "A Simplest Systematics for the Organization of Turn-Taking for Conversation." *Language* 50: 696–735.

Schegloff, E. A. 1990. "Discussion of Suchman and Jordan, 'International Troubles in Face-to-Face Survey Interviews'." *Journal of the American Statistical Association* 85: 248–250.

Schegloff, E. A. 1991. Reflections on Talk and Social Structure." Pp. 44–70 in D. Boden and D. H. Zimmerman (Eds.), *Talk and Social Structure*. Cambridge: Polity.

Schegloff, E. A. 1999. "Discourse, Pragmatics, Conversation, Analysis." *Discourse Studies* 1: 405–435.

Schegloff, E. A. (In press). "Conversation Analysis and 'Communication Disorders.' " In C. Goodwin (Ed.), *The Situation of Language in Brain Damaged Patients: Conversation and Language Impairment.* New York: Oxford University Press.

Suchman, L. A., and B. Jordan. 1990. "Interactional Troubles in Face-to-Face Survey Interviews." *Journal of the American Statistical Association* 85: 232–241.

PART 2

Recruitment to Participate in the Survey Interview

CHAPTER 7

Introductory Interactions in Telephone Surveys and Nonresponse

Mick P. Couper and Robert M. Groves

BACKGROUND

The introductory interactions between a telephone survey interviewer and householder (potential respondent) are short conversations between a stranger who has placed a call to a person who did not solicit it. These are the interactions during which the interviewer introduces himself or herself, describes the purpose in calling, answers the householder's concerns, often seeks information about all members of the household in order to select a potential respondent, and seeks cooperation with the survey request. In turn, the householder (or potential respondent) is trying to determine the reason for the call, evaluate the legitimacy of the caller, weigh the consequences of continuing the interaction, and consider the decision to agree with or decline the request.

We believe that these interactions have implications both for survey nonresponse and for data quality. In terms of nonresponse, because of the success of callbacks in obtaining householder cooperation, we believe householders do not generally have preformed decisions about participation. Such decisions are usually made during the course of these brief interactions, using the available cues to reach a decision. As far as data quality is concerned, the introductory interaction may set the stage for the subsequent interview, defining roles and expectations for the interaction to follow. We focus in this chapter on the implications of these interactions for nonresponse. An even more narrow focus of our

Standardization and Tacit Knowledge: Interaction and Practice in the Survey Interview
Edited by Douglas W. Maynard, Hanneke Houtkoop-Steenstra, Nora Cate Schaeffer, and Johannes van der Zouwen
ISBN 0-471-35829-0

investigation is an empirical evaluation of the concept of tailoring, developed from our work on survey introductions in face-to-face surveys.

This chapter differs from many others in this volume in several respects. First, where the focus of the other chapters is largely on the structured question–answer process during the body of the survey interview, we explore the unstructured interaction that precedes it (see also Maynard and Schaeffer, this volume, Chapter 8). Second, the data we use are from a different survey, the Survey of Consumer Attitudes, conducted by a different organization, the Survey Research Center at the University of Michigan. Third, whereas most of the other chapters use conversation analytic methods to explore survey interactions, we adopt a somewhat more quantitative approach, albeit on a small number of cases.

With few exceptions (e.g., Campanelli, Sturgis, and Purdon, 1997; Maynard and Schaeffer, 1997, Chapter 8; Morton-Williams, 1993; Pondman, 1993; this volume), relatively little attention has been paid to exploration of the nature of the interactions that occur during survey introductions. This is in sharp contrast to the growing body of research on interactions during survey interviews (see, e.g., Dijkstra and van der Zouwen, 1982; Schaeffer, 1990; Schwarz, 1995; Suchman and Jordan, 1990). There are two main reasons, we suspect. First, the difficulty of taping the introductory interactions makes the collection of data difficult. Second, the unscripted nature of the introductory interactions complicates the analysis of these data. However, the importance of these interactions in affecting both nonresponse (e.g., Groves and Couper, 1996) and measurement error (e.g., Couper, 1997) has been demonstrated.

While relatively little attention has been paid to observation or measurement of survey introductions, and particularly to the behavior of householders during the introduction, a number of studies have investigated the effects of different manipulations of the standard introductions used by interviewers (see, e.g., Gonzenbach and Jablonski, 1993; O'Neil, Groves, and Cannell, 1979; Pinkleton et al., 1994). However, by and large, these studies have focused on the scripted behavior of the interviewer only, rather than the interaction between the interviewer and householder. It is the latter that is of interest to us here.

The development of behavior coding (Cannell, Lawson, and Hausser, 1975; Dijkstra, this volume, Chapter 17; van der Zouwen, Dijkstra, and Smit, 1991) as a tool for examining survey interactions is based on the premise that much of the behavior is constrained and deviations from scripted behavior on the part of the interviewer (e.g., not reading the question as worded) or from expected responses on the part of the respondent (e.g., seeking clarification) can be reliably coded using a relatively simple set of codes. In contrast, conversation analytic transcription methods have more recently been applied to the survey setting (e.g., Schaeffer, 1990; Suchman and Jordan 1990; this volume), permitting detailed investigation of interactional activities, both scripted and unscripted, within the survey interview. However, this is a time-consuming process that has largely been limited to small-scale qualitative studies. We do not undertake a detailed conversational analytic transcription of these introductory interactions but instead develop our own coding system to test a particular set of hypotheses using a relatively large number of interactions.

Not only is the coding of the interview interactions less complicated than for introductory interactions, but the corpus of data can also be more readily generated. It is standard practice in many survey organizations to tape a subset of survey interviews, yielding an abundance of raw material for behavior coding or conversational analysis. In contrast, the problem of taping survey introductions is less tractable. A major difficulty lies in the need to begin recording before the interview begins. Morton-Williams (1993; see also Morton-Williams and Young, 1987) successfully did this in a face-to-face interview setting in the United Kingdom, and their work was recently replicated in that country by Campanelli, Sturgis, and Purdon (1997), but this has proved to be hard to do elsewhere. The logistics of such taping are less complex in a telephone survey setting, and taping and monitoring are more routine in such settings, suggesting that this mode may be more amenable to examination of the introductory interactions. Restricting our focus to telephone surveys may be both good and bad, for reasons we discuss later.

BEHAVIOR IN INTRODUCTIONS VERSUS INTERVIEWS

Given the difficulty of taping and analyzing these introductory conversations, relatively little is known about what goes on at this stage of the survey process. We speculate that there are some key differences between this part of the survey interaction and the structured interview that make the introductory interaction an interesting subject of inquiry. The interview itself is highly structured, with little deviation from the prescribed script permitted on the part of the interviewer (see, e.g., Fowler and Mangione, 1990). In contrast, the behavior of interviewers in survey introductions is largely unscripted (with the possible exception of the opening statement), and adaptive behavior in response to householder objections or concerns is encouraged. Within a set of general guidelines on what is permissible or not in this stage of the interaction, interviewers make heavy use of their tacit knowledge (see Maynard and Schaeffer, this volume, Chapters 8, 10), gained from training and experience, in their efforts to gain cooperation from sample householders.

The purpose of the interaction in the initial conversation is also different from that of the interview. During the interview, the interviewer is seeking information (by means of structured questions) from the respondent, whereas during the introduction the goal is to obtain cooperation, or an agreement to be interviewed. This may imply that the interview is more cooperative than the initial interaction. In the latter, the householder is seeking information to determine the purpose of the call and may be trying to find a polite way to decline the request or terminate the interaction. On the other hand, the interviewer is seeking information on the householder's likely willingness to cooperate and hoping to turn that information to his or her advantage in gaining cooperation. The householder is seeking rapid resolution of the interaction. In the words of one interviewer, this is a "behavioral dance of enticement." During the interview, on the other hand, both parties have agreed to cooperate and are assumed to be working toward a common purpose.

From the householder's perspective, different cognitive scripts may be evoked during these different parts of the process. During the initial interaction, the householder may be interpreting the interaction as telemarketing, fund raising, or other solicitation activities and may react accordingly. During the interview, in contrast, the householder may be applying interactional norms more akin to that of a job interview or similar situation. To what extent do these interactional norms differ across these two phases of the interaction? We suspect that interviewers are inclined to view these as two separate activities, with different goals and different norms operating. Does the householder see them in the same light or view them as parts of the same conversation? What implications do these differences have for data quality? While many have recognized the importance of the interviewer's behavior in the survey introduction, less attention has been paid to what the householder says and does at that point in the interaction. We believe that as survey introductions become more accessible to empirical investigation, we may learn much about both the nonresponse and data quality implications of this part of the survey process. We invite the reader to compare our findings in this chapter to those of others in this volume as a way to contrast interactions before the interview with those during the interview.

TAILORING

As noted earlier, we are interested in the persuasion-related strategies of interviewers during the introductory interaction. From our work on face-to-face surveys (see Groves, Cialdini, and Couper, 1992; Groves and Couper, 1998), we have observed that successful interviewers engage in a continuous search for cues about the person who is likely to open the door or answer the phone and their likely reaction to the interview request. Interviewers then adapt or customize the approach they use to the particular concerns of the householder. We have come to term this behavior *tailoring* (see Groves and Couper, 1996). The concept of tailoring developed from a series of discussions with interviewers from various organizations and from observations of interviewers in the field. Experienced interviewers are often aware of their tailoring behavior, as these statements from focus groups with interviewers working at various organizations suggest:

"I use different techniques depending on the age of the respondent, my initial impression of him or her, the neighborhood, etc."

"From all past interviewing experience, I have found that sizing up a respondent immediately and being able to adjust just as quickly to the situation never fails to get their cooperation."

"After 3 or 4 minutes you can almost tell what they are going to say, at that time you have to change your attack."

"I give the introduction and listen to what they say. I then respond to them on an individual basis, according to their response. Almost all responses are a little different, and you need an ability to intuitively understand what they are saying. Sometimes, the words they are saying are not the true reasons they are refusing. I attempt to discern what the situation is and respond to it."

From such observations, we can deduce that successful tailoring requires that a number of conditions be met. First, the interviewer must have a wide variety of techniques or strategies at his or her disposal. In some sense, expert interviewers have access to a large repertoire of cues, phrases, or descriptors corresponding to the survey request that can be called upon as appropriate. Second, the interviewer must be a good reader of verbal and (in the case of face-to-face contact) nonverbal cues from the householder and surroundings. The important cues are those indicating, for example, the householder's choice of script guiding their behavior, their likes and dislikes, salient issues, and time pressures. In short, any behavior that reveals decision factors the householder might use to decide about their participation can be valuable to the interviewer. Third, the interviewer must apply the appropriate strategy or technique in response to the cues received. Interviewers must be able to make fast and accurate judgments about the particular script reflected in the householder's initial response to the request and react accordingly. Fourth, the interaction (whether within a single contact or across multiple contacts) must be of sufficient duration for tailoring to be applied. An interviewer will not be able to tailor a response if no cues are forthcoming from the householder. Thus, there is a premium on maintaining interaction with the householder, even without apparent progress toward gaining consent to the survey request. The longer the conversation continues, the more information may be gleaned by the interviewer about the likely concerns of the householder regarding participation.

Tailoring need not necessarily occur only within a single contact. Many times contacts are very brief and give the interviewer little opportunity to respond to cues obtained from the potential respondent. Tailoring may take place over a number of contacts with that household, with the interviewer using the knowledge he or she has gained in each successive visit or call to the household. Tailoring may also occur across sample households. The more an interviewer learns about what is and is not effective with various types of potential respondents encountered, the more effectively requests for participation can be directed at similar others. This implies that the success of interviewer tailoring evolves with experience. Not only should experienced interviewers have acquired a wider repertoire of persuasion techniques, but they should also be better able to select the most appropriate approach for each situation.

The concept of tailoring may aid in explaining some of the puzzling empirical findings in the survey methodological literature. There is a relatively large literature studying the effects of different introductory scripts of interviewers on nonresponse rates (see, e.g., Cannell, Miller, and Oksenberg, 1981; Gonzenbach and Jablonski, 1993; Houtkoop and van den Bergh, 1995; Morton-Williams and Young, 1987; O'Neil, Groves, and Cannell, 1979; Pinkleton et al., 1994). In most instances, interviewers were instructed to deliver verbatim versions of alternative scripts. Over several years this research has not yielded a consistent set of guidelines, and failures to replicate have occurred. This result is exactly what would be expected if tailoring is a principal influence on the ability of interviewers to gain participation. Tailoring implies that there are few "main effects" of any particular interviewer's utterance. What addresses one respondent's concerns may be irrelevant or antithetical to another's concerns. Thus, as Morton-Williams (1993) found, allowing interviewers to apply adaptive behavior results in higher cooperation rates than does requiring interviewers to adhere to any single script.

Thus far our development of the tailoring concept has focused on face-to-face surveys and is based in large part on discussions with interviewers and observations of interviewers in the field. In addition, limited tests of the tailoring concept have been undertaken using a structured interviewer form and focusing on cross-contact changes in behavior in a face-to-face survey or interviewer self-reports of variation in strategies (Couper and Groves, 1992; Groves and Couper, 1996). Thus, to date we have not been able to directly test the concept of tailoring and its effect on survey cooperation. Given the difficulties of recording doorstep interactions in a face-to-face survey, we use a telephone survey to test some of these ideas. This has both disadvantages and advantages.

Focusing on telephone introductions makes the detection of tailoring and its effects more difficult. Telephone interactions are shorter than face-to-face ones; fewer cues are likely to be provided by the householder, and this affords less opportunity for tailoring. In face-to-face surveys, interviewers may have more information available for tailoring, including contextual visual cues (neighborhood, housing characteristics, appearance of person at the door), nonverbal cues (e.g.,"body language," such as the manner in which the householder opens the door, and facial gestures), and verbal cues (what the householder says at the door). In telephone surveys, the interaction is restricted to the oral medium only (what the householder says and how she or he says it). This may severely restrict the ability of interviewers to elicit cues for tailoring in telephone interactions, making tailoring harder to detect.

Given this, we expect that initial introductions ("Hi, my name is...") would be much more varied in face-to-face surveys than in telephone surveys, as more preintroduction cues are available to permit tailoring [Campanelli, Sturgis, and Purdon (1997) refer to this as *pretailoring*]. In telephone surveys, the interviewer has no visual cues about the respondent prior to their first utterance, and generally only a "hello" answer, which reveals little about the householder or his or her likely reaction to the request. For this reason, we expect much of the tailoring (to the extent possible) in telephone surveys to occur *after* the initial introduction and announcement of the purpose of the call by the interviewer.

While tailoring may be more difficult to detect in telephone survey introductions, the limitation of the interaction to the aural medium makes the coding of these interactions easier. We can focus only on the utterances of the participants, without attempting to code all the subtleties of the full face-to-face interaction. While restricting the range of behaviors examined, the coding task becomes more tractable.

With this in mind, our research had relatively modest goals. First, we explored whether it is possible to observe and measure interviewer tailoring behavior in telephone survey introductions. Second, we examined whether such tailoring is predictive of eventual cooperation with the survey request. We caution that this work should be viewed as exploratory in nature.

RESEARCH DESIGN

The survey vehicle for this work was the monthly Survey of Consumer Attitudes (SCA), conducted by the Survey Research Center at the University of Michigan. Each month's survey has both first-wave (RDD) cases and second-wave (recontact) cases. We focus on initial interactions with first-wave cases, who are receiving the request for participation for the first time. In April 1994, a random subset of interactions were taped. This was accomplished by randomly selecting interviewer shifts for taping. Interviewers were instructed to tape all interactions during the selected shifts. After excluding ineligible sample cases, this yielded 101 taped interactions, representing 87 unique sample cases (some repeat calls to the same sample unit are included). Of the 33 interviewers who worked on the SCA during that month, 23 are represented in our set of taped interactions.

A total of 453 eligible RDD sample cases were called in April 1994. Among these, 63.8 percent were interviewed, 20.8 percent refused to participate, and 15.4 percent could not be contacted. Our focus is limited only to interactions with contacted sample cases. Of the 383 contacted RDD cases, 75.5 percent cooperated with the survey request. Of the taped calls we have, 75.3 percent resulted in final interviews. This suggests we have a reasonable representation of the set of first-wave calls from the SCA in our set of taped interactions.

The call-level outcomes of the 101 taped contact interactions are distributed as follows: 28 interviews, 18 refusals, and 55 callbacks (e.g., appointments, call another time). Thus we have relatively few explicit refusals at the call level, limiting the conclusions we can draw about call-level outcomes from these data. Subsequent calls that were made to the 55 unresolved cases were made during shifts not selected for taping. Taping was stopped about halfway through the data collection period, resulting in fewer tapes than we had anticipated.

The coding of these 101 interactions proceeded through a number of different steps. First, the tapes consisted of much extraneous material, including ring-no-answers, contacts with ineligible numbers, full interviews with some respondents, and so on. Thus the interactions of interest were identified from the tapes and that subset placed onto a smaller number of tapes. These were then transcribed. The crude transcription was designed to capture only the utterances of the two participants, and in no way approximates a full conversational analytic transcription.

Based on a subset of the cases, a coding scheme was developed to identify tailoring behavior. This was done at the turn level within each interaction. A turn-pair was defined as any utterance by the sample householder, followed by the interviewer.[1] Thus the first turn-pair often consisted of the householder answering the phone (e.g., "Hello") and the interviewer's initial introduction ("Hi, my name is. . .").

We coded the householder's statements in the first part of the turn-pair into the following broad categories:

Positive statements

Negative statements

Time-delay statements

Questions

Other informative statements

Uninformative statements

The first four of these categories were based on earlier analyses from face-to-face surveys (see Groves and Couper, 1996, 1998). While these broad groupings sacrifice most of the nuances of these interactions, they nonetheless permit reliable coding of the critical features of the statements as far as the likelihood of cooperation is concerned. We have found in earlier analyses that these classifications are reliable predictors of cooperation in face-to-face surveys.

Positive statements are those in which a positive expression regarding participation in the survey is given. These include statements such as "I enjoy doing surveys" or "this sounds interesting." We expect utterances of this type to be extremely rare. Examples of negative statements we encountered include the following:

"I'm not interested."

"I don't like giving out information over the telephone."

"Find somebody else."

"I'm in too many surveys."

Time-delay statements are those in which the householder makes mention that the interviewer called at an inconvenient or inappropriate time or suggests that he or she has little time for participation in surveys. Examples of time-delay statements are as follows:

"You woke me up."

"I'm too busy."

"Now is not a good time."

Examples of questions include the following:

"How long will this take?"

"What is [the survey] for?"

"What are you selling?"

"Why did [person *x*] get selected?"

Whether positively or negatively stated, questions are treated as a separate category. By definition, questions facilitate continuation of the interaction, as they seek a response by the interviewer. We have found that questions (of whatever type) tend to be associated with a higher likelihood of eventual cooperation (Groves and Couper, 1996).

We classified as "other informative" those householder statements in which some information was volunteered by the householder beyond what was asked for by the interviewer. This information may reveal something about the household(ers) or their likely reaction to the survey request but did not fall into one of the above categories. For example:

"Well I think the economy stinks."

"I just retired from the army last year."

We coded as uninformative all other statements. These are uninformative only as far as providing relevant information for interviewer tailoring is concerned. Examples of this type include householder responses to interviewer questions regarding household composition or verification of the telephone number and "uh huh's" and "uhm's" during the interviewer's speech turn. Given the dominant role of the interviewer in these introductory interactions, we expect much of the householder's statements to be of this kind.

The distinction between other informative and uninformative statements can be illustrated with two examples.

```
1 I: ... and are any members of your household 17 years of age
2    or younger?
3 H: Um. No.
4 I: No.
5 H: They're all married
6 I: Umhm. So you get to babysit today?
7 H: Oh yeah. ((laughter))
```

In the first part of this interaction (line 3), the householder's (H) turn is uninformative to tailoring. She simply answers the questions posed by the interviewer (I). However, in line 5 she offers information that was not solicited by the interviewer ("They're all married") that allows the interviewer to make a tailored comment. This householder turn is coded as "other informative." The second example is a short extract from the telephone number verification process:

```
1 I: and because this number was randomly generated by a computer
2    I don't know whether I've reached a business or ... some
3    other type of residence
4 H: It's a home and an elderly couple
```

Here the householder has offered additional information, beyond that requested by the interviewer (is this a home or a business). The statement that the house is occupied by an elderly couple conveys some information to the interviewer that may permit a tailored response.

For the coding of interviewer behavior in the second part of each turn-pair, we employed a very narrow definition of tailoring. We considered tailoring to have occurred when (1) the householder provides some information in the first turn of the pair and (2) the interviewer uses that information appropriately in his or her response in the second turn of the pair. Appropriateness is not defined in terms of an evaluation of the efficacy of the interviewer's response, but simply whether the interviewer's response reflected information the householder revealed. An example of appropriate tailoring is as follows:

```
H: I personally feel the economy and everything in the United
   States as a whole suck... Just to live in the U.S. is
   something to be shameful about.
I: Umhm. Sounds like you have some really strong opinions and
   we'd really like to talk to you about those...
```

The following is an example of inappropriate tailoring (per our coding scheme):

```
H: This whole place is ripping the consumer off, that's all
   I'm going to tell you. All the politicians are making money
   and us poor blue collar workers we're getting screwed every
   day. Alright, that's my answer.
I: Oh I understand that sir, but you know you might not even
   be the person we wanted to speak to...
```

Another example of inappropriate tailoring is as follows:

```
H: ...what do you want?
I: I see... O.K. Have I reached you at a home or business?
```

In this last example, the householder's behavior would be coded as a question. The interviewer's response is coded as inappropriate tailoring as it does not address the question raised by the householder (ascertaining the purpose of the call), but rather deals with some other issue (verifying the telephone number).

Finally, a third code of "other" was used for all interviewer behavior that could not be classified as appropriate or inappropriate tailoring. This includes the introduction, much of the household roster, and verification of the telephone number. We expected that much of the interviewer's behavior in the introductory interaction would be classified in this "other" category.

Granted, this is a crude measure of only one narrow element of tailoring (within turn-pair behavior), but our goal was to develop reliable measures of tailoring behavior that could be easily measured on a relatively large number of cases, permitting more detailed statistical analysis of the correlates and consequences of interviewer tailoring.

The first 10 cases were coded independently by three coders. High levels of agreement were achieved, both on the identification of turn-pairs and on the classification of each turn-pair using the coding scheme outlined above. The remainder of the cases were then coded by a single coder, with independent review and verification. While we do not have measures of intercoder reliability, we are confident from this exercise that the measures as narrowly defined here can be reliably coded.

RESULTS

Some descriptive data on the nature of these interactions may be informative. For these, the unit of analysis could be considered to be the turn-pair or the interaction (a single telephone contact). We have a total of 101 interactions in our data and an average of 10.6 turn-pairs per interaction. This suggests, as we have found elsewhere, that these are typically relatively short interactions (Groves and Couper, 1998; Oksenberg, Coleman, and Cannell, 1986).

Looking at all the turn-pairs coded, the distribution of these turn-pairs in terms of householder behaviors are presented in Table 7.1 (first column). These behaviors are also presented in terms of their distribution across interactions in the second column in Table 7.1. As expected, positive statements about participation in the survey are extremely rare (on average less than 1 in 100 statements or less than 1 in every 10 interactions). The majority of turn-pairs (64 percent) can be described as uninformative (i.e., turns in which no information is revealed by the householder that would permit tailoring on the part of the interviewer). However, combining categories of householder behavior, we find that almost all (92 percent) of the interactions contain at least one statement by the householder that could be deemed informative. That is, only 8 percent of the interactions in our sample contain *no* information that could be used for interviewer tailoring. Put another way, an average of 3.8 turn-pairs per interaction (or about a third of all turn-pairs) provide at least some information for tailoring.

To what extent are these householder behaviors related to their (eventual) cooperation or otherwise with the survey request? We have found elsewhere (see Groves and Couper, 1996) that negative statements on early contacts with householders are positively associated with final refusals on the survey. To a lesser degree, the same was found for time-delay statements. In other words, those cases in which a time-delay statement was made were more likely to be final refusals than those in which no such statement was made, although the effect

Table 7.1 Occurrence of Various Householder Behaviors by Turn-pair and Full Interaction

	Percent of All Turn-pairs That Are...	Percent of Interactions Containing One or More Such Statements	Average Number of Such Statements per Interaction
Uninformative	64.3	99.0	6.84
Informative			
Positive	0.8	7.9	0.09
Negative	5.5	26.7	0.58
Other informative	8.7	42.6	0.92
Time delay	15.0	63.4	1.59
Questions	5.7	34.7	0.60
Any informative	35.7	92.1	3.80
Number	1074	101	101

Table 7.2 Cooperation Rates by Householder Behaviors

	Cooperation Rate (%)	Number of Cases
One or more positive statements		
Yes	75	8
No	75	93
One or more negative statements		
Yes	56	27
No	82	74
One or more time-delay statements		
Yes	77	64
No	73	37
One or more questions		
Yes	77	35
No	74	66

was not as large as for negative statements. We also found in previous work that questions were associated with a positive outcome (i.e., an interview).

Do we find the same results in our set of 101 taped telephone interactions from the SCA? Table 7.2 shows final cooperation rates (interviews/interviews + refusals) by various householder behaviors for the 101 cases in our sample. We see from this table that interactions (or contacts) in which a negative statement was made by the householder have a lower cooperation rate than those where no such statements were made. We detect few reliable differences in this set of interactions for the other householder behaviors.

It is interesting to note that over half (56 percent) of the householders who make some negative statement about participation in the survey nonetheless eventually cooperate with the request whether on that same contact or on a subsequent one. Even calls that appear on the surface to be hopeless can be successfully converted. Of the 18 call-level refusals, half involved hangups by the respondent in the middle of the interaction, probably the strongest form of refusal (see Maynard and Schaeffer, 1997; this volume, Chapter 8). However, 4 of these 9 were successfully converted, compared to the same proportion (4 out of 9) of non-hangup call refusals. This suggests some success on the part of interviewers in turning initial reluctance into a positive outcome. We believe that interviewer tailoring is one of the mechanisms by which this occurs.

Turning our attention to interviewer behavior in the turn-pairs, we find that appropriate tailoring occurred in 24 percent of all turn-pairs, while inappropriate tailoring was coded for only 6 of the 1074 turn-pairs coded (0.6 percent), and "other" interviewer behavior occurred in the remaining 75.1 percent of interactions. However, given that tailoring can only occur in instances where some information is provided by respondents (about 36 percent of the 1074 turn-pairs), we should focus our analyses of tailoring on these cases.

From a statistical viewpoint, tailoring implies interaction effects of statements that the interviewer might deliver to the householder. If the householder behaves in one way, there will be positive effects of an interviewer statement; if not, there could be negative or no effects of the statement on the likelihood of participation. In other words, the success or otherwise of the interviewer's tailoring behavior depends upon the particular householder behavior observed in that turn-pair.

On average, interviewers appropriately tailored their responses in about 62 percent of all turn-pairs in which information was provided (positive, negative, time delay, questions, or other informative). The percentage tailored for various types of householder statements is as follows:

Negative	48.6%
Questions	93.3%
Time delay	89.8%
Other informative	35.7%

It is not surprising that householder questions are more likely to be tailored, as by definition such questions both seek a response and permit such a response to occur. Similarly, a large proportion of time-delay statements are appropriately tailored. This, we believe, reflects the fact that interviewers are often well trained in the negotiation over appropriate times to call sample households and are encouraged to back off and suggest alternatives if it is apparent that they reached a householder at an inconvenient time. As an aside, we believe such actions have a positive effect on overall cooperation because they invoke reciprocal concessions on the part of the householder (Groves, Cialdini, and Couper, 1992). It also allows the interaction to be resumed at a later time. In about half the instances in which a householder makes a negative statement about participation in the survey, we find that the interviewer addresses that concern appropriately *in that same turn-pair*. In the balance of cases, the interviewer either tailored inappropriately or made some other unrelated comment. Given the few instances of inappropriate tailoring detected, we collapse these two groups in subsequent analyses.

Does interviewer tailoring increase the likelihood of cooperation? If we simply look overall whether any tailoring occurs, we find that when behavior was tailored in an interaction (79.2 percent of all interactions), the final cooperation rate was 73.8 percent. This cooperation rate is lower than when no tailoring occurred (81.0 percent). However, the question is not whether tailoring works, but whether *given a negative statement* (for example), tailoring increases the likelihood of cooperation over not tailoring. Once we condition on the type of householder statement made (e.g., negative or time delay), though, the number of cases in our analyses is even further reduced. Thus, our results can only be speculative at best, and hence estimates of statistical significance are not presented.

Table 7.3 presents the cooperation rates for tailoring given negative and time-delay statements. (With tailoring occurring on almost all turn-pairs in which questions were asked by the householder, we cannot examine the effectiveness of tailoring to questions.) These results suggest that, given the presence of a negative or time-delay statement, interviewer tailoring may increase the likelihood of cooperation over not tailoring to such statements. For example, given an interaction with a negative householder statement, when tailoring to the negative

Table 7.3 Cooperation Rates by Interviewer Tailoring, Given Negative and Time-delay Statements

	Cooperation Rate (%)	Number of Cases
No negative statements	82	74
Negative statements, no tailoring	50	8
Negative statements, tailoring	58	19
No time-delay statements	73	37
Time-delay statements, no tailoring	67	3
Time-delay statements, tailoring	77	61

statement was observed in the same turn-pair, the overall cooperation rate was 58 percent, compared to 50 percent when no such tailoring occurred. While it is clear we need to replicate these findings on a larger sample, these findings suggest that this is a fruitful line of inquiry.

DISCUSSION

Tailoring is a complex phenomenon that involves both statistical and conversational interactions, not only within turn-pairs as we have measured, but also across turn-pairs within an interaction, as well as across interactions or contacts. To test the utility of this construct empirically, we have severely restricted our definition of tailoring. Still, this has allowed us to reliably measure the tailoring construct as narrowly defined and code telephone interactions relatively cheaply. Our results suggest that the construct has some merit, and we are encouraged to pursue this line of investigation.

In operationalizing the construct to permit empirical testing of the tailoring notion, we have deliberately ignored the rich interactional substrate of the introductory interactions. We believe that telephone survey introductions yield much information that could be productively explored in an effort to understand why people agree to survey requests and what role the introductory interaction plays in that request. For example, we speculate that interviewers may begin tailoring earlier, varying their initial introductions based on information gleaned from the householder's initial "hello" (whether it be gender, age, race, or tone of voice). However, to address this, we will need to examine within-interviewer cross-household behavior, requiring a much larger set of interactions.

Maynard and Schaeffer (1997; this volume, Chapter 8) have focused their conversational analytic lenses on introductory interactions, and others are beginning to follow. As more attention is paid to these interactions, we believe these studies will yield rich insights into the process of decision making on cooperation and the role of the interviewer–householder interaction in that processes.

The results presented here are suggestive, based on a relatively crude coding of a small number of telephone introductions. Several extensions of this work are possible. For example, a larger set of interactions may allow the examination of variations in tailoring and their success in response to different types of householder comments (e.g., "I'm not interested" and "We don't do surveys" as distinct types of negative statements). In other words, are there some householder statements that are more impervious to tailoring or persuasion efforts than others, and what kinds of tailored responses work better than others? We assume, for example, that hangups provide little information to permit tailoring but that the timing of the hangup may be indicative of likely cooperation (see Maynard and Schaeffer, 1997).

We believe this line of inquiry to be of use not just to survey methodologists interested in reducing nonresponse through more effective persuasion. The nature of the householder–interviewer interaction may also provide insight into the quality of the data collected during the subsequent interview, among those who agree to cooperate (see Couper, 1997). Furthermore, these interactions may be of broader interest to those who study conversation and the compliance process. We believe these brief "request interactions" are an increasingly common feature of telephone interactions, and understanding the conversational features of these special types of interactions may yield much of interest to social scientists and survey researchers alike. Similarly, comparing and contrasting the character of the preinterview interaction with that during the structured interview itself (the focus of this volume) and how the two may interact could inform our understanding of both parts of the survey process.

ACKNOWLEDGMENTS

Earlier versions of this chapter were presented at the Workshop on Interaction in the Standardized Survey Interview, Amsterdam, November 1995, and at the Annual Meeting of the American Association for Public Opinion Research, Salt Lake City, May 1996. We thank Lisa Holland for her assistance in the collection of these data. We also gratefully acknowledge the assistance of Aaron Klein and Ashley Bowers in transcribing, coding, and analyzing the taped introductions. This research was supported in part by internal funds to the Survey Methodology Program at the Survey Research Center and by the Joint Program in Survey Methodology.

NOTES

1. While it is true that the interviewer's turns typically dominate the interaction and the householder is often reacting to the interviewer's utterances, we begin each pair with the householder because of our focus on interviewer tailoring in response to householder statements.

REFERENCES

Campanelli, P., P. Sturgis and S. Purdon. 1997. "Can You Hear Me Knocking: An Investigation of the Impact of Interviewers on Survey Response Rates." The Survey Methods Centre at SCPR, London.

Cannell, C. F., S.A. Lawson, and D. L. Hausser. 1975. *A Technique for Evaluating Interviewer Performance*. Ann Arbor: Institute for Social Research, University of Michigan.

Cannell, C. F., P. V. Miller, and L. Oksenberg. 1981. "Research on Interviewing Techniques." Chapter 11 in S. Leinhardt, (Ed.), *Sociological Methodology 1981*. San Francisco: Jossey-Bass.

Couper, M. P. 1997. "Survey Introductions and Data Quality." *Public Opinion Quarterly* 61: 317–338.

Couper, M. P. and R. M. Groves. 1992. "The Role of the Interviewer in Survey Participation." *Survey Methodology* 18: 263–277.

Dijkstra, W., and J. van der Zouwen. (Eds.). 1982. *Response Behaviour in the Survey-Interview*. London: Academic.

Fowler, F. J., and T. W. Mangione. 1990. *Standardized Survey Interviewing; Minimizing Interviewer-Related Error*. Newbury Park: Sage.

Gonzenbach, W. J., and P. Jablonski. 1993. "The Non-Solicitation Statement: A Methodological Consideration for Survey Introductions." Paper presented at the annual meeting of the American Association for Public Opinion Research, St. Charles, IL.

Groves, R. M., R. B. Cialdini, and M. P. Couper. 1992. "Understanding the Decision to Participate in a Survey." *Public Opinion Quarterly*, 56: 475–495.

Groves, R. M., and M. P. Couper. 1996. "Contact-Level Influences on Cooperation in Face-to-Face Surveys." *Journal of Official Statistics* 12: 63–83.

Groves, R. M., and M. P. Couper. 1998. *Nonresponse in Household Interview Surveys*. New York: Wiley.

Houtkoop-Steenstra, H., and H. van den Bergh. 1995. "Effects of Introductions in Large Scale Telephone Survey Interviews." Unpublished paper. Utrecht University, Utrecht.

Maynard, D. W., and N. C. Schaeffer. 1997. "Keeping the Gate: Declinations to Participate in the Survey Interview." *Sociological Methods and Research* 26: 34–79.

Morton-Williams, J. 1993. *Interviewer Approaches*. Aldershot: Dartmouth.

Morton-Williams, J., and P. Young. 1987. "Obtaining the Survey Interview — An Analysis of Tape Recorded Doorstep Introductions." *Journal of the Market Research Society* 29: 35–54.

Oksenberg, L., L. Coleman, and C. F. Cannell. 1986. "Interviewers' Voices and Refusal Rates in Telephone Surveys." *Public Opinion Quarterly* 50: 97–111.

O'Neil, M. J., R. M. Groves, and C. F. Cannell. 1979. "Telephone Interview Introductions and Refusal Rates: Experiments in Increasing Respondent Cooperation." *Proceedings of the American Statistical Association, Survey Research Methods Section*, 252–255.

Pinkleton, B., J. Reagan, D. Aaronson, and E. Ramo. 1994. "Does I'm Not Selling Anything Increase Response Rates in Telephone Surveys?" Paper presented at the Annual Conference of the American Association for Public Opinion Research, May, Danvers, MA.

Pondman, L. M. 1993. "The Influence of the Interviewer on the Refusal Rate in Telephone Surveys." Ph.D. dissertation, Vrije Universiteit, Amsterdam.

Schaeffer, N. C. 1990. "Conversation with a Purpose–or Conversation? Interaction in the Standardized Interview." In P. P. Biemer, R. M. Groves, L. E. Lyberg, N. A. Mathiowetz, and S. Sudman (Eds.), *Measurement Errors in Surveys*. New York: Wiley.

Schwarz, N. 1995. "What Respondents Learn from Questionnaires: The Survey Interview and the Logic of Conversation." *International Statistical Review* 63: 153–168.

Suchman, L., and B. Jordan. 1990. "Interactional Troubles in Face-to-Face Survey Interviews." *Journal of the American Statistical Association* 85: 232–241.

van der Zouwen, J., W. Dijkstra, and J. H. Smit. 1991. "Studying Respondent-Interviewer Interaction: The Relationship between Interviewing Style, Interviewer Behavior, and Response Behavior." In P. P. Biemer, R. M. Groves, L. E. Lyberg. N. A. Mathiowetz and S. Sudman (Eds.), *Measurement Errors in Surveys*. New York: Wiley.

CHAPTER 8

Opening and Closing the Gate: The Work of Optimism in Recruiting Survey Respondents

Douglas W. Maynard and Nora Cate Schaeffer

Elsewhere (Maynard and Schaeffer, 1997), we described a "requesting sequence" by which interviewers on the telephone solicit participation in a survey and recipients of the call grant or decline the solicitation. In all, we view call recipients' grantings and declinations of requests as a *gatekeeping* matter on the part of recipients, reflecting decisions about their willingness to sustain contact with anonymous interviewers who approach them over the telephone.[1] The *granting* of a request involves not just a single sequence. Rather, the recipient "goes along with" an interviewer's statements or questions regarding the nature of the call, the phone number reached, the gender and number of household residents, the confidentiality of the interview, and other matters at the "front end" of the interview until the substantive interview is commenced. In this chapter, we add a little flesh to this bare-bones statement about the interactive organization of successful requests for participation and show the various points at which call recipients could decline a request but instead provide for continuation to the substantive interview. We also discuss the interactive organization by which recipients may insert their questions about the length or purpose or other matters concerning the interview. Furthermore, just as requesting and *granting* an interview are interactively organized, so are requesting and *declining*. When recipients decline the request to participate, that does not necessarily conclude their interaction with interviewers. Rather, depending upon the declination device a recipient uses, and the interviewer's response to it, there are different routes

Standardization and Tacit Knowledge: Interaction and Practice in the Survey Interview
Edited by Douglas W. Maynard, Hanneke Houtkoop-Steenstra, Nora Cate Schaeffer, and Johannes van der Zouwen
ISBN 0-471-35829-0

to terminating the encounter. We examine these different routes as organized interactional phenomena.

In analyzing the requesting sequence and its two outcomes of granting or denying an interviewer's solicitations, our purpose is to depict the interactive anatomy of the survey interview's front end—that is, the opening of the call and the phase that can but may not lead to a completed interview.[2] Our research, as yet, does not offer prescriptions about how to maximize survey response rates. It does lay some analytical ground work for further research on this topic. Our most general finding is a propensity for interviewers, when dealing with apparently resistant call recipients, to be "optimistic" in their approach in ways that attempt to persuade them to stay on the line. Such optimism may succeed in a call recipient's opening the gate; other times it may work adversely and only reinforce the recipient's resolve to close the gate. We draw implications of this for interviewer training.

OPENING THE GATE

Granting a request to participate is a matter of call recipients going along with various portions of the "front end" of the survey as an interviewer conveys them. Call recipients, on an utterance-by-utterance basis, provide signals—either silences or "continuers" (Schegloff, 1981) or other utterances—that provide for the interviewer to go ahead with any next component to the front end. These go-ahead signals occur at junctures where call recipients could produce a declination. In the excerpt below, we identify eight points at which go-ahead signals, rather than resistance or declination, occur. Subsequently we will summarize research by Hadley (1999) that investigated patterns of acceptance and resistance at these eight points:

(1) [LSSC122:03:A11]

```
        ((phone ringing))
        (2.0)
  MR:   H'lo!
  MI:   ((tch)) .hh Uh hello: I'm Craig Johnson[3] calling from the
        University of Wisconsin (1) as: part of our national
        public opinion study? .hhhh (2)
        (0.2)
  MR:   An:d ↑we are trying to reach people at their home
        telephone numbers .hhh is this a residential number? (3)
        (0.2)
  MR:   Yes ((breathy))
  MI:   .hhhh An: to be sure that I reached the number I dialed
        .hh >is this area code< seven one five .hh nine two four
        four eight six three?   (4)
        (0.2)
  MR:   .h Yes=
  MI:   =O::kay .hh well because we'd like our results to
        represent the opinions of (.) people all over the country
```

```
        .hhh we need to select one adult to speak with in your
        household .hhh an: to make this selection scientifically
        we first need to know .hh how many adults eighteen or
        older live there (.) first how many m[en ]  (5)
MR:                                              [Two]
        (0.3)
MI:     O:kay?=
MR:     =>Husband and wife<
        (0.3)
MI:     O↓::ka↑y ((tch)) (0.5) well we've randomly selected the
        man of the household as the person that we'd like to
        interview for our important national study .hh and would
        that be you sir?  (6)
        (0.2)
MR:     Yep.
        (0.3)
MI:     O::↓kay .hhh well before begin I wanta assure you (.)
        that all the information you give us is confidential↑
        .hhhh an:that >none of it< will be released >in any way<
        that would permit identification of .hhhhh you or your
        family  (7) and your participation of course is voluntary
        hh .hhhhhhhh and in addition (.) we would like your
        permission to tape your interview for research we are
        doing on the way questions are asked .hhhh an:d ↑this
        too is kept completely confidential .hh an' the tapes
        will be:: .hhhh uh: used for:: research studies of
        interviews only. .h and your participation in this is
        also voluntary .hh uh >do we< have your permission to
        tape the interview?  (8)
        (.)
MR:     Yeah.
        (0.3)
MI:     O::↓kay. (0.2) .hh an::d uh first how many persons .hh
        live in your household counting all adults and children
        and including yourself ...
```

Positions 1–8 represent only the grossest points at which call recipients allow for progression instead of resisting or declining the interview. That is, it is safest to say that there are "at least" these eight points rather than that these are the only places where call recipients can accept or reject the interviewer's continuing solicitations. We call these eight points *marked* opportunities for declining or giving the go-ahead to an interviewer.

Other points at which recipients allow for continuation rather than declining the interview are 'transition relevance places' (TRPs) (Sacks, Schegloff, and Jefferson, 1974) and are relatively *unmarked.* Recipients can analyze potentially complete utterances in the interviewer's talk and take them as opportunities for turn transition. In excerpt 1, such TRPs include line 9, after "home telephone numbers," line 19 after "people all over the country," line 20 after "in your household," and so on. Below, excerpt 2 shows that, in contrast to what happens at line 8 in excerpt 1, the call recipient interjects a declination at the end of the interviewer's announcement about trying to reach people at home:

(2) LSSC248

```
MR:   Hello?=
FI:   =H:i I'm Mary: calling from the University of Wisconsin?
      .hh as part of our national public opinion study. .hh
      And we're try:in to reach people at their home telephone
      num[bers]
MR:      [I'm ] sorry you got the wro::ng guy.
```

MR's declining turn (line 6) starts in overlap with the finish of FI's turn, which, because it occurs at the projectable finish of FI's utterance, is considered to be a nonviolative or noninterruptive overlap (Sacks, Schegloff, and Jefferson, 1974). Our point is that, as the front end of the interview progresses, there are many places where declination can and does occur; as call recipients allow progression beyond each of these points, they may become increasingly committed to doing the interview, but it is rare to find a formal granting of the request for an interview, unless there is a question about taping or something similar and the recipient gives explicit permission for that. Otherwise, it is when the interviewer is allowed to start the substantive questions that a recipient's action of granting the request becomes fully realized.

Comparison of Grantings and Declinings

In her research on the eight marked opportunities for resisting or allowing for the progression of the interview, Hadley (1999) compared the recorded front ends of 12 successful or completed interviews with 35 cases in which a call recipient, at one of the eight opportunities or other TRPs, declined a request for participation.[4] The main pattern that emerged is very nuanced. After point 1 or interviewers' issuance of their institutional identification, recipients, whether they end up declining or granting the interviewer's request, rarely give a vocal go-ahead and instead "passively" allow for progression by remaining silent.[5] After point 2, when interviewers have produced the reason for the call (RFC) (Maynard and Schaeffer, 1997), there are also no discernable differences in the talk of eventual grantors (hereafter, "grantors") and eventual decliners (hereafter, "decliners") who remain on the line rather than declining at this point.[6] That is, both kinds of recipients use continuers, tokens such as "yes," or remain silent and passively allow for continuation.

It is after point 3, which is when they are first explicitly questioned, and after subsequent opportunities through 8 that something noticeable may occur. Whereas grantors who produce a go-ahead signal at these points often do so almost immediately, decliners are more likely to hesitate by pausing for 1 second or more before answering the interviewer's questions about a residential number, whether the correct number is reached, the number and gender of adults in the household, and so on. As an example, notice in excerpt 1 that the recipient pauses only briefly after the question about residential number and then answers straightforwardly with a "yes." Another call also shows how rapidly (without delay) a call recipient can answer the residential number question:

(3) 255L2204

```
FR:      Hello?
MI:      .hh Hello (.) my name is Bill Collins and I'm calling
         from the University of Wisconsin as part of our
         national public opinion study .hh we aren't selling
         anything at all it's purely public opinion research
         and we're trying to reach people at their home
         telephone numbers .hh ↑is this a: residential number?
FR: →    Yes.
```

As with excerpt 1, this fragment is from a call in which the interviewer succeeds in gaining the recipient's cooperation for the interview. Compare excerpts 1 and 3 with what happens in a declination case:

(4) LSSC123

```
R:       Hello?
MI:      .hh Hello:: I'm Greg Johnson calling from the
         U:niversity of Wisconsin  (1) .hh as part of our national
         public opinion study?  (2) .hh and we are trying to reach
         people at their home telephone numbers:, is this: a
         residential number?  (3)
         (1.5)
R:       Uh yeah it's a resident.
MI:      O::kay and to be sure that I reached the number I dialed
         .hh is this area code six oh two xxx xxxx?
         (0.5)
R:       Mm hmm
```

The recipient allows progression of the call at points 1 and 2 (lines 3 and 4, respectively). After point 3 and the residential question, not only is there a long silence but the confirmation, itself done in a lax form ("Yeah"), is preceded with an "uh" token. Overall, the answer is packaged in what conversation analysts call a *dispreferred* format that displays the recipient's reluctant answering and possible resistance to progression of the interview. Again at line 12, after the phone number verification question (lines 9–10) and a pause (line 11) is a lax token on the part of the recipient. In fact, this is the last time R talks at all; subsequently he or she leaves the phone[7]; eventually a male voice in the background can be heard to say, "Go hang it up, and" after 7 seconds the phone line goes dead.

In excerpt 4, the recipient's hesitation in answering is relatively subtle. Interviewers regularly disregard signals like those this recipient produces, even though in ordinary conversation participants are attuned to such subtle signals and may alter the course of their talk when discerning them. For example, in asking a question that suggests a particular answer, and hearing the recipient's pause, the questioner may project that a disagreeing answer is forthcoming and may quickly change the question so that the recipient can agree with it. This shaping of the course of talk to elicit a confirming rather than disconfirming response is one exhibit of the preference for agreement in conversation (Pomerantz, 1984a; Sacks, 1987). Notice in excerpt 5 how A, after asking a question that proposes a "good cook there" and hearing a silent response, revises the question in a way that then receives confirmation:

(5) (Sacks, 1987:64)

```
A:   They have a good cook there?
     ((Pause))
A:   Nothing special?
B:   No, everybody takes their turns.
```

In the front end of surveys, rather than dealing with displays of reluctance on the part of call recipients by altering the course of talk, interviewers wait for responses, sometimes for long intervals:

(6) LSSC269

```
MI:      Okay .hh um I was saying m::y name is Edward Price? an:d
         I'm calling from the University of Wisconsin? in Madi↑son
         Wisconsin? as part of a national public opin↑ion study we
         do? .hh uh we're dialing random numbers trying to reach
         people at their ho::mes? Is: this a residence?
     →   (3.0)
FR:      I:: don't think I'm interested thank you.=
MI:      =Well it's not a sales pitch or anything [ma'am
FR:                                               [((hang up))
```

Here, a long delay (line 6) in answering the question regarding home telephone numbers foreshadows a declination in the turn slot where an answer to the question is due.

The general issue is that call recipients who end up participating in the survey usually go along with the set of front-end statements and questions[8] until the substantive interview commences. Although some recipients who end up declining the request for participation do so immediately, many also allow progression of the front end by giving go-ahead signals. They initiate their declining moves somewhere along the way, the latest being just after the final front-end query and before the first substantive question. We have not found distinguishing features in the responses of the two sets of call recipients — grantors and decliners — except in a minimal way: More often than grantors, eventual decliners may offer subtle forms of resistance at the opportunity points identified in excerpt 1 and at other TRPs in the interviewer's talk (Hadley, 1999). By and large, interviewers disregard these forms of resistance by moving ahead, or attempting to move ahead with the next segment of the front-end. This may be one facet of interviewers' optimistic orientation in soliciting participation.

Recipients' Questions: Insertion Sequences

We said that call recipients *usually* go along with the interviewer during the front end. Instead of going along at each point, recipients sometimes produce queries about the interview. In our data, call recipients mainly ask about two matters: how long the interview will last and what the purpose of the interview is.

When such queries occur, they often operate as *insertion sequences* (Schegloff, 1972:109–110) between an interviewer's question and the recipient's own answer:

(7) LSSC249

```
FI:    ... an: I would just like to ask you now if we have
       permission to tape ( ) interview.
MR:    Uh how many minutes will this [take]
FI:                                  [.hh ] Um it takes:
       abou:t (0.3) fifteen probably at the most (0.3)
       depending on how quickly the: responses are
       (1.0)
FI:    tch=
MR:    =O:kay
```

The interviewer's question (lines 1–2) is a *first pair* part of an *adjacency pair* (Schegloff and Sacks, 1973), and it concerns permission to tape the interview. Instead of answering right away, MR asks about the duration of the interview (line 3). This query starts an insertion sequence and obtains FI's production of an estimated duration (lines 4–6). In our earlier report (Maynard and Schaeffer, 1997), we noted the tendency for optimistic rendering of answers to queries about timing, which characterizes FI's answer here. That is, she suggests 15 minutes as an outside estimate and implies that it might be shorter if the answers are done "quickly."

As noted by Sacks (1992:528), insertion questions propose "If you answer this one I will answer yours." Accordingly, after completion of the insertion sequence with FI's estimate of duration [and after a silence (line 7) and FI's tongue click (line 8)], MR provides the *second pair* part to the participation request by granting it (line 9). In an investigation of insertion sequences about timing, Woodbury (1999) suggests that optimistic proposals regarding the length of an interview may be more effective than neutral or *informational* responses, such as "about 15 to 20 minutes," in garnering a recipient's continuation in the front end. Optimism embellishes the informational response in a favorable way but whether it also affects overall response rates is not clear.[9]

Optimism is a device that can *persuade* call recipients to participate in an interview. For example, in the next excerpt, the call recipient asks two regular questions — first about the purpose of the call and second about its duration. The excerpt takes up after the interviewer has introduced himself, checked the call recipient's telephone number, and randomly selected (at lines 1–3 below) the "man of the household" (who is the call recipient):

(8) LSSC/Nov. 26/I2:A1

```
MI:   We've randomly selected th- the man of the household
      as the person we'd like to interview for our national
      survey. [.hh Would-   ]
MR:           [What is this] in regard to=
MI:   =.hh Sure .hh uh >like I said I'm calling< from the
      University of Wisconsin and we conduct national public
      opinion studies all the time on a variety of topics .hh=
MR:   =(>Is it gonna be<) long? or-
      (0.5)
```

```
10  MI:     Uh well it dep- it depends from person to person, the
11          shortest interviews I've done have taken about ten minutes
12          but (.) a lotta people ta[lk longer       be]cause they=
13  MR:                              [(on the phone? Or-)]
14  MI:     =enjoy the questions
15          (0.5)
16  MR:     On the phone you mean?
17  MI:     .h Yes
18          (0.4)
19  MR:     Okay go ahead.
```

In answering "what is this in regard to" (line 4), MI identifies himself again and says that they "conduct national public opinion studies all the time on a variety of topics" (lines 5–7), thereby suggesting that the request is a routine matter involving no particularly mentionable content. This may imply that other recipients routinely comply with the request and that the interview involves nothing objectionable. Then, MR asks about duration in a way that proposes the interview as "long" (line 8), and MI produces an answer proposing contingency (line 10), with a numerical estimate for a lower boundary ("ten minutes," line 11). He then characterizes upward departures from this boundary not in numerical terms but according to an enticing *reason* for their occurrence: "people talk longer because they enjoy the questions" (lines 12 and 14). This can put a positive cast both on what the interview is "in regard to" as well as how long it might take. After a further query about whether this would be a phone interview (line 16), and the interviewer's confirmation (line 17), MR agrees to the interview (line 19). The interviewer's optimism appears to have worked persuasively.

Such persuasiveness does not always work, and if an interviewer is more factual with a recipient, it does not necessarily mean that the recipient will decline the interview. But if optimism is to be deployed as a persuasive device, it appears that it should be done in and as the answer to a timing or purpose question. In our next excerpt, the recipient asks a purpose question in the middle of the interviewer's request for phone number confirmation:

(9) LSSC127

```
 1  MI:     ... an:: to be sure that I reached the number I dialed
 2          is this:: (.) area code five five seven .h four eight
 3          th[:ree ].
 4  FR:       [What']s this about?=
 5  MI:     =five oh four two?
 6          (0.5)
 7  FR:     What is this regarding?=
 8  MI:     =Well um this is our: uh national public opinion study
 9          that we're doing from the University of Wisconsin.
10          [.hh]
11  FR:     [ O ]kay I'm not interested.
12  MI:     Well um if [I    could] just mention ma'am
13  (FR):              [((hang up))
```

That is, FR's "What's this about?" (line 4), produced before MI has completely uttered the full 10-digit residential number, interrupts him—MI stops midturn (line 3). After FR's query (line 4), MI finishes uttering the number at line 5. At line 6, FR pauses and asks a different version of the purpose question, saying, "What is this regarding?" (line 7). This insertion works to obtain an answer (lines 8–9) to the query, but the answer simply repeats information from the interviewer's opening turn (in which he says he is "calling from the University of Wisconsin as part of our national public opinion study"). We call this *repeating the opening*; the answer contains the bare facts embodied in the initial request for participation and no optimism or other device displaying a persuasive effort to garner cooperation. At line 11, FR declines the interview, and now at line 12 MI starts a "mention" projecting a turn in which a persuasive device might have been deployed. In the midst of this utterance, FR hangs up. This may be a case of "too little, too late" on the part of the interviewer. Although we cannot know whether anything he might have said would have kept the recipient on the line, we can say he missed an opportunity to embellish his effort at answering the recipient's question.

CLOSING THE GATE: FURTHER OPPORTUNITIES FOR PERSUASION

When a call recipient produces a declination, that does not necessarily end the call. Rather it proposes to *move toward* terminating the encounter. Accordingly, even after a declining move by recipients, interviewers still have opportunities for keeping recipients on the line and for "converting" their resistance to cooperation. In the remaining part of this chapter, we analyze the interactive organization of closing the call and examine how interviewers may continue to work persuasively even as termination becomes imminent. Our data in this section involve a collection of declinations used in an earlier study and numbering 53 initial calls to households (Maynard and Schaeffer, 1997). For analysis, we concentrate mainly on this collection of initial calls and draw from a smaller set of household callbacks for specific illustrative purposes.

Our analysis shows there to be two categories of declinations of the request to participate in a survey interview—polite and impolite ones. *Impolite* declinations represent a rude or brusque way of terminating the phone call encounter (Schegloff and Sacks, 1973) and are relatively rare (11 percent of the corpus), while *polite* declinations, which constitute the large majority (89 percent), are of two subtypes whereby recipients give a reason for declining the request to participate: It represents "bad timing" and/or they are "not interested" (see Table 8.1).

Table 8.1 Types of Declinations

	Polite		
Impolite	Type 1: Bad Timing	Type 2: Not Interested	Total
6 (11%)	29 (55%)	18 (34%)	53 (100%)

Impolite declinations are those in which recipients, rather than employing turns of talk that enact procedures for mutually and systematically closing and terminating the call, engage in activities that unilaterally disable pursuit of the interview then and there. The most straightforward way of disabling pursuit of the interview is to hang up:

(10) LSSC276

```
FR:      Hello?
MI:      .h tch Uh: hello: ma'am my name's Edward Price? a:n I'm
         calling from the University of Wisconsin? as part of
         our national public opinion study?.hh uh were tryin
         to reach people: at their home numbers (.)
         I[s this a residence ma'am?
FR: →     [((hang up))
         (1.4)
MI:      <Don't hang up on me.>
```

After the hangup, interviewers may complain (line 9) in a manner that exhibits the objectionable and violative nature of the recipient's act.

Polite Declinations: Dispreferred Actions and Systematic Closing

By polite declinations, we refer to devices through which recipients (a) exhibit their declining action as *dispreferred*, and (b) observe the propriety of opening up closings or systematically terminating the call (Schegloff and Sacks, 1973). As discussed in Chapter 1, the *preferencing* and *dispreferencing* of actions in conversation have to do with the systematics by which participants fashion their utterances depending on which set of alternative responses they employ. Preferred responses are done immediately, often in short turns or at the beginning of longer ones, while dispreferred ones are delayed by silence and/or by mitigating components, such as accounts, excuses, preferred responses, repair sequences, and the like (Heritage, 1984; Schegloff, 1988). Participants in conversation regularly *accept* invitations, *agree*, and *deny* accusations (to cite some instances of actions) in a preferred manner, while they *reject* invitations, *disagree*, and *admit* accusations in a dispreferred mode (Atkinson and Drew, 1979; Davidson, 1984; Heritage, 1984; Pomerantz, 1984b; Sacks, 1987).

Excerpt 11 below illustrates how declinations of requests for survey participation often exhibit a dispreferred status. The interviewer initially identifies himself by name and institutional affiliation (lines 2–3) and provides a reason for the call ("as part of our national public opinion study?" lines 3–4) that constitutes the request to participate. Following this, MI produces a residential verification question (lines 4–6), and it is after this that MR fashions a declination by first answering the residential question and then announcing that he doesn't "have any time" (lines 8–10). Accordingly, an *agreeing* component appears early in MR's turn, contiguous with the question it answers, and the *disagreeing* declination is deferred to a later position in the turn and is not contiguous with the talk (the reason for the call and the request to participate, lines 3–4) to which it responds

(Sacks, 1987). Furthermore, MR pauses before talking (line 7), apologizes ("I'm sorry") after his answer to the residence question and before declining, and gives a reason for not participating rather than using a negative or rejecting term ("no"). All of these features are indicative of a dispreferred response:

(11) LSSC121

```
MR:      Hello:.
MI:      .hh Ah hello: I'm Greg Sanders calling from the
         University of Wisconsin .hh as: part of our national
         public opinion study? .hh a:n:d we are trying to reach
         people at their home telephone numbers is: this a
         residential number?
         (0.8)
MR: →    Uh yes it is but I'm sorry I d- I don't have any time
         to (.) answer any of your questions [right now]
MI:                                          [Oh : : : ] kay
         well we'd like to try you back at another time then
         would that be alright?
MR:      That'd be fine.
MI:      O::kay well thank you very much [sir.    ]
MR:                                      [Thanks.] Bye bye.
MI:      Bye bye.
```

Polite declinations, then, are those that exhibit the dispreferred status of the rejecting action they implement.

Besides being exhibited as dispreferred actions, polite declinations systematically work toward closing and terminating the conversation between interviewer and recipient. They present an opportunity for mutually discontinuing the encounter between interviewer and call recipient, representing a solution to the problem of how to arrive "at a point where one speaker's completion will not occasion another speaker's talk, and that will not be heard as some speaker's silence," or, recalling the impolite declinations discussed above, as a rude or brusque act (Schegloff and Sacks, 1973:294–295). When done in a relatively polite way, termination represents the lifting of *transition relevance*, a feature of any conversational turn obligating coparticipants to select another or self-select for a next turn of talk (Sacks, Schegloff, and Jefferson, 1974:237–238). *Terminal exchanges*—that is, the use of goodbyes—are prototypical devices for lifting transition relevance.

Terminal exchanges cannot occur just anywhere when two parties are talking, and require proper initiation, through what Schegloff and Sacks (1973) term a *closing section*. Closing sections, like terminal exchanges, regularly comprise two turns of talk. Thus, when closings occasion termination, it is possible to characterize the two events as an *archetype closing* consisting of four turns of talk (Button, 1987:102), as arrowed below:

(12) NB:III:5:10 (Button, 1987:102)

```
Guy:       I'll be down there, oh en you'll- you'll be aroun' then
           when I [(come in)
Emma:             [Yeah.
```

Guy: 1→ Okay.
Emma: 2→ Okay dear,
Guy: 3→ Buh bye,
Emma: 4→ Bye bye,

Turns 1 and 2, the closings, employ terms such as "okay" and "all right," which, as Button (1987:102) suggests, are "bereft" of specific topical implicativeness and provide a *free turn* and opportunity to move toward termination yet allowing for return to topical talk if a participant so chooses. In this respect, as Schegloff and Sacks (1973) argue, the initial "okay" (arrow 1) is a *preclosing* whose status as a "closing" as such is only accomplished when it is treated with a return closing item rather than with a device that reoccasions topical talk. When a preclosing is reciprocated, it provides for the relevance of a terminal exchange (arrows 3 and 4 above).

This analysis of archetype closings in ordinary conversation is relevant to an examination of survey interview declinations:

(13) LSSC222

FR: Hello
FI: ... my name's Susan and I'm calling from the University
of Wisconsin? as part of our national public opinion
study .hh we're tryin to reach people at their home
telephone numbers. Is: this a residential number?
(0.5)
FR: 1→ Yes it is [but] I do not have time for=
FI: [((tch)) Oka-]
FR: → this [( )]
FI: 2→ [Okay] we'll get back to you another time=
FR: 3→ =Bye.
FI: 4→ Tha:nk you bye bye.
(0.6)
((hangup))

Declinations (as in lines 7 and 9) can interlock two sequence types. They answer the request to participate and simultaneously constitute the first move in closing the conversation (as per arrow 1). Then, the turn that responds affirmatively to the opening is a second move (arrow 2) in the closing section and occasions turns 3 and 4, a terminal exchange (lines 11–12).

POLITE DECLINATIONS: "BAD TIMING" AND "NOT INTERESTED"

As we shall see, the organization of conversational exit in excerpt 13 has complexities that we have not yet analyzed. Nevertheless, the notions of conversational closing and terminal exchange are useful in appreciating the orderly fashion whereby polite declinations eventuate in ending an encounter between an interviewer and a call recipient. Now we explore how, depending on the reason a recipient gives for declining the request to participate—bad

Table 8.2 Types of Polite Declinations

Type 1: Bad Timing	Type 2: Not Interested	Total
29 (62%)	18 (38%)	47 (100%)

timing or not being interested—there are different and particular avenues toward closing and/or termination. Others (Groves and Couper, 1992) have also described the frequent use of bad-timing and not-interested declinations but have not distinguished them according to their *sequential implicativeness*—the range of consequences they project for subsequent talk and interaction. Also, as Table 8.1 shows, bad-timing declinations are over half of all (polite and impolite) declinations. When compared only with the other kind of polite declinations (not interested), they are 62 percent (Table 8.2). Accordingly, another task here is to explain, in interactional and structural terms, call recipients' preponderant use of the bad-timing device.

Bad-Timing Declinations

The predominant mode for recipients to decline participation in an interview is through announcing involvement in a competing state or activity. If they do not simply do such an announcement, recipients may *formulate* the matter as bad timing, for example saying, "I like what you're doing but you called at a real bad time" (LSSC177). Or an interviewer may follow a recipient's claim of being "busy right now" with the comment, "I'm sorry to catch you at a bad time" (LSSC169). Regularly in such calls, (1) the recipient invites the interviewer to call back or (2) the interviewer proposes to call back.

1. *Invitations to Call Back* Our first example occurs relatively deep in a phone call, which a child first answered. With the child, the interviewer verified the telephone number as residential, found out how many adult men and women lived in the household, and selected the "man of the household" to interview. The interviewer then asked to speak to him and the child assented to the request, after which there were 22 seconds of silence with an audible "click" near its end. Then the man came on the line:

(14) LSSC215

```
MR:      Hello?
MI:      ((tch)) .hh Hello I'm Marty Jameson calling from the
         University of Wisconsin? (0.2) .h as part of our
         national public opinion study?
MR: 1→   Okay you'll have to call back later I'm on another
         line right now.
```

```
MI: 2→   Okay.
MR: 3→   Okay.
MI: 4→   Thanks a lot. (0.3) Buh bye.
         (0.4)
         ((hangup))
```

At lines 5–6, declining the request to participate, the recipient's declination proposes that the interviewer "call back" and gives a reason, thereby initiating a closing section (thus arrow 1). MI accepts the proposal for a callback at line 7 with what Davidson (1990:161–163) calls a *finalizer*. That is, just as a request sets up acceptance/declination, so a declining move also occasions a display of (a) agreement/acceptance or (b) rejection (Davidson, 1990:160). Here, the interviewer's "okay" at line 7 accepts the declination and thus finalizes it. At the same time, the interviewer's finalizer is a return close component (arrow 2). Next, it appears that MR produces a first terminal component ("okay," arrow 3) that is followed by the interviewer's appreciation and return salutation, which constitute a second terminal component at arrow 4.

To capture the real-time, in-course production of the devices by which encounters are interactionally completed and, per our discussion of Schegloff and Sacks' (1973) concept of the preclosing, it is better to regard these items as possible close components and possible terminal components whose actuality is only accomplished according to the unfolding interaction. For example, consider that the interviewer's gratuity ("Thanks a lot," line 9) shows appreciation for the callback invitation[10] and might occasion an acknowledgement and then a terminal component (Button, 1987:123).[11] However, upon the completion of the "Thanks a lot" utterance—a place that makes the transfer of speakership relevant—a silence develops (middle of line 9), which the interviewer can interpret as the recipient's refusal to talk further and as indicating that the MR's line 8 "okay" was serving to initiate termination. The silence, accordingly, provides for the relevance of a terminal component, which MR produces after the silence ("Buh bye," line 9). Then, through the silence at line 10 and MR's hanging up after that, MR treats this "Buh bye" not as a component to be reciprocated with another "bye bye" but as itself a second part of a terminal sequence. Thereby, MR's "okay" at line 8 comes to be the first part of the terminal exchange.[12]

Foreshortening a closing section is to handle a problem to which these sections are vulnerable. Participants can occasion moves *out* of closings and thereby delay termination. This happens in the next excerpt, which is a callback from an earlier contact.[13] The recipient declines the request to participate—even though it is made in a callback—by announcing a competing involvement (exhibiting the "bad timing" of the request) and inviting yet another callback. The tape starts with the utterance at line 1; we do not have a recording of the earlier part of the call.

(15) LSSC161

```
MI:    ... ↑I'm Greg Sanders ↓calling from the University of
       Wisconsin .hh an:d on an earlier call to this household
       we selected you as the adult we would like to speak with
```

```
                there? for our national public opinion study? .h[hhh]
  FR: 1→                                                   [You]
                know what? (0.2) I'm just headed out do you mind calling
                another day?
  MI: 2→        N:o no that would be fine we could call back .hh=
  FR: 3→        =↑Okay
  MI:  ↑        An::d um if I can just verify the number then to make
   move         sure that I did reach the number I dialed .hh is it
   out          area code four oh eight? .hh uh seven six seven four
   of           three one two?
 closing        (1.5)
  FR:  ↓        <Yep>
                (0.4)
  MI: 2r→       tch O::kay well we'll try to reach you another: time
                then. .hh=
  FR: 3r→       =Thanks
  MI: 4→        Thank you bye bye.
                ((hang up))
```

Lines 5–7 comprise three interlocking actions: declining the request to participate, inviting a call-back *arrangement* (Button, 1987), and (as per arrow 1) initiating closing. The interviewer accepts the declination and invitation, which would also be a second close component (arrow 2). When FR produces the "okay" at line 9, then, it proposes to start a terminal exchange (arrow 3). At this point, however, MI moves out of the closing section by interjecting a query about his recipient's phone number (lines 10–13). With a silence at line 14, FR delays and then at line 15 confirms the number, after which (at line 17) MI reaccepts the earlier invited arrangement (Button, 1987:106–107),[14] thereby returning to closings. This turn is redoing a second close component (thus arrow 2r; compare to MI's turn at arrow 2). The recipient's turn at line 19 is an appreciation and, as in excerpt 7, appears to operate simultaneously as a terminal device—albeit in this case redoing (hence arrow 3r) and upgrading the earlier offering of a terminal component (arrow 3). That is, "Thanks" (line 19) is more termination implicative than "okay." The interviewer then returns the appreciation as part of a responsive terminal turn (arrow 4), thereby completing the closing section.[15] In short, the interviewer's move out of closing in this case substituted for a terminal device and delayed completing the call; foreshortening possibly could have occurred with an initial use of "Thanks" or some other device that is more termination implicative than "okay."

2. *Proposals to Call Back* Recipients often use bad-timing declinations without inviting a callback. Interviewers then respond with a declination finalizer and either request or claim the right to a callback. Because both requests and claims implicate an acceptance or rejection from the call recipient, we refer to them as *proposals*. A request occurs as in our next excerpt:

(16) LSSC121

```
  MR:          Hello:.
  MI:          .hh Ah hello: I'm Greg Sanders calling from the
               University of Wisconsin .hh as: part of our national
```

```
 4                public opinion study? .hh a:n:d we are trying to reach
 5                people at their home telephone numbers is: this a
 6                residential number?
 7                (0.8)
 8  MR:     1→    Uh yes it is but I'm sorry I d- I don't have any time
 9                to (.) answer any of your questions
10                [right now]
11  MI:           [Oh : : : ]kay well wed like to try you back at
12                another time then would that be alright?
13  MR:     1r→   That'd be fine.
14  MI:     2→    O::kay well thank you very much [sir.    ]
15  MR:     3→                                    [Thanks.] Bye bye.
16  MI:     4→    Bye bye.
```

We have again arrowed the four base components of a closing section. The declination (lines 8–9) initiates closing (as per arrow 1) and does not get a second close component until line 14. The request to call back (lines 11–12) starts an insertion sequence (Schegloff and Sacks, 1973), which is a move out of the closing section initiated with the declination. When the request is granted (line 13; arrow 1r), that is redoing the initiation of closing; the interviewer then responds to the initiation with a return close component and appreciation (arrow 2), which occasions a terminal exchange (arrows 3 and 4).

In the center where we collected data, interviewers are encouraged not to ask permission for calling back and, instead, to put the matter in a positive or affirmative fashion by announcing the intention to call back rather than requesting permission. That is, optimism, as displayed in the interviewer's work to elicit eventual cooperation, is partly a product of their training and work culture. Excerpt 17 below shows how claiming the right to call back operates as a proposal. In the turn subsequent to the announced claim, recipients can be heard to produce acceptances or rejections:

(17) LSSC271

```
 1  FR:         eHello::.
 2  MI:         ((tch)) uh hello ma'am my name is Edward Price and I'm
 3              calling from the University of Wisconsin as part of our
 4              national public opinion study? .hh uh we're trying to
 5              reach people at their home numbers is: this a residence
 6              ma'am?
 7  FR: 1→      .h Ye- uh yes it yes it is uh I'm sorry uh I won't be
 8              able to keep my phone tied up tonight oka::y?
 9  MI: 2→      Okay we'll just try another time then
10  FR: 3→      Okay thank [  you  ]
11  MI: 4→                 [((tch))] Su:re bye
12              ((hang up))
```

A bad-timing declination is constituted in the recipient's formulation that she is unable to keep her "phone tied up tonight" (lines 7–8). This is a first close component (arrow 1). While MI responds with an "okay" that can acknowledge FR's reason for declining, FR's preceding temporal formulation "tonight" and her appending of a tag question provide for the relevance of MI's also proposing the intention to "try another time then." This is accomplished as a second close component (arrow 2) in part through the answerer's next turn (arrow 3), which, in granting the request and using an appreciation, is also a possible terminal component. The interviewer's acknowledgment and salutation (arrow 4) is accomplished as second terminal component in part through the caller subsequently hanging up.

Largely, call recipients are affirmative in their responsiveness to callback proposals. Later in this chapter, however, is an instance in which a request to call back is made after several other types of appeals have been denied; the requested callback is denied.[16] However, such denials or rejections are not often so overt. In the excerpt below (18), the recipient's response to a callback proposal appears rejecting even though there is no blatant refusal. The interviewer's approach to the recipient's response embodies optimism:

(18) LSSC222

```
FR:       Hello
FI:       ... my name's Susan and I'm calling from the University
          of Wisconsin? as part of our national public opinion
          study. .hh an' we're tryin to reach people at their
          home telephone numbers .is:: this a residential number?
          (0.5)
FR: 1→   Yes it is but [I  ] do not have time
FI:                    [Ih-]
FR:       for this [(   )]
FI: 2→             [Okay ] we'll get back to you another time.=
FR: 3→   =Bye
FI: 4→   Tha:nk you bye bye
          (0.6)
          ((hangup))
```

After a bad-timing declination (arrow 1, lines 7 and 9) the interviewer proposes calling back (arrow 2), and the recipient responds with the terminal component "Bye" (arrow 3) rather than a token such as "fine," "okay," and/or "thank you," as in excerpts 14–17. Accordingly, as contrasted with the items used in slot 3 in these other excerpts, this may suggest a rejection of FI's proposal. Of course, whether FR intends to reject the proposal cannot be settled with finality here. Observe how the interviewer *treats* the line 11 terminal component by prefacing a return salutation ("bye bye") with an appreciation (arrow 4), thereby submitting that the proposal to call back *virtually* has been accepted. In this way, it is optimistic.

That interviewers may try ploys such as treating a recipient as *if* she accepts a proposal to call back is more evident in excerpt 19. After his request for participation (lines 1–3), MI begins another utterance (".hh Um we're not tryin-") that the call recipient interrupts with a declination (lines 5–6):

(19) LSSC146

```
FR:   Hello?
MI:   Hello I'm Joseph Lew↑is calling from the University of
      Wisconsin? as: part of our national public opinion
```

```
        study? .hh Um we're not [tryin- ]
FR:                             [I'm sor]ry we- we don't have
        the time right now.=
MI:     =Oh really? Okay we'll try you back thank you.
        (0.5)
FR:     [((hangup))]
MI:     [Bye bye   ]
```

This interruptive, bad-timing declination could indicate firm resistance to an interview. However, MI first receives FR's declination (lines 5–6) with a surprise marker ("Oh really?" line 7), then produces the callback proposal, and, finally appends an appreciation to his turn of talk. Optimism, then, marks the entire turn. The interviewer's display of surprise can question the recipient's resolve in declining the request for an interview. In addition, MI's lattermost action (thanking) presupposes acceptance of the just previous action (proposing the callback). Furthermore, by its positioning, "thanking" is the action that is sequentially implicative, specifically of an acknowledgment (e.g., "you're welcome" or "okay"), which would in a post hoc way not just acknowledge the gratuity but accept MI's proposal. Instead, the recipient here is silent (line 8) and then hangs up (line 9). MI orients to the absence of a response at line 10 with a terminal utterance that ends up occurring in simultaneity with the recipient's hangup. Although by being silent and hanging up the recipient may be rejecting the callback proposal, the interviewer has not heard an overt "no" or other rejection. Accordingly, the interviewer can avoid coding this call as a refusal and can retain the number for another attempt at recruitment by someone in the survey center. In a way, optimism pervades the treatment of the call recipient's actions and the case as a whole.

Not-Interested Declinations

Making arrangements in ordinary telephone calls, as Button (1987) has put it, represents a "minimal" movement out of a call's closing section. In survey calls, a more accurate term might be that making arrangements for calling back moves *away* from closing. That is, an invitation or proposal to call back often expands a turn that is otherwise preoccupied with closing and implicates acceptance or rejection in the next turn. The acceptance or rejection is often done with the same token that performs closing or termination. Accordingly, a closing section can be unexpanded even if one of the turns within it is expanded. Excerpt 18 above is typical; arrow 1 indicates a first close component, arrow 2 the answering close component with a callback proposal, arrow 3 a one-word ("Bye") initiation of a terminal exchange that can be optimistically treated as accepting the callback proposal, and arrow 4 the responding terminal component. Even in excerpt 16, an insertion sequence comprised of a request and its response can be heard as *on the way* to completion of a closing section. Clear movement *out of* closing occurs only in excerpt 15, when the making of arrangements occasions a different activity, the verifying of the recipient's telephone number. Because the call recipient has given permission for a callback, checking the phone number is related to closing and termination, and closings are reentered upon completion of the verification sequence. In terms of the structure of close and termination, then, bad-timing declinations, in a relatively polite way, effect efficient exit from the social encounter between interviewer and call recipient.

Not-interested declinations are not as efficient in politely exiting the interview, because interviewers often treat this reason for declination differently from the "bad-timing" one. Below, we reproduce excerpt 6; the recipient's declination is at line 7:

(20) LSSC269

```
MI:   Okay .hh um I was saying m::y name is Edward Price?
      an:d I'm calling from the University of Wisconsin? In
      Madison Wisconsin? as part of a national public opinion
      study we do? .hh uh we're dialing random numbers trying
      to reach people at their ho::mes? is: this a residence?
      (3.0)
FR:   I:: don't think I'm interested thank you.=
MI:   =Well it's not a sales pitch or anything [ma'am
FR:                                            [((hang up))
```

The interviewer responds to the recipient's declination with an *appeal* (line 8). Appeals consist of a *modification* or *addition* or some other *revision* to an original request, not taking a rejection as final (Davidson, 1990:165–166). Appeals also exhibit the speaker's analysis of possible inadequacies or troubles with the original request. In the survey interview, appeals are another way in which interviewer optimism manifests itself. Here, MI's line 8 utterance submits that FR's line 7 declination is not final and offers to make his request more *appealing* by suggesting that the original request is not to be heard as a "sales pitch or anything."

Recipients, for their part, *monitor* responses to their declinations, and when a declination is not initially accepted—an interviewer produces no finalizer—the recipient may take corrective action, such as revising the declination, to achieve finality (Davidson, 1984, 1990). Above, after hearing the interviewer's appeal (line 8), FR's corrective action is strong: she hangs up (line 9). Something similar happens in the next excerpt:

(21) LSSC265

```
MI:   ((tape cutoff) ... name is Mike Baker, I'm calling from
      the University of Wisconsin as part of our national
      public opinion study. .hh and we're trying to reach
      people at their home telephone numbers .hh i::s this a
      residential number sir?
MR:   Ah:: <no thank you not interested>
MI:   Uh well sir? [ We're    ] not tryin to sell=
MR:                [((hangup))]
MI:   =any°thing::
```

In this call, however, MI manages to issue only a preface to the appeal ("Uh well sir?") before MR hangs up, which suggests that recipients can *closely* monitor responses to a declination for some token of acceptance and that hangups occurring after a declination but with no terminal salutations are often an outcome of an interviewer's optimistically regarding declinations as *not final* when in fact call recipients evidently mean them to be just that. Indeed, in our category of not-interested declinations, 13 of 18 calls are terminated in this way, suggesting that a large number of initially polite declinations are transformed into something less than polite terminations because of the way the interviewers handle those declinations.

In only one instance out of 18 not-interested declinations does the interviewer immediately *accept* the declination and allow for an undelayed, systematic progression through closings and toward the terminal exchange:

(22) LSSC116

```
FI:   ... And (.) we're doing a national (0.4) public
      opinion study?
      (0.5)
FR:   Oh no I don- I'm not interested
FI:   OH: Okay=
FR:   =Thank you
FI:   Alright [bye bye.]
FR:           [Bye bye ]
```

It may be that FR's use of "Oh no" at the beginning of her turn, which is akin to the unhesitating way in which participants initiate their vocalizations of preferred actions, may signal a degree of firmness in this dispreferred action to which FI is responsive. Indeed, the "Oh no" is an example of an "oh-prefaced response to inquiry," which Heritage (1998:330, note 21) describes as a device for showing the "inapposite" nature of the inquiry it follows; it may be a "particularly powerful means to emphasize rejecting actions." FI issues a finalizer at line 5. Then FR initiates an appreciation sequence with the "thank you" at line 6, which occasions the interviewer's acknowledgment (line 7) and a terminal sequence whose components overlap one another (line 7–8).

Not-interested declinations, therefore, do not automatically engender appeals. They may be accepted or finalized and lead straightforwardly to a terminal exchange, particularly if the declination they implement is conveyed firmly. However, in part because of interviewers' optimistic approach to declinations, they do appear *vulnerable* to nonacceptance in a way that bad timing announcements do not. In our data, appealing a not-interested declination is much more frequent than accepting it, and there can be extended cycles of modifications or additions to the original request to participate, each one occasioning yet another opportunity for the recipient's response, and preferring acceptance (Sacks, 1987). In the next excerpt, an interviewer appears heroic in his effort to obtain acceptance; the recipient is equally stouthearted in the patience with which he responds to each appeal with another declination. This is a callback, and the transcript takes up where the interviewer has asked the first recipient for, and obtained, the person in the household who had been previously selected for the interview:

(23) LSSC163

```
 1  MR:      Yhello.
 2           (0.2)
 3  MI:      Uh: hello this is Greg Sanders calling from the
 4           University of Wisconsin, as: part of our national
 5           public opinion study? .hh
 6           (0.3)
 7  MR:      Uh huh? [uh I'm n]ot interested.
 8  MI:              [You::-  ]
 9  MI: A→   O::ka:y well um .hh it's awfully important that we:,
10           get the opinions of everyone in this sample otherwise
11           the results won't be very useful.=
12  MR:      =Well:: I: (.) don't wanna (0.2) give ya any (0.2) o'
13           my opinions.=
14  MI: B→   =O::ka:y well uh uh we think that the survey's very
15           important because the questions are ones that people
16           in .hh universities and in the goverment want to know
17           answers to so .hh [we'd-]
18  MR:                        [Oh:: ] I don't have any
19           opinions to give you so I don't wanna answer any o'
20           [your questions]
21  MI:      [O::ka::y      ] w- well [y'know ]
22  MR:                               [I don't] mean to be rude
23           about it that's the way I am.=
24  MI: C→   =Y:eah well well if you don't have oh- an opinion about
25           something we can jus:: s:kip it .hh y'know or or:: or
26           put down that you don't have an opinion.
27  MR:      Well: my dinner's getting cold so: [I don't] feel=
28  MI:                                         [°O:kay° ]
29  MR:      =like talking.
30  MI:      Yeah um well is it okay if we call you back at another
31           time?=
32  MR:      =N::o hh I'd- n- wish that you not call me back.
33  MI:      O::ka:y well [thank] you anyway=
34  MR:                   [Tha- ]
35  MR:      =Thank you.
36  MI:      Mm bye bye.
37           ((hang up))
```

Following MR's not-interested declination (line 7), there are three (arrowed) appeal sequences (Jefferson and Schenkein, 1978). Each starts with an acknowledgment of the declination and contains a formulaic response. The first one (arrow A) stresses the importance of the *sample*, to which MR replies by showing reluctance to give his opinions (lines 12–13). Second, at arrow B, the interviewer shifts ground slightly by characterizing the importance of the *survey* and the questions. MI then appears to start another piece of his appeal (line 17) when, in overlap (line 18), MR starts a turn exhibiting stronger reluctance about answering; that is, there are two interconnected components to his declination — (a) not having opinions to give the interviewer and therefore (b) not wanting to answer his questions. The interviewer appears to begin another appeal (line 21), producing, in overlap, an acknowledgment ("okay") and preface ("well y'know"); here, MR produces a mild apology or excuse (lines 22–23) and then MI finishes his third appeal (arrow C), proposing that they could "skip" or put no opinion for some questions. The recipient now (at line 27) produces a bad-timing announcement, thus shifting to a different type of declination and one that, as we already have seen, strongly implicates closing and termination while providing the grounds for making arrangements for another call. Thus, at lines 30–31, MI makes a request to call back, which is refused (line 32), and the participants return to closing and terminating the call (lines 33–36).

This excerpt shows that, in response to not-interested declinations, interviewers may optimistically engage *cycles* of appeals before an initial declination is finalized and the participants close and terminate the call. Put differently, not-interested declinations are vulnerable to nonacceptance and to radical rather than minimal moves out of closing. Also this call shows that the bad-timing announcement, in being the last one that the recipient uses, is the one that is most successful in reinvigorating movement toward closure and termination, even though the announcement, consistent with other uses of the device we have analyzed, occasions a request to call back.

Our suggestion is that the effectiveness of the bad-timing device for effecting closure and termination may be one reason that call recipients invoke it in the first place, independently of their actual involvements in activities (making dinner, going out the door) or states (not feeling well, for example) that compete with participation in the interview. In other words, bad-timing announcements are a *stronger* form of declination, whereas not-interested claims are *weaker*, in *bringing about a polite exit from the call.* Accordingly, there is an interactional–structural reason for recipients to invoke the bad-timing declination. Moreover, because bad-timing announcements contain temporal formulations as an embedded feature, they naturally occasion the effort at making arrangements for a callback, however mutual or unilateral those arrangements turn out to be. Because the request to call back itself prefers acceptance but no immediate commitment, recipients may accede to the request regardless of their intentions for eventual cooperation. Therefore, the bad-timing device as a form for declination may eventuate in spurious arrangements for the callback. High numbers of coded callbacks may reflect structures of interaction rather than substantive commitment on the part of call recipient.

CONCLUSION

In interview contexts, the *interactional substrate* refers to those skills of the participants that allow them to ultimately generate answers as accountable items (Maynard and Marlaire, 1992). That is, the proper recording of attitudes and opinions for social scientific purposes is an achievement that depends upon an organization of interviewers' and recipients' concerted practical actions. From this substrate both successful interviews and unsuccessful ones as well are consummated, in and through participants' actions for opening encounters, making requests, and accepting or declining such requests (Maynard and

Schaeffer, 1997). Opening the gate involves call recipients going along with the front end of the interview until substantive questioning begins. Recipients may interject questions along the way, especially about the length and purpose of the interview. We find that interviewers, often because of the training they receive, work optimistically in answering these questions, often to good effect. That is, optimism as a persuasive device used to handle recipient queries may be more effective in keeping recipients on the line than unembellished factual responses or those which repeat the opening.

However, when interviewers produce optimistic responses to bad-timing declinations, they may elicit spurious arrangements for callbacks. Also, optimism, embodied in appeals to recipients who declare they are not interested, often transforms a recipient's polite effort at closing into a more rude termination because a recipient will hang up when noticing that an interviewer is continuing to talk with an effort at persuasion. Training interviewers to use optimistic devices persuasively is no doubt important, but interviewers should not always do so during the front end of the survey interview. Our research suggests conditions under which persuasive devices such as optimism are appropriate. Optimism when answering questions about the length and purpose of the interview appear effective; optimism about callbacks when recipients are resistive with their bad-timing declinations, and optimism that claims of not being interested can be appealed and converted, not only seem ineffective but engender impolite—perhaps even hostile—retorts in the form of recipients hanging up the phone.

NOTES

1. The notion of gatekeeping, which can be tracked back to Lewin's (1947) discussion of social planning and effecting social change by influencing individuals in key institutional positions, has a venerable history in the social sciences. For discussion of the literature and also research on the gatekeeping that occurs in the context of newspaper editorial conferences, see Clayman and Reisner (1998). Much previous research deals with cultural, people-processing, and other public institutions, while, following Schegloff (1986), we use the term in relation to the more private domain of family households. It is "citizens" or household members, rather than the members of some institution or organization, who are keeping the gate.

2. For studies of ordinary telephone call openings, see Schegloff (1979, 1986), and for research on 911 emergency call openings, see Whalen and Zimmerman (1987); Whalen, Zimmerman, and Whalen (1988); and Zimmerman (1992).

3. We use pseudonyms for any personal names in our data.

4. The total number of cases examined was 47 (12 grantings of the request to participate, 35 declinations). This was a "convenience" sample: for the "grantings," Hadley used all transcribed completed interviews from an earlier data collection (described in Maynard and Schaeffer, 1997). For the declinations, Hadley used a subset of the declinations in that earlier data collection as well.

5. Here is an instance in which, rather than remaining silent, the recipient inserts a "yes" (line 5) after this identification (lines 3–4):

LSSC 117

```
 1  FR:  Hello:?
 2       (0.5)
 3  FI:  hh Hi:: (0.2) I'm Karen Smith (.) an' °I'm° (.) >calling
 4       from the University of< Wisconsin? [.hhh  ]
 5  FR:                                     [↑Ye↓s.]
 6  FI:  An:: we're ↑doing a national public opinion study, (0.4)
 7       an' we're try[ing to r-]
 8  FR:               [I'm just ] on my way out the door.
 9  FR:  [Thanks for calling]
10  FI:  [Oh    okay        ] I'm sor[ry          ]
11  FR:                              [((hang up))]
```

Although this "↑ye↓s," which the respondent speaks in a sing-song manner, allows for continuation, subsequent to FI's next bit of talk (lines 6–7), FR declines the request for participation. [Hadley (1999) finds that sing-song voice in a go-ahead signal regularly foreshadows subsequent resistance to progression of the interview.]

6. The pool of grantors is a constant 12, whereas the pool of decliners — initially 35 — decreases as there is movement through the front end of the interview. At each of the eight marked opportunity points and at other TRPs, that is, call recipients drop out of the decliners' pool.

7. The voice of the call recipient is high pitched and could be either a younger male or female child or an older female.

8. Sometimes, instead of going along with the front-end statements and questions, they may produce a query about the interview — see discussion below.

9. Woodbury's (1999) study is very preliminary in that the number of interviews was small (13 interviews), and she examined only momentary "going along with" decisions rather than full-fledged grantings of the request to participate. That is, her data show that optimistic responses to questions about timing are more successful than informational responses in getting the recipient to continue to the next front-end question. The data did not allow for assessing the effects of optimism on granting an interview per se.

10. Goffman (1971:142, fn. 33) argues that "thank you" is a *sign* of gratitude that is used for closing encounters.

11. See excerpt 17 for an instance in which a recipient's "thank you" is followed by an acknowledgment and then a terminal exchange.

12. Such an analysis is consistent with Button's (1987:135) arguments about how participants can *foreshorten* conversational closings.

13. Callbacks are not included in the enumerations in Tables 8.1 and 8.2. This excerpt is used here because it illustrates the way that a participant can move talk away from closing the encounter.

14. In Button's (1987:106–107) data and discussion thereof, it appears that the introduction or reintroduction of arrangements *precedes* the opening of closing sections, whereas in these data, arrangements talk coincides with closing sections.

15. It is in part by reference to the subsequent hangup that the component at line 19 can eventually be heard as having initiated termination rather than as having initiated a separate sequence of appreciation to be followed by a terminal sequence in which FR would have returned MI's line 20 "bye bye."

16. See excerpt 23 and analysis below.

REFERENCES

Atkinson, J. M., and P. Drew. 1979. *Order in Court: The Organisation of Verbal Interaction in Judicial Settings*. London: Macmillan.

Button, G. 1987. "Moving out of Closings." Pp. 101–151 in G. Button and J. R. E. Lee. (Eds.), *Talk and Social Organisation*. Clevedon: Multilingual Matters.

Clayman, S. E., and A. Reisner. 1998. "Gatekeeping in Action: Editorial Conferences and Assessments of Newsworthiness." *American Sociological Review* 63: 178–199.

Davidson, J. 1984. "Subsequent Versions of Invitations, Offers, Requests, and Proposals Dealing with Potential or Actual Rejection." Pp. 102–128 in by J. M. Atkinson and J. Heritage. (Eds.) *Structures of Social Action*. Cambridge: Cambridge University Press.

Davidson, J. A. 1990. "Modifications of Invitations, Offers and Rejections." Pp. 149–180 in G. Psathas. (Ed.), *Interaction Competence*. Washington, DC: University Press of America.

Goffman, E. 1971. *Relations in Public: Microstudies of the Public Order*. New York: Harper and Row.

Groves, R. M., and M. P. Couper. 1992. "Respondent-Interviewer Interactions in Survey Introductions." Paper presented at the *Third International Workshop on Household Survey Nonresponse*, September, Voorburg, the Netherlands.

Hadley, K. 1999. "Managing the Interactional Gate: Going Along with the Request to Participate in a Telephone Survey Interview." Unpublished Master's Paper, Department of Sociology, Indiana University, Bloomington, IN.

Heritage, J. 1984. *Garfinkel and Ethnomethodology*. Cambridge: Polity.

Heritage, J. 1998. "Oh-prefaced Responses to Inquiry." *Language in Society* 27: 291–334.

Jefferson, G., and J. Schenkein. 1978. "Some Sequential Negotiations in Conversation: Unexpanded and Expanded Versions of Projected Action Sequences." Pp. 155–172 in J. Schenkein (Ed.), *Studies in the Organization of Conversational Interaction*. New York: Academic.

Lewin, K. 1947. "Frontiers in Group Dynamics." *Human Relations* 1: 143–53.

Maynard, D. W., and C. L. Marlaire. 1992 "Good Reasons for Bad Testing Performance: The Interactional Substrate of Educational Testing." *Qualitative Sociology* 15: 177–202.

Maynard, D. W., and N. C. Schaeffer. 1997. "Keeping the Gate: Declinations of the Request to Participate in a Telephone Survey Interview." *Sociological Methods and Research* 26: 34–79.

Pomerantz, A. 1984a. "Agreeing and Disagreeing with Assessments: Some Features of Preferred/Dispreferred Turn Shapes." Pp. 57–101 in J. M. Atkinson and J. Heritage. (Eds.), *Structures of Social Action: Studies in Conversation Analysis*. Cambridge: Cambridge University Press.

Pomerantz, A. 1984b. "Pursuing a Response." Pp. 152–64 in J. M. Atkinson and J. Heritage. (Eds.), *Structures of Social Action*. Cambridge: Cambridge University Press.

Sacks, H. 1987. "On the Preferences for Agreement and Contiguity in Sequences in Conversation." Pp. 54–69 in G. Button and J. R. E. Lee (Eds.), *Talk and Social Organisation*. Clevedon, England: Multilingual Matters.

Sacks, H. 1992. *Lectures on Conversation*, Vol. II. Oxford: Blackwell.

Sacks, H., E. A. Schegloff, and G. Jefferson. 1974. "A Simplest Systematics for the Organization of Turn-Taking for Conversation." *Language* 50: 696–735.

Schegloff, E. A. 1972. "Notes on a Conversational Practice: Formulating Place." Pp. 75–119 in D. Sudnow (Ed.), *Studies in Social Interaction*. New York: Free Press.

Schegloff, E. A. 1979. "Identification and Recognition in Telephone Openings." Pp. 23–78 in G. Psathas (Ed.), *Everyday Language*. New York: Erlbaum.

Schegloff, E. A. 1981. "Discourse as an Interactional Achievement: Some Uses of 'Uh huh' and Other Things that Come between Sentences." Pp. 71–93 in D. Tannen (Ed.), *Analyzing Discourse (Georgetown University Roundtable on Languages and Linguistics 1981)*. Washington, DC: Georgetown University Press.

Schegloff, E. A. 1986. "The Routine as Achievement." *Human Studies* 9: 111–151.

Schegloff, E. A. 1988. "On an Actual Virtual Servo-Mechanism for Guessing Bad News: A Single Case Conjecture." *Social Problems* 35: 442–457.

Schegloff, E. A., and H. Sacks. 1973. "Opening Up Closings." *Semiotica* 8: 289–327.

Whalen, M., and D. H. Zimmerman. 1987. "Sequential and Institutional Contexts in Calls for Help." *Social Psychology Quarterly* 50: 172–185.

Whalen, J., D. H. Zimmerman, and M. R. Whalen. 1988. "When Words Fail: A Single Case Analysis." *Social Problems* 35: 335–362.

Woodbury, S. 1999. "Interviewer Strategies for Maximizing Acceptances in the Front End of the Survey Interview." Unpublished Master's thesis, Department of Sociology, Indiana University, Bloomington, IN.

Zimmerman, D. H. 1992. "The Interactional Organization of Calls for Emergency Assistance." Pp. 418–69 in P. Drew and J. Heritage. (Eds.), *Talk at Work: Social Interaction in Institutional Settings*. Cambridge: Cambridge University Press.

CHAPTER 9

Effects of Introductions in Large Scale-Telephone Survey Interviews

Hanneke Houtkoop-Steenstra and Huub van den Bergh

INTRODUCTION

An important issue in telephone survey research is the problem of nonresponse. Since nonrespondents may differ from respondents (Lin and Schaeffer, 1995), a large proportion of nonrespondents limits the generalizability of research results. It is said that response rates in surveys are falling and that this tendency is expected to continue (Groves and Lyberg, 1988). Groves and Lyberg find this especially problematic because "non-response in voluntary surveys is the result of behaviour of persons who are outside the researcher's control" (p. 207).

Of the several types of nonresponse, refusing to participate is the most important one (Bethlehem and Kersten, 1986; Cochran, 1977). Most refusals occur in the introduction of the interview. In a study of telephone interviews mentioned by Oksenberg and Cannell (1988), about 90 percent of the refusals occurred in the introduction and less than 10 percent after the interview had started. This implies that, in order to increase the response rates, special attention has to be paid to the quality of the introduction. Several attempts have been made to improve the quality of the introduction. Reviews of the literature (e.g., Groves, Cialdini, and Couper, 1992; Singer, 1993) show that many studies on response effects focus on information on content and purpose of the interview. Singer and Frankel (1982) studied the effects of four types of introduction, differing in the amount of information that was provided concerning the content and the

Standardization and Tacit Knowledge: Interaction and Practice in the Survey Interview
Edited by Douglas W. Maynard, Hanneke Houtkoop-Steenstra, Nora Cate Schaeffer, and Johannes van der Zouwen
ISBN 0-471-35829-0

purpose of the research. It turned out that an introduction with more information on both content and purpose slightly increased response rates. Oksenberg and Cannell (1988) observed that refusal rates differed significantly among telephone interviewers. To test whether the refusal rates vary with the content of the introduction, they provided a group of interviewers with a standard introduction, based on introductions used by the most successful interviewers. They found that refusal rates were not related to the content of the introductory statements. Singer (1993) concluded that information on content has no perceptible effect on response rates. Dillman, Gallegos, and Frey (1976) manipulated the content of the introductory text by stressing the social significance of the research. They did not find differences in response rate due to their manipulation. It has also been assumed that the assurance of confidentiality might influence response rates. Singer's review of the literature (1993), however, shows that assurance of confidentiality has a positive effect on response rates in a limited number of the studies. The reviews make clear that, in general, these factors only marginally influence response rates; whenever positive effects are reported, they tend to be small. Rather than wondering whether aspects of the content of the introduction influence response rates, we believe that the style of the introduction might be a more relevant factor.

A read-aloud scripted introduction may not be the most effective way to persuade people to participate in an activity with few or no benefits. If an interviewer reads a written text aloud, this fact will be perceived by the recipients (Blaauw, 1995). This will subsequently set the parameters for the rest of the interview: The formal interviewer will read out the questions, and the recipient will answer them in a subservient role.

Moreover, studies in the field of communication accommodation theory (Pittam, 1994) and conversation analysis (Sacks, 1995) suggest that a conversational introduction makes it easier for interviewers to adapt their conversational and speech style to that of the would-be respondent. Such an opening provides better possibilities for flexible interviewer behavior, or *tailoring* (Couper and Groves, this volume, Chapter 7; Groves and Couper, 1992). The interviewer will be perceived as a more likeable individual. This in turn makes it more likely for recipients to comply with the request to participate (Groves, Cialdini, and Couper, 1992; Morton-Williams, 1991, 1993).

We therefore hypothesize that the success of an introduction is related to the style in which it is presented, rather than to its content. We speculate that a standardized scripted introduction will be less successful than a conversational one, in which the interviewer has to construct an introductory text out of a list of content elements.

We believe that a standardized scripted introduction will be less successful than a conversational one, regardless of the number of content elements that the standardized scripted introduction contains. Our second research question therefore concerns the content of the introduction. We hypothesize that a standardized scripted introduction with relatively many content elements will lead to response and refusal rates that are no higher than those brought about by shorter standardized introductions.

METHOD

Materials

The study was carried out in cooperation with a Dutch market research company. Within the daily course of the interviewing we experimented with four different introductions. The company's standardized, scripted introduction of a particular recurrent national survey served as starting point. This text says:

> Good morning, this is INTERVIEWER'S NAME speaking, from COMPANY'S NAME in CITY. We are conducting a nation-wide research about reading magazines, watching television, shopping and some current affairs. Could I please first verify the phone number? Is your number... ? Then, I would like to have an interview with someone from your household about television and magazines and so on. In order to decide whom I should ask the questions, I would like to know how many people live in your household for at least 4 days a week. This, of course, includes yourself. Then, I would like to have an interview with....

Based on this introductory text, we made a conversational introduction (CI). Rather than having a script to read out, the interviewers only had some catchwords on their screens, which referred to those elements that had to be mentioned:

> your name; company's name; research topic; phone number check; recipient selection; "at least 4 days a week in household."

The interviewers were allowed to use the elements in any order, but without leaving out any of them.

Based on the company's standardized scripted introduction, we also constructed a shorter and a longer standardized scripted version. This, then, led to three standardized scripted versions, that we abbreviate MS (medium standard, the company's script), SS (short standard), and LS (long standard).

Whereas information about the topic of the research is provided twice in the medium version, it is given only once in the short standardized scripted version (SS). Furthermore, verification of the telephone number is put earlier in the text in the SS condition (see Table 9.1).

In the long standardized scripted version (LS) we primarily added elements that are theorized or occasionally found to positively influence response rates: information on the length of the interview (*"the interview will not take long"*) and the nature of the questions (*"the questions are quite simple")* and *"For example, we would like to know which television programs you watch, and which ones you don't. And, of course, why you don't watch these"*). Groves, Cialdini, and Couper (1992:477–478) suggest that when the research is done by an institution with authority, this might have a positive effect on the candidate's willingness to participate in an interview. This is the reason that we added a reference to the authority of the research company (*"You may know about us from television"*). It is also pointed out that the recipient's information is important (*"As you too, of course, do your shopping and watch television, we would like to hear your opinion.*

Table 9.1 Content Elements per Introduction

		Standardized Scripted		
	Conversation	Short	Medium	Long
Greeting		*	*	*
Interviewer's name	*	*	*	*
Company's name	*	*	*	*
Company's authority				*
Research topic	*	*	*	*
Reason for calling			*	*
Second mentioning of topic				*
Respondent's importance				*
Example of question				*
Length of interview				*
Confidentiality[a]				*
Respondent selection	*	*	*	*
Check phone number	*	*	*	*
Request to participate		*	*	*
Number of Dutch words	Varies	81	108	164

[a]The Netherlands differ from some other countries, in that there it is not obligatory to mention the length of interview and the respondent's confidentiality.

Your opinion is important because we try to get a clear view of the situation"). The long version also contains the phrase that "*it goes without saying that your answers will be used confidentially*" (see Table 9.1). Unlike in the United States, references to confidentiality and to length of an interview are not obligatory elements in Dutch interviews.

It needs to be pointed out that the decision for the final choice for the content elements was strongly influenced by the company. Since these interviews were also used for the interviewing project, the research company insisted on maintaining the crucial elements of their standardized introduction. Since the company is accountable to its clients for the quality of the interviews, we were restricted in the formulation of the variations of the original introduction. For the same reasons of accountability, these interviewers were subjected to the company's interviewer monitoring. Table 9.1 lists the content elements as they were used in each of the three standardized scripted versions and in the conversational introduction.

Data

In total 132 interviewers called 1831[1] candidate respondents (first contacts only). The interviewers were (systematically) randomly assigned[2] to one of the four conditions: a long standardized introduction (LS), a standardized introduction of medium length (MS), a short standardized introduction (SS), and a conversational introduction (CI).[3] Telephone numbers were randomly generated and assigned to interviewers as soon as a previous call was finished.

Table 9.2 Number of Interviewers, Number of First Contacts, and Number of Respondents' Reactions per Condition

	Number of				
	Interviewers	Respondents	Completes	Appointments	Refusals
Long standard	48	855	183	410	262
Medium standard	30	234	61	89	84
Short standard	34	381	79	172	130
Conversational	20	361	107	189	65
Total	132	1831	430	860	541

To eliminate carryover effects, each interviewer introduced the survey the same way to all recipients he or she contacted. Table 9.2 presents a summary of the numbers of interviewers and first contacts per version. Table 9.2 contains the number of interviewers and first contacts per condition stipulated and the results of these first contacts. We distinguish three types of recipients' reactions. First, the recipient may comply with the request for cooperation, and the interview may successfully be completed. We have labeled this type of reaction as "complete." Second, the recipient may indicate that the present time is unsuitable for holding an interview, and an appointment for a callback at a designated time is made. This reaction is called "appointment." The third type of reaction is a straight refusal for cooperation.

The figures in Table 9.2 seem to indicate a somewhat higher completion and appointment rate and hence a lower refusal rate for the conversational introduction. Unfortunately these differences are hard to interpret due to both the different number of first contacts and the different number of interviewers per condition.

Although interviewers are randomly assigned to conditions and telephone numbers are randomly assigned to interviewers, the number of first contacts differs between interviewers. First, the number of first contacts depends on the length of a call and the length of an introduction, the number of first contacts of an interviewer differs due to the results of the contacts, and the respondents' reaction to an interviewer depends on (voice) characteristics of the interviewer. Furthermore, one interviewer may be more persuasive than another (compare Oksenberg and Cannell, 1988). Therefore, first contacts cannot be considered as independent observations (in a statistical sense). Hence, in effect a two-stage sampling procedure is in operation (Cochran, 1977; Kish, 1965), in which first contacts with the candidate recipients are nested within interviewers.

It is generally known that such a clustering of data cannot be ignored, neither in the analysis of experimental data nor in the analysis of survey data (Fienberg, 1977; Goldstein, 1987; Holt, Scott, and Ewings, 1980). That is, treating data as if obtained under a simple random sampling scheme can lead to serious inferential errors (Cronbach, 1976; Goldstein, 1987). The treatment of data elements as if they were independent observations can only lead to an overoptimistic estimate of the standard errors and is therefore likely to produce type I errors (i.e., false rejection of H_0).

In the present study, the interviewers were randomly assigned to one condition. So all calls of one interviewer belong to the same condition and are nested within the interviewer. Therefore, in the analysis the mean numbers per condition (version), the variance between respondents within interviewers, and the variance between interviewers have to be estimated [see Houtkoop-Steenstra and van den Bergh (2000) for model details]. For each type of reaction of the respondents (completes, appointments, and refusals), the proportion per condition as well as both variance components have to be approximated.

RESULTS

In Table 9.3 the parameter estimates are presented per response variable. We will concentrate first on the estimates of the means. The mean proportion of completes varies from 0.21 to 0.30 in the LS condition and the SS and CI conditions respectively.

Of course, it has to be tested whether these differences are attributable to sample fluctuations. As the overall testing statistic for a simultaneous comparison of all 12 contrasts reached significance ($\chi^2_{12} = 55.52$; $p < 0.001$), we can without (too much) chance capitalization interpret the statistics per comparison. The results of these pairwise comparisons are presented in Table 9.4. Note that according to our hypotheses the differences in responses are pairwise compared for the three standardized introductions, whereas the conversational introduction is compared to all three standardized versions simultaneously.

As can be seen in Table 9.4, the logits of the mean proportions of completes in the LS and MS conditions do not differ ($\chi^2_1 = 2.29$; $p = 0.13$). Only the differences between the three standardized conditions on the one hand and the conversational introduction on the other are significant ($\chi^2_1 = 7.22$; $p < 0.01$). That is, the mean proportion of completes in the three standardized conditions (0.21–0.26) is lower than the proportion in the conversational condition (0.30; see Table 9.3).

Table 9.3 Proportions per Condition per Type of Reaction[a]

	Result of First Contact		
	Complete	Appointment	Refusal
Long standard	0.21	0.46	0.33
Medium standard	0.26	0.37	0.36
Short standard	0.21	0.44	0.34
Conversational introduction	0.30	0.51	0.18

[a]Note that the estimated proportions differ from the proportions to be calculated from Table 9.2, due to the between-interviewer variance.

Table 9.4 Testing Statistic χ^2 for Comparison of Means per Condition[a]

	Comparison			
	Long versus Medium	Long versus Short	Medium versus Short	Conversational versus Standard
Completes	2.29	0.07	2.34	7.22
Appointments	4.56	0.27	2.42	5.89
Refusals	0.56	0.26	0.09	21.04

[a]Note that significant differences are marked, i.e., df = 1; $\chi^2_{1;\ p\ \leq\ 0.05} \geq 3.84$.

The mean proportion of appointments varies from 0.37 in condition MS to 0.51 in the CI (see Table 9.3). For the standardized conditions a difference is shown between the proportions of appointments in the LS condition (0.46) and the MS condition (0.37; see Table 9.4). Also the differences between the three standardized introduction versions and the conversational version proved to be significant (see Table 9.4). That is, more appointments for a callback at a designated time (0.51) were realized in the latter.

It appears from Table 9.3 that the proportion of refusals also varies between conditions. The proportion of refusals varies from 0.34 in condition SS to 0.18 in the conversational condition. Especially the latter stands out with respect to the proportion of refusals. That is, the proportion of refusals is significantly less in the conversational condition than in all other conditions (see Table 9.4).

The differences between interviewers is not something to be neglected. The differences between interviewers is not the same for all three types of responses: The variance between interviewers is rather small for "completes" and substantially larger for both "appointments" and "refusals." In Table 9.5, the 95 percent confidence intervals for all three response variables are approximated.

As can be seen in Table 9.5, the 5 percent of interviewers with the lowest completion rates in the LS condition have a completion rate of 20 percent or less. The upper 5 percent show completion rates of 23 percent or more. In general, the difference between the completion rates of the lowest 5 percent of the interviewers and the upper 5 percent is only 2–3 percent. Although this difference

Table 9.5 Confidence Intervals (95 percent) for Differences between Interviewers per Condition per Type of Reaction[a]

	Result of First Contact		
	Complete	Appointment	Refusal
Long standard	0.20–0.23	0.43–0.50	0.29–0.36
Medium standard	0.25–0.28	0.34–0.41	0.32–0.39
Short standard	0.20–0.22	0.41–0.48	0.31–0.38
Conversational introduction	0.28–0.31	0.48–0.55	0.16–0.21

[a]Note that the mean proportion (see Table 9.3) does not lie in the middle of the confidence intervals because the parameters are estimated in terms of logits instead of the reported proportions.

is rather small, it is not to be neglected as the same interviewer makes many calls. If, for instance, interviewers make ten thousand calls during a month, the difference would amount to (at least) 200–300 completions. A second point worth noting is that the differences between interviewers is not larger in the conversational condition. Hence, we cannot show that more freedom for the interviewers inevitably leads to larger differences. Although this finding should be interpreted with care, as it concerns only well-trained interviewers, it suggests that there is no empirical support for the fear that only part of the interviewers profits from this kind of freedom.

For both the proportion of appointments and the proportion of refusals the differences between interviewers appear to be relatively large. That is, the difference between the bottom 5 percent and top 5 percent of interviewers is (about) 7 percent (or more); the top 5 percent of interviewers make (about) 7 percent more appointments for callbacks than the bottom 5 percent. The bottom 5 percent have about (roughly) 7 percent more refusals than the top 5 percent. Perhaps, needless to say, although these figures look small, they amount to large numbers if many calls are being made.

With respect to the number of appointments made by an interviewer, a distinction can be made between an appointment with the person to be interviewed and an appointment with another member of that household. Therefore, we explored whether the differences in mean proportions of appointments between the four conditions are to be attributed to either these two types of appointments. The results of this exploration are presented in Table 9.6.

As can be seen in Table 9.6, the mean proportion of appointments with the candidate recipient varies from 0.23 to 0.32; especially condition MS stands out in a negative way as the number of this type of appointments is remarkably low in this condition; however, only a difference between MS and CI on the one hand ($\chi_1^2 = 3.97$; $p = 0.047$) and MS and SS on the other hand ($\chi_1^2 = 3.92$; $p = 0.048$) could be shown.

For the proportion of appointments with another member of the household the differences between conditions are less marked. Due to the relatively low numbers of observations (and hence large standard errors) only the difference between SS and CI reached significance ($\chi_1^2 = 4.51$; $p < 0.01$).

Table 9.6 Proportions for Appointments with Respondent or Other Person of Same Household

	Appointment for Callback with	
	Respondent	Another Person
Long standard	0.31(0.17–0.48)	0.15(0.06–0.33)
Medium standard	0.23(0.12–0.39)	0.14(0.06–0.30)
Short standard	0.32(0.18–0.49)	0.13(0.05–0.28)
Conversational	0.32(0.18–0.50)	0.19(0.08–0.38)

Note: Numbers in parentheses are 95 percent confidence intervals.

CONCLUSIONS AND DISCUSSION

An experiment was carried out to test whether standardized scripted and agenda-based introductions in telephone surveys have a different effect on unit response rates. We examined the effects of a conversational, agenda-based introduction (in which only content elements were given) and three standardized introductions for reading out, differing in length, on three types of reactions of the recipients: completing the interview, making an appointment for a callback at a designated time, and refusing participation. Only reactions to first contacts with the recipients were analyzed.

It is shown that, ceteris paribus, the conversational introduction is more likely to evoke cooperative behavior from the recipients than the standardized introductions. That is, the mean proportion of completes in conditions with a standardized introduction is less than the mean proportion of completes in the conversational introduction.

With regard to the mean number of appointments, the conversational introduction leads to more appointments than the short, medium, or long standardized introduction. If a distinction is made between appointments with the candidate recipient or with another member of the household, this effect seems attributable to appointments for a callback at a designated time with the candidate recipient. We have to stress, though, that the data are not very clear in this last respect, due to the limited number of observations and therefore relatively large standard errors. Given these results, the conversational introduction has to stand out with regard to the number of refusals as well (as proportions sum to 1.0); the conversational introduction led to fewer refusals than each of the standardized introductions.

Between the three standardized versions of the introduction hardly any differences showed in either completion, appointment, or refusal rates. According to results in previous studies, no relation between the number of content elements and completion rate could be assessed; addition of content elements that appeared to be successful in some of the previous studies, such as stressing the authority of the research institute and confidentiality, did not increase the number of completed interviews.

In the introduction we suggested that the conversational introduction might be more successful because it provides the interviewers with the possibilities for tailoring their behavior, which may cause them to be perceived as more attractive to talk to. Now that the conversational introduction turns out to be relatively successful, we need to know what exactly characterizes it that may explain its effect on response rates. Since the concept of tailoring is a general one, a comparative analysis of tape recordings of scripted and conversational introductions is needed.

One future research topic based on the work by Oksenberg, Coleman, and Cannell (1986) and Oksenberg and Cannell (1988), showed a relationship between response rates and the prosodic and paralinguistic features of the interviewers' speech in the introduction. It may be that the specific speech characteristics of the two types of introductions correlate with the way the

Dutch judge these characteristics in terms of personality traits (compare Scherer, 1979). This possible alternative explanation calls for a further study comparing conversational and read-out introductions with respect to interactional, (para)linguistic, and prosodic aspects.

In addition, if individuals are addressed by an interviewer who presents himself or herself as a conversational partner, rather than a person who reads out a script, they will perceive the interviewer as a person instead of an "instrument," and they will be more likely to treat him or her accordingly (Dijkstra, 1987; Fowler and Mangione, 1990). Candidate recipients will then feel tempted to respect the face-keeping principle (Goffman, 1955) and accept the request to participate. Candidates who wish not to participate may feel reluctant to perform the dispreferred act of declining the request to participate (Pomerantz, 1984). They may make an appointment to be interviewed some other time and postpone a rejection in this way. This may have happened in our study, since the conversational introduction results in a higher appointment rate compared to the standardized versions. Unfortunately we were not in the position to collect research data for a study into a possible difference between the response rates for the callbacks in the conversational group and in the standardized groups.

When recipients participate to a higher extent only for reasons of face-keeping, this motivation may influence the quality of the answers in a negative way. This may especially be true if the interviewer shifts to a more formal interviewing style once the interview starts. There is little information about reluctant recipients providing data of similar quality compared to recipients who easily accept to be interviewed (Groves and Lyberg, 1988; Lin and Schaeffer, 1995). Further research may tackle questions related to the effects of face-keeping on various aspects of the survey interview. Also a study into differences between conversational and standardized introductions with respect to the types of declinations may be worthwhile (Maynard and Schaeffer, 1997). Candidates who are approached by a "conversational" interviewer may provide reasons for their declination that are easier to counter successfully (cf. Morton-Williams, 1993:94).

An important factor in the success of the conversational introduction may be that the interviewers who formulated their own introduction were better motivated because they especially liked to do this, as several of them told us after the experiment. Better motivated interviewers may be expected to be more successful than less motivated ones. Singer, Frankel, and Glasman (1983) showed that interviewers' expectations as to whether or not a telephone survey is going to be successful has a positive effect on the response rates. We think that the reason for this may be that interviewers who are better motivated may display a speech style that is perceived as more pleasant to listen to. They may put more effort in persuading reluctant candidate recipients.

We believe that genuine differences between the different types of introductions were found. Although the company where this experiment was carried out has switched from a standardized scripted introduction to a conversational introduction, we think that further experiments will be needed.

As our last point we would like to stress the applicability of multilevel models for this type of experiment. One advantage of multilevel models is that one circumvents the stringent assumptions of more traditional statistical models, but this is by no means the only advantage. In our view the main advantage of multilevel models resides in larger possibilities of greater interpretation.

ACKNOWLEDGMENTS

This chapter was previously published (copyright © Sage Publications Ltd.) in the journal *Sociological Methods and Research* 28: 281–300 (2000) and is reprinted according to the publication agreement with authors.

NOTES

1. Due to on-line coding errors 17 first contacts were coded twice. These cases were excluded from the analysis.
2. The (systematic) randomization was based on the last digit of the employee ID number of the interviewer (i.e., digits 0, 1, 2 to LS; 3 and 4 to condition MS; 5, 6, and 7 to condition SS; and finally 8 and 9 to condition CI). Note that the differences in number of interviewers between conditions are attributable to sample fluctuations ($\chi_3^2 = 4.62$).
3. Although all interviewers were subjected to the monitoring routines of the company (i.e., a supervisor listened in at random to calls), it needs to be pointed out that there is no guarantee that all interviewers have read out the texts at all times exactly as they were scripted. *Occasionally* they may have left out parts of the texts or they may have changed the wording (see, e.g. Dijkstra and van der Zouwen, 1987; Houtkoop-Steenstra, 1995; Schaeffer and Maynard, 1996; Suchman and Jordan, 1990). However, if interviewers deviate frequently from the script, they are corrected by the supervisor. Hence, in the standardized conditions there were not many possibilities to tailor the introduction to the recipients.

REFERENCES

Bethlehem, J. G. and H. M. P. Kersten. 1986. *Werken met Non-Response* [Working with Non-Response]. Amsterdam: University of Amsterdam.

Blaauw, E. 1995. *On the Perceptual Classification of Spontaneous and Read Speech*. Utrecht: OTS Publications.

Cochran, W. G. 1977. *Sampling Techniques*. New York: Wiley.

Cronbach, L. J. 1976. "Research in Classrooms and Schools: Formulation of Questions, Designs and Analysis." Occasional paper, Stanford Evaluation Consortium, Stanford.

Dijkstra, W. 1987. "Interviewing Style and Respondent Behavior." *Sociological Methods and Research* 16: 309–334.

Dijkstra, W., and J. van der Zouwen. 1987. "Styles of Interviewing and the Social Context of the Survey Interview." Pp. 200–211 in H. J. Hippler, N. Schwarz, and S. Sudman (Eds.), *Social Information Processing and Survey Methodology*. New York: Springer.

Dillman, D. A., J. Gallegos, and J. H. Frey. 1976. "Reducing Refusal Rates from Telephone Interviews." *Public Opinion Quarterly* 40: 66–78.

Fienberg, S. E. 1977. *The Analysis of Cross-Classified Categorical Data*. Cambridge, MA: MIT Press.

Fowler, F. J., and T. W. Mangione. 1990. *Standardized Survey Interviewing: Minimizing Interviewer-Related Error*. Newbury Park: Sage.

Goffman, E. 1955. "On Face Work: An Analysis of Ritual Elements in Social Interaction." *Psychiatry* 18: 213–231.

Goldstein, H. 1987. *Multilevel Models in Educational and Social Research*. London: Oxford University Press.

Groves, R. M., and M. P. Couper. 1992. "Respondent-Interviewer Interactions in Survey Introductions." Paper presented at the Third International Workshop on Household Survey Nonresponse, Voorburg, The Netherlands.

Groves, R. M., R. B. Cialdini, and M. P. Couper. 1992. "Understanding the Decision to Participate in a Survey." *Public Opinion Quarterly* 56: 475–495.

Groves, R. M., and L. E. Lyberg. 1988. "An Overview of Nonresponse Issues in Telephone Surveys." Pp. 191–213 in R. M. Groves, P. P. Biemer, L. E. Lyberg, J. T. Massey, W. L. Nichols, and J. Waksberg (Eds.), *Telephone Survey Methodology*. New York: Wiley.

Holt, D., A. Scott, and P. Ewings. 1980. "Chi-squared Tests with Survey Data." *Journal of the Royal Statistical Society* (Series A) 143: 303–320.

Houtkoop-Steenstra, H. and H. van den Bergh. 2000. "Effects of Introductions in Large Scale Telephone Survey Interviews." *Sociological Methods and Research* 28: 281–300.

Houtkoop-Steenstra, H. 1995. "Meeting Both Ends. Between Standardization and Recipient Design in Telephone Survey Interviews." Pp. 91–107 in P. ten Have and G. Psathas (Eds.), *Situated Order: Studies in the Social Organization of Talk and Embodied Activities*. Washington, DC: University Press of America.

Kish, L. 1965. *Survey Sampling*. New York: Wiley.

Lin, I-F., and N. C. Schaeffer. 1995. "Using Survey Participants to Estimate the Impact of Nonparticipation." *Public Opinion Quarterly* 59:236–258.

Maynard, D. W., and N. C. Schaeffer. 1997. "Keeping the Gate: Declinations of the Request to Participate in a Telephone Survey Interview." *Sociological Methods and Research* 26: 34–79.

Morton-Williams, J. 1991. "Obtaining Co-operation in Surveys — The Development of a Social Skills Approach to Interviewer Training in Introducing Surveys." Working Paper No. 3, Joint Centre for Survey Methods, London.

Morton-Williams, J. 1993. *Interviewer Approaches*. Cambridge: Cambridge University Press.

Oksenberg, L., and C. Cannell. 1988."Effects of Vocal Characteristics on Nonresponse." Pp. 257–273 in R. M. Groves, P. P. Biemer, L. E. Lyberg, J. T. Massey, W. L. Nichols, and J. Waksberg (Eds.), *Telephone Survey Methodology*. New York: Wiley.

Oksenberg, L., L. Coleman, and C. F. Cannell. 1986. "Interviewers' Voices and Refusal Rates in Telephone Surveys." *Public Opinion Quarterly* 50: 97–111.

Pittam, J. 1994. *Voice in Social Interaction. An Interdisciplinary Approach.* Thousand Oaks, CA: Sage.

Pomerantz, A. M. 1984. "Agreeing and Disagreeing with Assessments: Some Features of Preferred/Dispreferred Turn Shapes." Pp. 57–102 in J. M. Atkinson and J. Heritage (Eds.), *Structures of Social Action.* Cambridge: Cambridge University Press.

Sacks, H. 1995. *Lectures on Conversation* (2 Vol.) Edited by G. Jefferson. Cambridge: Blackwell.

Schaeffer, N. C., and D. W. Maynard. 1996. "From Paradigm to Prototype and Back Again: Interactive Aspects of 'Cognitive Processing' in Standardized Survey Interviews." Pp. 75–88 in N. Schwarz and S. Sudman (Eds.), *Answering Questions: Methodology for Determining Cognitive and Communicative Processes in Survey Research.* San Fransisco: Jossey-Bass.

Scherer, K. R. 1979. "Voice and Speech Correlates of Perceived Social Influence in Simulated Juries." Pp. 88–121 in H. Giles and R. N. St Clair (Eds.), *Language and Social Psychology.* Oxford: Basil Blackwell.

Singer, E. 1993. "Informed Consent and Survey Response: A Summary of the Empirical Literature." *Journal of Official Statistics* 9: 361–375.

Singer, E., and M. R. Frankel. 1982. "Informed Consent Procedures in Telephone Interviews." *American Sociological Review* 47: 416–426.

Singer, E., M. R. Frankel, and M. B. Glasman. 1983. "The Effects of Interviewer Characteristics and Expectations on Response." *Public Opinion Quarterly* 47: 68–83.

Suchman, L., and B. Jordan. 1990. "Interactional Trouble in Face-to-Face Interviews." *Journal of the American Statistical Association* 85: 232–241.

APPENDIX

Medium Standardized Scripted Version

Good morning, this is INTERVIEWER'S NAME speaking, from COMPANY'S NAME in CITY. We are conducting a nationwide research about reading magazines, watching television, shopping, and some current affairs. Could I please first verify the telephone number? Is your number 123456? Then I would like to have an interview with someone from your household about television and magazines and so on. In order to decide whom I should ask the questions, I would like to know how many people live in your household for at least 4 days a week. This, of course, includes yourself. Then, I would like to have an interview with SO AND SO.

Short Standardized Scripted Version

Good morning, this is INTERVIEWER'S NAME speaking, from COMPANY'S NAME in CITY. Did I dial telephone number 123456? Then I would like to have an interview with someone from your household about reading magazines, watching television, shopping, and some current affairs. In order to decide whom I should ask the questions, I would like to know how many people live in your household for at least 4 days a week. This, of course, includes yourself. Then, I would like to have an interview with SO AND SO.

Long Standardized Scripted Version

Good morning, this is INTERVIEWER'S NAME speaking, from COMPANY'S NAME in CITY. You may know about us from television. We are currently conducting a research about reading magazines, watching television, shopping, and some current affairs. As you too, of course, do your shopping or watch television, we like to hear your opinion. Your opinion is important because we would like to get a clear view on the situation. For example, we would like to know which television programs you watch and which you don't. And, of course, why you don't watch these. The questions are quite simple and the interview will not take long. It goes without saying that your answers will be used confidentially. In order to decide whom I should ask the questions, I would like to know how many people live in your household for at least 4 days a week. This, of course, includes yourself. And did I dial telephone number 123456? Then, I would like to have an interview with SO AND SO.

CHAPTER 10

Refusal Conversion and Tailoring

Douglas W. Maynard and Nora Cate Schaeffer

Survey research centers distinguish between the substantive portion of the interview and what is called the *front end*, where interviewers call potential respondents, introduce themselves, select a member of the household to question, and request participation from that member. Potential respondents decline these requests at notoriously high rates. The higher the refusal rate, of course, the more costly it is to administer the survey and the more threat there is of biasing the results. That is, the accurate measurement of attitudes and opinions itself may be compromised when success rates are low. The chapters in this volume not withstanding, there is little interaction-based research on the front end.[1]

In this chapter, we examine an episode of refusal conversion. This is a single-case analysis, and the case comes from videotaping interviewers at the Survey Research Center of Indiana University when a study concerning the role of television in two-party adult relationships (how much TV they watch, whether and what they watch separately or together, and so on) was underway. As in other studies, interviewers sometimes met with declinations of their initial requests to potential respondents. With some of these declinations, interviewers arranged call backs, but others were refusals. After an initial refusal, rather than dispense with the household and the respondent, the survey center may retain the phone number for calling back at a later time, when an interviewer will attempt to "convert" the former refusal into a complete interview. Here, we examine the second in a series of three telephone calls to a single household. In the first one, the male recipient firmly refused to participate in the study. In the second call (our focus), this same recipient agrees to a later callback. In the third call, his wife answered and, after conveying the request to her husband, relayed another refusal from the husband. However, if the initially selected respondent refused participation, the design of this study permitted asking the wife or partner. In this instance, the wife, after a

Standardization and Tacit Knowledge: Interaction and Practice in the Survey Interview
Edited by Douglas W. Maynard, Hanneke Houtkoop-Steenstra, Nora Cate Schaeffer, and Johannes van der Zouwen
ISBN 0-471-35829-0

persuasive appeal from the interviewer and subsequent to the husband's refusal, agreed to the interview. It was completed and the overall refusal conversion effort succeeded.

The question is, how, especially in the face of a firm previous refusal from the husband or male recipient in this household, was the interviewer able to reengage him in a subsequent call and obtain permission for yet another, later, call — during which the wife agreed to participate in the interview? What interactive skills were involved in this conversion effort?

TAILORING

Survey protocols for soliciting participation, like other aspects of interviewers' standardized instructions, are often not helpful when interviewers meet the individuality and particularity of respondents whom they attempt to recruit. Survey methodologists recognize this problem of particularity in suggesting that interviewers can *tailor* instructions for recruitment to specific concerns and objections that potential respondents may raise. Successful interviewers, Groves, Cialdini, and Couper (1992:487) suggest, engage in a continuous search for cues about a householder who answers the door or telephone. They should be able to "size up a situation quickly," and apply the appropriate persuasive messages (Couper and Groves, 1992:267). Experienced interviewers themselves are aware of their tailoring:

> I give the introduction and listen to what they say. I then respond to them on an individual basis, according to their response. Almost all responses are a little different, and you need an ability to intuitively understand what they are saying. Sometimes, the words they are saying are not the true reasons they are refusing. I attempt to discern what the situation is and respond to it.
>
> (Couper and Groves, this volume, Chapter 7)

Tailoring is facilitated at the front end of the interview because interviewers are not required to adhere strictly to a standard introduction. However, with computer-aided telephone instruction (CATI), they do have routines or devices from two sources for fashioning the opening request for participation. One is the computer screen and a scripted introduction it provides. Another is a document that interviewers tape to their interview carrels that has preformulated answers to expected questions from householders. That is, survey research personnel, including principal investigators (researchers), those who design the instruments, and those who administer a survey center, anticipate *queries* about the purpose of the call, its funding, and who is responsible, plus *declarations* of disinterest and not having enough time.

At the center we studied, possible answers to these queries and declarations are on a sheet called "What the Respondent Might Like To Know" (WRMLTK, referred to as the "rummeltick"). While phrases on the WRMLTK are meant to handle concerns that householders raise, how, when, and in what manner call recipients raise them are contingencies that the WRMLTK as another docile set of instructions does not itself address. Interviewers as employees of the center (rather than researchers, survey designers, or administrators) must handle such contingencies continuously in real time. They often use the messages on the WRMLTK in a mechanical fashion that cannot be regarded as tailoring as such. Consequently, our analysis explores the relation between various computer- and WRMLTK-scripted devices and the actualities of their implementation that are particular to the household, the respondent, and the actual course of interaction between interviewer and call recipient.

PREPARING FOR THE CALL

In our case study, the second call requesting participation is the pivot between the initial very firm refusal and the later callback that ended in a complete interview. Our focus is on how the interviewer displays in his talk the unique positioning of the phone call (calling back a firm refusal) according to unscripted devices that depart from a standard opening script, and how he engages a rhetoric of recruitment to persuade the reluctant call recipient to participate. When the interviewer makes the second phone call, he is sitting at his carrel in the Center for Survey Research (CSR), which has a computer, monitor, and keyboard (see Figure 10.1).

The computer is his primary source of information about the interview he will attempt to make and the household he needs to call and is also his data entry point. In addition to videotaping the interviewer from a camera above the carrel, we registered on another camera, and by way of a digital-to-analog converter, successive displays on the computer screen. The interviewer is Bob Roth (pseudonym), who is a senior interviewer at the center. *Refusal conversion* is a task that is assigned to such experienced employees, and on the night when we taped the call in question, he was provided with a number of households for attempted conversion. At the center, CATI involves use of a software

Figure 10.1

program called CASES; as interviewers make a call and (if possible) go through an interview, they also page through a number of screens that have instructions and questions on them. With this particular call, as he sits at his workstation and readies himself, Bob first views a computer screen that provides him with the case number assigned to the household, the current time of day, the number of previous contacts (one) and what happened (respondent refused). The second screen, after he enters his interviewer number, verifies his identity (the screen asks, "You are Bob Roth, right?").

By entering "1" ("yes") on his keyboard, and hitting the return key, Bob next arrives at screen 3 (Figure 10.2). This screen has the "caseid" or case identification number (4673v), which we have fictionalized, the time of day in the right-hand corner, repeated as "CSR Time" below, and the "Household Time" (were the household in a different time zone than the CSR, this number would reflect that). The "Call Grid" below the time listings allows for coded entries for up to 20 previous contacts with the household. With this case, there has been only one, and its date ("320" or March 20) and time ("1904" or 7:04 p.m.) are noted. The ID is a number (we have substituted "xy") that identifies who the previous interviewer was, and the outcome is selected from a table of codes that describe such things as having reached a business, a recorder, and household who asked to be called back, and so on. The code here ("refR") indicates that the previous call was a refusal done by the selected respondent.

Bob immediately moved to screen 4 (Figure 10.3), which has the case notes — a description of what happened in the first call — written by the previous interviewer ("dhoffer," the computer username for a survey center employee we

COMPUTER SCREEN 3

```
CASEID:              4673v                                        [18:28]

<            Call  Grid
             CSR Time:  1828  EST
             Household Time:  1828  EST

---------------------------------------------------------------------
| #| Date Time    ID   Outcome         | #| Date    Time    ID   Outcome
---------------------------------------------------------------------
| 1| 320   1904   xy   refR            |11|
| 2|                                   |12|
| 3|                                   |13|
| 4|                                   |14|
| 5|                                   |15|
| 6|                                   |16|
| 7|                                   |17|
| 8|                                   |18|
| 9|                                   |19|
|10|                                   |20|
---------------------------------------------------------------------

                 <1> PROCEED
                 <0> Mistake - CODE ONLY WITH SUPER/SI APPROVAL!
                 ====>1
```

Figure 10.2

COMPUTER SCREEN 4

```
CASEID:            4673v                                            [18:28]

|10|                                 |15|
|11|                                 |16|
|12|                                 |17|
|13|                                 |18|
|14|                                 |19|
|15|                                 |20|
----------------------------------------------------------------------

                 <1> PROCEED
                 <0> Mistake - CODE ONLY WITH SUPER/SI APPROVAL!
                 ====>1

3/20/97 tv 1904 refr dhoffer
fi brought male r to phone.  at "30 min" r said not interested b/c
he had to watch tv.  i asked if his wife would be willing to do the
survey and r said, no she's watching tv too.  tried to convert
saying we could cb at his convenience.  r ref'd.  i said it was
an import survey and most people enjoy it.  i said our funds limited
so import to speak with people we reach.  r said last year they had
gotten a call from new york and a week later they started getting th
things in the mail.  i assured r we wern't trying to sell anything.
r asked how would he know that.  i said i would be happy to give
him our # and he could speak with a supervisor but r again ref'd
and hu. ////1
```

Figure 10.3

designate with the pseudonym Diane Hoffer). Bob can be seen reading these case notes for 18 seconds and thereby obtaining a sense of history for the telephone call he is about to make. We refer to this as *preparing* for the call.

In the case notes, observe that Diane uses a number of abbreviations, including "fi" for female informant, "r" for respondent, "b/c" for because, "cb" for call back, "ref'd" for refused, "import" for important, "hu" for hung up, and so on. Although several of these are transparent in meaning, they also display the use of jargon and one way in which a community of understandings for mediated written communications among staff has developed in this center. Also of note is how firm, according to Diane's case notes, the refusal was. Often respondents will claim that they are presently "too busy," and agree to a later call back. Or they will state vaguely that they are "not interested," and the call will close (Maynard and Schaeffer, 1997). Here the respondent is reported to have refused four different times, and given reasons for each refusal. The first refusal occurred as he was being informed that the interview would take "30 min," when he said he was "not interested" because he "had to watch tv." Next, Diane asked if his wife would do the survey and the respondent refused that because she was "watching tv too." Third, Diane reports trying "to convert by saying we could cb [call back] at his convenience," after which the respondent is said again to have "ref'd [refused]." Finally, after Diane quotes herself as saying that it was an important survey, that "most people enjoy it," and that funds were limited so that it was important "to speak with people we reach," the respondent related having been called "from New York" and later "getting things in the mail." Even though Diane suggested that they "weren't trying to sell anything" and told him how he could verify this, the respondent still "refused and hung up." This, then, was the fourth rejection in this call of the interviewer's solicitations.

THE TELEPHONE CALL: REVIEWING THE SCRIPT

Recall that Bob reads these case notes for 18 seconds in a preparatory way. After this, Bob paged through two more screens (5 and 6), which involved setting the *reentry point*, or place at which the interview should begin if the respondent were to participate. Next, he arrived at screen 7 (Figure 10.4), which gives him the phone number to dial and a billing number to use for calls that are long distance.

COMPUTER SCREEN 7

```
CASEID:            4673v                                        [18:29]

4673v<             DIAL THIS NUMBER:

            Phone:  8+1+ xxx yyy zzzz   Billing #: 12345678

            R Name:         Current Household Time: 1828 EST

            Address:     Some Where, IN

            <1>  Someone answers

            <2>    No answer
            <3>    Busy
            <4>    Temporarily not in service, temp. line trouble
            <5>    Line problem - computer line, fast busy, no ring,
                   circuit problem
            <6>    Not in service
            <?>    Mechanical answering device
            ===>1
```

Figure 10.4

Bob now dials the number, gets his recipient on the line, and starts talking to him before he pages to the next screen. Here is a transcript of the conversation (MR refers to the male respondent, MI to the male interviewer):

(1) [CSR Refusal Conversion]

```
        ((Ring))
        (2.6) ((Another interviewer audible in background))
   MR:  He:llo?
   MI:  .hhhh Hi my name's Bob [Roth I'm calling from Indiana
                               [{hits ''1'' key = ''Someone answers''}
        Uni[versity in Bloomington? .hhhh >an'' we're doing uh<
           [{hits ''return'' and gets screen 8}
        research project with: Indiana residents fer their
        o↑pinions about tee vee in their lives .hhhh we'd like
        a chance tuh speak with any adult in yur househo:ld who
```

is ↓currently married (.) divorced >or living with a<
partner:.
(0.4)
MR: >I'm an adult?< (0.1) but I'm eating dinner right no:w?
[and I'd] rather not be inner-
MI: [Okay.]
(0.2)
MI: .hhh[h
MR: [(It's-) I understand this is gunna take a half
ho[ur:?
MI: [Yeah. well? >uh we c'n call yuh back ↑la↓ter:< at
yur convenience:. ·hhh A:n >I'd also like to stress that
we don't haftuh< do the whole interview at once, we c'n
break it up intuh segments .hhhh an' just do a little bit
at a ↑ti:me if that's more convenient for you .hhhh Uh:
it's very important we >get a chance speak with
you=why'on'< (0.2) >why don' I try givin' you a call
ba:ck in a little while:.<
(0.3)
MR: °All right.°
(0.2)
MI: Thank you very much sir.
(0.2)
MI: Buh bye.
(0.3)
((click))

After Bob gets an answer at line 3, he starts an introduction (line 4) whose beginning precedes his viewing of the screen that has an introductory script. The interviewer has, of course, performed such introductions frequently before, so that he knows the opening routine without necessarily having to read it. In the middle of giving his name (lines 4, 5), however, he hits the keyboard code (the numeral 1, per screen 7, Figure 10.4) for having gotten an answer to his call. And next, in the midst of identifying himself as being "from Indiana University" (transcript lines 6, 7), Bob strikes the return key, which brings up the introductory screen 8 (Figure 10.5) and provides a script for him so that he can continue his introduction by reading from that screen.

Bob has seen the script on screen 8 many times and does not read the screen exactly or otherwise enact the scripted introduction.[2] We argue that he simply *reviews the script*, which is a resource and not something to be enacted mechanically. Differences between the script and what Bob actually says to his call recipient, rather than being accidental or haphazard, are doing work in relation to *just this* respondent and household on behalf of the attempt at converting a refusal into a completed interview. In other words, the script is an embedded feature in interactional work that encompasses both Bob's efforts at garnering participation and the respondent's reactions to those efforts. Bob fashions his efforts with *devices* of talk and social interaction that are useable again and again within the real time of soliciting participation. The discriminating deployment of these devices exhibits his particularized knowledge of the respondent and the interaction so far.

COMPUTER SCREEN 8

```
CASEID:           4673v                                      [18:29]


4673v<                 Hello, I'm Bob Roth calling from Indiana University
            in Bloomington.  We're speaking with Indiana residents
            for their opinions about TV in their lives.

            May I please speak with any adult in your household who is
            currently married, divorced, or or living with a partner?

            (Phone: xxx yyy zzzz  Address: Some Where, IN)

            <1> Proceed (i.e. R is I, or I brings R to phone)

            <x> Call back
            <y> Refusal
            <z> Problem (Incapable, language, away, etc.)

            ===>x
```

Figure 10.5

TAILORING THE SCRIPT

Personalizing the Call

One difference from the script is that instead of saying "Hello, I'm Bob Roth calling from Indiana University," our interviewer says, "Hi my name's Bob Roth, I'm calling from Indiana University." The "Hi" form renders the greeting as more informal than "Hello," while "my name's Bob Roth, I'm calling from..." appears to parse the opening into two parts—the interviewer's introduction of himself, followed by a naming of the institution under whose auspices he is calling. The script, by contrast, blends the introduction and institution naming into a more seamless opening that rather suppresses the identity of the individual and asks for recognition of the institution (Maynard and Schaeffer, 1997:47). In short, these minute alterations of the script in the initial microseconds of Bob Roth's first utterance may exhibit the work of *personalizing* his attempt at refusal conversion.

Claiming Academic Auspices

Another difference from the script occurs with the second sentence on screen 8 (Figure 10.5). Rather than "We're speaking with Indiana residents for their opinions about TV in their lives," Bob says, at lines 6 and 8 on transcript 1, "we're doing a research project with Indiana Residents for their opinions about tv in their lives." The script suggests that the interviewer only identify himself as being from Indiana University, and in most of our other recorded introductions interviewers follow the script. However, in his actual introduction, Bob invokes "research project" and thereby can be more clearly offering an identity of the survey call as being under academic sponsorship, exhibiting his knowledge that during the earlier attempt, the respondent complained about getting a call from New York and a week later "getting things in the mail." Just as in the earlier call the interviewer had "assured" the respondent that he wasn't trying to sell anything, now, in the present call, interjecting that this is a "research project," the interviewer may be proposing to counter as early as possible any inference by the recipient that this solicitation is a marketing ploy.

Soliciting a Favor

Still another difference from the script is that instead of saying, "May I please speak with an adult in your household who is currently married, divorced, or living with a partner" (screen 8, Figure 10.5), the interviewer says "We'd like a chance tuh speak with any adult in your household..." (lines 9–12, transcript 1). In most "fresh" or initial calls, interviewers use the scripted "May I please..." format for requesting the interview, although there are series of screens and other questions that are invoked before this request takes place.[3] The crafting of his opening talk, and its departure from the more conventional script, can be working interactionally in at least two ways. First, "May I please..." occasions the relevance of a straightforward yes–no answer as a reply[4] to the request for an interview, while "we'd like a chance..." implicates different answering options, in particular the granting or withholding of an opportunity.[5] More expressly than the "May I please..." introduction, "we'd like a chance..." is a device that thereby may invoke a sense of being *solicited for a favor* rather than simply being *asked to comply* with a request. The "we'd like a chance" device, accordingly, can more strongly compel a positive response.

Requesting Ambiguously

A second way this device may work has to do with the position of this phone call as a second approach to the household. At no point has this interviewer provided an official recognition of this, although such recognition is regularly displayed in calls where interviewers and recipients have agreed to a return call:

(2) [TD30:2399:025:JL]

```
MR:   He↓llo.
      (0.2)
MI:   .hhh Hi:. (0.1) My name's Bo:b Ro:↓th I'm call↓ing
      ba:ck from India:na University. .hhh an we're >speaking
      with< adu:lts who are currently ma:rr↓ied (.) divo:rced
      or living with a p↑a:rtner. .hhh Doing some academic
      res↓earch to fi:nd ou:t the role that tee: vee: plays in
      peoples li:ves. .hhhh  [and we'd-       ]
MR:                          [(She's not here)] the one you
```

```
          wa:nta talk to is my w↓i:fe?=
    MI:   =Ok[ay    ]
    MR:      ['n sh]e's not here.
    MI:   All ri:ght...
```

In a previous call that same evening, the female respondent in this case had told the interviewer that she was leaving "right now" and asked for a callback on the next night. The interviewer told her that this was the last night for the study and asked to return the call later that evening, to which the respondent agreed. As excerpt 2 shows, when Bob performs the callback, he reaches the woman's husband, and now announces that he is "calling back." Unfortunately, the wife has not returned home.[6] [As in excerpt 1, Bob also says that the call represents "some academic research" (lines 6–7), which suggests that this is an unscripted but routine form that interviewers may invoke in a variety of circumstances.]

To return to excerpt 1 and our focal episode, Bob is in a very different position. Far from arranging a callback, the previous call involved a firm refusal. In this circumstance, it may be that calling back and officially pronouncing oneself to be calling back is something to be avoided, for respondents can take umbrage if, after having refused an initial request, they get such a return call:

(3) [td16:2268:011:AK] Revised

```
          ((ringing))
    FR:   °hell:o?°
    MI:   .hh H↑i: my name's Bo::b and I'm ca:ll:ing from Indiana
          University in (.) Bloo:mington (.) .hhh we're spe[ak-=
    FR:                                                    [You
          called me once before an' I told ju I wasn't interested
          in this.
    MI:   >Well we'd like a chance to speak with you< u::m or
          >someone in your household.<
          (0.2)
    MI:   hell:o:?
          (1.0)
    FR:   ((hangs up))
```

In this call, Bob (the same interviewer as in our other excerpts) does not get to make a full-fledged request because the woman he has phoned, after he identifies himself at lines 3–4, displays an understanding of who he is and what he is calling about and refuses him by claiming to have told them that she "wasn't interested" (line 6). The call recipient's turn, in announcing that "you" had called before (the previous interviewer was female, so the "you" here appears to refer to "Indiana University" and to be employed in a categorical rather than personal sense) and reporting her own previous response, also exhibits a complaining quality. The utterance can be implying a failure: that the Survey Research Center did not heed her "not interested" response to the previous solicitation, which they could have done by ceasing to recruit her.[7] It is notable that at line 4 with "we're speak-," the interviewer next appears to start a scripted introduction, rather than with the unscripted "we're doing a research project. . . ," which he employed in excerpt 1, or with "doing some academic research," as in excerpt 2.

Another feature of excerpt 3 is that, after the respondent has reformulated that she is not interested, Bob, at line 8, produces the "we'd like a chance to speak" device, after "well" (which regularly precedes utterances that disaffiliate from a previous one). By referring to this as a device, we mean it is a form that an interviewer uses across contexts, again and again. In excerpt 3, the device is used to plead with a respondent who has just firmly said that she is not interested and complained about having been approached again. As Sacks (1992:637-638) has observed about complaints, they set up a sequence in which some explanatory or other response is required.[8] "We'd like a chance to speak... ," in offering to account for the callback in a kind of minimal way,[9] is such a response, and now portrays the call as an appeal, not as a first request but as a second one that is justified by the need it formulates. This is further evidence that the device is being used to ask for a favor from a resistant recipient. Similarly, in excerpt 1, instead of requesting in the first place, the "we'd like a chance to speak" device can be appealing a request that has been turned down (but at a much earlier time). However, in the context of this call—our focal excerpt 1—it is working to avert a complaining announcement, such as "you called me once before." That is, as an account, it may propose to preempt a complaint to which it might otherwise have to respond.

However, the "we'd like a chance" device may also stand as a request in its own right, having its own integrity, for example, when there is a delay in the contact between a respondent and an interviewing organization. The call in excerpt 1 is taking place 5 days after the original request was refused. Accordingly, if the request is potentially hearable as *either* a second or a first request, the device may be *purposely ambiguous*, and be precisely fitted to the circumstance of calling back a firm refusal, where the interviewer may want to avoid announcing himself as calling back and yet potentially provide for the fact that he is doing so. As a device, therefore, it is cloaked with significations of the real-time field in which it operates.

Our interviewer appears to provide for the ambiguity—whether this is a request in its own right or a pleading appeal—to be resolved according to how his call recipient elects to respond. Initially, the man appears to treat the request in its own right, announcing himself at line 14 to be an adult, thereby responding to the "speak with any adult" formulation. Then he fashions a refusal,[10] using a "bad timing" claim. He says he's eating dinner (line 14), and by giving this excuse, still appears to treat the request as having its own integrity. Bob Roth receives this with an "okay" at line 16, while MR continues at line 15 in what appears to be a protest against being interrupted. But he stops in the midst of that utterance and in the middle of the projected word "interrupted" and restarts his turn at line 19, saying he understands "this is gunna take a half hour?" This is evidence that he now discerns the interviewer's request to be related to the earlier call. In the midst of his bad-timing refusal, that is, he appears to revise his understanding of this request, displaying a recognition of its relation to the earlier call and an orientation to the appeal embedded in the interviewer's "we'd like a chance" request.

COMPLETING THE CALL

As the respondent announces his recollection that the interview would take half an hour (line 19), and just at the point where the half hour is being articulated, in overlap with the latter part of the word "hour," Bob intersects his recipient's turn to acknowledge the time it will take ("Yeah" at line 20). So the case notes on screen 4 may have been informative in a way that partially motivates the interviewer to grab a turn of talk at just this point in the respondent's refusal, in an attempt to answer the respondent's objection immediately and before any other move toward *closing* and *terminating* the call (Maynard and Schaeffer, this volume, Chapter 8).[11] The acknowledgment "yeah" prefaces a long turn of talk containing a set of persuasive bids to garner the respondent's cooperation.

Persuading Verbally

At line 21, Bob starts a response that interviewers regularly use to announce the bad timing of a solicitation, *offering* a later callback at the respondent's convenience, which can address the recipient's concern about the length of the interview. At line 26, he produces another version of the "we'd like a chance" request, this one suggesting that it is important that they get this chance with the respondent. Recall that the previous interviewer had also mentioned the importance of the interview. The interviewer ends his turn in line 28 with an escalated version of the offer to call back with which he started his long turn of talk. That is, whereas at line 21, the interviewer says "we can call yuh back at your convenience," here he says, "why don't I try giving you a call back in a little while," thus *proposing* more specifically who would make the call and when.

The recipient's "all right" at line 30 appears resistive: it is delayed, and very muted. Nevertheless, it is enough to be heard as granting the offer/proposal and to be counted as a callback, which Bob Roth codes on screen 8 (Figure 10.5; see the "x" next to the arrow at the bottom of the screen). So the interviewer has been successful in this call; the household has been upgraded from a refusal to a callback.[12]

Performing the Rhetoric

Parts of Bob's talk are done in a kind of rush through or with fast-paced speech, exhibiting his manipulation of the prosody of his talk. Additionally, a topic we can only partially address is how his persuasive work is embodied. In systematic ways, he engages movements of his torso, head, and arms in conjunction with the talk and its pacing, tone, and other prosodic features to help render its rhetorical features. And although the respondent cannot see these gestures, they may be available at least in the prosody of the talk that he hears.

From the videotape of this telephone call, we have selected 12 frames because they help highlight the systematics of Bob's movement. Roughly the episode can be divided into two parts. The first part (frames 1–6; Figure 10.6) consists of the first six frames or so, during which Bob positions himself before the computer

(1) Line 1

((Ring))

(2) Line 4

MI: Hi my name's Bob Roth I'm
call~ing from Indiana
University

(3) Line 5

MI: ...we're doing a res~earch
project with Indiana
residents

(4) Line 6

MI: ...fer their o↑pinions
about tee vee in their
li:~ves

(5) Line 8

MI: ...currently married (.)
divorced >or liv~ing with a
partner:.

(6) Line 11

MR: I'm eating dinner right
no:w?~[and I'd] rather not=
MI: ~[Okay.]
MR: =be inner–

Figure 10.6 Tilde (~) in utterance indicates where video frame occurs in the speech stream.

in a way that allows him to gesture as he answers and deals rhetorically with the respondent's query about how long the interview will take. It is as if, having read the case notes, he is priming himself to deal with a resistive call recipient not only with his mind but with his body. The second part of the episode (frames 7–12; Figure 10.7) involves Bob's persuasive talk at lines 21–28. His head and arm movements during this persuasive talk are not haphazard occurrences but are systematic in relation to that talk (Erickson, 1982). In fact, the gesturing that occurs is *batonic*; particular gestures co-occur with points of emphasis or other prosodic aspects of the talk (Ekman and Friesen, 1969; Feyereisen, Van de Wiele, and Dubois, 1988; Kendon, 1994; Schegloff, 1984). But let us begin with frames 1–6 and how Bob positions his body during the opening request.

1. Frame 1 (Figure 10.6) shows how, after dialing the phone number, Bob positions himself in a kind of "rest" position as the phone is ringing. He is within reach of the keyboard, and with his hands in lap, he can initiate a variety of movements.
2. In frame 2, which occurs just after Bob has responded to his recipient's "hello?" by identifying himself, he is ready to hit the return button on the keyboard with his right hand, while his left hand is moving toward the phone to adjust the volume.
3. Then, after adjusting the volume and hitting the return key, and while saying "we're doing a research project," Bob moves his torso back, not quite to where it was before but in a slightly forward position, and his hands come back to his lap, as we see in frame 3.
4. Now while he is in this position, Bob starts rubbing his legs a little, and next, as he continues with the script about "tv in their lives," he adjusts himself in his chair, as frame 4 shows.
5. At this point, he pops himself off of the chair a few inches and then back on. Next, he swings himself to the right and back to the center again while rubbing the tops of his legs and saying, "we'd like a chance to speak with any adult who is currently married divorced or living with a partner." See frame 5. While doing this lateral movement, he keeps his head turned toward the computer screen.
6. Then, as frame 6 shows, during the brief silence after this request, and while the respondent is saying that he is "an adult," Bob once again moves his hands up to the table, with his right hand resting to the right of the keyboard, and his left hand momentarily grabbing the WRMLTK, which is the source of prepared answers to concerns that a respondent might raise.

Summary of Frames 1–6 (Figure 10.6)

These frames correspond to Bob's making the request to participate. Bob appears in a posture of readiness, insofar as his movements represent (metaphorically) a way of preparing his body for the response of someone who can be anticipated to resist the resolicitation for interview participation. As we move to frame 7 and Figure 10.7, the respondent is now talking in response to Bob's request.

(7) Line 12

MR: ...I'd rather not be inner--~
MI: ˙hhh[h]
MR: [It's-)] I understand...

(8) Line 17

MR: ...this is gunna take a half hour[:?
MI: [Ye~ah, well? uh we c'n call yuh back...

(9) Line 19

MI: a:n I'd <u>a</u>~lso like to stress...

(10) Line 19

MI: ...that we <u>don</u>~'t <u>haf</u>tuh do the whole interview at <u>on</u>ce.

(11) Line 20

MI: We c'n <u>brea</u>k it up intuh segments ˙hh~hh <u>a</u>n' just do a little bit ...

(12) Line 28

MI: Thank you very much~ sir

Figure 10.7 Tilde (~) in utterance indicates where video frame occurs in the speech stream.

7. Frame 7 is at the point where Bob, with his "okay" token, acknowledges the busy-right-now claim and begins self-grooming gestures. As Bob says "okay," and as the respondent is saying "I'd rather not," he lifts his left hand to his nose; then he moves his hand and grabs the left joint of his glasses and pulls them up.
8. Next, during the respondent's "take a half hour," Bob moves his left hand from his glasses hinge to his right brow, as seen in frame 8, so that from the camera perspective his face is shielded. While this can be another grooming gesture, it emerges as something different from that. From this point, Bob begins literally to *launch* his verbally rhetorical bid to elicit the respondent's participation, and he also launches gestures that coincide with the stress that he puts on key parts of his utterances.
9. As he starts to say "I'd also like to stress," his left hand moves out from his brow in a kind of hat tip gesture. The height or acme of that gesture is exactly at the point where he says "also" with stress on that word, as shown in frame 9.
10. Immediately, Bob lowers his left hand to the keyboard so that both hands are at rest position on the desk on either side of the keyboard momentarily. From there, both hands rather "bounce" up and then back down in shortened movements in correspondence with the next verbal stress on the "don't" of "don't haftuh do the whole interview at once." This is at frame 10. Bob's head also dips to the right as he does this. After this, his hands return to the rest position on the desk and his head moves back to center position.
11. Bob has a brief little gesture with the fingers of his left hand being extended upward when he says "We can break it up intuh segments," but then his hands are at rest position on either side except for two times when both hands flash up to a kind of *cocked* position. The reason for this term is that the hands go up during Bob's taking a breath before "an' just do a little bit at a time" and before "Uh: it's very important," so that the hands are then coming down during the utterance and correspond to the slight stress that occurs on the initial units in those utterances. One of these instances is at frame 11. Bob also does this hand flash during the "WHY don't-" that precedes "why don' I try givin' you a call ba:ck in a little while:" (lines 25–26 on the transcript).
12. Then Bob's hands are at rest position until at "Thank you very much sir" on frame 12, he reaches for the phone and hangs up after the respondent's "bye bye."

Summary of Frames 7–12 (Figure 10.7)

These frames are from the segment of the interaction where Bob is listening to his respondent's reply to the request to participate and where Bob makes a fast-paced persuasive effort to gain cooperation. The grooming gestures to his nose and to adjust his glasses and to hold his brow represent a kind of transition between frames 1–6 and 7–12 at the point where the respondent replies to Bob's request for participation. Grooming, however, emerges as the incipiency of batonic gestures that occur during Bob's next turn of talk. As frames 8–12 show, after getting resistance from his recipient, and when he verbally counters the respondent's bad timing declination and possible complaint about the half hour, he gestures in ways that seem to help with the emphasis and therefore the expressiveness with which he produces the rhetorical appeal to the respondent. Although the respondent has no visual access to these batonic gestures, they may help infuse his talk with its hearable emphasis.[13] In a sense, Bob is *performing the rhetoric* that is his persuasive talk.[14]

CONCLUSION

The situated and particular work whereby interviewers obtain respondents is not an incidental part of the survey enterprise. It is this work that enables obtaining a legitimate and accountable sample of cases for purposes of large-scale analysis in survey research. The numbers of those who accept or decline is a product, at least in part, of the effectiveness of interviewers in soliciting participation both at the initial call stage and in the attempt at refusal conversion. Moreover, the manner of solicitation can affect the type and quality of responses to interview questions that occur within successfully consummated interviews. That is, the tacit practices and personal or commonsense knowledge deployed to obtain completed interviews not only make it possible to assemble a sample of respondents; insofar as they intertwine with what survey methodologists refer to as nonresponse and measurement error, these practices and forms of knowledge affect the very content of those completed interviews and, accordingly, the ultimate quality of data and findings that interviews generate. For these reasons alone, it seems important to study these practices and forms of knowledge.[15]

The practices and forms of knowledge investigated in this chapter have to do with an interviewer's successful effort at refusal conversion, and another way of describing such practices and knowledge is in terms of the skill and competence this interviewer exhibits. In a sense, Bob Roth is a master at tailoring his callback to the particularities of the case, the household, and the interactive situation in which his callback is attempted. While using his computer-based instructions and the WRMLTK, Bob deploys unscripted devices that specialize and individualize his attempt to convert the respondent. These include preparing (reading the case notes carefully), personalizing the introduction, reviewing the script, claiming the academic auspices of the interview, soliciting a favor, requesting participation ambiguously, persuading verbally, and performing the rhetoric.

A final point about the *embodiment* of this refusal conversion effort—the rhetorical performance. We have examined other episodes for interviewer gesturing and movement. Although this inquiry is not yet systematic, it seems that the more that interviewers are tethered, via their gaze to the terminal and via their hands to the keyboard, the less they engage in gross arm and hand movements, and, significantly, the more their voices sound like "reading voice." Conversely, the less interviewers are tethered, the more they engage these movements, and the more their speech is informally articulate and also tailored to the circumstances of a specific respondent, including the here-and-now specifics of what the respondent says to just this request or just this interview question. Accordingly, what we have discussed as tailoring, on the one hand, and embodied action and batonic gesturing, on the other hand, sometimes go together in intricate and intimate ways.

ACKNOWLEDGMENTS

Much of the work for this chapter derived from the Sociological Research Practicum (SRP) in the Indiana University Department of Sociology. The SRP is supported by the Department of Sociology and by the College of Arts and Sciences at Indiana University. We thank Jeremy Freese, Bob Moore, and Robert Soto for valuable research assistance, and the participants in the SRP for stimulus and discussion. Particular gratitude goes to Dr. John Kennedy, the director of the Indiana University Center for Survey Research. Data collection was aided by the generous support of the Institute for Research on Learning, in Palo Alto, CA; Jack Whalen, Marilyn Whalen, and Erik Vinkhuyzen were instrumental and indispensable. A different version of this chapter was previously published as part of a longer work by Maynard and Schaeffer (2000), in the journal *Social Studies of Science*, 28:281–300, copyright © Sage Publications Ltd. Republication is according to the publisher's agreement with authors.

NOTES

1. Cicourel (1974:79, 84) observes that survey measurement reflects the *structure of interaction* in various ways. Such interactional structure includes "negotiating" the recruitment of subjects and "the need to convince the respondent to complete the interview." See also Maynard and Schaeffer (1997).

2. Because this is the front end and not the substantive interview, the ordinary requirement to read the wording from screens literally is relaxed. Speaking of a computer form that 911 emergency call takers must use when talking to callers, Whalen (1995:196) remarks that "the trajectory of the encounter cannot be treated as the straightforward realization of some standardized agenda, technology, or policy; instead, because contributions to interaction are subject to the independent actions of others, the actual course of a call (or any other social encounter) will be a fundamentally interactional achievement."

3. Here is one example, which we have excerpted from the interaction at a point when the interviewer is at screen 10 in her call to a "fresh case." The arrow below points to where the interviewer uses the scripted "May I please..." request:

[TC01:2505:001]

```
MI:     .h To find out which adult in your household I need to speak
        with .hhh I need to ask how many adults currently live in your
        household .hh who're at least eighteen years of a::ge (0.3) and
        are currently married (0.1) divorced, or living with a partner
               (0.4)
FR:     There are two: adults.
               (0.2)
MI:     Mhmm
               (0.1) {Interviewer changes screen}
MI: →   .hhh (0.3) For purposes of our re:search may I please speak
        with a male resident of your household .hhh who is at least
        eighteen and currently married (0.1) divorced (0.1) or living
        with a partner.
```

4. However, "yes" answers and "no" answers are not equivalent in terms of interactional organization. See Wootton (1981).

5. We are basing this assertion on the inspection of other materials in which participants invoke the notion of "chance" when making requests or describing how they or others have obtained a courtesy. See also the discussion of sales appeals in Jefferson and Schenkein (1978:169, fn. 3).

6. Bob attempted to recruit the husband for the interview, but he was "getting ready to go to work." After that, Bob asked when the wife would be returning, and the husband said that it would be in about 20 minutes. This would be after the end of the shift in the Center for Survey Research, and because this was the last night for the study, Bob said to the husband that they "probably won't be calling back," and the two parties hung up.

7. On the noticing of failures as a method for complaining, see Schegloff (1988:120). Also see Jacoby and Gonzales (in press).

8. See also Schegloff (1988:122).

9. That is, it does not propose to apologize for the callback, or explain it in so many words, or otherwise invoke a more elaborate remedy for the offense. See Goffman (1971:118–122).

10. Observe also that MR's utterance at line 14 exhibits features of a "dispreferred" (rejecting) response: the first part ("I'm an adult') agrees with part of the previous turn; that is followed by a brief pause, a contrast marker ("but"), and an account for a rejection rather than a rejection per se. See Pomerantz (1984).

11. Also see Schegloff and Sacks (1973).

12. As mentioned, Bob did call the household back an hour later, and the wife answered. When she checked with her husband about participating, he refused once again, which illustrates a point Maynard and Schaeffer (this volume, Chapter 8), make about the spuriousness of some callback arrangements, which are outcomes of interactional structures rather than substantive agreements.

However, in this case, the wife then agreed to participate and did complete the interview.

13. See Kendon (1994) for extended discussion of batonic gesturing. For experimental evidence on how "unseen" gesturing relates to a speaker's expressiveness, see Rimé and Schiaratura (1991). For a collection of studies that examine the relations among interaction, bodily comportment, and computer usage in work settings, see Engström and Middleton (1996).

14. For extended discussion of how rhetoric is designed for performance in scientific presentation, see Ochs and Jacoby (1997).

15. It is also the case that such study means opening for further inquiry an infinitude of orderlinesses in the concrete human activities that, in this case, happen to be situated in survey research centers as one utterly ubiquitous and pervasive institution of modern society. That is, aside from the practical questions emerging from the study of survey interviewing, there are topics of intrinsic ethnomethodological interest. See Maynard and Schaeffer (2000).

REFERENCES

Cicourel, A. V. 1974. *Theory and Method in a Study of Argentine Fertility*. New York: Wiley.

Couper, M. P., and R. M. Groves. 1992. "The Role of the Interviewer in Survey Participation." *Survey Methodology* 18: 263–277.

Ekman, P., and W. V. Friesen. 1969. "The Repertoire of Nonverbal Behavior: Categories, Origins, Usage, and Coding." *Semiotica* 1: 49–98.

Engström, Y., and D. Middleton. 1996. *Cognition and Communication at Work*. Cambridge: Cambridge University Press.

Erickson, F. 1982. *The Counselor as Gatekeeper: Social Interaction in Interviews*. New York: Academic Press.

Feyereisen, P., M. Van de Wiele, and F. Dubois. 1988. "The Meaning of Gestures: What Can Be Understood without Speech?" *European Bulletin of Cognitive Psychology* 8: 3–25.

Goffman, E. 1971. *Relations in Public*. New York: Harper Colophon.

Groves, R. M., R. B. Cialdini, and M. P. Couper. 1992. "Understanding the Decision to Participate in a Survey." *Public Opinion Quarterly* 56: 475–495.

Jacoby, S., and P. Gonzales. In press. "Saying What Wasn't Said: Negative Observation as a Linguistic Resource for the Interactional Achievement of Performance Feedback." In C. E. Ford, B. A. Fox, and S. A. Thompson (Eds.), *The Language of Turn and Sequence*. Oxford: Oxford University Press.

Jefferson, G., and J. Schenkein. 1978. "Some Sequential Negotiations in Conversation: Unexpanded and Expanded Versions of Projected Action Sequences." Pp. 155–172 in J. Schenkein (Ed.), *Studies in the Organization of Conversational Interaction*. New York: Academic.

Kendon, A. 1994. "Do Gestures Communicate?" *Research on Language and Social Interaction* 27: 175–200.

Maynard, D. W., and N. C. Schaeffer. 1997. "Keeping the Gate: Declinations of the Request to Participate in a Telephone Survey Interview." *Sociological Methods and Research* 26: 34–79.

Maynard, D. W., and N. C. Schaeffer. 2000. "Toward a Sociology of Social Scientific Knowledge: Survey Research and Ethnomethodology's Asymmetric Alternates." *Social Studies of Science* 30: 323–370.

Ochs, E., and S. Jacoby. 1997. "Down to the Wire: The Cultural Clock of Physicists and the Discourse of Consensus." *Language in Society* 26: 479–505.

Pomerantz, A. 1984. "Agreeing and Disagreeing with Assessments: Some Features of Preferred/Dispreferred Turn Shapes." Pp. 57–101 in J. M. Atkinson and J. Heritage (Eds.), *Structures of Social Action: Studies in Conversation Analysis*. Cambridge: Cambridge University Press.

Rimé, B., and L. Schiaratura. 1991. "Gesture and Speech." Pp. 239–281 in P. Ekman and K. R. Scherer (Eds.), *Studies in Emotion and Social Interaction*. Cambridge: Cambridge University Press.

Sacks, H. 1992. *Lectures on Conversation*, Vol. 1. Fall 1964–Spring 1968. Oxford: Blackwell.

Schegloff, E. A. 1984. "On Some Gestures' Relation to Talk." Pp. 266–296 in J. M. Atkinson and J. Heritage (Eds.), *Structures of Social Action*. Cambridge: Cambridge University Press.

Schegloff, E. A. 1988. "Goffman and the Analysis of Conversation." Pp. 89–135 in P. Drew and A. Wootton (Eds.), *Goffman: Exploring the Interaction Order*. Boston: Northeastern University Press.

Schegloff, E. A., and H. Sacks. 1973. "Opening Up Closings." *Semiotica* 8: 289–327.

Whalen, J. 1995. "A Technology of Order Production: Computer-Aided Dispatch in Public Safety Communications." Pp. 187–230 in P. ten Have and G. Psathas (Eds.), *Situated Order: Studies in the Social Organization of Talk and Embodied Activities*. Washington, DC: University Press of America.

Wootton, A. J. 1981. "The Management of Grantings and Rejections by Parents in Request Sequences." *Semiotica* 37: 59–89.

PART 3

Interaction between Interviewer and Respondent

CHAPTER 11

Questioning Turn Format and Turn-Taking Problems in Standardized Interviews

Hanneke Houtkoop-Steenstra

INTRODUCTION

The standardized survey interview is a tool for taking measurements that can be aggregated to describe a certain population. These measurements should be complete, comparable, and categorized. The interviewers should present each question to each respondent. With closed questions, interviewers should not only ask the questions but also deliver a limited set of response options from which the respondents must choose. The questions and response options are to be read verbatim and in the exact order in which they are written in the questionnaire (e.g., Fowler and Mangione, 1990).

As it turns out, interviewers often do not act according to these rules of standardized interviewing. For example, Brenner's (1982) analysis of tape-recorded interviews shows that more than 50 percent of all questions and probes are not asked as required. In a Dutch study, Dijkstra, van der Veen, and van der Zouwen (1985) find that interviewers alter the essential content or the meaning of the written questions to the effect that 40 percent of all answers cannot be trusted.

Oksenberg, Cannell, and Kalton (1991) who analyzed several question formats, found that the interviewers' reading of questions were sometimes interrupted by the respondents. This occurred frequently with one question that was followed by two response instructions and with another question that was followed by response options. Studies by Loosveldt (1995) and van der Zouwen and Dijkstra

Standardization and Tacit Knowledge: Interaction and Practice in the Survey Interview
Edited by Douglas W. Maynard, Hanneke Houtkoop-Steenstra, Nora Cate Schaeffer, and Johannes van der Zouwen
ISBN 0-471-35829-0

(1995) also suggest that scripted questions followed by response options are especially vulnerable to interruption.

Johnstone, Ferrara, and Bean (1992:408) write that "any deviation from the script by the interviewer can be assumed to serve the interactional functions that are most vital (from the interviewer's perspective) in this speech event: to keep the respondent from hanging up the phone and to ensure that the respondent hears and understands all the questions and answers them in appropriate ways. . . , and that the interview is completed." And Schaeffer (1991:368) states that "locating where standardization fails and suggesting solutions require a closer look at interaction in the interview." She formulates a series of conversation analytic (CA) research questions that deal with the sequential organization of talk and the *interactional substrate* (Maynard and Marlaire, 1992) of the interview.

Schaeffer (1991) and Schaeffer and Maynard (1996) agree with Suchman and Jordan (1990), who point out that from the moment the research interview starts it assumes and relies on a wealth of conventions and resources from ordinary conversation. In ordinary conversation, speakers are sensitive to the history of the current talk and accommodate to their recipients. This is called *local control* (Sacks, Schegloff, and Jefferson, 1974).

However, Suchman and Jordan (1990) point out that designing instruments to accommodate a large number of people goes against the principles of local control. For example, requiring interviewers to extensively specify questions before accepting answers limits their ability to accommodate to individuals who may start to respond before the entire question has been read. Furthermore, certain questions may turn out not to be relevant for a specific recipient, or may have become redundant during the process of the interaction. Interviewers, despite being aware of this, are supposed to read the scripted text verbatim in order to maintain standardization.

This practice is fundamentally contrary to the conversational maxims formulated by Grice and Sacks. Grice (1975) points out that conversationalists should be relevant, brief, and orderly, they should provide neither more nor less information than necessary, and they should avoid using unclear language. Sacks (1992, Vol. 2) formulates similar maxims: Speakers should orient their talk to their recipients (p. 438), design their talk by reference to what they know their recipients know (p. 565), and "if you've already told something to someone then you shouldn't tell it to them again" (p. 438). Also Clark and Schober (1992) formulate a number of maxims or principles, for example, the *principle of utterance design*, according to which speakers try to design each utterance so that their addressees can discern what they mean by considering the utterance against the current common ground.

When interviewers take such maxims into account, they violate the rules of standardized interviewing. However, interviewers have methods for dealing with the incompatibility of the interviewing rules and the conventions of ordinary conversation, and they are able to arrive at accountable answers (Houtkoop-Steenstra, 1995, 2000; Schaeffer and Maynard, 1996).

RESEARCH QUESTION AND DATA

The aim of this chapter is to provide an ethnomethodological–conversation analytic account of why some question formats seem to be more vulnerable to interruption than others. The focus is on the rules of conversational turn-taking. I will show that interview questions are sometimes structured such that they violate the standard organization of turn-taking. These violations of turn-taking may cause interviewers to breach interviewing rules by omitting specific parts of the question in order to maintain the interaction.

After discussing the organization of conversational turn-taking and the way speakers organize their turns, we will examine the structure of survey questions that contain one or more specifications. Then the focus is on questions that are followed by response options. I will also briefly discuss how proper respondent selection may be endangered by a problematic question structure.

The analysis is based on the Wisconsin data and on a Dutch survey instrument and a number of corresponding tape-recorded and transcribed telephone survey interviews.[1] The Dutch interviews were carried out by trained interviewers who work for a large market research organization. The respondents were questioned about their possible interest in adult education.

TURN-TAKING AND TURN-CONSTRUCTION

Before turning to the structure of survey interview questions, some points on turn-taking and turn-construction in ordinary conversation need review. Sacks, Schegloff, and Jefferson (1974) have described the organization of turn-taking in detail. The speaker of a current turn can allocate the next turn to a specific next speaker by producing the *first pair-part* of a sequence (e.g., the question of a question–answer sequence) and name or otherwise address a co-participant as the one who should produce the *second pair-part* (the answer) (Schegloff and Sacks, 1973). When the utterance is completed so that it can be heard as a question, the addressed recipient has the right and obligation to take the next turn and to produce the answer. The sequence may then be ended by the first speaker, who may produce a *third turn acknowledgment*[2] or a new sequence-initial action, for example a next question (Houtkoop-Steenstra, 1987), or both.

As Sacks, Schegloff, and Jefferson (1974) discuss, turns are constructed out of one or more *turn-constructional units* (TCUs). A speaker can use various unit types to construct a turn, such as lexical, phrasal, clausal, or sentential units. Familiar instances of the unit type allow the recipient to project an approximation of what the unit requires to be completed. For example, an utterance that begins with a verb or with "what" or "when," is likely to be a question. The first possible completion point of the unit employed constitutes an initial *transitional relevance place* (TRP).

When conversationalists take a multiunit turn, they instruct their recipients not to regard the first possible completion point as the intended one by marking the turn as a multiunit turn. For instance, they can project the minimal set of actions to follow. This projection may be explicit, for example, "two answers to that," or implicit, such as "on the one hand" which projects "on the other hand." Also, a speaker can use cataphoric terms, that is, forward-looking lexical items, with the referents still to be filled in, for example, "something" in "I'd like to ask you something." A TCU that cannot be heard as the projected turn completion component, that is, cannot be heard as the second answer, the punch line, the cataphoric fulfillment, for example, is incomplete (Ford and Thompson, 1996; Houtkoop and Mazeland, 1985; Schegloff, 1982).[3]

In interviews, however, multiunit turns are often not marked as such. Recipients cannot project that there is more to come after an initial TRP. Indeed, as will become clear later, some interview items tacitly instruct respondents to take a turn at a point that was not really meant to be the actual TRP.

TURN STRUCTURE OF SURVEY QUESTIONS

The following questions are designed either as single-unit turns or multiunit turns. An example of a single-unit turn is item 80: "Were you born in the Netherlands?" "Were" in turn initial position signals the onset of a question. The end of the utterance is then heard as the analyzable completion of a syntactic question. At this point the respondent can and should take the next turn to answer the question.

In some cases a question is prefaced by an *action projection* (Schegloff, 1980), a phrase that defines the action that follows as a question: "Could you tell me. . . ," or "May I ask you. . . ."

Dutch Survey, Item 71

```
First of all I would like to ask          ACTION PROJECTION
how many people, including yourself,
live in your household.                   QUESTION
```

Many items are even more complex, built as multiunit turns that constitute a *question delivery structure* (Heritage and Greatbatch, 1991; Heritage and Roth, 1995; Lerner, 1996):

Dutch Survey, Item 29

```
Now we would like to talk about
your possible future plans with respect to such
courses or education.


They may be either fixed plans or vague ideas.


Do you have any plans for the COMING YEAR to take
any course or education?
```

A multiunit turn may be designed so that each TCU accomplishes a different activity, for example, addressing someone ("Hey John") and asking a question

("Do you have the time please?"). Item 29 is built out of four questioning TCUs, each unit containing a different activity component. The turn begins with an action projection, as something "to talk about." This action projection is followed by a statement on what the question will be about: "your future plans with respect to such courses or education." Following Button (no date), I will call the second component the *question target component* (QTC). The next TCU defines or specifies the term "plans." This is the *question specification component* (QSC) (in short, *specification*[4]). The last component is the question itself. Button (no date) refers to this component as the *question delivery component* (QDC). Item 29, then, is structured in the following way:

Action Projection Component
Now we would like to talk about
+
Question Target Component
your possible future plans with respect to
such courses or education.
+
Question Specification Component
They may be either fixed plans or vague ideas
+
Question Delivery Component
Do you have any plans for the COMING YEAR
to take any course or education?

The item is thus systematically organized as a coherent multiunit turn in which the first three components work toward the question delivery component in final position. This question is designed to inform respondents about both the topic of the question and the type of answer that is expected.[5]

What questionnaire designers often miss, however, is how the components of a multiunit question delivery structure will function in the conversational interaction in which the question is ultimately actualized. It may be most effective to examine items that are less adequately structured than item 29 to see the consequences for the interaction and, subsequently, the information that is gathered.

Dutch Survey, Item 12

```
First of all I would like to ask you                  APC
what is the highest level of education that           QDC
you took in FULL-TIME daytime education.
By full-time daytime education we mean a course or    QSC
education that is taken at least three days a week.
```

This turn is composed of (1) an action projection, (2) the question as such, and (3) the specification of what is meant by "full-time daytime education." So item 12 is therefore designed as follows:

[Action Projection Component]
+
[Question Delivery Component]
+
[Question Specification Component]

This format is general practice in questionnaire design.[6] However, when such a question is used in interaction, it is vulnerable to conversational problems, as the fragments below will make clear.

In the first fragment the interviewer begins the question as it is worded in the instrument, but the respondent takes the turn to provide an answer immediately after the question delivery component. I will first show the scripted question again.

Item 12

```
First of all I would like to ask you                      APC
what is the highest level of education that               QDC
you took in FULL-TIME daytime education.
By full-time daytime education we mean a course or        QSC
education that is taken at least three days a week.
```

(1) Dutch Survey, Christine 12

```
I.     hh Uh first of all I'd like to ask you
       what is the highest level of education that
       you took in full-time daytime education?
C. →   In full-time day- a uh basic knowledge course.
I.     No, I  [mean the high[est-
C.            [(  )         [the school.
I.     Right, [the school,
C.            [the school.
I.     Right, daytime school right.=
C.     =No seven years of elementary school.
I.     That's- that's it?
C.     Yes.
```

Following a regular conversational turn-taking practice, the respondent, Christine, answers the question at the TRP following the completion of the question delivery component. The consequence is that she has not been informed about the specific meaning of "full-time daytime education" (taking place at least 3 days a week) for the purposes of this interview. The type of course Christine initially mentions generally takes place only a few hours per week for a limited period of time. Christine's answer, therefore, is very likely to be incorrect. Having noticed the (possible) problem, the interviewer produces a *third position repair*[7] in line 5, saying, "No, I mean the highest-" (Schegloff, 1992). This utterance is completed by Christine with "the school" in line 6. Although "school" is not the only thing the researchers refer to when they so carefully talk about "courses or education,"[8] the interviewer agrees on this term, saying, "Right, the school" in line 7, and "Right, daytime school right" in line 9.

The result is that not only is standardization at stake now, but also the original survey question is interactively transformed into a question with a different meaning. Whereas the question was meant to be about "a course or education that is taken at least three days a week," it is interactively changed, narrowed down to a question about "school." Furthermore, the respondent provides an answer that is appropriate, though not necessarily correct, in terms of what the researchers meant to ask. It is appropriate because "seven years of elementary school" is an instance of a form of education that takes place at least 3 days a week. But because the respondent was not informed about what was really meant by "full-time day time education," she may very well have given an incorrect, and therefore invalid, answer.[9]

When we look at item 14, we see the same problem. This multiunit turn is constructed out of (1) a question delivery component, (2) a question specification component explaining an important concept that is mentioned in the question, and (3) a second question specification component further specifying the question.

Item 14

```
Have you, for the last five years, during or after          QDC
this education you just mentioned, taken any courses
or education in so-called adult education?


You may think of part-time education,                       QSC 1
such as daytime, night or weekend education,
service training, courses at the School for Adult
Classes, adult centers, creativity centers, et cetera.
Also courses that you have not completed,                   QSC 2
are to be counted.
```

When this item is presented to respondents, various kinds of problems arise:

(2) Dutch Survey, Bert 14

```
I.  Have you <for the last five years> (.)         QUESTION
    during or after this education you just
    mentioned (0.7) uh taken any courses or
    education in so-called adult education?=
B.  =[ N o.   ]                                    ANSWER
I.  =[.hh You-] You >may think of                  SPECIF. 1
    part-time education, such as daytime,
    night or weekend education<, service
    training, courses at the School for
    Adult Classes, adult centers,
    creativity centers, et cetera.
B.  No:, I haven't taken any of these.=            ANSWER
I.  =Okay.
```

Again, immediately upon completion of the question delivery component, the respondent takes the turn to provide the answer. At the same time the interviewer takes an inbreath in line 6 (".hh") by which he holds the turn, and begins to read the first specification, to the effect that both speakers talk at the same time. The interviewer stops talking, and as soon as the respondent has finished his answer, the interviewer begins to read the specification again. When this specification is completed, and before the interviewer has started the third component of the question turn (the second specification), Bert again answers the question, saying, "No:, I haven't taken any of these" in line 12. With this reply, Bert shows that his answer is a *next-positioned new answer* (Jefferson, 1981) that responds to the now-specified question. Whereas Bert's first answer is left unacknowledged by the interviewer, the second one is acknowledged by the third-turn receipt "Okay" in line 13, by which the interviewer also finishes the question–answer sequence.

Though Bert's second answer is acknowledged and accepted by the interviewer, Bert has not been asked the complete question. He has not been informed that the courses and forms of education mentioned in the question include those that have not been completed. Because the prior question only concerned itself with completed forms of education, this distinction is important and bears mentioning. Thus Bert's answer is possibly invalid.

The same question is delivered to Ann in the following way:

Item 14

```
Have you, for the last five years, during or after      QDC
this education you just mentioned, taken any courses
or education in so-called adult education?

You may think of part-time education,                   QSC 1
such as daytime, night or weekend education,
service training, courses at the School for Adult
Classes, adult centers, creativity centers, et cetera.

Also courses that you have not completed,               QSC 2
are to be counted.
```

(3) Dutch Survey, Ann 14

```
I.      U:h have you, for the last five years,           QUESTION
        during or after this education you just
        mentioned, taken any courses or education
        in so-called adult education? .hh You
        may think of >part-time education,
   →    such as daytime, night or weekend                SPECIF.1
        education<, [sERvice training,
A. →                [U:h                                 ANSWER
I.      courses at the School for Adult                  SPECIF.1
        Classes, adult centers, creativity
        centers, e::t [cetera.
A.                    [U:h ((laughs)) No,
        No, [no.                                         ANSWER
I.          [No, y(h)ou ha(h)ven't. ((laughs))           ACKNOWL.
A.      No.
```

After the interviewer has completed the question, he takes an audible inbreath (line 5) and continues to produce the first specification. Halfway through his specification (his intonation indicates that he has not yet finished), Ann decides that the utterance is *completed-for-response* (Schegloff, 1980:150), and she begins to speak: "U:h" (line 8). But the interviewer continues and holds Ann off by increasing his volume for a short while, saying "sERvice" in line 7. This specification is completed with "e::t cetera" (line 11), as stated in the instrument. Since "et cetera" is a common last element of a list, a *generalized list completor* (Jefferson, 1981), it is not surprising that Ann begins to talk again in overlap with this possible turn-final word. The interviewer acknowledges the answer ["No, y(h)ou ha(h)ven't.," line 14], and shortly after he proceeds to the next item, omitting the second specification "Also courses that you have not completed, are to be counted."

If Ann has ever taken courses she did not complete, her present answer is invalid from the point of view of the researchers. The possibly invalid answer occurs because to the question turn is structured in such a way that it projects two turn completion points (TRPs) not intended by the questionnaire designer. When, according to turn-taking procedures, the respondent legitimately takes a turn at talk at such a point, she finds the floor still occupied by the interviewer. If the interviewer follows the interview protocol, the resulting situation is one in which the interviewer must compete with the respondent for turn space.

The next fragment shows a fierce competition for the floor. Although the respondent answers the question twice, the interviewer continues to read the question to her:

Item 14

```
Have you, for the last five years, during or after         QDC
this education you just mentioned, taken any courses
or education in so-called adult education?


You may think of part-time education,                      QSC 1
such as daytime, night or weekend education,
service training, courses at the School for Adult
Classes, adult centers, creativity centers, et cetera.


Also courses that you have not completed,                  QSC 2
are to be counted.
```

(4) Dutch Survey, Christine 14

```
I.       Have you for the last five years             QUESTION
         during or after this education (.) during
         this elementary school taken any courses
         or education in so-called adult education?
C. → 1 No:.                                           ANSWER
I.       You may thi[nk-                              SPECIF. 1
C. → 2               [No, not for the last the        ANSWER
         last five years, because u [h
I.                                  [>You
```

```
             may think of part-time                      SPECIF. 1
             education, such as daytime, night
             or weekend education<, or service
             training, or courses at the
             School for [Adult Classes,
    C  → 3              [No, no, not a[t all.            ANSWER
    I.                  [.hh
             Not, [no.                                   ACKNOWLEDGMENT
    C.            [no.
```

Christine provides an answer in line 5, immediately after the interviewer completes the question. When the interviewer continues to read the first question specification, Christine interrupts and expands her answer, saying, "no, not for the last the last five years." By repeating her initial answer in a more elaborate way, and also accounting for it ["because uh-" (line 8)], Christine may be proposing that she has understood the question and that there is no need for further explanation. The interviewer nevertheless interrupts the account and speedily continues to produce the specification. Before the specification is completed, the respondent interrupts him again and gives an answer that is formulated more strongly than the first one: "No, no, not at all" (line 15). The interviewer then stops reading the specification and acknowledges the receipt of the answer.

For the research on behalf of which these interviews are conducted, the omission of the second specification, is rather problematic. This specification, which indicates that incomplete courses are also to be counted, is very important. The omission of this specification could lead to invalid answers being collected.

In the next excerpt the interviewer leaves out the specifications altogether when the respondent provides the answer immediately after the question delivery component.

Item 14

```
Have you, for the last five years, during or after        QDC
this education you just mentioned, taken any courses
or education in so-called adult education?


You may think of part-time education,                      QSC 1
such as daytime, night or weekend education,
service training, courses at the School for Adult
Classes, adult centers, creativity centers, et cetera.


Also courses that you have not completed,                  QSC 2
are to be counted.
```

(5) Dutch Survey, Ernst 14

```
   I.    Have you (.) after (.) or during (.)        QUESTION
         this education,
   E.    Ye[:s?
   I.      [uh >so this VGLO school right<,
         .hh for the last (.) five years taken
```

any courses or education in (.)
so-called adult education?
E. No. ANSWER
I. No. ACKNOWLEDGMENT
((sound of typing))[10]

Having analyzed what happens in these interviews when a [question] + [specification] structure is used in interaction, it is clear that respondents tend to behave like ordinary conversationalists. They treat the first possible completion point of an utterance as a turn transitional relevance place. When the interviewer continues to produce a specification, thus treating the prior answer as premature, the respondent looks for the next possible completion point of the meanwhile specified question.

Furthermore, conversationalists do not necessarily wait for the end of an utterance before taking a turn. They may begin to respond as soon as the prior utterance seems "adequately complete" (Sacks 1992, Vol. 1:310; cf. Suchman and Jordan, 1990). Thus the first specification of item 14 is also vulnerable to interruption. When the interviewer has mentioned a few types of courses given as examples of what is meant by adult education, the utterance may be heard by respondents as being complete enough to respond to again. When the respondent answers the question near the completion point of the first specification, the interviewer still has a second specification to read. This means that to follow interview protocol he or she must define the respondent's answer as misplaced for a third time, though it is properly placed sequentially. Therefore, it not surprising that this second specification is omitted in all interviews.

This analysis makes clear that specifications following questions have a high chance of being omitted by interviewers. The [question] + [specification] format is fundamentally contrary to the principles of turn-taking and turn-construction in ordinary conversation.

[QUESTION] + [RESPONSE OPTIONS] TURN FORMAT

Oksenberg, Cannell, and Kalton (1991:355) suggest that "respondents probably often think that the question has been completed before the interviewer has finished reading it, or they feel that they have heard enough to answer it." When we look at the Wisconsin data, this perception seems to vary among respondents. While some wait for the interviewer to finish reading, others do not. For example, the respondent below provides her answer once she recognizes what sort of listing is underway, and she immediately identifies herself in terms of political party (line 5).

(6) AW01:364, {q56a}

IV. Generally speak↑ing do you usually
think of your<u>self</u> as a re↑publican
(0.4)
demo[crat
FR. → [republican

```
IV.     independent?
        or something else[:? re[publican?]
FR. →                   [      [republi  ] can
                        [#     [#
```

In this example, the question is formatted such that the response options are part of the question as such. Usually, however, the list of response options is structurally separated from the question, resulting into a multiunit turn that is formatted as a [question] + [response options] turn. The interviewer is supposed to read the entire list of response options. The respondent below, however, provides an answer before the list has been read completely:

(7) AW01:264 {q30}

```
IV.     And the next ↑few questio:ns concern your
        feelings about the taxes you have to pay.
        .hh[h For ea(h)ch ty(h)pe of ta(h)x (.)
FR.        [UH
IV.     please tell me whether you think the tax
        is .hh excessively high: (.) about
        right (0.2) or low. .hh[h ]
FR.                            [ok]ay
IV.     First
        (.)
        federal income taxes? Do you think
        federal income taxes are excessively high
FR. →   [: excessively high]
IV.     [(0.5) about right ] =
        =>or low.<
        (0.4)
IV.     [okay excessively high
        [#
```

Notice that the interviewer perseveres with the list of options, after which he acknowledges the respondent's answer by saying "okay excessively high" in line 17.

When my colleague Wil Dijkstra and I looked at a series of 98 Dutch interviews in which the reading of a certain question was to be followed by 5 response options, in 13 cases the respondents answered the question before the interviewer began reading the options. In 14 cases the interviewers were interrupted halfway through the list of options by the respondents providing an answer.

Analysis of survey interviews in which interviewers do not read (all) the response options (Houtkoop-Steenstra, 1996) shows that this procedure may lead to respondents providing answers that do not fit in the precoded response categories. The interviewers may then reformulate these answers in order to make them fit. In these cases, the interviewers are doing coding work, and, as the analysis makes clear, they do not always do this correctly. As van der Zouwen and Dijkstra (1995) have shown, one solution to this problem with fixed-choice questions is to present the respondent with a card that displays the response categories. However, this procedure can only be accomplished in face-to-face interviews.

QUESTIONING TURN STRUCTURE AND SAMPLING PROCEDURES

Methodologically, respondent selection is a critical aspect of the introduction portion of the interview. The following question in the introduction to a Dutch survey is meant to direct the interviewer to select the proper respondent. Notice that the question is followed by two specifications:

Dutch Survey

```
     In order to decide whom I should ask the questions,
     I would like to know how many people live in your
     family or household,
→    for at least 4 days a week.
→    This, of course, includes yourself.
```

The following fragment shows what may happen to postquestion specifications:

(8) Dutch Survey

```
I.   Bu::t in order to decide whom I should
     ask the questions, first of all I'd
     like to know how many people live
     in your household,
R.   Five.
I.   That's five people.
```

When the specifications are omitted, the person ultimately interviewed may be the wrong one. The respondent in fragment 8 for instance, may be producing an answer that does not take the qualification "for at least four days a week" into account. If so, then she may also be giving an incorrect number, which then alters the selection of the person to be interviewed. Further research is needed to study the extent to which this phenomenon occurs and what its effect is on sampling.

CONCLUSION

When questions need to be specified, the specification should precede any utterance that could possibly be heard as the end of a question-turn (cf. Oksenberg, Cannell, and Kalton, 1989; Schaeffer, 1990:20, 1991:386). This happens in the following questionnaire item, where the specification is sandwiched between two syntactic components of the question utterance:

```
     First of all I would like to ask how many people,
→    including yourself,
     live in your household.
```

This turn structure may prevent respondents from answering prematurely and may keep interviewers from omitting specifications and being forced to interrupt the respondents. As a consequence it may increase the standardization and validity of the research instrument.

It is difficult to prove that a question followed by specifications easily generates invalid answers. As long as respondents do not provide contradictory information during the course of the interview, invalid answers are seldom made explicit. However, as I have argued, answers generated in this way cannot always be trusted at face value.

Apart from the issue of validity, there is no doubt that this structure has a negative influence on standardization. It invites the interviewers to omit specifications. When the respondent begins to answer immediately after the completion of the actual question, one way for the interviewer to produce the specification is by interrupting the respondent. Another way is to continue to read the response options after the respondent's "premature" answer, and thus to ignore the answer. Both possibilities are unhappy interactional options in that they may offend the respondent. The interviewer then risks that the respondent will lose his or her motivation to fully cooperate.

NOTES

1. It goes without saying that the Dutch data were translated only after the analysis was finished. For the sake of the clarity it sometimes was necessary to keep the Dutch word order in the English translations. The Dutch fragments can be obtained from the author.

2. Jefferson's (1972:317) term *satisfactory termination* reflects the action that a first speaker performs: indicating that the question is answered satisfactorily. Heritage's (1984:317) term *sequence exit device* reflects another, related action: When the first speaker is satisfied with the response, he or she may terminate the sequence. See also Maynard and Marlaire (1992) on third-turn acknowledgments in the context of educational testing.

3. Ford and Thompson (1996) pay close attention to intonation as a resource for projecting the upcoming occurrence or nonoccurrence of turn completion.

4. I will use the term *specification* as a general term that comprises explanations and restrictions.

5. Questionnaire designers tend to pay attention to semantic aspects of the questions. However, as discussed in Houtkoop-Steenstra (1996), they often fail to take into account that the questions should also make clear to respondents what type of answer they are expected to provide.

6. Especially in written forms meant to be filled out, such as tax forms or application forms, we find the [question] + [explanation/specification] turn structure. Research shows that people tend to write down the answer without having read the specification or explanation (Jansen and Steehouder, 1992). Current research tries to find ways to redesign forms such that the explanation or specification is more likely to be read (Steehouder and Jansen, 1992).

7. Third-position repair is a repair in the speaker's turn after the recipient's (second turn) response to the speaker's (first turn) utterance.

8. In the Dutch educational system the notion "education" is associated with "school," whereas "course" is not.

9. Mazeland's (1992) study of nonstandardized social research interviews shows that interviewers work more toward appropriate answers than toward correct answers.

10. In this excerpt it is audible from the tape that the interviewer, immediately after his acknowledgment, enters something into the computer, which is almost certainly the respondent's answer.

REFERENCES

Brenner, M. 1982. "Response-Effects of 'Role-Restricted' Characteristics of the Interviewer." Pp. 131–165 in W. Dijkstra and J. van der Zouwen (Eds.), *Response Behaviour in the Survey Interview*. London: Academic.

Button, G. No date. "Some Design Specification for Turns at Talk in a Job Interview." Unpublished manuscript.

Clark, H. H. and M. F. Schober. 1992. "Asking Questions and Influencing Answers." Pp. 15–48 in J. M. Tanur (Ed.), *Questions about Questions: Inquiries into the Cognitive Bases of Surveys*. New York: Sage.

Dijkstra, W., L. van der Veen, and J. van der Zouwen. 1985. "A Field Experiment on Interviewer-Respondent Interaction." Pp. 37–63 in M. Brenner, J. Brown and D. Canter (Eds.), *The Research Interview: Uses and Approaches*. London: Academic.

Ford, C. E., and S. A. Thompson. 1996. "Interactional Units in Conversation: Syntactic, Intonational, and Pragmatic Resources for the Management of Turns." Pp. 134–185 in E. Ochs, E. A. Schegloff, and S. A. Thompson (Eds.), *Interaction and Grammar*. Cambridge: Cambridge University Press.

Fowler, F. J. and T. W. Mangione. 1990. *Standardized Survey Interviewing: Minimizing Interviewer-Related Error*. Newbury Park: Sage.

Grice, H. P. 1975. "Logic and Conversation." Pp. 41–58 in P. Cole and J. L. Morgan, (Eds.), *Syntax and Semantics*, Vol. 3: *Speech Acts*. New York: Academic.

Heritage, J. 1984. *Garfinkel and Ethnomethodology*. Cambridge: Polity.

Heritage, J., and D. Greatbatch. 1991. "On the Institutional Character of Institutional Talk: The Case of News Interviews." Pp. 93–138 in D. Boden and D. H. Zimmerman (Eds.), *Talk and Social Structure: Studies in Ethnomethodology and Conversation Analysis*. Cambridge: Polity.

Heritage, J., and A. L. Roth. 1995. "Grammar and Institution: Questions and Questioning in the Broadcast News Interview." *Research on Language and Social Interaction* 28: 1–60.

Houtkoop, H., and H. Mazeland. 1985. "Turns and Discourse Units in Everyday Conversation." *Journal of Pragmatics* 9: 595–619.

Houtkoop-Steenstra, H. 1987. *Establishing Agreement. An Analysis of Proposal-Acceptance Sequences*. Dordrecht/Providence: Foris.

Houtkoop-Steenstra, H. 1995. "Meeting Both Ends: Between Standardization and Recipient Design in Telephone Survey Interviews." Pp. 91–107 in P. ten Have and

G. Psathas (Eds.), *Situated Order: Studies in the Social Organization of Talk and Embodied Activities*. Washington, DC: University Press of America.

Houtkoop-Steenstra, H. 1996. "Probing Behavior of Interviewers in the Standardized Semi-Open Research Interview." *Quality and Quantity* 30: 205–230.

Houtkoop-Steenstra, H. 2000. *Interaction and the Standardized Interview: The Living Questionnaire*. Cambridge: Cambridge University Press.

Jansen, C., and M. Steehouder. 1992. "Forms as a Source of Communication Problems." *Journal of Technical Writing and Communication* 22: 179–194.

Jefferson, G. 1972. "Side Sequences." Pp. 294–338 in D. Sudnow (Ed.), *Studies in Social Interaction*. New York: Free Press.

Jefferson, G. 1981. "The Abominable 'Ne?': An Exploration of Post-Response Pursuit of Response." Pp. 53–89 in P. Schröder and H. Steger (Eds.), *Dialogforschung*. Düssseldorf: Schwann.

Johnstone, B., K. Ferrara, and J. Mattson Bean. 1992. "Gender, Politeness, and Discourse Management in Same Sex and Cross-sex Opinion Poll Interviews." *Journal of Pragmatics* 18: 405–430.

Lerner, G. 1996. "Finding 'Face' in the Preference Structure of Talk-in-Interaction." *Social Psychology Quarterly* 59: 303–321.

Loosveldt. G. 1995. "Interviewer-Respondent Interaction Analysis as a Diagnostic and Validating Instrument." Pp. 290–295 in *Proceedings of the International Conference on Survey Measurement and Process Quality*, Bristol, 1–4 April. Alexandria: American Statistical Association.

Maynard, D. W., and C. L. Marlaire. 1992. "Good Reasons for Bad Testing Performance: The Interactional Substrate of Educational Exams." *Qualitative Sociology* 15: 177–202.

Mazeland, H. 1992. *Vraag/antwoord Sequenties* [Question/Answer Sequences]. Amsterdam: Stichting Neerlandistiek.

Oksenberg, L., C. Cannell, and G. Kalton. 1991. "New Strategies of Pretesting Survey Questions." *Journal of Official Statistics* 7: 349–369.

Sacks, H. 1992 [1964–1972]. *Lectures on Conversation*. 2 vol. Edited by Gail Jefferson. Oxford: Blackwell.

Sacks, H., E. A. Schegloff, and G. Jefferson. 1974. "A Simplest Systematics for the Organization of Turn Taking in Conversation." *Language* 50: 696–735.

Schaeffer, N. C. 1990. "Conversation with a Purpose — or Conversation? Interaction in the Standardized Interview." Unpublished manuscript. University of Wisconsin-Madison.

Schaeffer, N. C. 1991. "Conversation with a Purpose — or Conversation? Interaction in the Standardized Interview." Pp. 367–391 in P. P. Biemer, R. M. Groves, L. E. Lyberg, N. A. Mathiowetz, and S. Sudman (Eds.), *Measurement Errors in Surveys*. New York: Wiley.

Schaeffer, N. C., and D. W. Maynard. 1996. "From Paradigm to Prototype and Back Again: Interactive Aspects of 'Cognitive Processing' in Standardized Survey Interviews." Pp. 65–88 in N. Schwarz and S. Sudman (Eds.), *Answering Questions: Methodology for Determining Cognitive and Communicative Processes in Survey Research*. San Fransisco: Jossey-Bass.

Schegloff, E. A. 1980. "Preliminaries to Preliminaries: 'Can I ask you a Question'." *Sociological Inquiry* 50: 104–152.

Schegloff, E. A. 1982. "Discourse as an Interactional Achievement. Some Uses of 'Uh Huh' and Other Things That Come Between Sentences." Pp. 71–93 in D. Tannen (Ed.), *Analyzing Discourse: Text and Talk*. Washington, DC: Georgetown University Press.

Schegloff, E. A. 1992. "Repair after Next Turn: The Last Structurally Provided Defense of Intersubjectivity in Conversation." *American Journal of Sociology* 98: 1295–1345.

Schegloff, E. A., and H. Sacks. 1973. "Opening up Closings." *Semiotica* 8: 289–327.

Steehouder, M., and C. Jansen. 1992. "Optimizing the Quality of Forms." Pp. 159–173 in H. Pander Maat and M. Steehouder (Eds.), *Studies of Functional Text Quality*. Amsterdam: Rodopi.

Suchman, L., and B. Jordan. 1990. "Interactional Troubles in Face-to-Face Survey Interviews." *Journal of the American Statistical Association* 85: 409, 232–241.

van der Zouwen, J., and W. Dijkstra. 1995. "Non-Trivial Question-Answer Sequences: Types, Determinants and Effects of Data Quality." Pp. 81–86 in *Proceedings of the International Conference on Survey Measurement and Process Quality*, Bristol, April 1–4. Alexandria: American Statistical Association.

CHAPTER 12

Occasions for Intervention: Interactional Resources for Comprehension in Standardized Survey Interviews

Nora Cate Schaeffer and Douglas W. Maynard

INTRODUCTION

In a typical survey interview, a freshly recruited respondent who may never have participated in a standardized interview before, is presented with a series of highly structured questions. These questions ask her to translate her experiences into the concepts and categories of a researcher as presented by an interviewer. The validity of the resulting data depends partly on how respondents understand key features of the task they are asked to perform, in particular how they interpret concepts presented in questions and apply them to their own experience. The importance of this understanding is highlighted in social–information–processing models of the response process. The first and last stages of these models — typically "understanding the question" and "mapping the answer onto response categories" (see, e.g., Cannell, Miller, and Oksenberg, 1981; Tourangeau, 1984) — focus on precisely this issue of the fit between the survey task and what a respondent might possibly report.

Social–information–processing models consider what goes on inside the respondent's head as he hears and then answers a question. But survey data are collected in interaction; they require *shared* understanding or intersubjectivity (e.g., Levine and Resnick, 1993), which is itself developed in interaction. The cognitive processing that leads to shared understandings is, in part, fundamentally

Standardization and Tacit Knowledge: Interaction and Practice in the Survey Interview
Edited by Douglas W. Maynard, Hanneke Houtkoop-Steenstra, Nora Cate Schaeffer, and Johannes van der Zouwen
ISBN 0-471-35829-0

interactive (e.g., Schaeffer and Maynard, 1996), and the task uncertainty (Schaeffer and Thomson, 1992) that results when these understandings fail to emerge may contribute to measurement error.

Survey researchers have long recognized that the activities of the interviewer could affect the validity and reliability of survey data (e.g., Hyman et al., 1975 [1954]; Cannell, Fowler, and Marquis, 1968). Studies of interviewer variance (e.g., Collins, 1978; Mangione, Fowler, and Louis, 1992), for example, consider that interviewer variance may be different for open and closed questions because they require different behaviors from interviewers. Methods that evaluate survey questions by coding the behavior of interviewers and respondents presume that some activities may affect the quality of the data (e.g., Oksenberg, Cannell, and Kalton, 1991; Dykema, Lepkowski, and Blixt, 1997). Despite their informed faith in the importance of the interviewer–respondent interaction for the resulting data, survey researchers have seldom looked at this interaction in detail. Yet to see how interviewers and respondents develop the intersubjectivity that both makes survey data possible and shapes its error structure, that is what we must do.

In this chapter we briefly describe three interactional resources that allow the participants in the interview to propose, display, review, and correct interpretations of survey questions or of how respondents' answers "fit" into survey categories. These resources — respondents' requests for clarification, respondents' reports of potentially relevant information, and a silence or delay after the reading of the question that is followed by the interviewer's preemptive probes (which we call "preprobes") — may reduce participants' uncertainty about the meaning of survey questions or about how the response categories fit the respondent's situation. These resources thus provide means to improve respondents' comprehension of the survey task and thereby to reduce task uncertainty. The tacit knowledge that interviewers and respondents bring to the survey interview includes how to deploy these resources appropriately and how to respond to them. We describe these resources as *occasions* for intervention because they consist of moments when the interaction is configured in a way that interviewers appear to routinely treat as an opportunity to intervene. This chapter presents a preliminary description of some such occasions rather than a detailed analysis, a comprehensive list of such occasions, or a comparison between occasions when the interviewer intervenes and those when she does not.

In an earlier work, we argued that if we examine the interactive aspects of the "cognitive processing" that interviewers and respondents jointly engage in, we can identify features of survey questions that could increase interviewer variability or reduce validity, and we may also see ways to improve the design of the instrument (Schaeffer and Maynard, 1996). The behaviors of interviewers and respondents that we describe in this chapter seem to address what a survey researcher might characterize as "comprehension problems." These behaviors illustrate how interviewers and respondents deploy in the survey interview the tacit knowledge available to them in other interactional contexts. We present findings from previous studies to show that these resources are not unique to survey interviews but occur in other sorts of talk.

It can be difficult for an observer to diagnose that a respondent does not understand a survey question in the way the researcher intended, because the respondent usually ends up providing an answer that the interviewer can record. If we can identify ways in which problems in comprehension appear and resolve in interaction, we may find behaviors that may prove useful either to identify questions that need further development or as indicators of measurement error in analysis [as measures of "response latency" or uncertainty may be (see Bassilli, 1996; Mathiowetz, 1998)]. In addition, what we observe when we listen to interviewers and respondents may suggest ways that we might augment the set of behaviors that we consider appropriate for standardized interviewers and may also suggest what the consequences of doing so might be for measurement error or how to train interviewers to use them in ways that further the purposes of measurement.

METHODS AND DATA

The interviewer's behavior in survey interviews is constrained by the rules of standardized interviewing, which are requirements of the interviewer's role. Several of those constraints are particularly relevant to this analysis: to read the question as worded, to obtain an answer that can be seen to be produced by the respondent and not the interviewer (compare Button, 1987), to probe nondirectively, and to maintain a cooperative relationship for the duration of the interview (see, e.g., Brenner, 1985; Fowler and Mangione, 1990).

Proper standardization sometimes requires that the interviewer ask a question with an "obvious" answer—an answer that uses information the respondent has already volunteered or that seems to be a reasonable inference from previous answers. Repeatedly asking such obvious questions may make the interviewer seem inattentive or socially inept and so threaten her cooperative relationship with the respondent. To help mitigate such role conflict, standardized interviewers at some institutions are allowed to use a practice called "verification." In this procedure, an interviewer verifies both that she has correctly heard and remembered information already provided by a respondent and that the information is the correct answer to the question. For example, if a respondent has already mentioned her age, the interviewer could, instead of asking a question about age, say something like, "I think you said you are 24 years old. Is that the age I should record for you?" or "How old were you on your last birthday? I think you said you were 24. Is that correct?" The situations in which a verification might be useful and appropriate are highly varied, and the techniques interviewers use for verifications are similarly diverse. Although we have not studied them closely, it seems that most verifications are constructed to show both that the interviewer has "heard" the respondent and that the answer is the respondent's own. In listening to actual interviews, however, it is sometimes difficult to determine whether the interviewer has changed the wording of a question as part of an acceptable verification or for some other reason that may have less justification under the rules of standardization.

Standardized survey questions constrain acceptable answers (compare Fowler and Mangione, 1990; Suchman and Jordan, 1990). Questions may be designed and formatted to obtain a "yes" or "no" answer, a specific number or piece of information, one of a set of response categories, or a relatively free-form description; the first three types of questions can be considered "closed" questions (see Dohwrenwend, 1965). These formats can be thought of as different question–answer "idioms" (Marlaire and Maynard, 1990), which are probably familiar to most survey respondents from other interactional contexts. Each of these idioms prefers an answer of a specific type (e.g., "yes" or "no"), which we refer to as a "formatted answer" because it matches the format of the question.

Our analysis presumes, but does not analyze in detail, the paradigmatic three-part structure of the question–answer–acknowledgment sequence (see this volume, Chapter 1; Cannell and Kahn, 1968:572; for parallels see Mehan, 1979:52–53 and Marlaire and Maynard, 1990:89). Often, in survey interviews — as in the testing sequences Marlaire and Maynard (1990) studied — the receipt is followed by a silence. During the receipt and this silence, the interviewer is often engaged in recording one answer and preparing to ask the next question. Participants can expand this basic three-part structure in several ways, for example, by commenting on the interview schedule, making remarks to others, and so on, or shorten it, as happens when the interviewer does not provide an audible receipt of an answer.

Our analysis uses the methods of conversation analysis, which provides a theory about how interactional sequencing makes utterances and associated interactional objects understandable (Heritage, 1984b; Maynard and Clayman, 1991; also see Chapter 1, this volume). Conversation analysis also offers a body of prior research and methodological procedures that enable systematic, close examination of interactional detail in the interview. Such examination permits identifying conversational practices — elements of the interactional substrate — that persist in the relatively constrained environment of the standardized interview. More generally, analysis of these practices shows how the organization of interaction enters into the production of social science data.

In negotiating an understanding of the concepts used in survey questions and of the structure of the task, interviewers and respondents draw on the interactional substrate, which includes the interactional resources we examine here. The interactional substrate consists of those basic interactional skills that interactants use to arrive at an *accountable answer* (Maynard and Marlaire, 1992). In standardized interviewing, an accountable answer is one that interactants can treat as reasonable, objective, verifiable, or properly achieved. Whether or not an answer is accountable, and how it is accountable, depends upon an organization of practical activities that constitute the interaction. These skills, which make up the interactional substrate of the interview, are presumed by and yet largely ignored in the rules of standardized interviewing (Marlaire and Maynard, 1990; Maynard and Marlaire, 1992).

The data we analyzed were drawn from two sources. One is transcripts of audiotapes of 24 computer-assisted telephone interviews (CATI) conducted by

U.S. Bureau of the Census interviewers for a test of redesigned questions about labor force participation in the Current Population Survey (CPS); the other is a set of 29 CATI interviews conducted at the University of Wisconsin Survey Center (UWSC/AW). The information available to us on the audiotapes approximates that available jointly to participants speaking on the telephone. The CPS tapes were not screened or selected using any criteria; they constitute a haphazard collection of interviews. The CPS tapes include only the portion of the interview concerned with labor force participation, and four different versions of the labor force participation questions appear on the tapes we analyzed. The UWSC/AW tapes were collected by asking 15 interviewers to each record two consecutive interviews on their shifts on the Wisconsin Continuous National Survey (WISCON) during approximately a two-week period in summer 1995. The WISCON fields a very small random digit dialing (RDD) sample of the continental United States every day. The tapes from the UWSC/AW include the entire interview.

In identifying segments of interaction for analysis, we consider that a question–answer segment begins with the turn in which the interviewer reads a question from the interview schedule and ends in the turn in which the next question is read. We partition the interview into overlapping segments, but the analysis considers interaction in other parts of the interview when it is relevant. We identified segments of interaction for analysis by examining the activity that occurs in the slot after the interviewer completes reading (or verifying) the question from the interview schedule; thus, we focus on the interactional slot in which the respondent's first turn could occur. In a preliminary analysis of the CPS interviews (Schaeffer, Maynard, and Cradock, 1993; Schaeffer and Maynard, 1995) we identified several interactional phenomena that seemed to occur in environments where there was some kind of uncertainty, either about the task or about the respondent's answer.

Based on this analysis, we identified different forms that the slot after the initial reading of the question could take and selected three for description here. These are the following: a question from the respondent (whether or not preceded by a silence or a particle), an unformatted answer that "reports" some potentially relevant facts (whether or not the report is preceded by a silence or particle), a silence or particle or other delay from the respondent (e.g., "uhhh," or "well") followed by a probe from the interviewer. The present analysis describes segments of these three types—requests for clarification (one type of question a respondent could ask), reports, and delays before answers by the respondent that are followed by preemptive probes—that occur following closed questions.

ACTIVE OCCASIONS: REQUESTS FOR CLARIFICATION

When a respondent does not understand something in the task—that is, when a situated ambiguity arises—an obvious recourse is to ask for clarification. When the interviewer treats the respondent's utterance as a request for clarification and responds to it as such, we can see the request for clarification as an "active" solicitation of help by the respondent. We begin by briefly examining a sequence of this type (for more detail about requests for clarification, see Moore and Maynard, Chapter 13). The following excerpts show several things: how even a question that appears to be "clear" can be rendered ambiguous by a particular respondent's situation, examples of how respondents actually ask for clarification (which is relatively rare in most standardized interviews), and how the structure of the request and the response to it unfolds.

(1) 001c0101, version C: 72

```
FI:    Do you operate your own business or a farm? hh
       (0.6)
FR:    Years ago?
       (0.3)
FI:    uh no currently (.) they want everything
       [currently ]
FR:    [No nothing] now no I jus::
       (2.0)
FI:    Okay uh- now I'm going to ask you a few (ch-)questions
       about work related activities for last week .hhh
```

When she was told the topic of the interview (not shown), the respondent in excerpt 1 volunteered that "at 82 years I don't think I'll be laboring any more," and that she is also "handicapped I... don't even get anywhere." When she was then subsequently asked if she operates her own business or farm (line 1), the respondent asks, "Years ago?" (line 3). The utterance could be a candidate answer or an invitation to the interviewer to clarify the time period to which the question refers. If "years ago" is heard as a candidate reference period, it proposes that the survey question was, despite appearances, responsive to the information she had already provided. Thus, the respondent might be proposing that the interviewer's question is consistent with the *common ground*—what participants mutually know, believe, and suppose—that is available to both parties to the interview because it has developed in the interview up to that point (Clark, 1985; Clark and Marshall, 1981). In addition, the respondent's question displays features of her understanding of the task: that she has been asked a question for which a time is relevant (possibly as a response, possibly as a reference period). The interviewer's statement clarifies that "they want everything currently" (lines 5–6), which establishes her "footing" as "not the author of the question" [see Goffman (1979) and Houtkoop-Steenstra (2000)]. The respondent then provides her formatted answer ("No") in overlap with the interviewer (line 7), adds an elaboration that provides an emphasis that suggests that the interviewer's disambiguation has been attended to, and reflects the interviewer's reference to the present ("nothing now" line 7), and projects the possibility of a further elaboration ("I jus::" line 7) before pausing (line 8). The respondent's question and the interviewer's response are shaped as an insertion sequence (lines 3 and 5) within the sequence of the survey question and its answer (lines 1 and 7) (Schegloff, 1980). The sequence has a predictably orderly trajectory in which the question that begins the insertion sequence is answered before the framing question is.

The respondent in excerpt 2 also requests a clarification—about the meaning of "you" in the question. [The interpretations of "you" documented by Belson (1981), for example, suggest that the ambiguity the respondent points to is a common source of misunderstanding in survey interviews.] As in excerpt 1, the respondent's question is answered immediately. The interviewer's statement that these questions are about the respondent "personally" and "then" the "next" questions will concern the respondent's husband projects a sequence of questions and specifies a contrast set (you personally vs. your husband) that further specifies the meaning of "you" (lines 9–14).

(2) 14c0402, version C: 38

```
FR:    =yeah=
FI:    =t- did okay alright (0.5) ((tch)) (0.5).hhh Naow
       (0.3)
       Last week
       (0.5)
FI:    did you (.) do any work at all including work for pay
       or other types of compensation
FR:    N:ow are you asking this of me? [personally?]
FI:                                    [You yeah ub]
       you personally .hh
       be[cause] I'm speaking with=
FR:      [(    )]
FI:    =you pers- and then the next question will be (.)
       concerning (.) your husband=
FR:    =Oh um
       (0.7)
FI:    [So do y- ]
FR:    [I'm on ma]ternity lea- maternity leave right now=
FI:    =Okay so nother words you are .hhh w- last week you
       actually did not work [right?]
FR:                          [No did] not work=
FI:    =O:kay (.) for any pay
FR:    Right
FI:    h Okay last week did you do any unpaid work in the
       family business ?
```

The respondent begins to speak in the position (line 15) at which she could be expected to answer the initial survey question following completion of the insertion sequence. Instead of producing a "yes" or "no" answer in the format that the question implicitly requests, however, she reports a potentially relevant fact about herself—that she is on maternity leave (line 18). This question about whether the respondent worked for pay in the previous week may be unambiguous for a large majority of respondents, but is ambiguous for someone on maternity leave. It is this respondent's strategy of reporting that we examine next. In the next section, we see that after a report, the interviewer may intervene as if in response to a request for clarification.

LATENT UPSHOTS: REPORTS

When asked if he owns his own business or farm, the respondent in excerpt 3 does not produce a formatted answer. Instead he begins with a particle "Weahh" (line 2) that sounds like a combination of "Well" and "Yeah" or perhaps a contraction of "We" and "yeah." As a combination of "well" and "yeah," this particle suggests a mitigated agreement or acceptance of the presumption of the question. There is then a full "Well" and a report that he is in business with his sister. The particle "well" is known to preface disagreements or to provide for sequentially weak agreement or acceptance. Turns at talk that do not affiliate with a prior turn properly (as in disagreements or refusals) are often marked as dispreferred by delays, hesitations, and the like (Pomerantz, 1984a). The particle "well" here may indicate that the respondent is about to produce a similarly dispreferred turn, in this case, an answer that is not in the format requested by the question. This recitation of potentially relevant information does not draw any implications, and the subsequent silence (line 4) can be heard as produced by the respondent waiting for a response from the interviewer.

(3) 008c0301, version C: 11

```
FI:     Do you have your own business or farm?
MR:     Weahh Well I'm (.) in partnership with
        my sister in the shoe repair business?
        (1.5)
FI:     O:kay so that would (.) uh qualify as your
        own business?
MR:     I guess so=
FI:     =uh huh
        (6.0)
FI:     Now I'm going to ask a few questions about
```

When the interviewer responds, it is in the form of a "probe" that proposes an implication for the respondent to confirm (lines 5–6) (Moore, 1999). The respondent's reply (line 7) confirms in a hedged way: it would be a "guess" on his part and displays some uncertainty.

Excerpt 4 has many similarities to excerpt 3. The respondent in excerpt 4 has already reported that a business he works for is "both of ours" (not shown) and that he has "one job and two businesses" (not shown). When asked if his wife is "an employee of a private company, business, or individual" (lines 11–13) in a way that ignores this information, the respondent reports "Just the businesses we have" (line 14). This utterance uses "the" to formulate "the businesses" as a backward reference to the businesses he reported earlier, with the implication that the interviewer already knows the information. The respondent treats the interviewer's "Scuse me?" (line 15) as a request to repeat his report. The respondent's repetition (line 16) intensifies the backward reference to the businesses by characterizing them as the "ones," possibly proposing that the backward reference may have been the source of the interviewer's problem. That is, the revised report at line 16 may be an attempt to mitigate the interviewer's trouble at line

15 by strengthening the proposal that the businesses the respondent is talking about are already known to the interviewer. The phrase "that's all" appended to the utterance emphasizes the suggestion that the business in which the wife is employed is already known, and may also suggest that the respondent has provided information about "all" the businesses or "all" the information he can.

(4) 009c0302, version B: 176

MR: Uh: Secretarial (.) answering phone typing
(.) word processing
(2.8)
FI: Okay::
(6.0)
FI: Alright we'll take a minute to make a nóte of that
(1.2)
FI: and then go on
(1.5)
FI: to ask the next question
which would be .hh is she an employee of a private
company business or individual for wages salary
or commissions?
MR: Just the businesses we have
FI: Scuse me?
MR: Just the ones that we have that's all
FI: Okay: and she is an employee of a private company then?
(2.2)
MR: Uh what do you mean by private com[pany]?
FI: [.hh]
Well as opposed to a federal government
employ[ee or a]
MR: [Oh yes] [yes]
FI: [Okay]
(.) very well .hhh and then we'll go on

The interviewer begins her turn with "Okay" (line 17). "Okay" may accept the respondent's report and, coupled with "and," it signals a return to the official agenda (Heritage and Sorjonen, 1994). The interviewer's abbreviated repetition (line 17) of the survey question then proposes a candidate answer ("employee of a private company") for the respondent to confirm. The shortened question at line 17 actually omits the word "businesses," which the respondent had used to describe his work. The long ensuing pause (2.2 seconds at line 18) contrasts with the immediacy of the respondent's initial answer (line 14); the delay and the particle ("Uh," line 19) that begins the respondent's utterance suggests the presence of a problem in understanding. This possibility is confirmed by the respondent's question, "What do you mean by private company?" (line 19), which begins an insertion sequence similar to that in excerpt 2. The interviewer defines "private company" by "opposing" it to the "federal government" (lines 21–22). The contrast appears to be instructive, because the "Oh" that begins the respondent's utterance (line 23, in overlap with the interviewer) typically indicates a change of state in the speaker's understanding (Heritage, 1984a).

Turning to excerpt 5 (which reproduces excerpt 2, for convenience), we can now see the respondent's utterance that begins at line 15 as the beginning of a report.

(5) 14c0402, version C: 38

```
FR:  =yeah=
FI:  =t- did okay alright (0.5) ((tch)) (0.5).hhh Naow
     (0.3)
     Last week
     (0.5)
FI:  did you (.) do any work at all including work for pay
     or other types of compensation
FR:  N:ow are you asking this of me?[personally?]
FI:                                 [You yeah ub]
     you personally.hh
     be[cause] I'm speaking with=
       [(   ) ]
FI:  =you pers- and then the next question will be (.)
     concerning (.) your husband=
FR:  =Oh um
     (0.7)
FI:  [So do y- ]
FR:  [I'm on ma]ternity lea- maternity leave right now=
FI:  =Okay so nother words you are .hhh w- last week you
      actually did not work [right?]
FR:                         [No did] not work=
FI:  =O:kay (.) for any pay
FR:  Right
FI:  h Okay last week did you do any unpaid work in the
     family business?
```

The report is positioned (at lines 15–18) after the completion of the insertion sequence, in a turn that is available for an answer to the original survey question, "did you do any work at all..." (lines 6–7). The recitation details a potentially relevant fact ("I'm on maternity lea- maternity leave right now") that has the effect of displaying the original survey question as ambiguous for this respondent. In view of the facts the respondent has stated, it is relevantly ambiguous whether the *goal* of the question "did you do any work at all..." is to determine whether the respondent had a job last week that she expected to return to once the leave was complete, received pay while on leave, or actually worked at a job for pay or other types of compensation. The interviewer's "okay" at line 22 accepts the respondent's report and returns to the agenda of the interview by formulating one possible upshot of the respondent's report as a restatement of the respondent's report ("so nother words" line 19), thereby selecting one of the possible interpretations of the survey question as relevant (lines 19–20). The respondent confirms at line 21, and the interviewer follows with another "okay" and an additional check about the respondent's employment status (line 22). The respondent again confirms (line 23), and then the interviewer continues with the next question.

There are important similarities among the reports in excerpt 2/excerpt 5, excerpt 3, and excerpt 4. In excerpts 3 and 4, the initial reading of the survey question is followed by a report; in excerpt 2, the report appears in a comparable turn following an insertion sequence. In all three cases (and in the other cases in the interviews we examined), the report does not provide a formatted answer to the survey question, nor does it formulate an upshot that would "answer" the question from the details presented in the report. Following the report, the interviewer produces a probe that asks the respondent to accept or formulate such an upshot; such a formulation would then stand as an answer to the survey question that could be seen as the respondent's own—an accountable answer. The probes in excerpt 2 ("...last week you actually did not work right?" lines 19–20) and excerpt 3 ("...that would qualify as your own business?" lines 5–6) are "leading" and propose an upshot as a formulation for the respondent to accept. In the slot that the interviewer has thus again provided for an answer that would be properly formatted for the initial question, the respondents in excerpts 3 and 4 produce utterances that display problems in understanding the question in the first place (i.e., "I guess so" and "What do you mean by private company?"), which may thus account for the original report. In all three cases, as in others we have examined, the report appears to accompany an uncertainty on the respondent's part—not about the facts that are detailed in the reports (which are produced with no signs of hesitation or markers of uncertainty)—but about how to translate those details into the categories of the survey question. That is, the problem seems to be a kind of task uncertainty, and not state uncertainty.

Reports in Other Settings

We have described the recitations by respondents that we have been examining as *reports*, borrowing the term from Drew (1984), where he examined talk in contexts other than the survey interview and described how speakers may use reports to convey unstated upshots of various kinds. In excerpt 6, speaker C, who was going to take I to Syracuse, conveys the bad news that he is not going to Syracuse after all, by reporting a relevant circumstance (line 1). I's elongated "oh" at line 2 shows that she understands the upshot, which she then produces in line 4 and has confirmed in line 6. [See also Heritage (1984a) and Maynard (1992:332) for further discussion of this segment.]

(6) Simplified from Drew (1984:130)

```
1   C:  Yihknow I really don't have a place tuh stay
2   I:  hh Oh:::::.hh
3       (0.2)
4   I:  hhh So yih not g'nna go up this weeken?
5       (0.2)
6   C:  Nu::h I don't think so
```

Features of reports include the following: A report details an activity without making the upshot explicit; it is up to the recipient of the report to gather the upshot; the recipient may address what the upshot is taken to be in a subsequent turn; a person providing a report is responsible for the report, but not for what the recipient makes of it; because of the recipient's role in determining what is made of a report, the role of the speaker's intentions is mitigated (Drew, 1984:137, 139). The various potentials present in a report make it well adapted to delicate interactional environments. In particular, because a report leaves it to the recipient to gather the upshot, the recipient can calibrate the level of delicacy with which the upshot is formulated.

The reports we find in our data occur in situations in which some *goal* of the question is ambiguous. Almost any survey question can be rendered ambiguous by the particular circumstances (such as being 80 years old or a new mother) that a respondent has to report about. In the terms of social–information–processing models, the respondent has a kind of comprehension or interpretation problem. By displaying facts or details that show how the meaning of the survey question is ambiguous for her situation, the respondent displays *an* understanding of the question and the task. This information may be used by the interviewer to "diagnose" or "correct" problems in understanding or applying survey questions.

Based on the interviews we have examined, it is our impression that reports such as those described here are more likely to occur for questions that ask the respondent to classify themselves or to classify events or behaviors in her own life as relevant to the survey or not than for other types of questions, such as questions about subjective phenomena. This observation is compatible with the way respondents commonly use their reports: to solicit the interviewer's help in applying survey concepts to their own experiences.

Like the reports Drew (1984) describes in conversational settings, a respondent's report in a survey interview occurs in a situation that might require some "delicacy" in interactional terms. When an interviewer asks a survey question, she may be seen as proposing that the question is both intelligible and relevant. Thus, by presenting information as a report, the respondent avoids overt (and dispreferred) challenges to the assumptions underlying the survey question. Reporting is also instrumental in that it provides information that may be relevant to achieving a properly formatted answer. The structure of the report, however, leaves drawing the upshot to the interviewer—it proposes that the upshot is the interviewer's "job." Because subsequent proposals by the interviewer that the respondent determine the meaning of survey concepts (e.g., "we want your interpretation") violate this structure, such proposals can engender awkward moments between interviewer and respondent. Although a respondent who gives a report is offering diagnostically relevant information and is thereby being a "good" respondent, an interviewer who does not draw the upshot is not behaving as recipients of reports have been observed to behave in other interactional contexts.

TENTATIVENESS: DELAYS AND PREPROBES

We have seen that requests for clarification explicitly solicit intervention by the interviewer and that respondents' reports also occasion intervention by deploying a mechanism that leaves the upshot of a report to the interviewer. We next examine sequences in which the reading of the question is followed by a silence and then a preemptive probe (or preprobe) by the interviewer. In such a sequence, the initiative for the intervention appears to be the interviewer's, although the intervention is clearly occasioned by the respondent's delay in filling the turn provided for her answer. Interviewers often treat such delays as indicating that the respondent has some problem. In the interviews that we have examined, when the respondent delays answering, the interviewer routinely produces a preprobe.

Before examining examples from survey interviews, we note that the situation in which there is a "delay" in filling the space provided for the respondent's answer is similar to a situation that occurs in other social settings. Both interviewers and respondents have experience with situations in which the assertions of speakers are met with silence or delay. Pomerantz (1984b) notes several techniques that speakers use to "pursue a response" from listeners. Speakers may clarify potential problems in understanding, check on presumed common knowledge, or shift their position, and these techniques are part of the tacit knowledge participants bring to the survey interview. An example from talk in a setting other than a survey interview serves as illustration. In excerpt 7, which is from talk in a situation other than an interview, B's statement "And that went wrong" is followed by silence and then particles from A. B then resumes talking with a phrase, "that surgery, I mean," which diagnoses the referent for "that" as possibly being the source of A's delay in responding.

(7) (Pomerantz 1984b: pp. 156–157; simplified)

```
B:     And that went wrong.
       (1.0)
A:     Well, uh ---
B:     That surgery, I mean.
A:     I don't -
B:     Isn't she the one who- I think I heard about it-the
       daughter in law told me- wasn't she playing golf at the
       Valley Club?
```

Turning now to the survey interview, we can see that in excerpt 8, the asking of the question (line 3) is followed by a long silence. This silence exceeds the 1-second "maximum" that is commonly observed in conversation (Jefferson, 1989). This silence can be heard as produced by the respondent, whose "turn" it is to answer, and who further delays the start of that turn with the particle "uh::." Although the respondent has yet to answer the question, the interviewer follows the delay with a preemptive probe (line 6). The interviewer appears to use the silence and the delay diagnostically as indicating that the respondent has some trouble in understanding the question or formulating an answer. The interviewer's preprobe provides a contrast with "private company" that may embody the interviewer's analysis that the respondent's trouble is understanding this survey concept.

(8) 019c0502, Version C: 149

```
FI:   Okay
      (4.0)
FI:   And are you an employee of a private company?
      (2.5)
FR:   Uh::=
FI:   =As opposed to government puh- (.) company?
FR:   It i:s a private company we are doin some
      (0.5)
FR:   contract work for the goverment[(      )]
FI:                                  [Uh  huh]
      (2.5)
FI:   .hh O:kay (.) now uh last week hhh did Mark Roland do
```

The respondent replies immediately with a report (lines 7–9) that incorporates both terms from the question ("private company") and terms from the preprobe ("government"). By moving to the next question without providing additional probes (line 12), the interviewer appears to accept the report as sufficient, although the respondent has not provided a properly formatted answer (see Hak, this volume, Chapter 19, for a discussion of coding by interviewers).

Two features of the situations in the interviews we have examined in which preprobes occur are notable. First, the slot that follows the initial reading of the survey question can be quite complex. It might include various combinations of silence, particles such as "well" or "uhh," expressions of uncertainty such as "I dunno," false starts, and so forth. This complexity can make it difficult to determine how long it was from the time a question was asked until the beginning of the next turn (whether the "turn" is the respondent finally "answering" or the interviewer delivering a preprobe). Such complexity is relevant, for example, for the measurement of response latency, which may indicate problems in comprehension, retrieval from memory, or attitude salience or strength (e.g., Bassili, 1996:321). The delays observed when preprobes are produced—delays that are jointly produced by the interviewer and respondent—are frequently rather substantial, approaching 2 seconds (e.g., the discussion of excerpts 9, 10, and 12). Second, the interviewer's preprobe routinely displays some diagnosis of where the reason for the delay might lie. The proposals the interviewers make in the preprobes we have examined include problems in comprehension or interpretation, in retrieval of information from memory, and in mapping an answer onto the response categories. Thus, in excerpt 8, the preprobe that provides "government" as a contrast for "private company," proposes that the respondent might have trouble interpreting the survey concept "private company."

Standardized interviewers do not have available to them the range of options Pomerantz (1984b) describes for pursuing a response, but they use preprobes to propose a range of diagnoses or solutions as we see in the next series of excerpts. In excerpt 9, when the silence after the completion of the question reaches 2.5 seconds, the interviewer repeats the question with a different emphasis, possibly proposing her initial reading or a hearing problem as the source of the respondent's delay in answering.

(9) 012c0305, version C: 111

```
 1  FI:    Now is his business incorporated?
 2         (2.5)
 3  FI:    Is the business incorporated?
 4  FR:    I don't know
 5         (5.0)
 6  FI:    Last week (.) did Kathy Gimbel do any work for
 7         pay or profit?
```

In excerpt 10, the silence is ended after 1.8 seconds by the interviewer's providing (again) the categories the respondent should use for answering.

(10) AW01:330

```
 1  FI:  ↑Okay please tell me whether you agree: or disagree:
 2       with each of the following statements?.hhh The first
 3       statement is (0.4) people like me don't have any say
 4       about what the government does.
 5       (1.8)
 6  FI:  And it's just agree or a disagree
 7       (1.2)
 8  FR:  uh::m uh-
 9       (0.5)
10  FR:  Agree
11  FI: .hhh Government agencies (.) usually
```

In terms of social–information–processing models of the response process, the proposal in excerpt 10 might address a problem in mapping the answer onto the response categories, while that in excerpt 11 might attempt to reduce the difficulty of retrieval. The delay is much briefer (0.5 seconds) than the delay that precedes the other preprobes in our examples; it is possible that the interviewer produces this preprobe quickly because it is a routine solution to a routine problem with this question [compare other solutions that appear to be routine in Schaeffer and Maynard (1996)].

(11) 004c0202, version B:44

```
 1  FI:    When did you last work at a job or
 2         business for two weeks or longer
 3         (0.5)
 4  FI:    Has it been more [than a] year ago? or within the=
 5  MR:                     [(    )]
 6  FI:    =last year?
 7  MR:    Seventeen years ago
 8  FI:    Oka::::y?
 9         (2.5)
10  MR:    hhhh=
11  FI:    =And last week {q20} did Rhonda King do any work
```

Excerpt 12 presents a more complex delay. After the question (lines 7–9) there are both silence and tokens of hesitancy (lines 10–13). Eventually, the respondent issues laughter (line 15) that may invite reciprocation from the interviewer (who declines to laugh by issuing a preprobe, lines 16–17) (see Lavin and Maynard, this volume, Chapter 15). The interviewer's preprobe ("well this would be a number between one and seven...") proposes that the respondent might need information about the type of answer requested.

(12) AW22:651

```
FI:   .hh Now I'm going to read a list of the ways
      you might have felt or behaved during the past
      week .hh please tell me how many days you've
      felt this way
      (0.3)
MR:   mhmm
FI:   On how many days during the past week did
      you feel bothered by things that didn't usually
      bother you
      (0.7)
MR:   hh
      (0.9)
MR:   huh (0.2) .hh[hh
FI:                [uhuh
MR:   uhuh hah hih hih hah
FI:   .hh Well this would be a number between
      one and seven for heh [heh for days of the week
MR:                         [GO:D
FI:   .hh
MR:   God we're gonna have some re:al you're gonna
      ge-really get into some philosophical questions
      (I think) .hhhh u::m (0.2) ooh-ah-uh-eh howm
      hh I I I don't even (0.3) say that again
FI:   .hhh How many on how many days during the pa[st
```

The information the interviewer provides is not, however, sufficient to allow the respondent to give a properly formatted answer. His subsequent talk is a turn that begins with an expletive, a possible complaint about the question (lines 20–21), and additional hesitations before a request that the question be said again (line 23).

CONCLUSION

Even within the constraints of the standardized interview, both interviewers and respondents can be seen deploying resources from the interactional substrate toward solving problems in producing an accountable answer to a survey question. The "cognitive processing" required to produce an answer is often observable in the interaction. The three configurations that occasion intervention by the interviewer that we have described here vary in how "explicit" the solicitation of help by the respondent can be seen to be. Although explicit requests for clarification have been previously recognized as indicating that a question may be difficult to understand (e.g., Oksenberg, Cannell, and Kalton, 1991), such explicit questions are relatively rare for most survey questions. This can make it difficult for survey researchers to identify questions that are associated with comprehension problems during testing of questions. The reports, hesitations, and the like described here may be more common. In characterizing both reports and delays as possibly indicating comprehension problems (see Schaeffer, Maynard, and Cradock, 1993; Schaeffer and Maynard, 1995), we drew on the actions and displays of the participants. The insights suggested by examining these interactions in detail can be used to identify questions that may need further developmental work (e.g., in interaction coding as part of question testing), to develop indicators for comprehension or processing problems in experimental and quantitative analyses (such as those used in Bloom and Schober, 1999, and Belli, Lepkowski, and Kabeto, in press), to consider how the repertoire of behaviors for standardized interviewing might be augmented, and to identify issues that are relevant when training interviewers.

ACKNOWLEDGMENTS

The analysis of reportings, delays, and preprobes in this chapter was presented at the International Conference on Survey Measurement and Process Quality in Bristol, England, April 1995 and at the Vrije Universiteit Amsterdam in November 1995.

We thank Frederick Conrad and Michael Schober for comments on an earlier draft of this chapter.

This research was partially supported by a Joint Statistical Agreement (90-45) with the U.S. Bureau of the Census. Supplementary support was provided by the Research Committee of the Graduate School of the University of Wisconsin, Madison, and by the Ameritech Foundation Fellowship Program. Computational support was provided by the Center for Demography and Ecology, which receives core support from the National Institute for Child Health and Human Development (HD05876). The views expressed in this chapter are the authors' and not those of the funding agencies.

We thank the University of Wisconsin Survey Center for collecting interviews for analysis and the U.S. Bureau of the Census for permission to analyze the Current Population Survey pretest interviews. In the excerpts from the interviews, all names of persons, businesses, and locations are fictitious.

REFERENCES

Bassili, J. N. 1996. "The How and Why of Response Latency Measurement in Telephone Surveys." Pp. 319–346 in N. Schwarz and S. Sudman (Eds.), *Answering Questions: Methodology for Determining Cognitive and Communicative Processes in Survey Research.* San Francisco, CA: Jossey-Bass.

Belli, R. F., J. M. Lepkowski, and M. U. Kabeto. In press. "The Respective Roles of Cognitive Processing Difficulty and Conversational Rapport on the Accuracy of Retrospective Reports of Doctors' Office Visits." In *Seventh Health Survey Research Methods Conference Proceedings.* Chicago: University of Illinois.

Belson, W. A. 1981. *The Design and Understanding of Survey Questions*. Aldershot, England: Gower.

Bloom, J., and M. Schober. 1999. "Respondent Cues That Survey Questions Are in Danger of Being Misunderstood." Paper presented at the annual meetings of the American Association for Public Opinion Research, May, St. Pete Beach, FL.

Brenner, M. 1985. "Survey Interviewing." Pp. 9–35 in M. Brenner, J. Brown, and D. Canter (Eds.), *The Research Interview: Uses and Approaches*. London: Academic.

Button, G. 1987. "Answers as Interactional Products — Two Sequential Practices Used in Interviews." *Social Psychology Quarterly* 50: 160–171.

Cannell, C. F., F. J. Fowler, Jr., and K. H. Marquis. 1968. *The Influence of Interviewer and Respondent Psychological and Behavioral Variables on the Reporting in Household Interview*. Washington, DC: U.S. Department of Health and Human Services.

Cannell, C. F., and R. L. Kahn. 1968. "Interviewing." Pp. 526–95 in G. Lindzey and E. Aronson (Eds.), *The Handbook of Social Psychology*, Vol. II. Reading, MA: Addison-Wesley.

Cannell, C. F., P. V. Miller, and L. Oksenberg. 1981. "Research on Interviewing Techniques." Pp. 389–437 in S. Leinhardt (Ed.), *Sociological Methodology*, Vol. 11. San Francisco: Jossey-Bass.

Clark, H. H. 1985 "Language Use and Language Users." Pp. 179–231 in G. Lindzey and E. Aronson (Eds.), *Handbook of Social Psychology*, 3rd ed., Vol. 2. New York: Random House.

Clark, H. H., and C. R. Marshall. 1981. "Definite Reference and Mutual Knowledge." Pp. 10–63 in A. Joshi and B. L. Weber (Eds.), *Elements of Discourse Understanding*. Cambridge: Cambridge University Press.

Collins, M. 1978. "Interviewer Variability: A Review of the Problem." *Journal of the Market Research Society* 22: 77–95.

Dohrenwend, B. S. 1965. "Some Effects of Open and Closed Questions on Respondents' Answers." *Human Organization* 24: 175–184.

Drew, P. 1984. "Speakers' Reportings in Invitation Sequences." Pp. 129–51 in J. M. Atkinson and J. Heritage (Eds.), *Structures of Social Action: Studies in Conversation Analysis*. Cambridge: Cambridge University Press.

Dykema, J., J. M. Lepkowski, and S. Blixt. 1997. "The Effect of Interviewer and Respondent Behavior on Data Quality: Analysis of Interaction Coding in a Validation Study." Pp. 287–310 in L. Lyberg, P. Biemer, M. Collins, E. deLeeuw, C. Dippo, N. Schwarz, and D. Trewin (Eds.), *Survey Measurement and Process Quality*. New York: Wiley.

Fowler, Jr., F. J., and T. W. Mangione. 1990. *Standardized Survey Interviewing: Minimizing Interviewer-Related Error*. Newbury Park: Sage.

Goffman, E. 1979. "Footing." *Semiotica* 25: 1–29.

Heritage, J. 1984a. "A Change of State Token and Aspects of Its Sequential Placement." Pp. 299–345 in J. M. Atkinson and J. Heritage (Eds.), *Structures of Social Action: Studies in Conversation Analysis*. Cambridge: Cambridge University Press.

Heritage, J. 1984b. *Garfinkel and Ethnomethodology*. Cambridge, MA: Polity.

Heritage, J., and M. L. Sorjonen. 1994. "Constituting and Maintaining Activities across Sequences: And Prefacing as a Feature of Question Design." *Language in Society* 23: 1–29.

Houtkoop-Steenstra, H. 2000. *Interaction and the Standardized Survey Interview: The Living Questionnaire*. Cambridge: Cambridge University Press.

Hyman, H. H., W. J. Cobb, J. J. Feldman, C. W. Hart, and C. H. Stember. 1975 [1954]. *Interviewing in Social Research*. Chicago, IL: University of Chicago.

Jefferson, G. 1989. "Preliminary Notes on a Possible Metric Which Provides for a 'Standard Maximum' Silence of Approximately One Second in Conversation." Pp. 166–196 in D. Roger and P. Bull (Eds.), *Conversation: An Interdisciplinary Perspective*. Clevendon, Avon, England: Multilingual Matters, Ltd.

Levine, J. M., and L. B. Resnick. 1993. "Social Foundations of Cognition." *Annual Review of Psychology* 44: 585–612.

Mangione, T. W., F. J. Fowler, Jr., and T. A. Louis. 1992. "Question Characteristics and Interviewer Effects." *Journal of Official Statistics* 8: 293–307.

Marlaire, C. L., and D. W. Maynard. 1990. "Standardized Testing as an Interactional Phenomenon." *Sociology of Education* 63: 83–101.

Mathiowetz, N. A. 1998. "Respondent Expressions of Uncertainty: Data Source for Imputation." *Public Opinion Quarterly* 62: 47–56.

Maynard, D. W. 1992. "On Clinicians Co-implicating Recipients' Perspective in the Delivery of Diagnostic News." Pp. 331–358 in P. Drew and J. Heritage (Eds.), *Talk at Work: Social Interaction in Institutional Settings*. Cambridge: Cambridge University Press.

Maynard, D. W., and S. E. Clayman. 1991. "The Diversity of Ethnomethodology." *Annual Review of Sociology* 17: 385–418.

Maynard, D. W., and C. L. Marlaire. 1992. "Good Reasons for Bad Testing Performance: The Interactional Substrate of Educational Exams." *Qualitative Sociology* 15: 177–202.

Mehan, H. 1979. *Learning Lessons: Social Organization in the Classroom*. Cambridge, MA: Harvard University Press.

Moore, R. 1999. "Achieving Understanding in the Standardized Survey Interview." Unpublished Ph.D dissertation, Department of Sociology, Indiana University, Bloomington, IN.

Oksenberg, L., C. Cannell, and G. Kalton. 1991. "New Strategies for Pretesting Survey Questions." *Journal of Official Statistics* 7: 349–365.

Pomerantz, A. M. 1984a. "Agreeing and Disagreeing with Assessments: Some Features of Preferred/Dispreferred Turn Shapes." Pp. 57–100 in J. M. Atkinson and J. Heritage (Eds.), *Structures of Social Action*. Cambridge: Cambridge University Press.

Pomerantz, A. M. 1984b. "Pursuing a Response." Pp. 152–163 in J. M. Atkinson and J. Heritage (Eds.), *Structures of Social Action: Studies in Conversational Analysis*. Cambridge: Cambridge University Press.

Schaeffer, N. C., and D. W. Maynard. 1995. "Occasioning Intervention: Interactional Resources for Comprehension in Standardized Survey Interviewing." Paper presented at the International Conference on Survey Measurement and Process Quality, April, Bristol, England.

Schaeffer, N. C., and D. W. Maynard. 1996. "From Paradigm to Prototype and Back Again: Interactive Aspects of Cognitive Processing in Standardized Survey Interviews." Pp. 65–88 in N. Schwarz and S. Sudman (Eds.), *Answering Questions: Methodology for Determining Cognitive and Communicative Processes in Survey Research*. San Francisco: Jossey-Bass.

Schaeffer, N. C., D. W. Maynard, and R. Cradock. 1993. "Negotiating Certainty: Uncertainty Proposals and Their Disposal in Standardized Survey Interviews." Unpublished manuscript, Discussion Paper 93-25, Center for Demography and Ecology, University of Wisconsin, Madison.

Schaeffer, N. C., and E. Thomson. 1992. "The Discovery of Grounded Uncertainty: Developing Standardized Questions about Strength of Fertility Motivation." Pp. 37–82 in P. V. Marsden (Ed.), *Sociological Methodology 1992*, Vol. 22. Oxford: Basil Blackwell.

Schegloff, E. A. 1980. "Preliminaries to Preliminaries: 'Can I Ask You a Question?' " *Sociological Inquiry* 50(3/4): 104–152.

Suchman, L., and B. Jordan. 1990. "Interactional Troubles in Face-to-Face Survey Interviews." *Journal of the American Statistical Association* 85: 232–253.

Tourangeau, R. 1984. "Cognitive Sciences and Survey Method." Pp. 73–100 in T. B. Jabine, M. L. Straf, J. M. Tanur, and R. Tourangeau (Eds.), *Cognitive Aspects of Survey Methodology: Building a Bridge between Disciplines*. Washington, DC: National Academy Press.

CHAPTER 13

Achieving Understanding in the Standardized Survey Interview: Repair Sequences

Robert J. Moore and Douglas W. Maynard

The production of valid statistical knowledge about societies using sample surveys depends on several conditions. First, the questions in the survey instrument must elicit the type of information the researcher intends, which means respondents must understand the questions in the same way as the researcher (Czaja and Blair, 1996:94). Under such conditions, valid inferences about *individuals* can be made. Second, inferences about *populations* are made possible by the aggregation of inferences about individuals. Such aggregation is valid if the sample surveyed is representative of the population of interest and if the answers collected are reliable and comparable (Hahn, 1990). Survey researchers implement a variety of standardization procedures intended to ensure the reliability of the respondents' answers "on the assumption that 'the same' responses to 'the same' stimulus questions are equivalent" (Hahn, 1990:245). Thus, interviewers are required to read survey questions exactly as written, and respondents are often required to answer by choosing from a set of prespecified options. This helps to achieve two types of intersubjectivity, or mutual understanding, that enhance validity in survey research: that between respondent and researcher over the meaning of the survey questions and that between respondent and interviewer regarding the procedure for asking and answering questions in a standardized way.

However, in the course of particular survey interviews, the participants will inevitably encounter various interactional troubles. Where these troubles hinder the interviewer's and respondent's attempts to achieve intersubjectivity over the

Standardization and Tacit Knowledge: Interaction and Practice in the Survey Interview
Edited by Douglas W. Maynard, Hanneke Houtkoop-Steenstra, Nora Cate Schaeffer, and Johannes van der Zouwen
ISBN 0-471-35829-0

survey questions or the standardization procedures, the quality of the resulting data is potentially jeopardized. In dealing with such troubles, either party may initiate generic repair practices. For example, a respondent may request the interviewer to clarify the meaning of a question or an interviewer may request the respondent to redo an answer that is formulated inadequately. Each of these are examples of *other-initiated repair*, in which a speaker attempts to remedy a trouble with an utterance produced by another speaker (Schegloff, 1997; Schegloff, Jefferson, and Sacks, 1977). In this chapter, we will show what other-initiated repair looks like, in its concrete details, in the standardized survey interview. In so doing, we will specify the ways in which the unique requirements of standardized interviewing sometimes conflict with generic conversational repair practices.

ACHIEVING UNDERSTANDING IN TALK-IN-INTERACTION

In ethnomethodology and conversation analysis, *intersubjectivity* (or mutual understanding between two people) is conceived as an inherently interactional phenomenon, rather than as a cognitive one (see Coulter, 1989; Garfinkel, 1967; Garfinkel and Sacks, 1970; Sacks, Schegloff, and Jefferson, 1974; Schegloff, 1992; Schegloff, Jefferson, Sacks, 1977; Schutz, 1962; Wittgenstein, 1953).[1] Any claim that a person "understands" another ultimately rests on the ability to demonstrate that understanding in and through social interaction. As such, mutual understanding is always a local achievement between participants.

Studies in conversation analysis (Sacks, Schegloff, and Jefferson, 1974; Schegloff, 1992; Schegloff, Jefferson, and Sacks, 1977) show that participants organize interactions on a turn-by-turn basis. That is, they construct any given utterance with reference to previous and next utterances: "A turn's talk will display its speaker's understanding of a prior turn's talk, and whatever other talk it marks itself as directed to," according to Sacks, Schegloff, and Jefferson, (1974:728). Utterances also have implicativeness: They project a fairly defined range of next turns (Schegloff and Sacks, 1973). Overall, then, conversation is sequentially organized. In addition, Sacks, Schegloff, and Jefferson, (1974:43) argue that speakers design utterances specifically for their particular recipients, displaying an orientation and sensitivity to what *that* recipient knows, or might know (Sacks, 1992:564). In other words, actors custom fit actions to their recipients, thus maximizing the probability of their understanding. This is known as *recipient design*.

Although sequential organization and recipient design make the practical achievement of understanding in interaction possible, their use by no means ensures that an utterance or action will be understood the first time through. In the event that the subsequent action by a recipient displays a *mis*understanding or trouble in understanding, there are a variety of *repair* practices available to the participants to remedy these troubles. Schegloff, Jefferson, and Sacks (1977) outline the basic features of an *organization of repair* that is available to speakers in conversation. "Repair" is a term for the removal of a variety of interactional obstacles that in some way prevent a speaker from continuing.

Such obstacles may consist, for example, of a failure to find a word, to hear another speaker's utterance, or to understand a reference. Schegloff, Jefferson, and Sacks, (1977) observe that repairs can be *initiated* by self (i.e., the producer of the trouble source) or by other (i.e., a recipient of the trouble source) and can be *completed* by self or other. In cases in which the trouble source constitutes an *error* or *mistake*, its repair is referred to more specifically as "correction." In the standardized survey interview, two classes of other-initiated repair appear to be somewhat common: respondent-initiated repairs on survey questions and interviewer-initiated repairs on respondent answers.

Under current survey practice, researchers must assume that respondents understand the questions more or less adequately unless they display some kind of trouble, such as requesting clarification or providing an "inadequate" answer. Rather than depending on recipient design to help individual respondents understand particular questions (standardized interviewers are prohibited from particularizing the questions they ask), survey researchers attempt to construct questions that will be clear to an entire population (see Houtkoop-Steenstra, 1995:92 on *audience design*; Suchman and Jordan, 1990). Indeed, survey questions are often pretested in order to identify and repair recurrent understanding and answering troubles encountered by respondents. Questions are usually considered good to the extent that they result in the ideal or paradigmatic question–answer sequence, that is, one in which the interviewer *only* reads the question as written and the respondent *only* provides a properly formatted answer (Fowler and Mangione, 1990; Czaja and Blair, 1996; Schaeffer and Maynard, 1996). For example:

(1) SPPA110:530

```
FI:   .hh During the last twelve months did you
      play any jazz music.
      (0.2)
FR:   Oh yes.
      (0.6)
FI:   .hh Did you play any jazz in a public. . .
```

From the survey researcher's perspective, the ideal standardized survey would consist of a series of paradigmatic sequences for every question and for every respondent. That way interviewer behavior across questions and across different interviewers can be held constant and, thus, the resulting data would be reliable (because the component of the variation due to the interviewers would be zero). However, deviations from the paradigmatic sequence appear to be unavoidable. Deviations often consist of attempts to repair troubles in hearing or understanding the questions on the part of the respondent as well as attempts by interviewers to repair answers that do not match one of the answer options. Pretests and behavior coding studies attempt to identify questions that elicit such deviations so they can be rewritten (common practice at one of the university survey centers we studied; see also Fowler and Cannell, 1996). In conversation analytic terms, the task of question design is to obviate the need for interviewers or respondents to employ repair practices. However, questionnaire designers can never anticipate every

Table 13.1 Description of the Data Set

Survey	Survey Center	Number of Interviewers	Number of Interviews	Approximate Number of Questions
Economy and Government	University 1	7	14	1346
Neighborhood	University 2	5	7	1134
Television	University 2	3	5	845
The arts	Government	4	6	450
Labor force	Government	7	7	105
Total		26	39	3880

contingency or design a single question that is equally clear to all respondents (Suchman and Jordan, 1990).

THE DATA

Before we turn to a more detailed analysis, a brief description of the data is in order. The data used in this study consist of audio-taped recordings of telephone survey interviews from three survey centers. Two of the centers are located within large midwestern universities, while the third is located within a government agency. The three *university* surveys dealt with the following topics: opinions about government and the economy, neighborhood satisfaction and community ties, and television viewing habits of married couples. The first is a national study, while the latter two are regional. Data from the *government* center include a pretest of a survey on "participation in the arts" and a set of questions on "labor force participation" from a national survey. The numbers of interviewers, interviews, and questions for each survey are listed in Table 13.1. Because the vast majority of the questions in all of these interviews are closed or fixed-choice questions, the few open-ended questions are mostly excluded from these analyses.

RESPONDENT-INITIATED REPAIRS ON SURVEY QUESTIONS

We observe that respondents tend to initiate two types of repair on survey questions: repeat requests and clarification requests. The former seek a repeat of the survey question and officially propose trouble in *hearing* it, whereas the latter seek a clarification or elaboration and officially propose trouble in *understanding* the question. Sequences containing either type of repair exhibit the following general format:

1. Interviewer: Survey question (trouble source)
2. Respondent: Repair initiation/candidate repair
3. Interviewer: Repair/confirmation
4. Respondent: Answer
5. Interviewer: (Acknowledgment) + next survey question

The survey question is the source of the trouble (1), which is followed by the repair sequence (2 and 3). The repair sequence constitutes an *insertion sequence* (Schegloff, 1972) that is nested between the survey question (1) and its answer (4), after which the interviewer proceeds to the next question (5). Although requesting a repeat or clarification of a prior turn-at-talk is a routine practice in most forms of talk, offering a *clarification* — generally a paraphrase or elaboration of the trouble source turn — of a survey question is often unacceptable in a standardized interview. Because the aggregation of answers is based on "the assumption that 'the same' responses to 'the same' stimulus questions are equivalent" (Hahn, 1990:245), any modification of the question by the interviewer, either during its initial reading or during subsequent attempts to clarify it, may threaten the reliability of the data. Therefore, when respondents request a clarifying paraphrase or elaboration of a survey question, researchers face a dilemma: if they allow interviewers to clarify questions, the respondent's understanding may be achieved, but at the expense of reliability; if they do not allow clarifications, standardization may be achieved at the expense of the respondent's understanding and, thus, of the validity of the datum.[2]

Although most survey researchers would agree that *standardized interviewing* consists of reading the questions exactly as they are worded and in the same order as they appear on the survey instrument, there appears to be some disagreement about whether it is acceptable for interviewers to clarify questions or offer prescripted definitions (Fowler and Mangione, 1990). Fowler and Mangione (1990:38) take perhaps the most strict and logically consistent position on standardization, arguing that interviewers should *never* clarify questions, not even when definitions are prescripted by the researcher, because only those respondents who display some trouble in understanding the question will receive the special definition. In this chapter, although we often refer to Fowler and Mangione's (1990) position, we recognize that it is not *the* standard for survey research. We turn now to our analysis of respondent-initiated repair in which we array examples on a continuum from those that are least problematic for standardization to those that are most problematic.

Repairing Troubles in Hearing a Survey Question

Respondents use a variety of techniques to initiate repair on troubles in hearing and understanding survey questions. These techniques vary in terms of how specifically they identify the source of the trouble. Repeat techniques may solicit a repeat of the entire question (as in excerpt 2), a portion of the question, or a confirmation of a candidate repeat proffered by the respondent. We can see the general format in this example of a repeat technique (arrows indicate parts of the question–repair–answer format identified above):

(2) PC07:389

```
1  FI:  1→   .hhh ON the who:le (0.3) how satisf↓i:ed (0.1) are you
2            (0.3) with yer neighborhood.
3            (1.6)
```

```
 4  MR:  2→   Wha:t?
 5            (0.4)
 6  FI:  3→   .hhh ON the wh↓o:le (0.2) how satisfi:ed (0.2) are
 7            y↑ou (0.3) with yer neighborhood?
 8            (2.4)
 9  MR:  4→   ((cough)) (1.0) u:m: (0.1) (°Some°) (0.9) some (0.1)
10            satisfaction.
11            (0.5) #
12  FI:  5→   I see
13            (0.2) # (0.4)
14  FI:   →   .hhh (0.2) What o:ne thing (0.3) IF any. . .
```

In this excerpt, the interviewer reads the survey question (arrow 1), and the respondent pauses (line 3) before saying "Wha:t?" (arrow 2). This question word done with rising intonation proposes difficulty in hearing the prior utterance and is heard as requesting a repeat of the entire survey question (arrow 3). The respondent's request constitutes a type of repair initiation in that it locates, in the preceding turn, some obstacle to the production of an answer, although it does not identify that obstacle specifically. After receiving the repeat,[3] the respondent proceeds to give an answer (arrow 4), followed by acknowledgment and the next question (arrows 5).

In addition to using "What?," respondents can solicit global repeats of survey questions with explicit requests, as in the following excerpt at arrow 2:

(3) AW01:888

```
 1  FI:  1→   .hhh What do you think is the percent
 2            chance that your own total income before
 3            taxes will be under thirty five thousand
 4            dollars
 5            (0.7)
 6  FR:  2→   Say that one again please?
 7  FI:  3→   What do you think is the percent chance
 8            that your own total income before taxes
 9            will be under thirty five thousand dollars?
10            (1.8)
11  FR:  4→   Zero per[cent
12                    [##
13  FI:  5→   .hh And what is the percent chance that. . .
```

The interviewer repeats the question virtually verbatim (arrow 3) except for a lack of emphasis on the word "before" (line 8). The repeat of the survey question, in this case, succeeds in occasioning the respondent's answer (arrow 4), after which the interviewer records the answer (line 12) and moves to the next question (arrow 5).

Instead of requesting global repeats of a survey question, respondents also use a technique, found in ordinary conversation, to occasion a repeat of a particular portion of the survey question. The technique consists of a partial repeat of the trouble source turn, plus a question or question word (Schegloff, Jefferson, and Sacks 1977:368); the following example comes from ordinary conversation:

(4) Schegloff, Jefferson, and Sacks, 1977:368

Sue:		Yeah we used to live, on the highway, too. And when we first moved up there, it was terrible sleeping because all these semis were going by at night.
		((short silence))
Bob:	→	All the what?
Sue:		Semis.
Bob:		Oh

By saying "All the what?" (arrow), instead of simply "What?," Bob indicates that he is having difficulty with the word that immediately follows "all these" in Sue's utterance. This enables Sue to repeat only that particular word instead of the entire utterance. Respondents in survey interviews frequently use this technique as well.

(5) AW19:747

MI:	1→	. . . An' is your race white black asian: american indian or what=
MR:	2→	=Is my wha:t?
MI:	3→	.hh Oh is your ra:ce?
MR:	4→	I'm caucasian
		((##))
MI:	5→	°'kay° .hh An' are you of hispanic origin. . .

By repeating a portion of the survey question (arrow 1), "Is my" (transforming the pronoun), followed by a question word with rising intonation, "wha:t?" (arrow 2), the respondent locates the word immediately following "Is your" as a source of trouble. The interviewer repeats, "Oh is your ra:ce?" (arrow 3), from the survey question (the *oh* preface appears to display surprise at the request). The respondent then provides an answer, "I'm caucasian" (arrow 4), that can be heard as a synonym for one of the answer options, "white," and the interviewer treats it as such by acknowledging and proceeding to the next question (arrow 5). Similarly, in the next excerpt, the respondent uses a partial repeat of the survey question to occasion a repeat of a particular word:

(6) AW15:547

MI:	1→	. . . .hhhh An:d how 'bout (.) >how many days during the past week< did you <feel (.) ↑fear↓ful>
		(0.9)
FR:	2→	Feel what?
MI:	3→	Fearful?
		(0.8)
FR:	2′→	FEARful?
MI:	3′→	↓Yeah huh::-[huh- huh-] (.)
FR:	4→	[None.]
MI:	5→	[Okay great]
FR:		[huh- huh- huh-]

The respondent says, "Feel what?" (arrow 2), locating trouble with the word immediately following "feel" in the survey question (arrow 1). The interviewer promptly repeats only that particular word (arrow 3). After a pause (line 6), however, the respondent repeats the trouble source (arrow 2′), perhaps displaying surprise. The interviewer confirms this candidate hearing (arrow 3′) and laughs, and the respondent provides an answer (arrow 4).

Rather than asking the interviewer to repeat a portion of the question, the respondent may perform a *hearing check*, reproducing a part of the trouble source turn using questioning intonation (Schegloff, Jefferson, and Sacks, 1977:368). A hearing check operates as a *correction invitation device* (Sacks, 1992, I:380), with which a speaker "implicitly asks the recipient to either confirm the guess as correct or provide the correct answer if the guess is wrong" (Pomerantz, 1988:366). The respondent in excerpt 6 above (arrow 2′) employs this technique, as does the respondent below (arrow 2):

(7) AW01:805

```
FI:  1→   .hhh And what do you think is the percent
          chance that someone will break into your
          home? and steal something? during the next
          twelve months
          (0.6)
FR:  2→   (h)eh uh uh:: In my home?
FI:  3→   Yes
FR:       Uh::
          (0.7)
FR:  4→   Ten percent
```

The respondent offers a candidate hearing of a portion of the survey question, "In my home?" (arrow 2), which exhibits *what* she heard by repeating that portion (transforming the pronoun). The candidate invites the interviewer to confirm it if it is accurate or to *correct* it if it is a mishearing. The interviewer then confirms it (arrow 3), and the respondent proceeds to produce an answer (arrow 4). Thus, with a partial repeat, a respondent can check his or her hearing of a particular portion of the survey question without requiring a repeat of the whole question.

Although the respondent in excerpt 7 could be using the partial repeat to deal with an actual trouble in hearing, she could also be using it simply to "fill a pause." She produces the repeat (arrow 2) after a pause (line 5) and two "uh's," which Sacks (1992, II:410) argues are used to fill pauses in one's own utterance. After the confirmation of the partial repeat (arrow 3), the respondent produces another "uh" (line 8) and pause (line 9). Thus, the fact that the partial repeat "In my home?" (arrow 2) is produced in an environment of pauses and pause fillers supports the contention that it too is being used to help fill the space left in the interaction by her delayed answer.[4] This may be a common use of partial repeats in survey interviews since survey questions often require concentrated thought that may cause some delay in answering.

Further evidence that partial repeats are used by respondents to fill pauses as they attempt to answer survey questions comes from the fact that such repeats do not always receive a response from the interviewer. In these cases, the partial repeat is coproduced as the respondent's "thinking out loud."

(8) AW02:354

```
FI:   1→   . . . And how about sales taxes?
           (1.7)
MR:   2→   Sales tax? uh::m
           (3.3)
MR:   4→   Oh about right
FI:   5→   .hhh ↑Oh kay: A:::nd state income taxes. . .
```

After the interviewer reads the survey question (arrow 1), the respondent pauses (line 2) before repeating a portion of the question with rising intonation, "Sales tax? uh:m" (arrow 2). Although the respondent's utterance takes the initial form of a hearing check, he appends an immediate "uh:m" token, which may be a bid to continue. A substantial silence develops (line 4), and since no correction of the questioned phrase occurs, the respondent can take it that his hearing was adequate. He then produces an answer (arrow 4). Thus, the partial repeat is coproduced as a kind of thinking out loud, which, along with "uh:m," partially fills the pause preceding the respondent's answer.

As Schegloff (1997:531) points out, whether or not a partial repeat is doing repair or something else can be an issue not only for analysts of interaction but also for the participants themselves. In the following excerpt, the participants encounter such an ambiguity in the respondent's use of a partial repeat of the survey question.

(9) PC07:1349

```
FI:  1→   .hh Over the pa:st year (0.4) how o:ften (0.3)did you:
          (0.1) git together (0.4) with the people (0.4) from
          yer place of worship? (0.5) Would you s:ay (0.5)
          almost ev'ryday (0.5) at least (0.3) once a week (0.8)
          at least (0.5) once a month (0.7) less (0.4) than once
          a month (0.7) or (0.4) not at all?
          (1.1)
MR:  2→   Get together?
          (1.4)
MR:       Is that what you said?=
FI:  3→   =Mm hm
          (0.7)
MR:  4→   I would say uh::m (0.7) ev'ryday.
```

In this case, the respondent pauses (line 7) after hearing the survey question (arrow 1) and then repeats a portion of it with rising intonation, "Get together?" (arrow 2). Subsequently, a lengthy silence develops (line 9) that may indicate that the interviewer is waiting for the respondent to continue, thus treating the repeat as a thinking out loud that does not require a response. However, the respondent indicates that a confirmation of the candidate hearing was indeed requested by formulating her prior utterance *as* a hearing check: "Is that what you said?" (line 10). The interviewer then promptly confirms it (arrow 3), and the respondent can proceed to give an answer (arrow 4). Thus, whether or not partial repeats of survey questions require a response can be a question for interviewers as well as analysts.

To summarize, in each of the preceding cases, the repair involves some type of *repeat* of all or part of the survey question, while its initiation indicates a trouble in hearing on the part of the respondent or a display of thinking out loud in order to formulate the answer. Because interviewers merely repeat parts of the survey question or confirm the respondent's repeat, they do not modify the wording of the original question and therefore do not appear to violate standardization.

REPAIRING TROUBLES IN UNDERSTANDING A SURVEY QUESTION

Although they officially propose troubles in *hearing*, repeat techniques may also be involved in handling troubles in *understanding*. Sacks (1992, II:412–413) shows that in ordinary conversation, the utterance "What?" is often treated as a request for clarification. He writes, " 'What?' says at least 'I didn't hear what you said, say it again,' but it's employed and is well known to be employed as a substitute for 'I don't understand what you said' " (Sacks, 1992, II:413), and it often occasions a simpler version of the utterance, instead of merely a more clearly enunciated repeat. For example:

(10) Sacks, 1992:413

```
1   Dr:   Do you have some church affiliation, now?
2   Pt:   What?
3   Dr:   Do you belong to a church now?
```

By offering a paraphrase in response to "What?," instead of a verbatim repeat, the doctor treats the patient's nonspecific request for a repeat as possible evidence of trouble in understanding the word "affiliation."[5]

However, this practice of treating a request for a repeat as a request for a paraphrase becomes problematic in a standardized interview. Because standardization requires that each respondent be asked the "same" question, interviewers should not offer a simpler *paraphrase* of a question when a respondent officially requests a *repeat*. Although there are no cases in our data in which the interviewers do this, there is some evidence that *respondents* may attempt to use this technique to solicit a paraphrase. In the following excerpt, the respondent explicitly requests clarification after she first requests and receives a repeat of the survey item.

(11) SPPA110:599

```
1  FI:        .hhh I'm going to read you a list of some types of
2              music as I read the list tell me .hh which of these
3              types of music you like to listen to. Yes or no...
               ((Sequence omitted.))
4  FI:   1→    .hh Latin (0.2) Spanish salsa.
5              (1.1)
6  FR:   2→    What was that?
7              (0.3)
8  FI:   3→    Latin (.) Spanish an' salsa
9              (1.1)
```

FR: 2′→ What's that.
(0.4)
FI: 3′→ Tha- (.) uhm (0.7) Spanish people they have that salsa
beat (0.5)
FR: 4→ Oh (.) .hh No I guess not.
(0.9)
FI: 5→ .hhh Big band

In response to the survey question about musical taste (arrow 1), the respondent replies with a request for a global repeat, "What was that?" (arrow 2). The interviewer then repeats the question with only a slight modification (arrow 3), inserting an *and* ("an'") into the list, perhaps to make it clearer (the questionnaire item reads, "Latin/Spanish/Salsa"). In reading the item the first time (arrow 1), the interviewer pauses between the first and second types of music, but not between the second and third. As a result, the question could be heard as asking about "Latin" or "Spanish salsa," rather than about "Latin," "Spanish," or "salsa." By inserting the *and* between "Spanish" and "salsa" in the repeat (arrow 3), the interviewer repairs her earlier reading.[6]

The repeat occasions a second repair initiator which this time seeks clarification. Although the clarification request, "What's that." (arrow 2′), is similar to the first (arrow 2), it may be produced as a request for clarification, rather than a repeat, by virtue of its falling intonation (indicated by the period) and its sequential position (i.e., following a repeat request and the requested repeat itself). In addition, although the request does not explicitly signify to which type of music "that" refers, it can be heard as referring to the last one, "salsa," by virtue of the "preference for contiguity" in conversation (Sacks, 1987:60). "Salsa" is the most immediately prior or contiguous item. In addition, the interviewer exhibits her understanding of "What's tha:t" as a request for clarification of the term "salsa" by offering a kind of definition of it in her next turn (arrow 3′). Then, using a change-of-state token, "oh" (Heritage, 1984), the respondent marks the clarification as informative and produces an answer that displays uncertainty (arrow 4). The point here is that this respondent may have been attempting to solicit a clarification with her very first repeat request (arrow 2) and using the technique that Sacks (1992, II:413) describes.

The next excerpt provides another instance in which a respondent first requests a repeat and then requests a clarification. However, in this case, the interviewer deals with the request in a more "standardized" manner:

(12) PC07:1182

FI: .hh How import'nt is it (0.5) to yer place of wo:rship...
((Sequence omitted.))
FI: 1→ Tuh foster (0.1) a sense (0.1) of patriotism (0.4) an'
good c:itizenship?
(0.9)
MR: 2→ I din (0.3) I din catch 'at
FI: 3→ n'kay (0.1) tuh foster (0.1) a sense of (0.1) patriotism
(0.4) an' good citizenship?
(2.2)

```
 9   MR: 2′ →  What's that mean (°I mean°)
10             (0.6)
11   FI:       .tch
12             (0.1)
13   MR: 2″ →  Well I mean father (0.4) means take it on?
14             (0.9)
15   FI: 3″ →  It would be whatever it means tuh you.
16             (0.2)
17   FI:       °Um°
18             (0.1)
19   FI:       Uh jus (0.1) tuh foster a sense of patriotism
20             (0.2) an' good citishenship. (0.3) eh (0.3)
21             [what]ever  [it
22   MR: 4→    [ uh ]      [I would sa:y uh: probly be very import'nt.
```

After hearing the survey question (arrow 1), the respondent requests a repeat without specifying a particular source of a hearing trouble, "I din, I din catch 'at" (arrow 2). After receiving a verbatim repeat from the interviewer (arrow 3), he pauses (line 8) and then explicitly requests clarification with a *wh* question, "What's that mean" (arrow 2′), which appears to refer to the entire question. However, before he receives a response from the interviewer, the respondent reformulates the request (arrow 2″), this time displaying trouble with the word "foster." He repeats it incorrectly as "father" and then proffers a candidate definition of the term, "means take it on?" (arrow 2″). The interviewer responds immediately with silence (line 14), which can be heard as foreshadowing a *dispreferred* response (Sacks, 1987; Pomerantz, 1984). Rather than confirming (the preferred response) or correcting the candidate definition, the interviewer employs the phrase "It would be whatever it means to you" (arrow 3″). This is a routine device in standardized interviewing, familiarly called nondirective probing, which works to turn a clarification request back to the respondent. When a response by the respondent is not forthcoming (lines 16–18), the interviewer continues on to say "Uh jus" and repeating the survey question again (lines 19–20), another routine practice for handling clarification requests in a standardized way. The repeat, which contains articulation of and emphasis on "foster," could be an *embedded correction* (Jefferson, 1983), which gives the respondent an opportunity to catch his error without explicitly calling attention to it. However, the respondent displays no realization of the error (silences at line 20).[7] As the interviewer appears to start a repeat of her nondirective probe (line 21), the respondent, in overlap, offers a tentative answer (arrow 4) to the question.

Clarification requests, like repeat requests, vary in terms of how precisely they identify the source of trouble. They may request a clarification of the entire survey question, as in the following excerpt, or they may indicate the particular word or phrase within the question that is causing the trouble (as in excerpt 12 above).

(13) AW13:185

```
1   MI:  1→   . . . .hhh Sharing information (.) between diffrent
2             agencies ah goverment (0.3) when both want thuh same
3             information (0.4) sa::ves time an' money (0.6) but it
```

```
                also means some loss in privacy for thee individual .t
                .hh do you think thuh benefits, of saving time an'
                money (0.2) outweigh thuh costs in privacy, (0.2) .hhh
                or do you think loss of privacy outweighs th' benefits
                of saving time an' money.
                (3.3)
FR:   2→        I: don' understand that one
MI:             Oh kay I can read it a↓gain if you ↑want
                (.) #
MI:   3→        .hh Sharing information (.) between diffrent agencies
                uh goverment .t .hh when both want thuh same
                infor↓mation (.) ↑ sa::ves time an'money. (0.3) .t
                .hhhhh but it also means some loss in privacy for thee
                individual. (0.2) .t .hhhh do you think. (0.6) thuh
                benefits of saving time an' money, (0.1) .t outweigh
                thuh costs in privacy? (0.5) or do you think. (0.5)
                loss of privacy (.) outweighs thuh benefits of savin'
                time an' money.
                (0.4)
FR:   4→        Prob'ly tha (0.2) uh:: (0.2).hh priva↑cy
```

After hearing a lengthy and complex survey question (arrow 1), the respondent pauses for a long time (line 9) before she requests clarification by stating that she does not "understand that one" (arrow 2). The request initiates repair on the survey question without identifying which part of the question poses the trouble in understanding. The respondent thereby requests a paraphrase or simplified version of the entire question (Sacks, 1992, II:412–413). However, following standardization procedure, the interviewer instead responds to the request with an offer to *repeat* the entire question (line 11) and proceeds to do so (arrow 3). After hearing the repeat of the question, the respondent provides an answer that she marks as uncertain (arrow 4), suggesting that the repeat may not have been helpful in formulating a definitive answer. Thus, the interviewer's repair (arrow 3) is misaligned with the respondent's repair initiator (arrow 2) in that it is of a different pair type—that is, it repeats rather than offers clarification. In this way, these *clarification request–repeat* sequences are similar to the *repeat request–clarification* sequences discussed above (excerpt 10), except that in the latter, the second-pair part moves in the direction of being potentially more helpful than explicitly requested, while in the former it is less helpful. This may be an instance of what Suchman and Jordan (1990:238) suggest as a pattern whereby "the interviewer manages to engage in the appearance of an interaction over the meaning of the question without ever really offering an alternative, clarifying version." Indeed, treating a respondent's request for clarification as a request for a repeat is a routine interviewer technique for avoiding clarification, as is the "It's whatever it means to you" device in excerpt 12.

In our data, a more common method of requesting clarification than soliciting a paraphrase or elaboration from the interviewer is to produce a *candidate* clarification and to invite the interviewer merely to confirm or correct it. By producing a candidate clarification, respondents more precisely locate the source of their understanding trouble than by saying "What's that mean" or "I don't understand." A clarification proffer is also known as an *understanding check* (Schegloff, Jefferson, and Sacks, 1977:378), a generic repair device, which is similar in structure to the hearing check.

(14) AW13:547

```
 1  MI:  1→   .hh D'you support? (.) or oppose (.) the following
 2            measures tuh ↑deal with ↓AIDS. (0.4) .t (0.3)
 3            Prohibit students with thee AIDS virus (.) from
 4            attending public school. (0.2) .t (0.3) D' you support
 5            or oppose that ma'am.
 6            (0.5)
 7  FR:       I sup- (0.3) let's see.
 8            (2.1)
 9  FR:  2→   In other words sh- should they go to school or not?
10  MI:       Lemme lemme read you the question again.
11            (0.7)
12  MI:  3→   Prohibit students (0.2) with thee AIDS virus from
13            attending public school. (0.2) An' do you support or
14            oppose that.
15            (0.9)
16  FR:  4→   Ah:::: °uh° oppose it.
17            (0.4) ##
18  MI:  5→   .hhh Develop uh goverment information program. . .
```

In response to the question (arrow 1), the respondent begins to provide an answer (line 7) but then cuts it off. After a delay (lines 7–8), she proffers a candidate paraphrase of the question, "In other words sh- should they go to school or not?" (arrow 2) with rising intonation, which invites the interviewer to confirm or correct it. However, the interviewer resists the invitation, in accordance with standardized interviewing procedures, and offers to reread the question instead (line 10). The respondent does not receipt the alternative proposal (e.g., with *okay*), as indicated by the 0.7-second silence at line 11, perhaps indicating some misalignment with the projected course of action. Nonetheless, the respondent allows the interviewer to continue on with the repeat (arrow 3) and eventually provides an answer (arrow 4). Thus, with a clarification proffer, it is the respondent who produces the clarification for the interviewer merely to confirm or correct, yet due to standardization constraints, the interviewer may avoid evaluation of the candidate all together.

Although the respondent in excerpt 14 provided an answer, despite the fact that her candidate clarification was neither confirmed nor corrected, respondents do not always do so. Withholding evaluation of the candidate clarification by the interviewer may cause the respondent to withhold a substantive answer.

(15) AW04:635

```
1  FI:         .hh And do you have health insurance coverage for
2              hospitalization?
               ((Intervening talk omitted))
3  FI:  1→     .h And long term care
4              (0.3)
5  FR:  2→     Long term care=
```

```
 6  FI:  3→       =Yes
 7  FR:  2′ →     I don't know what you mean by that you mean like
 8                if I was an invalid?
 9                (.)
10  FI:  3′ →     U:m a:ll it says here is for long term ca[re it]=
11  FR:                                                    [heheh]
12  FI:  3′ →     =doesn't explain the question to [me
13  FR:                                            [ah Then I don't
14                know how I can answer it=
15  FI:           =I can put down don't know (.) And now. . .
```

After the question (arrow 1), the respondent initiates a hearing check (arrow 2), which the interviewer confirms (arrow 3). Then the respondent initiates further repair by claiming trouble in understanding, followed by a candidate elaboration, "you mean like if I was an invalid?" (arrow 2′). However, instead of confirming the candidate repair, the interviewer claims an inability to offer clarification because the questionnaire does not provide her with any information beyond the question (arrow 3′).[8] The respondent then proposes her own inability to provide an answer (lines 13–14), indicating the contingency of her answer on a confirmation or correction of her candidate elaboration. The interviewer resolves the impasse by affirming "don't know" as an appropriate answer (line 15) in this situation. Thus, the survey question does not receive a substantive answer.

The interviewers in excerpts 14 and 15 both handle the respondents' clarification proffers in a standardized way—by offering to repeat the question and claiming a lack of additional information. However, interviewers in our data sometimes did offer clarifying help when respondents proffered their own candidate clarifications.

(16) SPPA148:180

```
 1  FI:           Okay..hhh An did you listen to...
                  ((Intervening talk omitted))
 2  FI:  1→       .hhh A reading of novels or books either live
 3                or recorded
 4                (.)
 5  MR:  2→       Uhm: (.) Do short stories count?
 6                (0.3)
 7  FI:           .tch Uh(h)m
 8                (0.9)
 9  FI:  3→       Yes:
10                (.)
11  MR:  4→       Yes okay ↓yes
12                (0.2)
13  FI:  5→       .hhh .tch An during the last twelve months...
```

In response to the survey question (arrow 1), which projects a "yes" or "no" answer, the respondent proffers a candidate clarification, "Uhn: do short stories count?" (arrow 2), that indicates trouble in determining whether "short stories" qualifies as "novels or books." After a pause, an "uhm," and a longer pause (lines 6–8), during which the interviewer could be consulting a training manual or other notes about the interview, she then confirms the candidate with "Yes:"

(arrow 3). This enables the respondent to proceed immediately to produce an answer (arrow 4). Thus, the wording of the original question, "Did you listen to a reading of novels or books either live or recorded?" becomes collaboratively modified as "Did you listen to a reading of novels or books, *including short stories*, either live or recorded?" According to Fowler and Mangione (1990:38), this is an "unstandardized" way of handling the clarification request because not every respondent will receive the modified version of the question. Furthermore, because the *respondent* produces the candidate clarification (arrow 2), the interviewer completes the repair sequence merely by saying "yes" (arrow 3), rather than by providing a paraphrase or elaboration herself (as the interviewers would have to do in excerpts 11–13). In a sense, then, the clarification proffer encourages help. As compared with other clarification techniques (requests without candidate clarifications), this may explain why the clarification proffers in our data are more likely to receive an unstandardized response from an interviewer than the other types of clarification requests (see Table 13.2).

In Table 13.2, clarification requests are divided into four types: "I don't understand" avowals (e.g., excerpt 13), *wh* questions (e.g., excerpts 11 and 12), clarification proffers (e.g., excerpts 14–16), and procedural queries.[9] The table also reflects Fowler and Mangione's (1990) strict conception of standardization according to which *any* introduction of terms from outside the original question constitutes unstandardized interviewer behavior. "Standardized" responses consist of *repeats* of all or part of the survey question, refusals to provide clarification, or variants of the "It's whatever it means to you" response. Table 13.2 shows that, compared to "I don't understand" avowals and *wh* questions, clarification proffers in our data are about twice as likely to receive unstandardized responses (in the form of confirmations) from interviewers.[10] Indeed, in half of the instances with clarification proffers, interviewers produced confirmations in an unstandardized manner. Thus, the way a respondent designs a clarification request may influence the likelihood with which the interviewer will grant the request.

To summarize, when initiating repair on survey questions, respondents attempt to deal with troubles in hearing and understanding. Although requests for repeats are usually used to occasion a repeat of a preceding utterance, they are sometimes used to occasion a clarification of the utterance without officially admitting ignorance (Sacks, 1992, II:413). Given this possibility, recipients of repeat requests in ordinary conversation must decide whether to offer a repeat or a clarification

Table 13.2 Type of Clarification Request by Type of Interviewer Response

Type of Clarification Request	Standardized Response	Unstandardized Response	Total
I don't understand	10 (71%)	4 (29%)	14 (100%)
Wh questions	9 (75%)	3 (25%)	12 (100%)
Clarification proffers	28 (51%)	27 (49%)	55 (100%)
Procedural	20 (100%)	N/A	20 (100%)
Total	67 (66%)	34 (34%)	101 (100%)

without knowing for certain which is being solicited. However, in the standardized survey, interviewers do not enjoy this discretion: Because paraphrasing is not allowed, they should always treat a request for a repeat at face value. Furthermore, if the respondent requests clarification explicitly, the interviewer should avoid granting the request. This can be accomplished by providing a repeat of all or part of the survey question, by saying "It's whatever it means to you," or by refusing the request in other ways. However, by using a clarification proffer, which requires only a "yes" or "mm hmm" rather than a paraphrase or elaboration from the interviewer, respondents make it easier for interviewers to participate in the modifying of survey questions.

INTERVIEWER-INITIATED REPAIRS ON RESPONDENT ANSWERS

Having examined respondent-initiated repair on the questions that interviewers ask, we now turn to another kind of other-initiated repair. In this section, it is the interviewer who is "other" relative to the respondent who is the "speaker" of answers to survey questions. Sequences containing interviewer-initiated repairs on respondent answers exhibit the following general format:

1. Interviewer: Survey question
2. Respondent: Answer (trouble source)
3. Interviewer: Repair initiation/candidate repair
4. Respondent: Repair/confirmation
5. Interviewer: (Acknowledgment) + next survey question

The repair sequence, consisting of items 3 and 4, occurs in the turn following the trouble source, the respondent's answer (2), and before the interviewer acknowledges and/or proceeds to the next question (5). Interviewers use other-initiated techniques to repair troubles in hearing respondents' answers, apparent inconsistencies across answers, and inadequacies (for the purposes of recording) of answers. Where these repair techniques potentially influence answer content, they may violate standardization procedures; however, conforming to standardization in particular situations may conflict with the organization of repair in ordinary conversation.

Interviewer Hearing Checks

Like respondents, interviewers may initiate hearing checks on respondents' answers. These types of repairs consist of a repeat of the answer, often done with questioning intonation. We can follow the five-part format in this example:

(17) AW03:188

FI: 1→ .hh How would you feel about a national health
care system providing information about you to
the ↑census bureau .hh for use in the population

```
 4                 census .hhh would you favor it strongly? (0.3)
 5                 favor it somewhat .hh oppose it somewhat or oppose
 6                 it strongly
 7                 (2.3)
 8   FR:  2→       (Oh) I would favor it somewha:t=
 9   FI:  3→       =Favor it somewha:[t?
10                                   [#
11   FR:  4→                         [°Ye:ah°
12                 (0.4)
13   FI:  5→       (.tch) And would you be willing to give your. . .
```

After the respondent answers (arrow 2) the question (arrow 1), the interviewer produces a questioning repeat[11] (arrow 3) or hearing check that proffers a candidate repair. The fact that the interviewer records the answer (keystroke, line 10) *before* the respondent confirms it (arrow 4) may indicate that the "hearing check" is not actually being used to check her hearing, although that is what it officially proposes. Repeating the respondent's answer could also be used to acknowledge the answer and/or to fill the silence that might otherwise transpire while the interviewer works at the computer (i.e., recording the answer and finding the beginning of the next question on the next screen).[12] In any case, the interviewer, after the respondent's confirmation, proceeds to the next question (arrow 5).

The following is another case of an interviewer checking her hearing of the respondent's answer; however, in this case the respondent offers a correction.

(18) AW21:618

```
 1  FI:  1→       .hh What was your job last week that is what
 2                kind of work were you doing
 3                (1.0)
 4  MR:  2→       I'm a service tech fer a telephone company
 5                (0.5)
 6  FI:  3→       ((typing)) Service ex↑pert?
 7  MR:  4→       Service TECH
 8  FI:  5→       Oh=
 9  MR:  4′ →     =Service technician
10                (0.5)
11  FI:  5′ →     ((typing)) Service technician .hh A-and um (0.4)
12                what-what sort of things do you service?
```

The interviewer initiates a hearing check of the respondent's answer (arrow 2) by offering a candidate hearing with questioning intonation, "service expert?" (arrow 3) while typing the answer into the computer. However, the candidate does not match the answer ("service expert" instead of "service tech"). The respondent then corrects the candidate (arrow 4), and the interviewer receipts the correction with a freestanding "oh" (arrow 5), indicating a change in her state of knowledge (Heritage, 1984) (line 8). Next, the respondent restates his earlier repair in complete rather than abbreviated form (arrow 4′), which the interviewer records while proceeding to the next question (arrow 5′). Thus, interviewers, like respondents, also use the generic hearing check to repair their hearing of a prior utterance, sometimes receiving a confirmation and other times a correction.

Interviewer Detection of Inconsistencies across Answers

In addition to requesting repeats and clarifications, respondents may display trouble in understanding survey questions or the prespecified answer options by producing inconsistencies across their answers. Although they are not required to do so, interviewers sometimes detect such inconsistencies and initiate their correction. In these situations, they can use the turn immediately following the answer to initiate its correction.

(19) AW17:808

```
FI:  1→    An::d what is the percent chance that your own
           total income before taxes will be under thirty
           thousand dollars in the next twelve months.
FR:  2→    Well ↑zero. It won't be: [huh huh.
FI:  3→                             [Okay. So then it would
           be::- percent chance would be- it wouldn't be z-
FR:  4→    Uh- Be u:h
           (0.6)
FI:  3′→   One hundred then?
FR:  4′→   Yeah.
FI:  5→    Okay. [hhhhh]  [.hhh.t And what was your age. . .
                 [# # #]  [#
```

In answering the survey question (arrow 1), the respondent provides an answer that contains the answer option, "zero" (arrow 2). However, in response to prior questions, the respondent has indicated that she expects her income to be between "$24,000" and "$26,000," which would intimate an answer here of 100 percent instead of zero. Detecting this apparent inconsistency, the interviewer begins to initiate its correction (arrow 3). She appears to display some reluctance to do so by restarting the utterance twice before proposing, "it wouldn't be z-" (line 6), cutting it off just before completion. However, the respondent fails to produce the correction (arrow 4). The interviewer attempts to initiate repair again by proffering a candidate answer and inviting the respondent to confirm it (arrow 3′). The respondent confirms the correction (arrow 4′) and the interviewer proceeds (arrow 5).

Although the interviewer in this case appears to have helped the respondent provide a correct answer, interviewers are generally instructed neither to monitor respondents' answers for apparent inconsistencies nor to repair misunderstandings, because they could potentially introduce error into the data (Fowler and Mangione, 1990:33–34). However, by prohibiting such uses of interviewer-initiated correction, survey researchers lose a potential resource for achieving the respondent's adequate understanding of the questions.[13] Once the interview is over, there is no more opportunity to reconcile discrepancies between inconsistent answers, and the accuracy of both answers is cast into doubt (Suchman and Jordan, 1990:239).

INTERVIEWER PROBING

With closed or fixed-choice questions, respondents must answer using one of a prespecified set of answer options, which may consist of a set of categories, a yes and no, or a range of numbers. If the respondent produces an answer that does not match one of the options, it is considered inadequate for the purposes of standardized interviewing, and the interviewer should initiate repair on it. Of course, in the survey enterprise, this type of interviewer-initiated repair is officially known as *probing* (see Schaeffer, 1991:388; Houtkoop-Steenstra, 1996). As is well known, probing should be done *nondirectively* or in a neutral way that does not *lead* the respondent to choose any particular answer option. If a probe utterance is designed such that it suggests one answer option over the others, it is considered *directive* and improper probing because it potentially can introduce interviewer-related error into the resulting data (Fowler and Mangione, 1990:33; Czaja and Blair, 1996). The data for the following analysis consist of 176 instances of probing collected from the two university survey centers studied (26 interviews containing approximately 3325 question–answer sequences).

As stated above, whenever a respondent's answer fails to reproduce one of the prespecified answer options, interviewers should probe it, that is, initiate repair and invite the respondent to redo the answer by selecting a single option, as in the following excerpt:

(20) AW17:197

```
FI: 1→   . . . An:d how willing would you be to give your
         social security number to the Census Bureau (.) if
         this would increase the accuracy of items in your
         files. Would you be ↑very willing (.) somewhat willing
         (0.2) not very willing or not at all willing.
FR: 2→   .t I'd be willing?
FI: 3→    Okay. Ver↑y somewhat (0.5) not [very or not at all.]
FR:                                      [Ah:::::           ]
    4→    I'd ↑be (0.4) a- very willing.
FI: 5→    Okay. And how willing would you be: to give your. . .
```

The respondent's answer (arrow 2) fails to distinguish between two or three of the answer options—very willing, somewhat willing, and not very willing (lines 4–5)—and therefore is inadequate for enabling the interviewer to record a single option. The interviewer thus probes the response by listing the four modifiers of "willingness" and thereby invites the respondent to choose one (arrow 3). The probe is done correctly, or nondirectively, because it includes *all* of the relevant answer options (Fowler and Mangione, 1990:33; Czaja and Blair, 1996), although some researchers might fault her for not repeating the word "willing" for each option. The respondent then selects "very willing" (arrow 4), which corrects her earlier answer and enables the interviewer to acknowledge and proceed to the next question (arrow 5).[14]

Not all survey questions require a selection from a set of categories. Instead, the answer options may consist of a range of numbers or a yes or no. Still, if a respondent's answer does not reproduce one of these options, interviewers generally should probe:

(21) PC08:42

```
FI: 1→     HOW many years have you lived in the Indianapolis
           area?
FR: 2→     We've lived here since (0.2) June of nineteen eighty
           six.
FI: 3→     O:ka::y, .hh And can you just give me a (.) a number
           of years,
           (0.6)
FI:        that you've lived in the area,
           (1.8)
FR: 4→     °Eleven°
           (0.6)
FI: 5→     All:ri:::ght?
           ### (0.6)
FR:        In June, hhuh it'll be eleven
```

In response to the survey question (arrow 1) that projects a number of years as the preferred answer format, the respondent answers with a date (arrow 2). Although the answer provides adequate information for the interviewer to determine the appropriate number to record, it requires further calculation. The interviewer turns the work of calculating back to the respondent by probing with a request for a number of years: "And can you just give me a, a number of years," (arrow 3). The interviewer's request is immediately met with silence (line 7), which projects a dispreferred response by the respondent.[15] Before the respondent can give an answer, the interviewer appends the phrase "that you've lived in the area," (line 8) onto her original request. An even lengthier silence then develops (line 9), after which the respondent returns a number, "eleven" (arrow 4), in a flat and quiet tone of voice, which corresponds to the correct subtraction of "nineteen eighty six" from the current year (this survey was conducted in 1997). Hence, the interviewer accomplishes a nondirective probe by requesting that the respondent redo the answer in the form of a number, but the respondent, in her delays and in the prosody of her talk (voice tone), appears to display dissatisfaction with the probing procedure.

Instead of a request for a number, an interviewer probe may also take the form of a request for a yes or no. The question in the next excerpt refers to occasions on which the respondent and his wife "watch different shows on different television sets at the same time":

(22) TC02:395

```
MI: 1→     Does this ever bother you
           (1.7)
MR: 2→     Occasionally
MI: 3→     'kay I jus need a yes or a no ta that
MR: 4→     Y:es
MI: 5→     Thanks.
           (0.4) ##
MI:        Does it ever bother your wife
```

Here, the survey question projects a yes or no as the preferred answer format (arrow 1). Instead of offering one of these options, the respondent uses a format that implicates a yes (arrow 2), although in a somewhat mitigated form. Instead of recording a yes, the interviewer initiates correction on the form of the answer by reporting a need for a yes or a no, ' "kay I jus need a yes or a no ta that" (arrow 3). This occasions a prompt yes (arrow 4) from the respondent, and the interviewer proceeds (arrow 5). Thus, although "occasionally" would appear to be adequate for enabling the interviewer to record an answer of "yes," she treats it as inadequate or in need of repair. The inadequacy of the answer in this case appears to be a kind of formal or technical one instead of a more substantive inadequacy, as in excerpt 20.

So far we have examined instances of nondirective probing, but interviewers sometimes fail to follow standardized probing procedure. Fowler and Mangione (1990:40) define a *directive* probe as one that "can be answered with a 'yes' or 'no.' "[16] In conversation analytic terms, directive probing consists of employing what we discussed earlier as a *correction invitation* device, a single candidate answer for the respondent to confirm or correct. For example, the following (arrow 3) would be considered a directive probe:

(23) AW19:382

```
MI: 1→      An' would you call yourself a (.) strong Republi↑ca:n
            or a not very strong Republican
MR: 2→      Moderate
            (1.0)
MI: 3→      °'kay° (.) So would=ya say not very strong:?
MR: 4→      Yeah
```

Here the respondent provides an answer (arrow 2) that appears to fall in between the prespecified answer options—"strong Republican" and "not very strong Republican"—and thus inhibits the interviewer from recording one of them. After a pause (line 4), the interviewer receipts the answer and initiates its repair with a directive probe (arrow 3). With the *so* preface, the interviewer designs the probe as a candidate upshot or consequence of the respondent's answer.[17] Given that the respondent's answer does not appear to correspond to either answer option, by proposing "not very strong" over "strong," the interviewer leads the respondent toward choosing that option. It is this kind of leading that survey researchers seek to avoid out of a concern for interviewer-related error (Fowler and Mangione, 1990; Czaja and Blair, 1996).

Respondents sometimes provide answers that, although technically formatted incorrectly, nonetheless clearly imply one of the prespecified answer options (e.g., excerpts 21 and 22) and, therefore, are adequate for enabling the interviewer to record a single option. Such answers are thus "inadequate" in only a formal sense, for the purpose of standardized interviewing, rather than in a substantive sense (e.g., excerpts 20 and 23). Consequently, offering the answer option indicated as a candidate answer, in the form of a probe, does *not* necessarily lead the respondent. See excerpts 24 and 25:

(24) AW02:518

```
1   FI:          . . . .hhh Do you sup↑port or op↑po:se the following
2                ↑measures to deal with aids .hh
                 ((Sequence omitted.))
3   FI:   1→     . . . .hhh Permit insurance companies to test
4                ↑applicants (.) for thee ai:ds virus:
5                (1.3)
6   MR:   2→     .hhh I would uh:: be for that
7   FI:   3→     .hhh So you would su[pport it?
8         #→                         [#
9   MR:   4→     Yes
```

(25) AW01:464

```
1   FI:          . . . Do you support or op↑po:se the following
2                measures to deal with AI:DS. . .
                 ((Sequence omitted.))
3   FI:   1→     . . . .hhh (0.1) Ah:: >How about< (0.2) develop uh
4                goverment infer↑mation ↑program to ↓promote safe
5                sex practices (.) ↑such as ↓the use of condoms:
6                (0.4)
7   FR:   2→     Yer talking to uh health teacher=abso↓lutely?
8                (0.2)
9   FI:   3→     Uh Supp[ort?
10        #→            [#
11               (0.6)
12  FR:   4→     °Mhmm?°=
```

In answering the survey questions (arrow 1), the respondents in both excerpts offer answers that indicate a single answer-option using other words[18] (arrow 2): In their sequential positions immediately following each question, both "I would be for that" and "You're talking to a health teacher, absolutely" can be heard as indicating the option "support." Thus, the answers are "inadequate" in only a formal sense. Rather than recording support, the interviewer in both cases initiates repair by probing (arrow 3), offering "support" as a candidate answer. Although the probes would be considered directive according to Fowler and Mangione's (1990:40) definition, they do not appear to *lead* the respondents. Because the respondents have already indicated "support" using other words, proffering that answer option moves to *verify* it. In fact, the interviewer in both cases records the answer (arrow #) *before* the candidate answer is confirmed, indicating that it was indeed substantively adequate for enabling her to record a single answer option. In light of this analysis, these probes should perhaps be considered a kind of *verification* probe[19] rather than *directive* or *leading* probes.

Initiating correction on formally inadequate answers in standardized interviews appears to conflict with repair practices in ordinary conversation. Schegloff, Jefferson, and Sacks (1977:380) discovered a *preference for self-correction* in ordinary conversation, one feature of which is that speakers tend to *withhold* correction of another speaker's error if the error does not prevent them from proceeding to the sequentially implicated next turn. Thus, in line with ordinary conversational practice, probing should be withheld when a respondent uses a formally inadequate answer format because the error does not prevent the interviewer from recording the answer, as we saw in excerpts 24 and 25. Offering the answer option in these situations appears to be a kind of compromise between the total withholding of probing and probing nondirectively.

This conflict between the practice of nondirective probing and the preference for self-correction may account to a significant degree for why interviewers appear to probe in a directive way. In Table 13.3 we categorize each instance of probing for whether it was directive or nondirective, according to Fowler and Mangione's (1990:40) criterion, and for whether the inadequacy of the answer being probed was *substantive* or merely *formal*.

First, of all of the probes, 20 percent were done in a directive or incorrect manner.[20] Second, when probing answers with a substantive inadequacy, interviewers probed correctly 93 percent of the time. However, when probing answers that were inadequate in only a formal or *technical* sense, according to standardized interviewing procedures, interviewers probed correctly only 53 percent of the time. Furthermore, on every occasion in which an interviewer offered a single candidate answer as a replacement for a formally inadequate answer, he or she offered *the answer option implied or indicated by the respondent*. If our data are representative of most surveys, then the incidence of directive probing may not constitute as much of a compromise of data quality as survey researchers fear. Although interviewers may break the rules of standardized interviewing when they employ a *verification probe on a formally inadequate answer*, they do not appear to violate one purpose of standardization, which is to ensure the collection of answers (or choices of answer options) that were accountably produced as the respondent's own—that is, without systematic interviewer behavior inducing certain answers.

In summary, interviewer-initiated repairs on respondents' answers may involve troubles in hearing, apparent inconsistencies across answers, or the adequacy of the answer for the task of recording an accountable answer. Correcting apparent inconsistencies across answers and offering a single candidate answer are activities through which the interviewer can potentially introduce error into the resulting answers. However, as we argue, the latter do not appear to threaten standardization when used to verify an answer that the respondent has already given using a technically inadequate format.

Table 13.3 Probe Type by Type of Inadequate Answer

	Type of Inadequacy		
Probe Type	Substantive	Formal	Total
Directive	9 (7%)	26 (47%)	35 (20%)
Nondirective	112 (93%)	29 (53%)	141 (80%)
Total	121 (100%)	55 (100%)	176 (100%)

DISCUSSION

We have examined several types of question–answer sequences in which things go wrong, that is, when respondents or interviewers attempt to repair some kind of trouble in hearing a question, understanding a question, or recording an answer. But repair sequences can also tell us something about paradigmatic question–answer sequences in which there is no observable trouble in the interaction. Recall excerpt 1, in which neither respondent nor interviewer initiate repair, at least initially. The expanded excerpt below shows what follows the initial sequence (which is in lines 1–4).

(26) SPPA110:530 expanded

```
FI:        .hh During the last twelve months did you play any
            jazz music.
            (0.2)
FR:         Oh yes
            (0.6)
FI:         .hh Did you play any jazz in a public performance or
            rehearse for a public performance.
            (0.5)
FR:         No
            (0.6)
FI:        .hhh During the last twelve months did [you-
FR:   →                                            [Oh now
           wait a second when you said do I pla- did I
           p[lay (any   instrument)] as an instrument?
FI:         [Jus like an instrument]
FI:        [Right ]
FR:        [.hh No] no I don't play any instruments
           (0.3)
FI:        °O:↑kay:°
           (0.3)
FR:        I'm sorry.
           (0.5)
FI:        °Let me change this°
           (0.4)
FR:        °Sorry°
```

At the arrow, the respondent displays a realization that she misinterpreted the first question (lines 1–2).[21] She appears to have interpreted "play" as playing recordings of music and then realized that it should have been interpreted as playing a musical instrument instead. The respondent's realization is made possible by the second question (lines 6–7), which introduces "play or rehearse for a public performance" as a stimulus for retrospectively reinterpreting the prior question. Although the first question–answer sequence (lines 1–6) has the compact structure of a paradigmatic sequence, "Oh yes" actually *conceals a misinterpretation* of the question, which may never have been detected if it were not for the second question. Hence, paradigmatic sequences, which contain no repair, may be desirable for the purpose of standardizing the survey instrument, but they cannot reveal whether the respondent has interpreted the question as the researcher intended.

How frequently respondents misunderstand survey questions when they give paradigmatic answers is an open question. Using *comprehension probes*, Oksenberg, Cannell, and Kalton (1991:357) discovered *substantial misinterpretation* in a pretest of a health survey. By asking respondents who had given paradigmatic answers then to elaborate on the answers, to interpret concepts in the questions, and to report any comprehension difficulties, they (Oksenberg, Cannell, and Kalton, 1991:357) found that "fifteen percent or more of the respondents were found to have problems with 12 out of the 18 questions probed." For example, after respondents answered a question on exercise, they were asked, "You said that in the past 4 weeks you had not done any exercise, sports, or physically active hobbies. Did you get any exercise at all during that time?" In response to this probe, about one-third of the respondents who initially reported "no exercise" now mentioned "walking," which they may not consider "real exercise" (Oksenberg, Cannell, and Kalton, 1991:358). Oksenberg, Cannell, and Kalton (1991) advocate the use of comprehension probes in the pretest phase of a survey (in conjunction with behavior coding) to help identify understanding problems that might otherwise go unnoticed and then to rewrite the questions for the final version of the instrument. While this will not remove the structural problem of concealed misinterpretation with paradigmatic sequences, it may reduce its occurrence. Our point here is to show that repair sequences, although a nuisance for the purposes of standardization, can be informative about whether respondents understand survey questions.

CONCLUSION

In the course of a survey interview, respondents and interviewers on occasion encounter interactional troubles in attempting to achieve intersubjectivity regarding survey questions. Consequently, they must employ a variety of other-initiated repair techniques. However, the interview participants sometimes encounter conflicts between the standardization procedures — designed for obtaining reliable data — and generic conversational repair practices. For example, while recipients of repeat requests are generally free to diagnose the cause of the other speaker's trouble, that is, they may offer the officially requested repeat *or* a clarification of the problematic turn (Sacks, 1992, II:412–413). In the standardized interview, interviewers are constrained in their ability to diagnose the trouble, being sanctioned only to repeat a problematic survey question verbatim. This may preserve the reliability of the answers by reducing interviewer variability but may sometimes do so at the price of compromising the validity of those answers. Furthermore, respondents' requests for clarifications of survey questions vary in terms of how specifically they identify the source of the trouble. As a result, they vary in terms of what they solicit from the interviewer. Requests that seek only a minimal response from an interviewer, such as a confirming "yes" or "mm hmm," according to our data, appear to be more likely to be granted than those that seek a paraphrase or elaboration of the survey question.

In addition, recipients of utterances that contain an error of some kind in ordinary conversation may withhold attempts to correct it if the error does not prevent them from producing the sequentially implicated next turn (Schegloff, Jefferson, and Sacks, 1977). However, in the standardized interview, interviewers should initiate correction on the form of an answer even if it is adequate for enabling them to record a single appropriate answer option. In practice, interviewers at one center we studied tended to employ a kind of "verification probe" that, although it takes the form of a directive probe (Fowler and Mangione, 1990; Czaja and Blair, 1996), does not *lead* the respondent in any biasing way. If our data are representative of probing practice more generally, then some instances of directive probing in the survey interview may pose less of an actual threat to standardization than is currently believed.

Although the goal of question design in survey research is to avoid repair, repair sequences can be useful for revealing how well respondents understand the questions in a way that paradigmatic sequences cannot. Repair sequences expand the compact structure of the paradigmatic sequence and in so doing can reveal respondents' interpretations or misinterpretations of the questions. Thus, repair sequences may constitute a resource for question design, rather than simply a nuisance for standardization.

NOTES

1. See section on ethnomethodology and cognition in Maynard and Clayman (1991:392–396)

2. Valid inferences about populations do not of course depend on *every* respondent understanding the question adequately, but only on *enough* respondents understanding.

3. We are using the word "repeat" in a commonsense manner. Technically, a repeat of an utterance is never identical with the original; in fact, it may differ considerably. A more precise term might be "redoing" or "reperforming," but we avoid these because they are somewhat cumbersome.

4. Corsaro (1977) identifies a similar use of repeats in adult–child interaction. He (Corsaro, 1977:199) argues that when faced with utterances from children that are hard to respond to in a substantive way, adults sometimes simply repeat the child's utterance, with questioning intonation, and thereby "fill in" their conversational turn.

5. Requesting a repeat without locating a specific source of a hearing trouble may be a means for a participant to solicit a simpler paraphrase without having to admit ignorance (Sacks, 1992, II:413).

6. The survey item would probably be less confusing in this case if it were written the way that the interviewer repairs it: "Latin, Spanish and salsa."

7. An example of a display of realization would be something like, "Oh fos<u>ter</u>," where "oh" marks a change in the speaker's state of knowledge (see Heritage, 1984).

8. At this survey center, special definitions or examples are sometimes included with a question on the interviewer's screen but should only be read in the event that a respondent requests clarification. If no scripted clarification is provided, interviewers should avoid offering their own interpretations.

9. Requests for clarification of the answering procedure often inquire about the appropriate answer format, e.g., "Whaddaya want a number?" or about how to respond to a series of items, e.g., "And I should answer for each one?"

10. We offer this result cautiously because it could be caused by other things such as the types of questions asked (factual vs. attitude) or the differing guidelines at the three centers regarding the offering of clarification. A study in which *question type* and *interviewing guidelines* are held constant would probably be necessary for confirming it.

11. In a paper on probing in Dutch survey interviews, Houtkoop-Steenstra (1996) found that whether or not interviewers produce these repeats with rising intonation, respondents will respond to them, usually by confirming them.

12. We only have audio data and cannot confirm this. However, see Maynard and Marlaire's (1992) analysis of how educators, during one-on-one testing, use evaluative comments after examination question–answer sequences in simultaneity with their recording of the child's answer and looking up the next question. In the survey, interviewers are not permitted to use evaluative comments after receiving an answer to a question and may therefore use other devices to display interactional orientation during their recording of answers and locating a next question.

13. With computer-assisted interviewing, logical inconsistencies between two answers could be detected by the computer, and the interviewer could be alerted and could attempt to reconcile them. This type of programmed consistency check is used at some survey centers, but not at the three we studied.

14. On the use of "okays" as both completing a prior sequence and anticipating a next action, see Beach (1993).

15. The pause could reflect the time it takes the respondent to make the calculation; however, responding in a preferred format would consist of some kind of acknowledgement of the request, such as "okay," in the absence of an immediate answer.

16. By yes or no they mean a confirmation or rejection of candidate answer rather than the answer options of a yes/no questions.

17. The interviewers we studied commonly used the conjunction, so, to preface probes. At one center, 54 percent of all probes were designed with *so* prefaces (Moore, 1999). Initiating correction on the answer by proffering an upshot(s) may be another way of downgrading the initiator—in addition to marking it as uncertain or formatting it as a question—thereby orienting it to the *dispreference for other correction* (Schegloff, Jefferson, and Sacks, 1977).

Furthermore, the most common probe–utterance design the interviewers employed was, *Would you say X?* It was used at all of the centers studied, accounting for 24 percent of all probes. It is also one of the more versatile designs in that it can be used on answers to all of the three types of survey question—category, numerical, and yes/no—(unlike, e.g., *Can you give me a number?*) and can be used in both a nondirective and directive manner [unlike, e.g., *I (just) need X*] (Moore, 1999).

18. Answers such as "the first one," referring to one of the listed answer categories, or "same," referring to an answer to a prior question, are other types of answers that indicate a particular answer option using other words.

19. This is somewhat similar to the practice, known as a verification probe (Schaeffer and Maynard, 1996), used at the U. S. Bureau of the Census, by which if a respondent has already provided an answer to a current question in response to a prior question, the interviewer may ask a truncated version of the question. For example, see arrows:

SPPA156:273

FI: .hh During the last twelve months did you watch a
ballet program on television or a video tape?
(1.9)
FR: °hm:. .tch Have to think about (0.3) ( )°
(5.7)
FR: I (.) I can't be sure. Probabl(heh)y I(heh) did(heh)
once(heh) or(heh) twice(heh)
(.)
FI: .hh Alright. Was that on T V, V C R or both.=
FR: =uh: that'd be T V
(0.9)
FI: And during the last twelve months about how many times
→ did you do this?, You said once or [twice.]=
FR: [Yeah]
FI: → =Is that corre:ct?
(.)
FR: Right

20. Thirty-three of these directive probes came from one center, compared to only two for the other center.

21. This respondent clarification proffer is an example of "fourth position repair" (Schegloff, 1992:1321), while the ones analyzed earlier in this chapter constitute "second position repair."

REFERENCES

Beach, W. A. 1993. "Transitional Regularities for 'Casual' 'Okay' Usages." *Journal of Pragmatics* 19: 325–352.

Corsaro, W. A. 1977. "The Clarification Request as a Feature of Adult Interactive Styles with Young Children." *Language in Society* 6: 183–207.

Coulter, J. 1989. *Mind in Action.* Cambridge: Polity.

Czaja, R., and J. Blair. 1996. *Designing Surveys: A Guide to Decisions and Procedures.* Thousand Oaks, CA. Pine Forge Press.

Fowler, F. J., and C. F. Cannell. 1996. "Using Behavioral Coding to Identify Cognitive Problems with Survey Questions." Pp. 15–36 in N. Schwarz and S. Sudman (Eds.), *Answering Questions.* San Francisco: Jossey-Bass.

Fowler, F., and T. Mangione. 1990. *Standardized Survey Interviewing: Minimizing Interviewer-Related Error.* Newbury Park: Sage.

Garfinkel, H. 1967. *Studies in Ethnomethodology.* Englewood Cliffs, NJ: Prentice-Hall.

Garfinkel, H., and H. Sacks. 1970. "On Formal Structures of Practical Actions." Pp. 337–366 in J. D. McKinney and E. A. Tiryakian (Eds.), *Theoretical Sociology.* New York: Appleton-Century-Crofts.

Hahn, R. 1990. "Comment on Suchman and Jordon: "Interactional Troubles in Face-to-Face Survey Interviews." *Journal of the American Statistical Association* 85: 248–250.

Heritage, J. 1984. "A Change-of-State Token and Aspects of Its Sequential Placement." Pp. 299–345 in J. M. Atkinson and J. Heritage (Eds.), *Structures of Social Action.* Cambridge: Cambridge University Press.

Houtkoop-Steenstra, H. 1995. "Meeting Both Ends: Between Standardization and Recipient Design." In P. ten Have and G. Psathas (Eds.), *Situated Order: Studies in the Social Organization of Talk and Embodied Activities.* Washington, DC: International Institute for Ethnomethodology and Conversation Analysis and University Press of America.

Houtkoop-Steenstra, H. 1996. "Probing Behavior of Interviewers in the Standardised Semi-Open Research Interview." *Quality and Quantity* 30: 205–230.

Jefferson, G. 1983. "On Exposed and Embedded Correction in Conversation." *Studium Linguistik* 14: 58–68.

Maynard, D., and S. Clayman. 1991. "The Diversity of Ethnomethodology." *Annual Review of Sociology* 17: 385–418.

Maynard, D., and C. Marlaire. 1992. "Good Reasons for Bad Testing Performance: The Interactional Substrate of Educational Exams." *Qualitative Sociology* 45: 177–202.

Moore, R. 1999. "Achieving Understanding in the Standardized Interview: A Conversation Analytic Study of Talk in Telephone Surveys." *Unpublished dissertation*, Department of Sociology, Indiana University, Bloomington.

Oksenberg, L., C. Cannell, and G. Kalton. 1991. "New Strategies for Pretesting Survey Questions." *Journal of Official Statistics* 7: 349–365.

Pomerantz, A. 1984. "Agreeing and Disagreeing with Assessments: Some Features of Preferred/Dispreferred Turn Shapes." Pp. 299–345 in J. M. Atkinson and J. Heritage (Eds.), *Structures of Social Action.* Cambridge: Cambridge University Press.

Pomerantz, A. 1988. "Offering a Candidate Answer: An Information Seeking Strategy." *Communication Monographs* 55: 360–373.

Sacks, H. 1987/1973. "On the Preferences for Agreement and Contiguity in Sequences in Conversation." Pp. 54–69 in G. Button and J. R. E. Lee (Eds.), *Talk and Social Organisation.* Clevedon, England: Multilingual Matters.

Sacks, H. 1992. *Lectures on Conversation*, Vols. I and II. Cambridge, MA: Blackwell.

Sacks, H., E. A. Schegloff, and G. Jefferson. 1974. "A Simplest Systematics for the Organization of Turn-Taking for Conversation." *Language* 50: 696–735.

Schaeffer, N. C. 1991. "Conversation with a Purpose–Or Conversation? Interaction in the Standardized Interview." Pp. 367–391 in P. P. Beimer, R. M. Groves, L. E. Lyberg, N. A. Mathiowetz, and S. Sudman (Eds.), *Measurement Errors in Surveys*. New York: Wiley.

Schaeffer, N. C., and D. W. Maynard. 1996. "From Paradigm to Prototype and Back Again." Pp. 65–88 in N. Schwarz and S. Sudman (Eds.), *Answering Questions*. San Francisco: Jossey-Bass.

Schegloff, E. A. 1972. "Notes on a Conversational Practice: Formulating Place." Pp. 75–119 in D. Sudnow (Ed.), *Studies in Social Interaction*. New York: Free Press.

Schegloff, E. A. 1992. "Repair after Next Turn: The Last Structurally Provided Defense of Intersubjectivity in Conversation." *American Journal of Sociology* 97: 1295–1345.

Schegloff, E. A. 1997. "Practices and Actions: Boundary Cases of Other-initiated Repair." *Discourse Processes* 23: 499–547.

Schegloff, E. A., G. Jefferson, and H. Sacks. 1977. "The Preference for Self-Correction in the Organization of Repair in Conversation." *Language* 53: 361–382.

Schegloff, E. A., and H. Sacks. 1973. "Opening Up Closings." *Semiotica* 8: 289–327.

Schutz, A. 1962. *Collected Papers*, Vol. 1: *The Problem of Social Reality*. The Hague: Martinus Nijhoff.

Suchman, L., and B. Jordan. 1990. "Interactional Troubles in Face-to-Face Survey Interviews." *Journal of the American Statistical Association* 85: 232–241.

Wittgenstein, L. 1953. *Philosophical Investigations*. Translated by G. E. M. Anscombe. New York: Macmillan.

CHAPTER 14

Ad Hoc Inquiries: Two Preferences in the Design of Routine Questions in an Open Context

John Heritage

In the United Kingdom, community nurses visit all newly delivered mothers within a few days of their return from hospital. These visits occur without regard to the social or economic status of those visited and, though they occur at the mother's homes, they are often initiated without prior consent. They are nonetheless rarely refused. During these initial visits, the nurses — known as health visitors — begin to build a relationship with the mother that, ideally, will be deepened over the course of many others. The health visitors will attempt to establish a "befriending" relationship in which their professional expertise will be acknowledged and put at the disposal of the mother. They will also collect standardized information about the family and its circumstances, which they record on the spot. This chapter looks at how the health visitors design their *routine* questions in search of standardized information.

The conduct of the health visitors in this situation may offer us a glimpse of how social surveys may have been administered in the past. The origins of the health visitor service are rooted in the same historical and ideological context as the school board visitors, whose information formed the basis of Charles Booth's *Life and Labour of the People in London* (Booth, 1889–1903). Moreover, like the social surveys conducted before World War II, the routine information solicited in these visits is not collected by means of questions whose wording is determined in advance. Like the prewar "schedule" or "blank" (Converse, 1987:33–34), the

Standardization and Tacit Knowledge: Interaction and Practice in the Survey Interview
Edited by Douglas W. Maynard, Hanneke Houtkoop-Steenstra, Nora Cate Schaeffer, and Johannes van der Zouwen
ISBN 0-471-35829-0

form that the nurse must complete is a printed ("clinic") form that contains spaces, or slots, for relevant items of information concerning other family members, age, occupation, nationality, and siblings, and basic information concerning the course of the pregnancy and labor. And, like the blanks, the clinic form does not contain any specific wording for the questions to go with those slots. This circumstance offers an opportunity to see how the nurses, charged to administer routine questions on standard topics, design those questions when they have the latitude to vary them in ways that are not constrained by the requirements of survey methodology.

This chapter begins with a discussion of the context in which these questions are asked and answered. Specifically, it looks at how the task of completing the child health form is constructed by the nurses, and how it is positioned relative to the other business that is normally transacted in the nurse–mother encounter. Second, focusing on questions about (a) the child's birth and health postpartum and (b) the parents' age and occupation, it examines two preferences that the nurses consistently deploy in designing questions: (i) to embody a preference for normalized or *optimized* states of affairs and (ii) to design the questions in the light of information and other interactional dimensions of the interaction-in-progress so as to exhibit *recipient design*.

The chapter's central observation is that, regardless of the specific aims of questioning, the ways in which questions are designed unavoidably serve to index particular relationships between questioner and respondent. At one extreme is the kind of questioning found in the survey interview, designed to minimize bias and maximize the objectivity and standardization of results. This style of questioning, if reciprocated in the respondent's answers, constitutes a social relationship in which both parties treat their specific identities for one another as irrelevant for all practical purposes. This relationship may be termed "essentially anonymous." If questioners want their relationships with respondents to depart from this condition, then departures from the effort to maximize the neutrality and objectivity characteristic of survey question design will have to be made. This chapter discusses the departures that are recurrent when medical personnel, concerned with building personal relationships with new mothers, engage in *face sheet* and related standardized questioning.

DATA AND ITS CONTEXT

In the occasions that we will focus on, first-time mothers and their new babies have been discharged from hospital in the previous several days and are receiving their first visit from the health visitor. The visit is normally unannounced and, on one or two occasions, the mothers were caught somewhat unprepared for the visit.

With the knowledge and permission of all parties, 5 health visitors (HVs) audiotaped a series of 6 visits to a total of 9 first-time mothers, and another 10 mothers of second or subsequent babies.[1] The recordings were made in a large British midlands city and in households varying significantly in occupation and income. This is, of course, not a large sample of the hundreds of thousands of visits made by HVs to new mothers each year. Nor is the sample representative. These weaknesses would be problematic if we were attempting to assess the distribution or frequency of the activities going on in these visits. However, the task in this chapter is a more restricted one: to identify features of question design in this context and to suggest underlying principles that organize them. While the data are definitely adequate for this purpose, it is clear that they cannot yield anything like a full sampling of these features, and this chapter makes no pretense at any kind of exhaustive catalog.

BACKGROUND: THE HEALTH VISITORS AND THEIR ROLE

The British health visitor service is the largest single element of the UK community nursing program, comprising some 9300 qualified nurses (Cumberlege Report, 1986:10). The health visitor's role, as described by the Health Visitors' Association (1985), is to be "fully and completely involved in the giving of advice and support but only indirectly in the treatment of illness, environmental control and the provision of practical help." As this broadly worded description suggests, health visitors have very wide-ranging professional responsibilities comprising the detection and prevention of ill health in the community, the identification of health needs in the community, health teaching, and advice and guidance in cases of illness and in the care and management of children (Council for the Education and Training of Health Visitors, 1977). These responsibilities — in which advice giving plays a primary role — are discharged through verbal interaction with members of the community. Unlike community nurses in other advanced countries, British health visitors do not perform routine nursing tasks and concentrate instead on illness prevention, giving advice on health and social problems and case finding for other more specialized agencies.

The British health visiting service is distinctive in that, unlike other medical services in the United Kingdom and elsewhere, its provision is supply — rather than demand — driven. This characteristic is particularly prominent in relation to the health visitor's work with children. Health visitors have a statutory obligation to perform routine visits to all mothers with children under five regardless of whether these visits are requested or not.[2] The supply-driven character of the service reflects the origins of the health visiting service in the municipal sanitation movement of the nineteenth and early twentieth century. This movement was highly interventionist and was anchored in a sense of mission that bordered on a moral crusade.[3] While this explicit moral focus has all but evaporated with the health visitor's incorporation into an instrumentally oriented national health service, there remains a — perhaps irreducible — residue of ambivalence concerning the dual role of advisor and evaluator that is commonly attributed to the health visitor role (McIntosh, 1986).

During the home visits, HVs generally act as baby experts while also striving to construct a warm, befriending relationship (Heritage and Sefi, 1992; Heritage and Lindström, 1998; Robinson, 1982; Sefi, 1988). However these efforts are not by any means always successful. A substantial proportion of mothers, particularly those in poorer socioeconomic circumstances, see the HV service largely in terms of social control and surveillance and attempt to minimize contact with its representatives. Most surveys suggest declining levels of satisfaction with the HV service over time and indicate that mothers tend to remain unclear about the role and value of HVs. These studies also indicate that mothers have a strong preference for friendliness and informality in HV conduct and that HVs are quite commonly perceived as excessively authoritarian and didactic (McIntosh, 1986).

GENERAL CHARACTER OF THE FIRST VISITS

Of the visits that make up our database, most represented the first occasion on which the mother and HV had met one another and, although the mothers were expecting a visit from the HV at some time during the week of the first visit, all were unannounced.[4] In six of the nine visits making up our database, the new mothers were not alone when the HV called. In two cases, a female companion was also present: a young female friend in one case, and the mother's mother (with whom the couple were living) in the second. In the other four cases, the new father was present. In most of the visits, these other persons were significant, sometimes even dominant, participants. The visits varied in length from 15 minutes to just under an hour.

Although the visits are quite diverse, their content contains a number of common features. A large majority of topics were initiated and terminated by the HV in a basically *segmented* process of topical progression (Button and Casey, 1984; Sefi, 1988). Substantial parts of these first visits were occupied with three significant bureaucratic tasks: (1) getting face sheet data about the mother and baby for the records of the clinic to which mother and baby will be attached; (2) getting consent signatures for immunization injections for diphtheria, whooping cough, tetanus, polio and measles; and (3) explaining clinic procedures and the various subsequent health checks that mothers and babies will go through in the ensuing months and years. The visits were substantially diverse in terms of their overall trajectories, the type and ordering of the topics raised, and in terms of the tone or rapport achieved between the parties.

GETTING ROUTINE DATA: THE QUESTIONNAIRE SECTION OF THE ENCOUNTER AS OBSTACLE AND RESOURCE

The main body of routine information that the health visitor must gather is laid out as a block on the second page of a form labeled Child Health Record, which is conveniently color coded, pink or blue. The main block that the health visitors

Parents

	Nationality	Year of Birth	Occupation
Mother	**Item # 1**	**Item # 2**	**Item # 3**
Father	**Item # 4**	**Item # 5**	**Item # 6**

Siblings[1]

Name	**Item # 7**				
Sex					
D.O.B.					

Birth information

Pregnancy normal/abnormal, specify **Item # 8** ______________________

____

Type of delivery **Item # 9** ______________________

In S.C.B.U. Yes/No **Item # 10**

Neonatal problems 1 **Item # 11** ______ 2 ______ 3 ______

Feeding on leaving hospital Breast/Bottle/Mixed **Item # 12**

Family planning advice given Yes/No

Figure 14.1

have to complete in the visits we recorded is represented in Figure 14.1. There are 12 information items in the form that we focus on, which are numbered for reference purposes.

In principle, this information can be quite straightforward to obtain, as the following extract illustrates. The father was not present during this interaction.

(1)[5A1 : 9][6]

```
HV:  And Andrew's (.) j[ob?                                Item # 6
M:                        [Painter and decorator.
     (1.7)
HV:  So his- (.) he hasn't come into his own yet'as he=
M:   =No:=
HV:  =in the hou:se.
M:   No[:.
HV:    [((co[ ugh)) ('is:)
M:     [(Not) ye:t.
     (0.2)
M:   His ti:me will come (.) yeah.
     (1.0)
HV:  Has he got plenty of wo:rk on?
M:   He works for a university college.
HV:  O:::h.
M:   So: (.) he's in full-time work all the ti:me.
HV:  °Yeh.°
     (0.4)
HV:  And this is y'r first baby.                           Item # 7
M:   Ye(p).
HV:  And you had a normal pregnancy.=                      Item # 8
M:   =Ye:h.
     (1.5)
HV:  And a normal delivery,                                Item # 9
M:   Ye:p.
     (1.8)
HV:  °That's (great)°
     (0.7)
HV:  And sh'didn't go into special ca:re.                  Item # 10
M:   No:.
     (1.8)
HV:  °A:nd she's bottle feeding?°                          Item # 12
     (1.2)
HV:  °'nd° (0.5) and uh you're going to Dr.White
     for your (0.6) post-natal?
```

Here, in a very few turns at talk, the HV addresses 6 out of the 12 main items of information she is required to obtain.

However, while this *questionnaire section* of the visit can be managed quickly and easily, it was often treated as a source of difficulty (Heritage and Sorjonen, 1994). Set within a visit that is substantially aimed at relationship making with the family — and, in particular, befriending the mother — the questionnaire section is often portrayed as embodying the intrusion into the visit of an alien set of relevancies. For example, the health visitors may initiate the questionnaire section as an onerous task to be disposed of at a convenient moment:

(2) [4B1:2]

```
Well whi(le) she's asleep we might as well fill in these forms and get
them over and done with.<
```

(3) [4A1:7]

```
He's terrific. .hh Well while he's so settled and quiet (0.6)
maybe it would be a good idea to get (.) a:ll the paperwork
over and done with.
```

Or they may give the forms they are completing a functional gloss:

(4) [4B1:17]

```
This is his baby clinic card >you could keep it< (.) it's uhm
we keep at the health center its just (0.5) we put the dates
of the injections (0.7) uh:m any information there that's relevant,
(1.2) and the weight, (2.2) his date of birth where he was born
how many weeks gestation, (1.0) and on this side, (2.5) uh:m
> information about the parents< you're both British?
```

Or simply attempt to minimize the intrusiveness of the questions:

(5) [1C1:25]

```
Okay so that's that's your clinic fo:rm....An' all I put on here
is you:r (0.7) there's a bit about you: (0.7) it sa:ys here that
you're twenty one is that ri:ght?
```

Finally, they may distance themselves from questions that seem particularly bureaucratic or outside the range of relevancies of the visit:

(6) [4A1:13]

```
These details (.) I don't know why they want to know but
father's a:ge.
```

(7) [4B1:5]

```
Uh:m (0.5).hh now fi:rst th' particulars they want to know th' baby's
father's a:ge.
```

In these and other ways, the health visitors treated the *clinic form* section of the encounter as problematic and as in contrast to the main purpose of the visit.

However, in some of the home visits the clinic form represented a resource. In these cases, the forms were invoked early and their contents were treated as an agenda of inquiries that could be elaborated upon, or departed from and returned to in due course. This use of the clinic form was prominent in cases where fathers were present. In this context, the health visitors used the clinic form as a resource to legitimate questions, and associated topical lines, which mainly focused on the mother's experience and tended to exclude fathers from participation. A return to the clinic form, and its bureaucratic details, was also used as a means to reinclude fathers in the conversation in ways that were legitimately provided for and did not seem forced or unnatural.

This use of the form is illustrated by the following extended sequence:

(8) [1A1:12-22]

```
1 HV:    So this'll be her clinic ca:rd.
2        (1.0)
```

```
HV: → (.hh) And on here we have a record (0.8) of how you
      were in- (0.3) did you have a normal pregnancy:?    Item #8
 M:   Yes.
      (0.7)
HV: → I'll ring no:rmal. And what about your delivery,
      (1.6)
HV:   Did you (0.6) uhm have a normal delivery,=did       Item #9
      you wat[ch it.
 M:          [Ye::s
 F:   Mm hm,
HV: → Did you?=What did you thi:nk. ((smile voice))
      ...
      ... [ Approx. 2.5 minutes of talk that focuses on difficulties
      ...   in getting the baby's shoulders out during the delivery.]
      ...
HV: → =So you had a- uh: (1.0) You didn't- Did you-
      You didn't have forceps you had a:                  Item #9
 M:   =Oh [no:: nothing.
 F:       [(  )
HV: → An- and did she cry straight awa:y.                 Item #11
 M:   Yes she did didn't sh[e.
 F:                        [Mm hm,
      (1.0)
HV: → Uhm (.) you didn't go to scbu: you know the
      spe[cial care unit.                                 Item #10
 M:      [Oh: no: no:.
      (0.8)
HV:   °Good.°
      (0.5)
HV:   So the hospital: are quite pleased with her.
      (0.5)
 M:   Mm::.
      (2.0)
HV: → And you breast fed straight away did you:?          Item #12
 M:   Ye(p).
      (1.2)
HV: → °And she hasn't had a bottle yet.°                  Item #12
      (0.7)
HV:   °Good.°
      (0.7)
HV:   S[o-
 M:    [She'ad- (.) she'ad water the first night,
      (0.8)
 M:   uh-=
HV: → =Are you working.                                   Item #6
 F    Nope.=
 M    =No.
      (1.5)
HV:   (    )
      (1.0)
HV:   So you're unemplo:yed,
 F:   Mm hm,
HV:   .tch Poor you. What did you used t'do.
      ...
```

... [Approx. 4.5 minutes on F's employment problems]
...
405 HV: Ri:ght.
406 (4.5)
407 HV: Uh::m (1.5) >Where did we get to.<
408 (2.0)
409 HV: → How old are you father? Item #5
410 (0.7)
411 F: Twenty si:x.
412 (4.7)
413 HV: → Were you a uhm (1.5) what (1.0) uh: (.) you were a
414 nurse at the Randolph. Item #2
415 F: Ye[h
416 M: [Yea:h.
417 (5.2)
418 HV: Uhm (.) an S.E.N..
419 M: Ye(p).
420 (2.2)
421 HV: Are you going to go ba:ck.
422 F: Mm hm
423 M: Yea:h
424 F: ((sniff))
425 M: I've taken maternity leave.
426 HV: ([ )
427 M: [I'm due to go back in Ma:rch.
428 HV: And who will look after (.) Ann-Marie.
429 M: I was goin' back on nights.
430 HV: I see.
431 M: I wanted to'ave a go at a couple of nights to
432 see if we could (.) you know manage between us,

In this case, the health visitor takes about 11 minutes to cover the same ground that was covered in about 30 seconds in the first example. But, in the interim, she has gone out of her way to involve the father in a discussion of the delivery (starting at lines 9–10), has addressed the father's employment difficulties and the family's associated financial problems (lines 197–404), and explored the mother's employment plans and provision for baby care (lines 417–432).

In sum, routine data collection in these interactions represented both a source of difficulty and of opportunity, depending on the context. In interactions with mothers on their own, it was treated as an unequivocal intrusion into the encounter, and as something to be minimized as far as possible. In interactions where others were present, the form filling was sometimes treated as problematic, but it was also deployed as a resource in the management of the interaction to include both mothers and fathers, while legitimating lines of questioning that might otherwise be treated as excluding fathers from the interaction.

QUESTION DESIGN

Regardless of whether the nurses treated routine data collection as an intrusion or as a resource, the questions by which they implemented the process did not resemble traditional questionnaire questions. Two fundamental principles underlay the nurses' questions. These principles are drawn from the normative order of everyday life and clearly depart from the principles informing questionnaire construction and implementation. They are: (1) the principle of optimization, which yields questions designed to prefer "best-case," normal, or "no-problem" responses;[7] and (2) the principle of recipient design, which yields questions adapted to the specific circumstances of a recipient, and the state of the interaction between questioner and recipient that is current at the moment of the question.

BIRTH DETAILS: QUESTIONS EMBODYING THE PRINCIPLE OF OPTIMIZATION

Questions that embody the principle of optimization display a respect for what Maynard (in press) terms the benign order of everyday life. They do so by incorporating presuppositions and preferences that are biased toward best-case or no-problem outcomes, and they are a characteristic feature of medical questioning in a great variety of routine contexts (Boyd and Heritage, in press). For example, in questioning patients about their family history, doctors will rarely spell out a full range of potential options. Thus a question about parental mortality could be asked in at least the following ways:

Is your father alive?
Is your father dead?
Is your father alive or dead?

However, it is overwhelmingly asked in the first, best-case form, which is designed for, and thus prefers a confirming "Yes" response. The second method of asking this question — "Is your father dead?" — which would favor a confirming bad news response, is not ordinarily used outside of some aspect of the recipient's circumstances that would make its production specifically accountable. But neither is the third, objective, or bureaucratic, form of the question ordinarily deployed in medical contexts (Boyd and Heritage, in press).

In the data we are examining, the nurses' questions about the process of childbirth are almost uniformly constructed so as to favor optimized no-problem responses. A very clear case is the following in which, while a series of yes/no declarative questions varies in terms of the polarity preferred, each is constructed in search of a no-problem response:

(9) [5A1]

```
HV:      And this is y'r first baby.               Item #7
 M:      Ye(p).
HV: →    And you had a normal pregnancy.=          Item #8
 M:      =Ye:h.
         (1.5)
HV: →    And a normal delivery,                    Item #9
 M:      Ye:p.
```

```
 8          (1.8)
 9 HV:      °That's (great)°
10          (0.7)
11 HV: →    And sh'didn't go into special ca:re.           Item #10
12  M       No:.
13          (1.8)
14 HV:      °A:nd she's bottle feeding?°                   Item #12
15          (1.2)
16 HV:      °'nd° (0.5) and uh you're going to Dr.White ...
```

A similar pattern of questions favoring no-problem responses emerges in the following sequence, at lines 1, 4, 11, 14, 16, 19 and 20:

(10) [4A1:17]

```
 1 HV: →    Uh::m (.).hh So your pregnancy was perfectly     Item #8
 2     →    normal.
 3  M       Yeh.
 4 HV: →    And did you go into labor (.) all by yourself?   Item #9
 5  M:       No: I was started o[ff because uh:m (0.8) the blood
 6 HV:                          [Induced.
 7  M:      pressure (0.7) went up in the last couple of weeks.
 8          ...
 9          ... [Segment dealing with why mother was induced]
10          ...
11 HV: →    And was he alright when he was born.             Item #11
12  F:      Mm[:.
13  M         [Yeah.
14 HV: →    He came down head fi:rst.                        Item #9
15  F:      Mm h[m,
16 HV: →        [No:rm- no:rmal delivery?=                   Item #9
17  M:      =Ye:h.
18          (2.2)
19 HV: →    And did he stay with you all the time.=          Item #10
20     →    =He didn't go to special care baby unit.
21  M:      No:.
```

Here each question thematizes the medically and socially desirable outcome of labor as the focus for a yes/no question. In each case except the last, the question design favors a "yes" response, sometimes abandoning interrogative for declarative syntax in pursuing this objective (lines 14, 16, 19, and 20), and/or deploying negative polarity when a "no" response is the best-case, no-problem one (e.g., line 20).

It is very rare that any of these questions are asked in a form that provides explicit alternatives (the "alive or dead" option). The only such case in this data set is the question "And didju: (0.6) push her out yourself or did you have to have forceps (then)" in the following sequence (at lines 11–12):

(11) [1C1:25]

```
 1 HV:      Didju have a normal pregnancy:.                  Item #8
 2  M:      Yes a very goo:d pregnancy:.=I carried her we:ll.
 3          (1.2)
 4 HV:      A:nd uh: (1.5) how long were you in labor for the:n,
 5          (1.0)
```

```
   M       Uhm thirteen to fourteen hou:rs.
           (1.0)
   HV:     W'll that's pretty: average actually for a first
           baby:,
           (1.2)
   HV: →   And didju: (0.6) push her out yourself or did      Item #9
       →   you have to have forceps (then).=
   M:      =Pushed her out mese:lf.
           (4.2)
   HV:     And she: (0.5) cried straight awa:y_               Item #11
           (0.5)
   HV:     D[id she:?
   M:        [No:
   M:      She didn't no (it took'er (0.3) couple'v minutes
           actu'lly)=
   HV:     =Yeh,
   M:      They had to give'er uh:: (.) she come out blue:
           (0.4) the:y gave he:r oxygen I think it was a
           minute before she cried,
   HV:     Yeah.
           (0.6)
   HV:     But she didn't have to go to: (.) special ca:re    Item #10
           unit or any[thing.
   M:                 [No:: no:.
```

Subsequent to this question, the nurse reverts to declaratively formed questions shaped in favor of no-problem responses. Notably the "B-event" question (Labov and Fanshel, 1977) at line 15 "And she: (0.5) cried straight awa:y_" is revised with an incremental, tag question (line 17) in the face of deferred response by the mother. Here the deferral of the response projects a "No" answer that aligned against the design of the question, and the later description of a problem that the question was not built to favor. However, even after the mother's account of her baby's initial postpartum difficulties (lines 19–24), the nurse's question about the special-care baby unit is still declaratively shaped to favor a no-problem response (lines 27–28).

As already noted, these question designs carry forward a more general tendency in ordinary conversation for good news to be favored over bad news, and for interaction to be conducted under the auspices of the "benign order of everyday life" (Maynard, in press). As Houtkoop-Steenstra and Antaki (1997) have also suggested, questions designed to favor problematic responses are "marked" forms that are accountably asked only "for cause" and must be licensed by information from prior talk or its context (see also Boyd and Heritage, in press). We can see this kind of license emerging in the following sequence, where the nurse picks up a hesitation in the mother's response to an earlier no-problem inquiry:

(12) [4A1:19]

```
   HV:     And you're feeling well.
           (0.7)
```

```
 3  M:   Yeah.
 4       (1.5)
 5 HV:   And you're- (.) You didn't ha- Did you have stitches?
 6       (0.8)
 7  M:   Ye[:es
 8 HV:     [You did. [('N) are you so:[re=
 9  M:               [(nh hnhn )       [I had a third degree tea:r=
10 HV:   =O::::::h. ↑Did you::?
11  M:   Yeah. (0.2) It's uh (.) they think what happened'is
12       chin must'ave caught me.
13       (0.3)
14  M:   .hhh as'e w'[z coming ou:t.
15 HV:               [O::::h,
```

While it is open as to what the term "feeling well" might address in the inquiry that begins this sequence, the declarative question itself is clearly designed for an optimized no-problem outcome. However, the mother's response is both minimal and quite delayed (line 2). Perhaps because of these features, the nurse abandons a next question "And you're-" (line 5), that was begun as a move to a new *agenda* item (Heritage and Sorjonen, 1994) and instead moves to what is hearable as a contingent inquiry addressed to the whether the mother had stitches. This question is initially designed toward a no-problem outcome "You didn't ha-." However, this declarative question form is abandoned, and revised to an interrogative form "Did you have stitches?" which, relative to the declarative at least, favors what will be in this context a more problematic "Yes" response.

A similar case is the following. Here the nurse's question follows an extensive sequence in which both parents described the difficulties the mother experienced during labor with the child's shoulders. While the delivery process they describe was lengthy, they did not mention the deployment of any unusual measures (such as forceps). The nurse's question resumes her work of completing the child health card, and addresses item 9, "Type of delivery," in which the recording of a forceps delivery would be clearly relevant—see excerpt 11, lines 11–12. The initial form of the nurse's question at line 1 would probably have eventuated as "So you had a normal delivery" as the upshot of that account.

(13) [1A1:14]

```
 1 HV:   =So you had a- uh:
 2       (1.0)
 3 HV:   You didn't- Did you- You didn't have forceps          Item #9
 4       you had a:
 5  M:   =Oh [no:: nothing.
 6  F:       [(    )
 7 HV:   An- and did she cry straight awa:y.                  Item #11
 8  M:   Yes she did didn't sh[e
 9  F:                        [Mm hm,
10       (1.0) ((Wood cracking))
11 HV:   Uhm (.) you didn't go to scboo: you know the         Item #10
12       spe[cial care unit.
13  M:      [Oh: no: no:.
```

However, instead of taking her initial formulation of this question to its likely conclusion ("So you had a normal delivery"), the nurse abandons it at just the point at which the optimized word "normal" would be due, in favor of a question design that will come to address the possibility of forceps more explicitly. This new question, however, with its invocation of a problematic circumstance that was not previously referenced, causes her notable difficulty. She begins with a declarative form "You didn't-" that would likely favor a no-problem report. Abandoning this, she reinitiates her question with an interrogative form "Did you-" that, as in except 12, would be more favorable to a problematic response. Finally she reverts to a fuller declarative form favoring a no-problem outcome: "You didn't have forceps you had a:" Here it seems likely that the full question that the nurse was heading toward was "You didn't have forceps you had a normal delivery." However, once again, the nurse abandons the question at the point at which the word "normal" was due. In this case, the nurse was clearly in genuine doubt as to whether her question should be framed so as to favor a no-problem response, or whether in the light of the parents' description of a difficult birth process, she should depart from this norm toward a less upbeat question design.

OCCUPATIONAL DETAILS: QUESTIONS EMBODYING THE PRINCIPLE OF RECIPIENT DESIGN

At the most fundamental level, the prospect of standardizing the questioning of patients is interdicted by the principle of recipient design. This term refers to the "multitude of respects in which the talk by a party in a conversation is constructed or designed in ways which display an orientation and sensitivity to the particular other(s) who are the coparticipants" (Sacks, Schegloff, and Jefferson, 1974:727). It is, of course, considerations of recipient design and, in particular, the issue of whether a no-problem question design would be appropriate given what the mothers had conveyed or stated, that impacted the somewhat vacillating question designs we have already seen in excerpts 12 and 13.

The principle of recipient design is very apparent in the nurses' questioning about the details of parental nationality and occupation. For example, in the following case, the nurse (who visited the mother briefly just before she had her baby) is clearly oriented to the impropriety of asking redundant questions. Thus at line 2, her question about the parents' nationality (items 1 and 4) is built to presuppose that they are British nationals. Similarly, having asked about the mother's occupation (lines 27–35), she acknowledges having been told this on a previous occasion at line 37. It can also be noticed that, having asked the mother her boyfriend's name (line 7), the nurse "volunteers" his family name at line 9, based on previous knowledge that the mother is taking the boyfriend's family name (see also line 14):

(14) [5A1:8]

```
1 HV:  (pt) °N(h)o:w° jist need a little bit of (0.3)
2      information on the insi:de.=You're both British.  Item #1 & 4
3      (0.6)
```

```
   M:  Yeah.=
  HV:  ° Yeh.°
       (2.8)
  HV:  What's your boyfriend's name?
   M:  Andrew.
  HV:  Andrew.(0.2) Beetham.
   M:  Yeah.
       (3.0)
  HV:  You're Mary.
   M:  Yeah.
  HV:  °Beetham°.
       (1.8)
  HV:  And uh (.) his date of birth=his a::ge?                 Item #5
       (1.0)
   M:  He's twenty-one so what would his date of birth be.
 (?):  hhhhhh=
   M:  =six-
  HV:  Twenty one'll do:.=
   M:  =Yeah.
  HV:  =They can work it out from the[:re huh
   M:                                [Yeah
       (0.2)
  HV:  fr'm there.
  HV:  Were you working?                                       Item #3
       (.)
  HV:  Previous[ly?
   M:          [Yeh I wa:s.
  HV:  What were you doing?
   M:  Uhm General Infirmary.
       (0.7)
   M:  Housekeeping.=
  HV:  =Housekeeping.
       (.)
  HV:  That's ri:ght °(I remem-)°
       (2.8)
  HV:  And Andrew's (.) j[ob?                                  Item #6
   M:                    [Painter and decorator.
```

The nurse's treatment of these two questions embodies an orientation to the mother as someone with whom she is forming a relationship. In such a context, failure to remember significant details about the mother is accountable. The nurse treats herself as accountable for remembering them, and in this way she indexes an ongoing, if only nascent, relationship with the mother. Her treatment of these questions contrasts quite markedly with her pursuit of face sheet data about the mother's boyfriend (lines 7, 16, and 39) whose details are treated in a more anonymous fashion characteristic of the survey interview. Comparing the questions about the mother and the boyfriend suggests that the latter embody what might be termed 'legitimate ignorance' about the subject matter. Thus, the boyfriend's details are treated in a more anonymous fashion, which is characteristic of the survey interview.

In other cases, nurses find other ways to use questions to assert an incipient relationship with the mothers. For example, they question them by supplying candidate answers to their questions:

(15) [3A1:8)]

```
 1 HV:    Ehm, so (I-) anyway I'll fill'is- I'll finish filling
 2        this card in..hhh Eh:m (0.7) father's age at bi:rth.
 3        (0.5)
 4  M:    He's twenty nine.
 5 HV:    °Twenty nine.°
 6        (1.2)
 7 HV: →  And he is a builder.
 8         (1.5)
 9  M:     °Y:up°
10         (1.3)
11 HV: →  And um: (.) you were (.) anaesthetic's nurse weren't
12         you.
13  M:     Yep.
14 HV:     And dj- (0.2) Do you think you'll (.) be a housewife
15         now for a bit.
```

(16) [1A1:20]

```
1 HV: →  Were you a uhm (1.5) what (1.0) uh: (.) you were a
2     →  nurse at the Randolph.
3  F:    Ye[h
4  M:      [Yea:h.
5        (5.2)
6 HV:    Uhm (.) an S.E.N..
7  M:    Ye(p).
```

The preference for this form of question design shows up in the fact that it carries the risk of error:

(17) [3B1:24]

```
1 HV:     >I'd better fill in the rest of this ca:rd then,
2      →  .hh ↑Mothers occupa:tion it- that was catering
3          wasn't [it,
4  M: →           [No no::.=(I'm in) pottery teaching.
5         (.)
6 HV:     [Ah:
7  M      [Pottery teacher.
```

Nonetheless the imperatives of relationship making mandate this form of recipient design, notwithstanding the risks.

Some of the nurses' questions are, of course, less presumptive, but this does not mean that the significance of recipient design is thereby diminished. In the next two cases, the nurses approach the question of the husband's occupation in distinctive ways:

(18) [4A1:12]

```
1 HV:   These details (.) I don't know why they want to know
2       but father's a:ge.
3       (0.9)
4  M:   heh Twenty eig[ht
5  F:                 [Twenty eight [I (should think)
6  M:                               [>heh heh heh [heh huh
```

```
 7 HV:                                                    [Twenty eight.
 8   M:     .hhuh
 9   F:     (Short a week (    ))
10          (1.0)
11 HV: →   And your occupation?
12   F:     °(P'lic' f's)°
13 HV:      Uh::(h)::r hhh[eh heh heh huh huh
14   M:                   [heh heh heh heh
15   F:     Don't say it like that.
```

(19) [1A1:16]

```
1 HV: →  =Are you working.
2   F     Nope.=
3   M:    =No.
4         (1.5)
5 HV:     (     )
6         (1.0)
7 HV:     So you're unemplo:yed,
8   F:    Mm hm,
9 HV:     .tch Poor you. What did you used t'do.
```

While the nurse's question in excerpt 18 presupposes that the husband is currently employed, the question in excerpt 19, which also looks more like a survey question, does not. As it turns out the design of both these questions reflects the operation of recipient design and, in both cases, the presumption (or the lack of it) that the nurse makes turns out to be correctly fitted to the local environment. In except 18, the father has already made reference to the fact that he is on paternity leave for 2 weeks and, in this context, the question "Are you working?" would evidently be redundant. In the second case, the father has made no such reference and a more cautious, and less presuming, approach to the topic of employment is the more appropriate one.

DISCUSSION

The British health visitor service, from which these data come, is a state-sponsored remnant of the municipal sanitation movement of the nineteenth century. Developing as part of a middle-class response to the squalor of working-class housing in Britain's mid-Victorian cities, the movement culminated in the development of a national sanitory inspectorate early in the twentieth century. The women inspectors' duties of a century ago are clearly evoked in the following quotation from the rule book of the Manchester and Salford Sanitory Association (c.1880):

> They must visit from house to house, irrespective of creed or circumstances, in such localities as their superintendents direct. They must carry with them the carbolic powder, explain its use and leave it where it is accepted; direct the attention of those they visit to the evils of bad smells, want of fresh air and impurities of all kinds; give hints to mothers on feeding and clothing their children; where they find sickness, assist in promoting the comfort of the invalid by personal help they... must urge the importance of cleanliness, thrift and temperance on all possible occasions. They are

> desired to get as many as possible to join the mothers' meetings of their districts: to use all their influence to induce those they visit to attend regularly at their places of worship, and to send their children to school. (Clark, 1973:11)

It is just these kinds of women, including most importantly the school board visitors of the period, who were the source of most of the information recorded in Charles Booth's (1889–1903) monumental and pioneering exercise in survey research, *Life and Labour of the People in London.* As these women moved from house to house, gathering information in the neighborhoods to which they were assigned and with which they were familiar, it is difficult to believe that they did not employ questioning of the type we have seen in this chapter, embodying principles of optimization and recipient design. Furthermore, it seems likely that this questioning was interlinked with sequences in which they built rapport with their subjects, supporting their reported decision making and experiences, as in the following sequence:

(20) [4A1:17]

```
 1 HV:     Uh::m (.).hh So your pregnancy was perfectly
 2         normal.
 3  M:     Yeh.
 4 HV:     And did you go into labor (.) all by yourself?
 5  M:     No: I was started o[ff because uh:m (0.8) the blood
 6 HV:                        [Induced.
 7  M:     pressure (0.7) went up in the last couple of weeks.
 8         (2.2)
 9  M:     I was going in the G.P. unit until two weeks
10         befo[re he was bo:rn,
11 HV: →       [O:h what a pity.
12         (.)
13  M:     and then uh.hh the blood pressure went up,
14 HV:     M[m.
15  M:      [so I had him in the consultant's,
16 HV:     Yes.
17         (0.2)
18  M:     a week early.
19 HV: →  Oh well that was rather nice you didn't have to wait
20     →  quite s[o    [long.
21  M:              [N(h)[o(h):::.
22  M:      .hhh hnh (.) mm[:
23 HV:                     [And was he alright when he was born.
```

In this kind of conversational interviewing, the interviewer morally supports the interviewee, endorsing decisions she has made, sympathizing with unfortunate outcomes, and otherwise "looking on the bright side."

A similar kind of interactional process is recorded for survey interviews in societies where social surveys are not yet commonplace and institutionalized. In a recent study of Czech interviewing, for example, Jana Hoffmannová (1998) identifies a range of strategies that interviewers use to maintain rapport with their subjects, including praising, sympathizing, agreeing, and identifying with the interviewee. In the following sequences, concerning the interviewee's (IE's) vacation travel plans, the same interviewer (IR) explicitly supports each IE's standpoint on foreign travel, though these standpoints are somewhat opposed:

(21) [Hoffmannová, 1998:163]

```
1 IR:  Where are you going this year?
2 IE:  This year we're planning to go to Greece.
3 IR:  Greece is beautiful. I've been there before.
4 IE:  Well I hope it works out.
5 IR:  It's good to travel and look at different countries.
```

(22) [Hoffmannová, 1998:163]

```
1 IE:  We've been to the West two or three times to have a
2      look. But now the excitement's beginning to fade.
3 IR:  Not when we've already seen it before.
4 IE:  No, not when we've already seen it.
5 IR:  The shops are the same everywhere.
6 IE:  Exactly.
```

One sees here something of the same effort at relationship building and above all personalizing the interview relationship, as is readily apparent in the British health visitor data.

It is clear, I think, from these examples that the principles of optimization and recipient design as features of question construction in these data are fundamental to the process of making and sustaining personal relationships with particular other individuals. There is no "time out" from this process for anyone who is intent on sustaining such a relationship, regardless of whether its basis is professional (as in these data) or more personal. From this vantage point, we can begin to see what a peculiar social relationship is instantiated in the survey interview. It is a relationship from which optimization is excluded and in which, save for the use of *filter questions*, recipient design is systematically avoided. In the survey interview, both interviewer and interviewee must collaborate in the construction of an essentially anonymous relationship from which all personal characteristics that could be acknowledged in the design of a turn are rigorously excluded. This is a social relationship that most of us have little or no experience of outside the special province of the survey interview. It takes effort to sustain and, as several contributors to this volume document, can easily break down. Indeed it has taken upwards of 50 years for the survey interview as a form of social interaction to become institutionalized to the point that most of the population can collaborate in its co-construction without the kind of discomfort or collapse into conversation that Hoffmannová and others have documented.

The distinguished physician Eric Cassell (1985:86) observes that medical questioning must address much more than the patient's current condition: The doctor must also find out

> who the patient is, and how the kind of person the patient is, along with how he or she behaves, interacts with the pathophysiology to produce this specific illness. Next must come an attempt to discover whether other factors, environmental, familial, social, occupational, or personal habits, have played a role in making the patient sick.

To accomplish this, physicians employ a relatively standardized list of questions and become, as Cassell (1985:89) puts it, "a fixed measuring instrument," a kind of living questionnaire, neutral and consistent across patients. However, despite Cassell's observations, it is clear that medical personnel do not question in the style preferred in social surveys. While more formal or objective question designs might, in principle, minimize response bias, this consideration is offset by the need to put patients at ease and permit relationship building. These latter considerations mandate the implementation of norms of questioning from the everyday world, particularly optimization and recipient design. Thus in medicine, as in other walks of life, there may be a trade-off between objectivity on the one hand and the effort to build human bonds on the other. In this process, the effort after objectivity, at least in the strictest sense, is not often the winner. Indeed it may be hazarded that, as Weber (1948) contended, the effort at objectivity is only a dominant impulse in bureaucratically administered activities of which, perhaps, the survey interview is a supreme expression.

NOTES

1. This work was done in collaboration with Sue Sefi.
2. Although the health visitor has a statutory obligation to cater to the health needs of all children, she does not have a statutory right of entry into the parental home. In practice, however, the accountability of denying entry to the health visitor renders such a right unnecessary.
3. See Robinson (1982) for a summary account of the historical background to this unusual division of labor between health visitors and other community nurses and Donzelot (1980) for an account of the ideological background of health visiting.
4. Although the HV service stresses the value of establishing contact with mothers before the birth of a baby and efforts are normally made to do so, only one HV in our sample had in fact made contact with her two first-time mothers prior to the first recorded visit.
5. This item, while potentially multiple, is treated as a single item because all the mothers in the data set discussed here were first-time mothers. The issue of siblings could, therefore, be disposed of in a single question, as in excerpt 1.
6. In these transcripts all proper names, including those of towns, corporations, and other institutions, are pseudonyms.
7. For accounts of preference organization, see Heritage (1984), Pomerantz (1984), Sacks (1987), and Schegloff (1988).

REFERENCES

Boyd, E., and J. Heritage. In press. "Analyzing History-Taking in Primary Care: Questioning and Responding During Verbal Examination." In J. Heritage and D. W. Maynard

(Eds.), *Practising Medicine: Structure and Process in Primary Care Encounters*. Cambridge: Cambridge University Press.

Booth, C. 1889–1903. *Life and Labour of the People in London*. London: Macmillan.

Button, G., and N. Casey. 1984. "Generating Topic: The Use of Topic Initial Elicitors." Pp. 167–190 in J. M. Atkinson and J. Heritage (Eds.), *Structures of Social Action: Studies in Conversation Analysis*. Cambridge: Cambridge University Press.

Cassell, E. 1985. *Talking with Patients, Volume 2: Clinical Technique*. Cambridge MA: MIT Press.

Clark, J. 1973. *A Family Visitor: A Descriptive Analysis of Health Visiting in Berkshire*. London: Royal College of Nursing.

Converse, J. M. 1987. *Survey Research in the United States: Roots and Emergence 1890–1960*. Berkeley, CA: University of California Press.

Council for the Education and Training of Health Visitors (CETHV). 1977. *An Investigation into the Principles of Health Visiting*, London: CETHV.

Cumberlege Report. 1986. *Neighbourhood Nursing: A Focus for Care*. London: Her Majesty's Stationery Office.

Donzelot, J. 1980. *The Policing of Families*. London: Hutchinson.

Health Visitors' Association (HVA). (1985). *Health Visiting and School Nursing: The Future*. London: HVA.

Heritage, J. 1984. *Garfinkel and Ethnomethodology*. Cambridge: Polity.

Heritage, J., and A. Lindström. 1998. "Motherhood, Medicine and Morality: Scenes from a Medical Encounter." *Research on Language and Social Interaction* 31: 397–438.

Heritage, J., and S. Sefi. 1992. "Dilemmas of Advice: Aspects of the Delivery and Reception of Advice in Interactions between Health Visitors and First Time Mothers." Pp. 359–417 in P. Drew and J. Heritage (Eds.), *Talk at Work*. Cambridge: Cambridge University Press.

Heritage, J., and M. L. Sorjonen. 1994. "Constituting and Maintaining Activities Across Sequences: *And*- Prefacing as a Feature of Question Design." *Language in Society* 23: 1–29.

Hoffmannová, J. 1998. "How the Interviewer Accommodates His Respondent (Agreement, Praise and Identification as a Co-operative Strategy)." Pp. 159–166 in S. Cmejrková, J. Hoffmannová, O. Müllerová, and J. Svetlá (Eds.), *Proceedings of the 6th International Congress of IADA*, Volume 2 (International Association for Dialog Analysis). Tubingen: Niemeyer.

Houtkoop-Steenstra, H., and C. Antaki. 1997. "Creating Happy People by Asking Yes-No Questions." *Research on Language and Social Interaction* 30: 285–313.

Labov, W., and D. Fanshel. 1977. *Therapeutic Discourse: Psychotherapy as Conversation*. New York: Academic.

Maynard, D. In press. *Bad News, Good News, and the Benign Order of Everyday Life*. Chicago: University of Chicago Press.

McIntosh, J. 1986. "A Consumer Perspective on the Health Visiting Service." University of Glasgow, Social Paediatric and Obstetric Research Unit.

Pomerantz, A. 1984. "Agreeing and Disagreeing with Assessments: Some Features of Preferred/Dispreferred Turn Shapes." Pp. 57–101 in J. M. Atkinson and J. Heritage (Eds.), *Structures of Social Action: Studies in Conversation Analysis*. Cambridge: Cambridge University Press.

Robinson, J. 1982. *An Evaluation of Health Visiting*. London: Council for the Education and Training of Health Visitors.

Sacks, H. 1987 [1973]. "On the Preferences for Agreement and Contiguity in Sequences in Conversation." Pp. 54–69 in G. Button and J. R. E. Lee (Eds.), *Talk and Social Organisation*. Clevedon, England: Multilingual Matters.

Sacks, H., E. A. Schegloff, and G. Jefferson. 1974. "A Simplest Systematics for the Organization of Turn-Taking for Conversation." *Language* 50: 696–735.

Schegloff, E. A. 1988. "On an Actual Virtual Servo-Mechanism for Guessing Bad News: A Single Case Conjecture." *Social Problems* 35: 442–457.

Sefi, S. 1988. "Health Visitors Talking to Mothers." *Health Visitor* 61: 7–10.

Weber, M. 1948. "Buraucracy." Pp. 196–244 in H. Gerth and C. Wright Mills (Eds.), *From Max Weber: Essays in Sociology*. London: Routledge and Kegan Paul.

CHAPTER 15

Standardization vs. Rapport: How Interviewers Handle the Laughter of Respondents During Telephone Surveys

Danielle Lavin and Douglas W. Maynard

Survey researchers seek to measure subjects' past and intended future behaviors, their attitudes, beliefs and values, and their membership in social categories (Schuman and Kalton, 1985:643–652). Furthermore, as Fowler and Mangione put it (1990:14), the "key defining part of a measurement process is standardization."[1] Of course, both practitioners and critics of the survey recognize that interviewers and respondents construct answers during interaction, and that standardization cannot be uniform or absolute (Maynard and Schaeffer, this volume, Chapter 1). Practitioners and critics differ, however, in what the constructed nature of interview data means, with practitioners asserting that validity is not necessarily compromised. When standardization is unable to screen nonsampling error, for instance, researchers are able to investigate and remedy normative *context effects* and other influences on response patterns (Schuman, 1982; Schuman and Ludwig, 1983). Critics, on the other hand, argue that the survey interview is inherently distorting and that we need altogether alternative forms of interviewing (Cicourel, 1982; Mishler, 1986; Suchman and Jordan, 1990), including those that are more *conversational* in structure (Conrad and Schober, 2000; Schober and Conrad, 1997).

Although this controversy has an interesting history (Beatty, 1995; O'Muircheartaigh, 1997), we do not take a position on it. Rather, taking for granted that survey interviewing involves *construction* and some *negotiation* between

Standardization and Tacit Knowledge: Interaction and Practice in the Survey Interview
Edited by Douglas W. Maynard, Hanneke Houtkoop-Steenstra, Nora Cate Schaeffer, and Johannes van der Zouwen
ISBN 0-471-35829-0

interviewer and respondent, our question is how, in a context where an organization or institute maintains a strong orientation to standardization, workers can deal with behaviors not covered by official procedures. As observed in Chapter 1 of this volume, the survey interview, with its procedures for the collection of data through scripted interviews, is a paradigmatic case of *mechanical objectivity*, which is Porter's (1995) phrase for all forms of inquiry in which rules for measurement discipline the use and variability of local metrics and interpretations. Tacit or commonsense knowledge is necessary to any regulated work task because rules can neither anticipate nor prespecify the handling of unpredictable and contingent elements that invariably arise in the concrete, particular situations that rules are meant to encompass (Garfinkel, 1967:20–22).[2] Despite the ad hoc way in which practitioners use commonsense knowledge to attend to the contingencies of work tasks, their use is also a site of orderliness in social interaction.

In the survey interview, *laughter* on the part of a respondent is one contingent element, and it raises what is perhaps the central dilemma of the survey: how interviewers are to maintain both standardization and a proper relationship, or what we will call rapport, with their respondents. While rapport has long been recognized as a problematic concept to define and operationalize (Beatty, 1995; Goudy and Potter, 1975; Weiss, 1970), it is acknowledged to be an important component of the relationship between interviewer and respondent (Fowler and Mangione (1990:55). More than enabling completion of the interview, it has been suggested that "the relationship which the interviewer succeeds in establishing with the respondent to a large extent determines the accuracy of the information the latter will supply" (Cannell and Axelrod, 1956:177), and there seems to be some evidence supporting this assertion. That is, the quality of the relationship — partly effected through what van der Zouwen, Dijkstra, and Smit (1991) refer to as the *style* of an interviewer — may have an effect both on completion rates and on the quality of the data collected during the question–answer process.[3]

As an aspect of the interviewer–respondent relationship, we define *rapport* in a limited or restricted way — as the occurrence of reciprocal laughter between respondent and interviewers. In previous treatments, as Weiss (1970) observes, researchers have regarded rapport as a more global concept and measured it not by the analysis of interactive behavior, but largely through (a) counting such things as eye contact, smiles, nods, gestures, nontask conversation, "no" answers, and reinforcement tokens or (b) through ex post facto ratings. In her summary of the research, Weiss (1970) observes that studies are difficult to reconcile because they use different measures. Furthermore, measurement issues aside, findings about the effects of rapport sometimes conflict with one another. Consequently, Weiss (1970:19) suggests that "we abandon the concept of rapport at this point and concentrate on the specific attitudes and behaviors of which — in some formulation or other — it is constituted." Our aim is to follow this suggestion and concentrate on specific behaviors: When we use the term rapport and discuss reciprocated laughter, we mean that an interviewer responds positively to laughter that a respondent initiates during the course of a telephone survey interview.

We are interested not just in survey methodology and the effect of this kind of rapport on data quality, however. By analyzing the tacit, orderly social practices associated with laughter in the survey interview, we contribute to a *sociology of social scientific knowledge*. As Maynard and Schaeffer (2000) argue, survey research, when performed through computer-aided telephone interviewing, is a system of "laboratory-based" inquiry occurring in what Latour (1987:232) calls "centers of calculation." That is, by the use of measurement instruments employed in particular locales to collect, stabilize, and combine discrete observations, survey research shares with natural science the work of generating abstract knowledge of worldly entities. Such work, as situated in laboratories, and whether involving biology, astronomy, chemistry, or physics or the attempt to understand human behavior, is always concrete and particularistic. In the constructivist and ethnomethodological sociology of (natural) scientific knowledge (SSK), investigators have increased our understanding of how the concreteness and particularity of scientific work are socially organized (Lynch, 1993). However, a concomitant appreciation of social scientists and their assistants at work — a constructivist/ethnomethodological sociology of *social* scientific knowledge (SSSK) — is not as well-developed (Maynard and Schaeffer, 2000). Survey research, as a preeminent form of social science (Converse, 1987; Platt, 1996; Wilner, 1985), offers a setting ripe for study along these lines.

As a contribution to the sociology of social science, our study aims to understand the manifestation of affect in the generation of such knowledge. Although feminist sociologists of science (Keller, 1985; Traweek, 1988) have pointed out the emotional and affectual bases to the objectifications of science, little empirical work is devoted to the analysis of these bases. Our study demonstrates the variable ways that respondents, asked to submit to the uniformities of social measurement, sometimes find and show a humorous response to the impersonal, anonymous questioning and formatted answering through which measurement is conducted. Indeed, when respondents laugh, interviewers are momentarily confronted with the task of keeping the exchanges on the track of detachment and disinterest considered central to the mechanics of survey measurement even though (as our data show) that track may be the very feature of interviewing that respondents find humorous or laughable. How interviewers interact with respondents to keep things on track, while attending to relational ramifications of the respondent's laughter, reveals aspects of situated interactional order in the disciplined collection of social data.

This concern with the organization of interviewer-respondent interaction also adds to the growing literature on "talk in institutional settings" (Boden and Zimmerman, 1991; Drew and Heritage, 1992), which documents various dimensions to such talk that help distinguish it from that which occurs in more ordinary or mundane settings. Heretofore, as Drew and Heritage (1992:28–53) discuss, investigations of institutional talk have attended mostly to such dimensions as lexical choice, the ways that turns are designed, the organization of sequences (e.g., restrictions on turn-taking), the overall structure of interaction, as in 911 emergency phone calls (Whalen, Zimmerman, and Whalen, 1988; Zimmerman,

1992), and social relations and epistemology.[4] Our research, in analyzing the specificity of laughter as a problem in survey interviews as compared with laughter's relatively free-flowing status in ordinary conversation, fits with the dimension of social relations and epistemology. Other than Heath's (1988) study of embarrassment, Whalen and Zimmerman's (1998) examination of hysteria in 911 emergency calls, and investigations of laughter in the doctor–patient interview (Haakana, 1999; West, 1984), research on talk in institutional settings has neglected the management of emotion displays (Peräkylä, 2000). The consequence has been that analysis has overemphasized the rational cast to such talk. Finally, previous research on talk in institutional settings has not much compared different settings of the same type. By documenting variation in the distribution of orderly social practices across worksites, we show that while practices of talk and interaction are generic and therefore transportable from setting to setting, there is more to the story. We can appreciate local cultures and "communities of practice" (Lave and Wenger, 1991) in their handling the unpredictable and contingent elements endemic to those settings. These cultures and communities appear to provide differing normative environments that condition the use of practices. Accordingly, research on talk in institutional settings will benefit from increased comparative inquiry.

RAPPORT AND LAUGHTER IN INTERACTION

Early works suggest that laughter helps generate a common definition of the situation for the participants (Coser, 1959). More recently, investigators suggest that laughter helps coordinate talk in conversation, as a turn-taking cue signaling that the current speaker has finished (Gavioli, 1995; Jefferson, 1979, 1984, 1985; O'Donnell-Trujillo, Adams, and Adams, 1983; Poyatos, 1993). Laughter can also function as a signal whereby participants in conversation show "troubles resistance" (Jefferson, 1984:351) and that some experienced problem is not getting the better of them. As stated, we are interested in reciprocal laughter and how it ties participants in a conversation to one another (Glenn, 1989, 1991/1992, 1995; Norrick, 1993). This bonding or affiliation may occur in the accomplishment of "intimacy" (Jefferson, Sacks, and Schegloff, 1987; McAdams, Jackson, and Kirshnit, 1984) or in achieving an alliance as when some parties concertedly "laugh at" something or someone (Glenn, 1991/1992, O'Donnell-Trujillo, Adams, and Adams, 1983). In circumstances, such as the survey interview, in which participants are previously unacquainted, laughter may not indicate intimacy or any kind of strong alliance as such, but it can show affiliation or momentary rapport between the individuals (Glenn, 1995). All of these positive functions of social laughter suggest the converse point: Nonreciprocation of laughter can create relational distance, negate the formation of an alliance between parties, and undermine rapport.

For analysis, we draw on the work of Jefferson (1979), who details a sequence in ordinary conversation for inviting laughter. The laughter invitation sequence has two parts. First, a current speaker or initiator of the laughter episode signals the other conversant indicating that laughter is appropriate in the interaction. That is, a speaker laughs and may solicit reciprocal laughter as a possible, and perhaps desired, response (Glenn, 1989; Jefferson, 1979). Second, in a next-turn response "slot," the recipient may accept the invitation by providing laughter. For example (see Appendix 1 for transcription conventions):

(1) (Jefferson, 1979:81)

```
Ellen:        He s'd well he said I am cheap he said, .hh about the
         1→   big things he says but not the liddle things, hhhHA
          →   HA [HA HA HA
Bill:    2→      [heh heh heh
```

At the end of her turn of talk (arrow 1), Ellen appends several laughter tokens, which constitute the invitation to laugh. Bill begins to laugh (arrow 2), and thereby accepts Ellen's invitation, before she has finished her laughter. The participants thereby end up "laughing together." Laughter invitations are distinguished from *candidate laughables*, such as jokes, puns, and other items that are not themselves embedded in or with laughter but that make a recipient's laughter relevant.[5] If a recipient does laugh after such items, technically the laughter is *volunteered* rather than *invited*, even when a speaker of the laughable may have been seeking laughter in response. Accordingly, Jefferson (1979:83) observes that silences after a current speaker's turn may indicate that a recipient finds an utterance equivocally laughable and "may specifically be *awaiting* an invitation to laugh:"

(2) (Jefferson, 1979:80; Modified)

```
Dan:          I thought that wz pretty outta sight didju hear me say you a
              junkie
         s→        (0.5)
Dan:     1→  hheh heh=
Dolly:   2→  =hhheh-heh-heh
```

The silence here (s arrow), which indicates the joke recipient (Dolly) withholding a turn that could involve volunteered laughter, provides an opportunity for the speaker (Dan) to issue a laughter invitation (arrow 1) and obtain Dolly's laughter (arrow 2) as an acceptance of that invitation.

In the response slot, a recipient may also *decline* the invitation to laugh. As Jefferson (1979:83) observes, to decline a laughter invitation, a recipient must do more than "not laugh" or refrain from laughing. The recipient must act to "terminate the relevance of laughter." One way to do this is to *talk*, since talk replaces either laughter or silence as a response to a laughter invitation:

(3) (Jefferson, 1979:87; Modified)

```
Gene:    1→  So that shook the old (h)house(h)hold up fer a(h)whi(h)le
             heh =
Patty:   2→  =Oh yes I c'n imagine.
Gene:        Y'know, a:nd uh I think Jill's, realized thet hell maybe...
```

Serious pursuit of topic diverts the focus from the laughter episode. Notice here that Patty's declination (arrow 2) appears to intersect Gene's turn-ending pursuit of laughter, which follows within-utterance laugh particles that Patty has also bypassed as laughter invitations. Furthermore, after Patty's serious pursuit of topic, Gene follows with more topical talk (Jefferson, 1979:87). However, a recipient's declination itself may be countered if a speaker continues to laugh; that is, speaker and recipient may compete with regard to the serious or laughing implicativeness of the interaction.

In other research on laughter in interaction in an institutional setting, West (1984) and Haakana (1999) study the doctor–patient relationship. Both investigators find that patients laugh more during medical interviews than physicians do. Like patients, respondents in the survey interview initiate much more laughter than do the professionals with whom they interact. This may reflect the fact that respondents, like patients, are the ones who are asked to introduce "personal" materials whereas interviewers, like doctors, are the ones simply asking questions (Haakana, 1999:132–134). Moreover, the nature of the interview is more anonymous and abstract than the doctor–patient encounter. Questions often appear "laughable" to respondents as do their answering and their answers; a respondent's laughter often has to do with the constrained nature of the interaction, which is partly designed to preserve that anonymity and abstractness.

DATA AND METHODS

The first and principal data set for this chapter includes audiotapes, videotapes, and transcripts of interactions between survey interviewers and respondents recorded at the Central State University Survey Research Center (SRC) during March of 1997. The resulting collection is not systematic or random, but represents an effort to gather as much recorded interactional data as possible during a concentrated period of time. In all, 13 complete interviews were taped and transcribed. These interviews involved two different surveys, conducted by a total of 7 different interviewers; one survey involved television viewing habits and the other sought respondents' opinions about their neighborhood. We focused on the substantive portion of the interview, and not the "front-end" or request to participate, and selection of a household member for an interview, which is allowed to be more "conversational" than the interview itself. We collected all instances of respondent-initiated laughter that occurred in the 13 completed interviews.[6] The total number of laughter instances in our collection is 35, an average of not quite 3 per interview. At the Central State University SRC, interviewers are proscribed, both formally[7] and informally[8] from engaging in any emotional displays including laughter. If respondents deviate from the task at hand (e.g., by joking or other humorous asides) interviewers are to "remind them of the current question" (per an instruction on a computer screen for training). The overall rule is to "maintain seriousness."

A second set of data are audiotapes and transcripts from the University of Midstate center, which, unlike Central State, does not prohibit laughter on the part of interviewers. To match our Central State collection, we selected 13 interviews from the Midstate data, again from 2 surveys. In the Midstate data, there were five interviewers conducting the surveys. Both Midstate surveys measured public opinions about the government, politics, and public health issues. The interviews had been recorded in the early 1990s and were comparable in length to the Central State surveys.[9] We collected all instances of respondent-initiated laughter that occurred during the substantive portion of the interview in the 13 Midstate cases, and found 103 instances, or an average of about 8 per interview. That there were only 35 respondent initiations of laughter in the Central State collection, as compared with 103 in the University of Midstate, is a disparity we discuss later.

In the Central State data, after transcribing instances of respondent-initiated laughter (instances where the respondent produces laughter), we analyzed the data using conversation analytic procedures. This involves comprehensively examining all instances in the collection. The practices we identify for dealing with respondent laughter, as *generic* devices (see note 13), can be presumed to inhabit other survey research centers, although (as we shall see) the distribution of such devices from center to center may vary.

Schaeffer (1991:389) anticipates a study of this kind when she writes that "the work laughter does for either participant in an interview has not yet been explored." Our research extends that of Cannell, Fowler, and Marquis (1968:19–20), which coded interviews for interviewer and respondent laughter, finding that on average, interviewers initiated laughter 0.81 times during an interview, whereas respondents initiated laughter an average of 7.26 times. Their research, however, does not provide information on the contexts in which participants were laughing (or what they were laughing at) or exactly how, in real time, the receiving party (the interviewer) reacted to the initiations of laughter. Also, unlike our study, Cannell, Fowler, and Marquis (1968) do not compare participants' laughter across work settings.

There are three parts in the Analysis section. First, we analyze the sequential positioning of respondents' laughter invitations and hence what these invitations are related to. Second, we examine the tools interviewers use to handle respondent-initiated laughter. Third, we compare the practices of interviewers in our two survey centers.

ANALYSIS

Sequential Context of Respondents' Laughter Invitations

Before examining interviewer response to laughter, we examine and identify when and how, in the course of the interview, laughter emerges. Respondents initiate laughter during the survey interview in a number of different sequential positions. The laughter can be responding to or "commenting on" (1) a question they have been asked, (2) their answer, or (3) some aspect of a completed question–answer–acknowledgment "interviewing" sequence.[10]

Question-Oriented Respondent Laughter

When respondents initiate laughter, they often do so after an interview question. Excerpt 4 contains one question in a sequence of questions regarding the role of television in the respondent's relationship with her husband. A response of "ten" indicates complete agreement with the statement, whereas a "zero" indicates no agreement at all. The question is in lines 1–2 (interviewers and respondents are designated as male or female—MI is a male interviewer; FR is a female respondent):

(4) [TC01:2505:001:KD:500-508]

```
1  MI:  Someti:mes (0.1) you are more interested in watching
2       teevee: .hhhh than in ↑being ↓with (0.1) your husband.
3  FR:  Heh heh heh .hhhh u::::h hh (0.4) a one
4       (0.2)
5  MI:  Uh [↑huh]
6           [ #  ] #
7  MI:  .hhhh teevee is a time for you (0.2) and your husband .hhh
8       to be physic'lly close to each other
```

Directly following the interviewer's question, and prior to providing an answer, the respondent initiates laughter (line 3). We regard such laughter as a humorous comment on the question, since the laughter is produced just after it and prior to the provision of an answer.

Because of coding concerns and survey design, respondents are not always given the option to answer the question as they would like. For example, questions that require "yes" or "no" answers can be problematic and occasion laughter from the respondent:

(5) [TC04:2451:011&018:LF:245-256]

```
 1  MI:   ↑Ohkay, (0.8) # °thank you°. (0.2) .hhhhh (.) Does yer
 2        husband (.) still wa:tch this sho~w.
 3        (1.9)
 4  FR:   Uhh, =
 5        (2.0)
 6  FR:   =↑Sometimes.
 7  MI:   ↑Mkay >so would ↑you say yes or no.<
 8        (0.3)
 9  FR:   Heh heh! I can't say yes or no, (.) actually.
10  MI:   ↑Hmkay h. ((smile voice))
11  FR:   I mean (.) sometimes he will sometimes he won't=
```

A trouble in answering the question (lines 1–2) is initially evident at line 3 in a 1.9-second pause, after which the respondent utters "uhh," and then pauses again for 2 seconds. Instead of providing a yes or no answer, the respondent answers "sometimes" in line 6. The interviewer probes the respondent's answer to get a properly formatted answer (line 7). After another pause (line 8), the respondent, in line 9, laughs ("Heh heh!") and then emphasizes that she "can't" use these categories (line 9). After MI's acknowledgment (line 10), FR offers a reason for her inability (line 11).

Answer-Oriented Respondent Laughter

At times, when respondents initiate laughter in the interview, it is displayed as a reaction to their *own* answers or the situations depicted by their answers (boldface indicates when the interviewer's typing is audible):

(6) [PC07:2308:029-030:JL:100-112]

```
FI:   .hhhh Please tell me (0.2) the ma:in reason (0.3) you
      decided (0.1) to mo:ve (0.2) to the pla:ce (0.3) WHERE
      (0.1) you are currently living.
      (2.3)
MR:   Only place they had heh heh heh at the time
      ((FI continues typing, 2.6 seconds))
MR:   [(°it was open°)]
      [    #     #    ]
      ((typing, 6.0))
FI:   [okay]
```

The interviewer produces the question in lines 1–3, and after a 2.3-second silence (line 4), the respondent answers (line 5). Embedded within the answer are laughter tokens. It can be observed that the question presumes choice on the respondent's part as to the "place" he "currently" lives, whereas his answer claims there was no choice. The laugh tokens, placed midturn, may suggest the laughability of the lack of choice and the way this lack counters the question's presumption; in line 7 MR goes on to suggest that the place "was open," and in the context of the previous utterance this may mean it was the "only" place open. Other instances of answer-oriented laughter can be characterized as self-deprecating (Glenn, 1991/1992),[11] in which respondents somehow poke fun at themselves or the situations they experience.

Post-sequence Laughter

Respondents' laughter also can follow the completion of a question–answer–acknowledgment sequence. Laughter that occurs in this position is less common than laughter in the other two positions and may serve in the refinement of a respondent's answer.

(7) [PC08:2344:032:SS:41-54]

```
FI:   First, .hhhh I would like to find out more about WHERE
      you are living now. HOW many years have you lived in
      the Indianapolis area?
FR:   We've lived here since (0.2) Ju:ne of nineteen eighty six
FI:   O:k::ay, .hh and can you just give me a (.) a number of
      yea:rs (0.6) that you've lived in the area?
      (1.8)
FR:   °Eleven°
      (0.6)
FI:   All:ri:::ght↑
      ((# # #)) (0.6)
FR:   In Ju:ne, heh heh it'll be ele(hh)ven
FI:   .hhhh O:kay. Hhh How many years have you lived in your
      CURRENT neighborhood
```

In excerpt 7, FR's answering the question spans several turns. The first answer that the respondent produces (line 4) is not properly formatted for the survey instrument. The interviewer, in lines 5–6, explains to the respondent that she requires an answer that is framed in "number of years." The respondent then produces a numerical answer in line 8, which FI accepts with an acknowledgment token[12] "all:ri:::ght↑." However, the respondent reopens the exchange in line 12, self-correcting the accepted answer, while the interviewer types in the answer code. The correction offers a more precise answer and appears as a way of proposing that the correction is not serious. With "O:kay" and continuation in line 13, FI both accepts the correction and indicates that it is not consequential enough to inhibit progression to the next question.

Overall, the interview resembles any other social situation, in that it may furnish occasions that are ripe for laughter. Here, as we have observed, the questions as read by the interviewer can do just that, as respondents laugh in response to the questions. Or respondents themselves may say something that they construct as laughable. Their answers may contradict a presumption in the question or otherwise comment, sometimes in a self-deprecating way, on their performance in relation to the requirements of the interview. Or, after completing a question–answer–acknowledgment sequence, respondents may reopen the sequence and do so with laughter. All of these occasions can invite the interviewer to reciprocate with laughter. We now look at how interviewers respond to invitations to laugh.

Reciprocation: Accepting the Invitation to Laugh

In our main data set, we have only one instance of an interviewer accepting a respondent's laughter invitation. Reciprocation of laughter, or the acceptance of the respondent's invitation to laugh, aligns the interviewer with a momentary move away from the interviewing task at hand. As mentioned, previous research has suggested that "laughing together" or reciprocally is an achievement and signifies at least a momentary display of rapport between the involved parties.

(8) [PP09:2476:038:JH:292-303]

```
FI:   <How satisfied (0.1) are ↑you: (0.3) with the repu↑tatio:n
      (0.2) of your neighborhood.>
      (3.0)
FI:   [You-] (0.4) you ken just use the responses that we've used=
FR:   [Um- ]
FI:   =in the last few ques[tions?  ]
FR:                        [↑Oka:y.=]
      (0.2)
FR:   =um (·) fai:rly satisfied.
      (0.9)
FI:   .hhhh ↑OH: (·) ↑UM (0.4)>I- I'm sor↑ry whadid you say?<
      (0.6)
FR:   ↑Fai::rly satisfied? (0.2) °what I said for the first one°
      (0.7) is- (0.3) or ↑somewhat satisfied?=
FI:   O[kay (0.2) okay  ]
FR:   =[Yeah (0.2) ↑okay] >↑somewhat satisfied th(hh)at's
      ri(hh)ght du(h)h< °eh huh huh°
```

```
18   FI:   °hah heh°
19   FR:   [ >I'm sorry<   ]
20   FI:   [.hhh All ↑right] (0.7) the availability (.) of convenient
21         shopping.
```

In lines 1 and 2, FI reads FR the question. Not getting an answer (line 3), FI instructs FR (lines 4 and 6) to use the same options as she had for the previous set of questions, which include: very satisfied, somewhat satisfied, somewhat dissatisfied, very dissatisfied. However, FI does not repeat the response options themselves, perhaps because FR acknowledges the instructions by saying "okay" in overlap (line 7). The respondent then answers the question with "fairly satisfied" (line 9) — an answer that also does not fall into the proper answer categories. FI asks for a repeat of the FR's answer in line 11, which is a way of asking for a correction (Schegloff, Jefferson, and Sacks, 1977). In her next turn (line 13), FR does not correct and merely repeats her original answer, then pauses, and in a manner suggestive of "thinking out loud," quietly asserts that her answer is the same as what she "said for the first one" (line 13). After this, she pauses, starts up ("is-"), stops again, and changes her answer to "somewhat satisfied" (line 14), a self-correction that now fits into the response categories provided by the interviewer at the beginning of the question set.

In line 15, FI provides two acknowledgment tokens; in overlap (line 16) FR reaffirms her just previously articulated answer. Continuing in lines 16–17, FR produces laughter tokens in the midst of her talk and appends more tokens at utterance end. FR's laughter is characteristically self-deprecating. Notice the word "du(h)h" whereby the respondent can be commenting on her mistake as a "dumb" one or as having missed the obvious. This is because slightly earlier, the interviewer had read the response options ("very satisfied, somewhat satisfied, somewhat dissatisfied, or very dissatisfied") that would be used for the question sequence. Besides this, at other times in the interview, FI had interrupted the reading of responses and explained to the respondent that she is *required* to read all of the options before the respondent may select one. Accordingly, the respondent's laughter at lines 16–17 may be laughing at not only the present forgetting of correctly formatted response options but at her overall performance as a respondent during the interview. The interviewer accepts the invitation to laugh in line 18 with the two quiet laugh tokens ("°hah heh°"). After the laughter reciprocation, the respondent offers an apology (line 19). In overlap with this, FI produces an upwardly intoned "All ↑right," (line 20), accepting the respondent's answer, and then continues with the scripted questions.

If the laughter in excerpt 8 is what Glenn (1991/1992) terms self-deprecating or even apologetic, it may be difficult for the interviewer to avoid accepting this particular invitation to laugh. Reciprocation can align with the respondent and allow the interviewer to avoid portraying herself as rude and unresponsive, whereas not laughing could be interpreted by the respondent as a reprimand for her prior performance as a respondent during the interview. Hence, the one instance in our Central State data of reciprocal laughter on the part of an interviewer—a deviation from survey center policy—may occur for good interactional reasons.

Nonreciprocation

At Central State interviewers decline respondents' invitations to laugh more often than they accept them (see Table 15.1). Interviewers can decline a respondent's invitation to laugh by responding with speech rather than laughter (see Jefferson, 1979). In the following three instances, besides using talk, the interviewer relies on the properties of the survey interview to deal with the respondent's laughter. The structure of the survey interview, that is, allows the interviewer to follow the respondent's utterances with long pauses, audible key clicks, as well as the initiation of a new question–answer sequence, without disrupting the interaction. In excerpt 9, after FI asks a question (lines 1–2), FR provides an answer (line 3) that does not fit into the instrument's answer categories.

(9) [PC08:2344:032:SS:762-774]

```
FI:   .hhh Over the pa:st ye::ar (0.2) ↑how of:ten, if at
      all:, did you atte:nd religious services
FR:   Every Sunday.
      (0.8)
FI:   O:↑k::ay. (0.2) Um I'll go ahead and read [these.   ]
FR:                                             [Fifty two]
      weeks. If you want to say it or a whatever
FI:   [Ehhhhhhhuh
FR:   [Weekly eh heh heh heh
FI:   Okay would you say-
FR:   Uh heh I [du(h)(h)nn(h)o
FI:            [hhhhhhhhh .hhh several times a week, once a
      week, nearly every week, two to three times a month,
      about once a month, sev[eral  ]
FR:                          [(Once)] a week.
```

FI offers to "go ahead and read" the acceptable responses (line 5), but in overlap with FI's utterance, FR at lines 6–7 acknowledges the problem and self-corrects by providing an answer ("fifty two weeks") different from her first one ("Every Sunday"). This second answer still does not fall into the correct response categories. FI emits a long outbreath in line 8, and then FR provides a third candidate answer by saying "weekly" in line 9. In a self-deprecating way (i.e., we assume that the laughter is treating the series of revised answers humorously), she appends laughter to this utterance, inviting the interviewer to laugh with her. However, FI starts to probe the respondent (line 10), and when she momentarily stops, FR produces more laughter in a disclaiming utterance (she says "I dunno" with laughter tokens interspersed, line 11). In overlap (line 12), FI emits an outbreath and an inbreath, and when FR stops her talk, FI presses on with a nonlaughing reading of the response options (lines 12–14). The serious pursuit of talk on FI's part is finally taken by FR as a declination. FR produces no more laughter, allows FI to repeat the response categories (lines 12–14), and provides an answer at line 15. Here, not only does the interviewer overlap the respondent's laughter with a breathy prelude to serious talk, but also her talk is "on-task." Rather than pursuing a joint laughter interlude, FI proposes to realign FR to the job at hand. This realignment is achieved when FR provides her correctly formatted answer.

In the survey interview, silence or pausing after a laughter invitation can signal a declination. Unconstrained conversation, where a silence allows the initiator of laughter to reinvite the laughter (see excerpt 2 and discussion), can be contrasted with the survey interview where such a silence or pause more regularly signals the pursuit of the interviewing task at hand, as in excerpt 10:

(10) [TC01:2505:001:KD:239-247]

```
1  MI:   Do you wish your husband wa:tched less teevee?
2        (0.4)
3  FR:   Huh heh heh hah hah hah hah .hhhhh o::::h (0.3) ye::s.
4        (0.4)
5  MI:   Okay. ##
6        (0.8)
7  MI:   Have you talked with your husband about this.
```

In line 1, the male interviewer (MI) provides a question about the husband's television habits. After pausing (line 2), and before granting an answer to the question, FR in line 3 produces extended laughter, which displays the question as laughable. If this invites the interviewer to laugh, then the inbreath (.hhhhh) after the laughter may *cover* lack of interviewer uptake or acceptance. Lack of uptake may also indicate that the interviewer is waiting for an answer, which FR produces at the end of line 3. The "yes" is a correctly formatted answer to the interviewer's question, but the utterance in which the "yes" occurs is an "oh"-prefaced response treating the question that MI poses as self-evident or overly obvious (Heritage, 1998:31). That is, by suggesting that the question is "inapposite," the answer can extend the invitation for the interviewer to laugh. However, after a pause (line 4), the interviewer provides an acknowledgment (line 5) and enters the respondent's answer into the computer, as evidenced by the audible key strokes (also line 5). By refraining from laughter, MI seeks a serious pursuit of the topic at hand both during and directly following the respondent's laughter. When he finishes typing, and after further pause, he resumes by asking a new question (line 7).

The next example has more extended typing by the interviewer because he is dealing with the answer to an open-ended question. He asks the question at lines 1–3, waits (line 4), and reports to FR that it is "open ended" (line 5). Then, after MI repeats the question (lines 6, 8), at lines 10–12, FR produces an answer. The interviewer begins typing the answer at lines 11–12.

(11) [TC01:2505:001:KD:360-387]

```
1  MI:   .hhh Please tell me why: (0.3) you describe the ro::le
2        (0.2) teevee has played in your relationship (0.4) as
3        small a~nd neutral
4        (0.8)
```

MI: >This is open ended °so-° <
FR: Okay u::::m
(1.9)
FR: Why do I- why do I describe it as neu↓tral
(1.1)
FR: Becuz I- I mean it's really not a b- (0.2) I mean it's
not a big deal ta watch **teevee I mean we can (0.2)**
we're go:ne a lot
((typing for 0.5 seconds))
FR: **So if we sidown and I don't wanna watch something**
I kin (0.2) you know leave the roo:m or he kin
leave the roo:m
(0.7)
FR: an' go to another room an' (0.4) we don't have a problem
with tha::t
(0.4)
MI: 'kay let me [type all-]
FR: [does that] make se:nse
MI: yeah let me type all that ↑i::n↓
FR: h(h)h o(h)ka(hh)y heh heh
(1.2)
((typing resumes for 11.7 seconds))
MI: .hhh hh Okay s- I ha::ve (0.4) it is not a big deal to watch
teevee (.) we can do other wor::k or go to >another room and
watch< teevee or do something .hhh 'zat sound r↑ight?

As FR continues to talk (lines 14–16, 18–19), MI stops typing, but at a possible completion point, asks FR to let him "type all that" (lines 21, 23). After this, FR initiates laughter (line 24). The laughter, which may occur in response to MI's "all that" gloss of FR's long answer, occurs initially as breathy within-speech laughter and later extends beyond the utterance as postutterance completion laughter; thus, the entire turn can be considered a laughter invitation. Directly following FR's laughter, however, there is a 1.2-second pause (line 25), during which there is no audible typing, when MI may be reviewing (on screen) what he has typed so far. During the silences, MI does not respond to the previous laughter, nor does FR attempt to reinitiate any laughter. Audible typing recommences (line 26) for 11.7 seconds. In line 27, then, after the typing stops, MI reads the respondent's answer back to the respondent to verify that he has the correct answer typed into the computer.

The interviewer in this case announces that he has to type (before the laughter invitation) and, through audible keystroking, displays that he is doing so. Through such displays of orientation to task, an interviewer declines laughter invitations from respondents.

From these excerpts, it is apparent that part of interviewers' tacit knowledge includes conversational competence—knowing how to decline respondents' laughter invitations by pursuing serious, on-task talk. The interviewer proposes to complete a question–answer sequence already underway or to initiate a new one. In addition, using a device that is literally at hand in the survey interview, the interviewer can announce, or audibly engage in, the task of typing. In short, this collection of devices—talking, attending to survey questions, and typing—represents an interviewer's *repertoire* for declining laughter invitations in the survey interview.[13]

Pseudolaughing Responses: Smile Voices and Quasilaughs

So far, we have seen that interviewers can either accept or decline a respondent's invitation to laugh. Accepting the invitation is affiliative, but at our primary research site, this violates rules of standardization, and most often interviewers decline the invitation to laugh. However, they do this not by overt statement ("that's not funny") or other explicit forms of rejection but through a repertoire of tacit practices notifying the respondent that the survey, rather than respondent's momentary effort at humor, are the interviewer's primary mode of orientation. Respondents largely appear to go along with this orientation, which suggests that even though interviewers may refuse to laugh, they manage to keep the working relationship intact. Therefore, it is not as if achieving standardization and maintaining rapport are mutually exclusive. Interviewers, in the face of potentially disruptive laughter, seem to be skilled at keeping as fully on task as possible without jeopardizing their relationship with the respondent. Now the question is, can interviewers do more than this? Can they adhere to standardization protocols and not just maintain their relationship with the respondent but also show affiliation? That is, is it possible to attend to standardization and, in the context of respondent-initiated laughter, achieve rapport at the same time?

Interviewers rely on the use of a *smile voice* or quasilaugh to respond to, and yet avoid complete reciprocation of respondent-initiated laughter. That is, while it can be interpreted as responding positively to respondent-initiated laughter, smile voice does not violate interviewing procedure in the SRC and therefore averts admonishment from supervisors who may be listening to the call.[14] For example:

(12) [PC08:2344:032:SS:606-621]

FI: What is the na:me of the other religious group that
contacted you.
(2.0)
((click))
(0.6)
FR: Just a second.
(13.6) ((speaking in background))
FR: I believe it's the Congregational Church.
(0.6)
FI: O:: **k:::↑ay**
(2.0) ((audible typing))
FR: **I have no idea rilly of the na:(h)(h)me.**
FI: Oka:::↑y. ((smile voice))
(1.8)
FI: N:::OW I have some questions about your neighborhood
and HOW it fits into the community of Indianapolis.

In lines 1–2, the FI asks the FR a scripted question. Then, after pausing and hesitating (lines 3–5), FR asks for a "second" (line 6), and can be heard talking to someone in the background (line 7). After almost 14 seconds, FR returns to the phone and provides an answer (line 8) to the question. FI provides an acknowledgment (line 10) and proceeds to enter the respondent's answer into the computer during this acknowledgment and after. The typing continues during the 2-second pause in line 11. Next, FR comments on her answer and inserts laughter tokens at the end of the utterance. Then, the interviewer provides another acknowledgment, the elongated "okay," in a smile voice (line 13). On the videotape, the shape of the interviewer's mouth indicates a purposeful smile. This viewing, of course, is unavailable to the respondent, but there is a sound quality—a "cheery" resonance—that indicates this smiling.[15] Smile voice thus allows the interviewer to acknowledge the respondent's laughter and still avoid any overt engagement in laughter. That is, because smile voice occurs simultaneously with the speech stream, it does not require divergence from the interview script and, in this way, helps promote standardization.

Along similar lines, a quasilaugh provides interviewers with yet another way to appreciate laughter without fully engaging in it. The quasilaugh is a breathy within-word laugh particle that may have purposeful ambivalence: Respondents can interpret it as reciprocating their laughter while supervisors can hear it as a breathy talk that merely elongates an acknowledgment token. In the following example, the respondent's laughter invitation occurs after a completed question–answer sequence. FI poses the question at lines 1–4; FR almost immediately provides an answer (line 5), and FI acknowledges (line 6) and types in the respondent's answer (audible key clicks in line 7).

(13) [PC08:2344:032:SS:184-197]

```
FI:   Over the past ye::ar HOW ↑oft↓en did you talk on the
      telepho:ne with any of your neighbors. Wouldju sa:y
      (0.2) almost every day, at least once a ↑week, at least
      once a month, le:ss than once a month, or not at all.
FR:   Not at all.
FI:   Uh ↑h::uh?
      ((##)) (0.4)
      (0.4)
FR:   We see each other over the fence eh huh huh huh huh huh
      [huh huh huh heh heh heh]
FI:   [O:↑k:hhh:ay       .hhh   ] How ((smile voice)) well: do you
      think the people in your neighborhood know each hh other?
      Would you say very well, fairly well, not very well or not
      at all well:
```

In line 9, after a brief pause, the respondent reopens the exchange with a joke. She displays her utterance as laughable by engaging in extended laughter directly after the completion of the utterance. In response to and overlapping with the respondent's laughter, the interviewer engages in an instance of quasilaughter in line 11, embedding breathy sounds ("hhh") that are not quite laughter particles (see Appendix 1, item 13) within the "O:↑k:hhh:ay," which is thereby elongated and also has a rise in tone, and is followed by an inbreath. Occurring in overlap with FR's laughter, this semiserious utterance also signals pursuit of the interviewing task. And, at the precise moment that FR stops laughing, FI initiates the next question. As noted at line 11, she does this with a smile voice. Thus, this interviewer uses two devices — the quasilaugh and the smile voice — to show appreciation of, without fully accepting, the respondent's laughter invitation.

In contrast to the overlapping quasi laughing response used in excerpt 13, the next instance of such a response (excerpt 14) occurs in the clear and is also more defined than the one in excerpt 13.

(14) [TC01:2505:001:KD:219-232]

```
 1  MI:   About how many hours of teevee .hhh do ↑you ↓watch on a
 2        typical weekday
 3        (0.6)
 4  FR:   .hhhhhh hhhhhhhhh I would say'n-
 5        (1.4)
 6  FR:   Two hours
 7        (0.2)
 8  MI:   Two hours? #
 9  FR:   Say yea:::h [but] just be honest hih hih
10                    [ # ]
11  MI:   'ka(h)y(hh) # .hhhh ab(h)out how many hours of teevee
12        (0.3) do you watch on Sa::turdays
```

After the question–answer sequence is completed in lines 1–6, and after a short pause (line 7), the interviewer, with a questioning repeat, asks to verify the respondent's answer in line 8. FR confirms (line 9), and then makes a remark about being "honest" to which she appends laughter tokens. Again, this may be a self-deprecating laugh that is responsive to MI's questioning repeat, interpreting it as indicating surprise at the amount of television being watched. In any case, the appended laughter in line 9 elicits quasilaughter on the part of the interviewer in line 11. That is, the interviewer, at line 11, places breathy particles within his okay token and the succeeding "ab(h)out," which initiates the next question sequence. After this, MI's utterance is cleared of any breathiness or quasilaughter, once again suggesting serious pursuit of the interview.

As interactive practices, the smile voice and the quasilaugh can be construed as minimal reciprocation of respondent-initiated laughter and are *achieved* as ambiguous (Jefferson, 1978). Both are inserted into the speech stream, thus requiring minimal divergence from the interview script and the instrumental task that it embodies. These practices do not officially or formally accept a laughter invitation, although they exhibit, without violating procedures, some appreciation and awareness of the respondent's laughter. Accordingly, smile voice and quasilaughter indicate momentary affiliation or rapport between the interviewer and respondent.

HOW INTERVIEWERS RESPOND TO LAUGHTER AT TWO SURVEY CENTERS

Survey centers employ highly specialized rules to ensure standardized interviewing conduct. However, all centers are not uniform; many rules and guidelines for interviewing are site specific (Viterna and Maynard, this volume, Chapter 16). Our main site at Central State prohibits interviewer laughter, but the survey center at the University of Midstate does not. At the latter site, interviewers are instructed to be "natural" and to speak in a conversational manner at a normal pace. Given that there are no restrictions placed on interviewer laughter at the University of Midstate center, we would expect that the practices for tacitly declining respondent-initiated laughter would be less prevalent and that reciprocation would be more frequent. Our counting of acceptances, declinations, and pseudolaughing techniques in the two data sets confirmed these expectations. Table 15.1 provides the results from the comparison of the two survey centers.

The percentage of acceptance is much higher at the University of Midstate center. In half of the instances in which laughter was initiated by the respondent, the University of Midstate center interviewers accepted the invitation to laugh, as compared with the 3 percent (one instance) at Central State. By contrast, the percentage of declination or nonreciprocation is much higher at Central State (71 percent compared to 40 percent of the cases at the University of Midstate). Interviewers at Midstate have less of a need to avoid reciprocation actively because their training instructs them to be conversational. Nonetheless, they sometimes decline invitations to laugh. Finally, use of pseudolaughing responses is more frequent at Central State (26 percent) compared to University of Midstate (9 percent), suggesting that Central State interviewers work more often to straddle the fence between achieving rapport and maintaining standardization.

Another striking difference between the two survey centers is in the total number of laughter invitations. At Central State, there are only 35 instances of respondent-initiated laughter, whereas at the University of Midstate center there are 103 such instances. It is possible that respondents involved in the surveys conducted at the University of Midstate are reacting to the openness of the survey environment to laughter and the reciprocal nature of that laughter by

Table 15.1 Interviewer Response to Respondent-initiated Laughter[a]

	Reciprocation	Pseudolaughter	Nonreciprocation	Total
University of Midstate	53 (51%)	9 (9%)	41 (40%)	103
Central State University	1 (3%)	9 (26%)	25 (71%)	35

[a] $N = 13$ completed interviews for each survey center. Recall from earlier in the chapter that the number of questions differs depending on the respondent's answers to specific questions. The interviews chosen as representative had a median number of questions asked. The Central State University television study has approximately 131 questions. For the neighborhood opinion study at Central State, a representative interview is 96 questions. In the University of Midstate interviews, the average length of a survey on alcohol consumption and traffic laws is 114 questions and the survey on politics is approximately 90 questions long.

offering laughter more freely. It may be, in fact, that how interviewers handle the first laughter invitations teaches respondents in the course of interaction whether laughing at questions, answers, or other matters is acceptable.

In any case, the environment for laughter at Midstate does appear more permissive in a variety of ways than at Central State. For comparing the two sites, we provide an example below from the Midstate data. Recall that laughter reciprocation occurs only once in the Central State data, and that it is very minimal. The laughter in this Midstate excerpt is much more elaborate and spans a greater portion of the exchange between respondent and interviewer:

(15) [LSSC112:02:A3:11/14/91:109-127]

```
FI:  =Okay # great .hh all right? and on a scale of one to
     ten one being very poor and ten being excellent .hhhhh
     how would you rate the job the president is doing hhhhh
     (3.0)
FR:  Oh::: I think President Bush is:: probably about a::
     eight an a half heh heh [heh heh heh heh] heh
FI:  ((smile voice))         [Ei- heh heh .hh] an eight and
     a half [huh heh ] .hh alright =
FR:         [.hh Yeah]
FI:  ((smile voice:)) = Should I round up and put (in) a nine?
     [or hh heh heh ]
FR:  [oh- heh hh    ] Okay  [yeah eh ] heh heh heh heh=
FI:                         [.hh Okay]
FR:  =.hh eh heh
FI:  ((smile voice)) hh an eihhght n a half .ehhh =
FI:  =all right and how would you rate the job that Congress
     is doing this is on the same scale of one to ten
```

In answering the question posed by FI (lines 1–4), FR produces an answer (lines 5–6) that does not fit the units of the 1-to-10 scale that she is given by the interviewer. After providing the answer, she designates her utterance as humorous/joking by engaging in laughter (line 6). In overlap with FR's laughter, FI reciprocates the laughter (line 7) and repeats the joke back to FR (lines 7–8), showing affiliation and appreciation of the humorous comment. Instead of rereading the response options, FI furthers the humor, using smile voice to ask FR if she should "round up and put in a nine" (lines 10–11) and then engages in laughter herself. Note that this is a *directive* probe (Fowler and Mangione, 1990:37–45), leading the respondent to her answer. FR reciprocates the interviewer's laughter in (line 12) as she assents to the proposed rounding up to "nine" (line 12). The laughter sequence that emerges here is extensive in that the interviewer does not act to cut the laughter short but instead engages in her own laughter invitation, which occasions even further laughter from FR. Finally, FI offers another rerun of the punch line using smile voice (line 15). In contrast to excerpt 8 from Central State where the interviewer, responding to invitations, produces two small quiet laughter tokens, the laughter in this example is fully reciprocal, overt, and extensive. Many such examples exist in the Midstate data.

The interviewer's and respondent's actions in this excerpt exemplify an open laughter environment.

Another pattern supports the interpretation of the openness of the laughter environment at Midstate. Although our focus in this chapter is on *respondent*-initiated laughter, we also observed instances of *interviewer*-initiated laughter at both sites. Central State interviewers initiated laughter twice. In both instances, the laughter appeared to be what is vernacularly called "nervous" laughter because it occurred when each interviewer was having difficulty with an interview. For example, when an interviewer is having problems with her computer, trouble entering in a respondent's answer, or difficulties getting the respondent to fit the answer into one of the response options (or any combination of these problems), she may produce laughter in what Jefferson (1984) has described as displaying *troubles resistance*. As we noted earlier, such laughter generally does not invite reciprocation.

In contrast, at the University of Midstate center, interviewers initiated laughter 36 times in a number of different situations, most of which were a form of humorous commentary, joke, or as in excerpt 15, a reaction to respondents' comments inviting collaborative laughter. In another example, the respondent provides a relatively innocuous answer to a question about the military. She then produces an aphorism (line 7), asking for agreement ("right?") and in response FI does agree and offers laughter (line 8).

(16) [LSSC112:02:A3:11/14/91:147-157]

```
1   FI:   O:kay ## .hh an: how would you rate the job the military
2         is doing
3         (2.8)
4   FR:   Oh I think the military is doing very well hh probably::
5         (.) well let's go with an eight
6   FI:   Oka:y=
7   FR:   =There's always room for improvement right?
8   FI:   Yeah ehh heh [heh .hh hh YEAH heh .hh O:kay .hh]=
9   FR:                [heh  heh  ehh  heh  heh  heh  .hh]
10  FI:   =generally speaking do you usually think of yourself as:
11        a Republican? a Democrat an independent or: what.
```

This laughter as well as her agreement with the respondent's assessment (line 8) operate to invite the respondent to laugh with her. The respondent accepts (line 9) and hence engages in laughter in overlap with the interviewer's laughter. This is a case in which the respondent's original comment, although not necessarily designated as humorous or inviting of laughter in itself, functions to generate an offer of laughter from FI. Frequently in the Midstate interviews, interviewers initiate laughter in response to a respondent's comment or answer. Further, the laughter is generally affiliative (as above), as opposed to the interviewer "laughing at" a respondent.

Accordingly, standardization protocols appear to shape more than specific interviewer actions. In prohibiting or affirming interviewer laughter, for instance, protocols may affect the entire *context* of the interview. The interviewers at

Central State are in general less "mirthful" and those at Midstate are more "mirthful," suggesting that there are different local cultures and communities of practice (Lave and Wenger, 1991) across survey centers (Viterna and Maynard, this volume, Chapter 16). A more mirthful environment may affect quality of data adversely, as in excerpt 15 where the interviewer uses a directive probe along with laughter to propose a joke and elicit further laughter from the respondent. It is at least possible that the mirthful community of practice at Midstate entails other undesired effects, such as respondents' sense that their answers or attitudes are approved, or that "agreement" is a goal of the interaction (Schaeffer, 1991). On the other hand, a mirthful environment might increase respondents' motivation to complete the interview or their cooperation in answering difficult or sensitive questions (van der Zouwen, Dijkstra, and Smit, 1991). What "house" effects (Hill, 1991; Smith, 1982; Turner, 1984) there may be from differing mirthful environments awaits further investigation, and perhaps experimental manipulation.[16]

DISCUSSION

Respondent-initiated laughter is a contingent development in the survey that interviewers handle in various ways on the occasion of its occurrence. At Central State, our main research site, interviewers are constrained from laughing because it is perceived to threaten standardization and thus bias the data. And, although the survey center's protocols prohibit laughter on the part of interviewers, they do not specify how interviewers should handle *respondent* laughter, including that which is invitational, when it does occur. Nevertheless, in only one instance at Central State did we find an interviewer *accept* a respondent's invitation to laugh by reciprocating, and this instance was rather minimal in that the interviewer produced two very quiet tokens. Most often, interviewers deal with respondent laughter by adhering to the structure of the survey interview. That is, they *decline* invitations to laugh through a repertoire of practices that involve the deployment of both conversational and workplace competencies — they talk seriously, pursue interview-based question–answer sequences, and type responses into the computer in an audible fashion. Accordingly, interviewers are able to decline invitations to laugh without necessarily being seen as crassly ignoring the respondent; they are attending to the business of the interview.

At times, however, interviewers reciprocate respondent laughter by using smile voice or quasilaughter. These pseudolaughing responses to laughter invitations potentially communicate different messages depending on the listener. A supervisor in the Central State survey center is unable to discern any direct or literal violation of the protocol for standardized survey interviewing, whereas respondents can construe reciprocation to their offer of laughter. In all, interviewers appear to maintain both standardization and a good working relationship with their respondents through momentary episodes of rapport. Interviewers achieve this mixture of goals according to how participants in many work settings enact rules and procedures: They invoke a variety of underground or tacit methods and practices whereby a sense of rule adherence is achieved.

CONCLUSION

If the survey interview is a paradigmatic case of "mechanical objectivity," then, as Porter (1995) argues, participants in the interview will necessarily use their tacit knowledge to make its procedures for standardization work. Indeed, respondents answer questions in a variety of nonconforming ways, including with laughter, and interviewers react to breaches of survey order and maintain the orderly progression of the interview. Interviewers' responses to the situated contingencies that arise during the interview are dealt with in the interview through skills that standardized protocols do not formally specify.

The practices for handling laughter in the survey interview, and their variable frequencies at the two centers we studied, raise two sets of issues, one theoretical and the other methodological. In theoretical terms, Porter (1995) has conjectured about the importance and relevance of tacit knowledge in the achievement of *mechanical objectivity* in varied academic, government, business, and other settings in which participants seek to promote unity and consensus, while meeting the scrutiny and suspicions of regulators and other outsiders. However, little has been done demonstrating the use of such knowledge *empirically*. In the standardized survey interview, interviewer competence in handling a multiplicity of situated problems and contingencies (laughter being just one) involves considerable taken-for-granted conversational and work-related skills that are, as this volume attests, coming under increased investigation. We will benefit from research on other forms of tacit knowledge that lend to the various forms of workplace objectivity (Button, 1993; Engström and Middleton, 1996; Greif, 1988; Lave and Wenger, 1991). Such research, as an enterprise in its own right, contributes to an understanding of the social organization of interaction in work settings. For example, it augments the constructivist and ethnomethodological sociology of science by treating survey centers like natural scientific laboratories, as sites where practitioners, through their practices, generate knowledge of disciplinary phenomena and subject matter.

An additional theoretical issue is that, by investigating laughter in particular, it brings under the empirical umbrella of SSSK the role of one kind of affect in survey-based social science. During administration of the survey, respondent laughter copes with the constrained nature of the interview. Respondents appear to laugh at the formal questions, the categorical ways they are required to answer, and the process itself. The conundrum this presents is that interviewers using strictly standardized protocols are prevented from acknowledging displays of affect on the part of respondents that are related to the stringency of the task they face. Interviewers who are allowed to reciprocate laughter face a different puzzle: how to maintain professional distance from their respondents. Whichever strategy is allowed, survey centers settle by fiat what feminist scholars have proposed as a central theme for the sociology of scientific knowledge. In the pursuit of scientific knowledge this theme is what the proper relationship should be between subjectivity and objectivity, between feeling and thinking, between being personal and apersonal (e.g., Keller, 1985). The interview provides a rich site for studying in defalt the de facto resolution of this issue. However, humor and laughter involve only one kind of emotion and its display. Given the array of topics with which surveys are occupied — from states of health, to employment, to marital status, to income level, and so on — questions can no doubt evoke a variety of emotional responses and displays. Investigations of these displays and their socially organized management in the survey interview[17] will increase our understanding of how this form of social science constructs its knowledge base.

A similar theoretical issue involves conversation analytic research on talk in institutional settings, which has neglected the display and management of affect. Talk in institutional settings is more than a rational enterprise; it may involve an array of emotion displays that are consequential for the talk's organization and substance. In only one type of work setting — the survery interview — we have examined only one type of display — laughter — and how interviewers attempt to regulate it, even as the laughter is occasioned by other regulative procedures peculiar to the standardized interview. How talk in diverse institutional settings manifests emotion and how talk is organized around it is an area of inquiry in its infancy. We have seen how there is variation in the management of interview laughter according to local cultures, communities of work practice, and the notions of standardization they uphold. Conversation analytic emphasis on the generic character of interactional practices can be complemented by investigation of the differential distribution of these practices across both time (Clayman and Heritage, 2000) and space.

Methodologically, our research on laughter in the survey interview has implications for data quality. As interviewers manage laughter in the interview, they artfully maneuver through the dilemma of adhering to standardization protocols while maintaining rapport with respondents that retains participation and continues to elicit answers. Of significance is that survey centers approach this dilemma differently; some prohibit laughter and others allow its expression. Whichever pattern is operative, it appears to influence the course of specific question–answer sequences as well as the overall interview environment in its relative mirthfulness. A question for future research is how differences in these local sequences and in the larger interview context affect the retention of respondents and the accuracy of the data that survey centers generate. Among other matters to investigate is whether permitting or prohibiting laughter affects the length of interviews,[18] the percentage of questions answered,[19] the substance of respondents' answers,[20] or any other aspect of the data.

ACKNOWLEDGMENTS

A previous version of this chapter was published in the *American Sociological Review* and is reprinted here by permission of the American Sociological Association. For comments on earlier versions of this text we thank Chas Camic, Nora Cate Schaeffer, Rob Robinson, and Franklin Wilson.

NOTES

1. As Schaeffer (1991:368) puts it: "The standardized survey interview takes measurements on a large number of persons, for a set of topics selected by a researcher, in such a way that measurements can be aggregated to describe a population."

2. For a wide-ranging discussion of the issues here, see Maynard and Schaeffer (2000).

3. Hyman (1954:138–150) argues that when respondents are *task*-involved, validity is enhanced, whereas *social* or *interpersonal* involvement may create bias. As Fowler and Mangione (1990:64) put it, the kind of relationship interviewers can pursue with respondents is constrained by the need to standardize the interview and preserve a businesslike, professional relationship that does not create expectations for the respondent to "produce the right answers, or sophisticated answers, rather than responding openly to the question." Indeed "overly friendly" behavior has been found to produce negative effects on data quality (Weiss, 1968).

4. The subject of social relations and epistemology includes topics such as how professionals may be cautious in broaching sensitive topics—bad news, for instance (Maynard, 1996). It also encompasses how interaction in institutional settings exhibits systematic asymmetries, as in the responses to political speeches (Clayman, 1993).

5. As conversation analysts remark (e.g., Glenn, 1989; Haakana, 1999), the notion of a laughable is vague and glosses many different objects that may engender a laughing response. Participants produce and orient to particular utterances as laughables partly through the sequencing of conversation. Thus, the "humor" or "funniness" of a particular utterance is not an innate quality, but is established in situ.

6. In the Central State University data, we found only one instance of interviewer-initiated laughter, when an interviewer laughingly apologized for not hearing a respondent's answer. For another example of interviewer-initiated laughter, see excerpt 15 in Houtkoop-Steenstra (1996:220–221). After a respondent answers "I think it was four or six times" to a question about how often her elementary school class has been to a museum, the interviewer says "I'll just pu(h)t si(h)x time(h)s," laughing in the midst of her utterance. The respondent also laughs in return.

7. The CASES (Computer-Assisted Survey Execution System) training program, used at many survey research centers including our two sites, instructs interviewers to avoid laughing or otherwise discussing issues unrelated to the survey. Our main survey site (Central State) combined the CASES program, including module tests, with a Microsoft PowerPoint computer program that displayed and explained the information necessary to complete the CASES module tests. One screen of the training module in the PowerPoint instructional program, under the "Professionalism" category, instructs, "Be serious during the interview. Don't laugh or discuss issues not related to the study."

8. Both authors participated in interviewer training a Central State Survey Research Center (pseudonym) where the present research was conducted. This involved the CASES program (see note 7), instruction from supervisors, and simulation of interviews using center staff in the role of respondent. Here we draw on field notes from these training experiences. For example, the first author's notes document a supervisor addressing the problem of laughter by saying that interviewers "really" were not supposed to laugh, even though sometimes it seems natural to do so. She said to be pleasant when someone makes a joke, but to abstain from laughing. Additionally, after the first author laughed at a respondent's joke during a monitored call, the supervisor reminded her that she "shouldn't laugh" in such circumstances.

9. The number of questions in the Central State University television study, taken from a representative recorded interview, is 131. For the neighborhood opinion study at Central State, a representative interview is 96 questions. In the University of Midstate interviews, the first public opinion study on local and national politics is 114 questions and the second government/political survey is 90 questions. Note: The actual number of questions in any interview varies depending on how respondents answer the first question in a given series. The interviews chosen as representative had a median number of questions asked.

10. On the interviewing sequence, see Maynard and Schaeffer (this volume, Chapter 1). Some laughter initiations appear not to be invitational or are not placed in one of the three positions we have characterized. For example, respondents sometimes offer *apologetic* laughter that is not invitational and is more like laughter that shows troubles resistance (Jefferson, 1984).

11. The initiator of self-deprecating laughter pokes fun at her/himself in a teasing way that proposes the initiator or speaker as "butt" of the joke or object of laughter (Glenn, 1991/1992). See also excerpt 8 and discussion.

12. Interviewers are instructed to provide *short feedback* after every closed, short-answer question. Acceptable, standardized, neutral feedback tokens include: Uh huh, Um hmm, I see, Thank you/Thanks, OK, and All Right. The purpose of short feedback, according to the training program, is to encourage the respondent to provide "further thoughtful participation." We use the term *acknowledgments* to refer to such feedback.

13. We are agnostic about how *conscious* the use of this repertoire may be. That is, in attending to the computer, recording answers, asking questions, and so on, interviewers may sometimes not "hear" a respondent's laughter. Other times, they may be more strategic and purposeful in their manipulation of talking and typing to override laughter invitations. Either way, the practices they engage operate to *accountably* refrain from laughing. By this we mean that the respondent can hear the interviewer as reasonably declining an invitation to laugh rather than just ignoring that invitation.

14. Recall that supervisors at our main research site (Central State) monitored each interviewer at least once per night.

15. Shor (1978) and Tartter (1980) provide technical specification of a smile voice: The unintended effect of smiling is an audible alteration of the vocal tract. Individuals involved in speaking usually place their mouths in a neutral position. When the vocal tract is moved from this position, as in the case of smiling, the "mouth orifice widens, shortening the vocal tract, and it's opening enlarges," thereby altering voice quality (Shor, 1978:82). The frequency and the pitch of the speech are also noticeably heightened or increased when individuals smile.

16. Especially with survey questions on "threatening" topics, the conduct of the interview may have an effect of respondents' reports. For an overview, see Schaeffer (2000). For a discussion of how a person-oriented or socioemotional style of interviewing facilitates respondents' provision of delicate information in the survey interview, see Houtkoop-Steenstra (1997).

17. For suggestive research on how interviewers engage in face-saving practices on behalf of respondents who are asked about delicate issues in their lives, see Houtkoop-Steenstra (1997).

18. It is our impression that permitting interviewer laughter lengthens the interview because it also promotes more casual talk between interviewer and respondent. This impression is consistent with Belli, Lepkowski, and Kabeto's (in press) finding (in a factor analysis) that laughter seems to go along with other kinds of *digression behaviors* in interviews. In our two data sets, the attempt to measure overall lengths of interviews and a time-per-question rate was confounded by normative differences in pacing, types of questions (closed vs. open), and probing practices at the two centers.

19. As a preliminary effort, we counted question completions in the surveys at our two research cites and coded completions in two different ways. We did not find any significant differences; completions at both sites were close to 100 percent.

20. For a pertinent recent study, see Belli, Lepkowski, and Kabeto (in press), who report "rather surprising" results regarding rapport and response accuracy. As they remark, the expectation among survey methodologists is that certain verbal behaviors, including laughter, would bias survey data. In their study comparing responses to a health and health care utilization instrument (administered face to face) with data from a health maintenance organization, however, they found that verbal expressions in the (audio-recorded) interviews were not associated with response accuracy. Such verbal expressions included laughter by both interviewers and respondents, plus inappropriate interviewer feedback, directive probing, and digressions. (The authors did find expressions of cognitive difficulty to be negatively associated with response accuracy.)

REFERENCES

Beatty, P. 1995. "Understanding the Standardized/Non-standardized Interviewing Controversy." *Journal of Official Statistics* 11: 147–160.

Belli, R. F., J. M. Lepkowski, and M. U. Kabeto. In press. "The Respective Roles of Cognitive Processing Difficulty and Conversational Rapport on the Accuracy of Retrospective Reports of Doctor's Office Visits." In M.L. Cynamon and R.A. Kulka (Eds.), *Seventh Conference on Health Survey Research Methods*. Washington, DC: U.S. Government Printing Office.

Boden, D., and D. H. Zimmerman (Eds.). 1991. *Talk and Social Structure: Studies in Ethnomethodology and Conversation Analysis*. Berkeley: University of California Press.

Button, G. (Ed.). 1993. *Technology in Working Order: Studies of Work, Interaction, and Technology*. Cambridge: Cambridge University Press.

Cannell, C., and M. Axelrod. 1956. "The Respondent Reports on the Interview." *American Journal of Sociology* 62: 177–181.

Cannell, C., F. Fowler, and K. Marquis. 1968. "The Influence of Interviewer and Respondent Psychological and Behavioral Variables on the Reporting in Household Interviews." *Vital and Health Statistics* 2(26): 1.

Cicourel, A. V. 1982. "Interviews, Surveys, and the Problem of Ecological Validity." *American Sociologist* 17: 11–20.

Clayman, S. E. 1993. "Booing: The Anatomy of a Disaffiliative Response." *American Sociological Review* 58: 110–130.

Clayman, S. E., and J. Heritage. 2000. "Questioning Presidents: Journalistic Deference and Adversarialness in the Press Conferences of Eisenhower and Reagan." Paper presented at the annual meeting of the American Sociological Association, August, Washington, DC.

Conrad, F. G., and M. F. Schober. 2000. "Clarifying Question Meaning in a Household Telephone Survey." *Public Opinion Quarterly* 64: 1–28.

Converse, J. M. 1987. *Survey Research in the United States: Roots and Emergence 1890–1960*. Berkeley, CA: University of California Press.

Coser, R. L. 1959. "Some Social Functions of Laughter." *Human Relations* 12: 171–182.

Drew, P., and J. Heritage (Eds.). 1992. *Talk at Work: Interaction in Institutional Settings*. Cambridge: Cambridge University Press.

Engström, Y., and D. Middleton (Eds.). 1996. *Cognition and Communication at Work*. Cambridge: Cambridge University Press.

Fowler, F., and T. Mangione. 1990. *Standardized Survey Interviewing: Minimizing Interviewer Related Error*. Newbury Park, CA: Sage.

Garfinkel, H. 1967. *Studies in Ethnomethodology*. Englewood Cliffs, NJ: Prentice-Hall.

Gavioli, L. 1995. "Turn Initial vs. Turn Final Laughter: Two Techniques for Initiating Remedy in English/Italian Bookshop Service Encounters." *Discourse Processes* 19: 369–384.

Glenn, P. J. 1989. "Initiating Shared Laughter in Multi-Party Conversations." *Western Journal of Speech Communication* 53: 127–149.

Glenn, P. J. 1991/1992. "Current Speaker Initiation of Two-Party Shared Laughter." *Research on Language and Social Interaction* 25: 139–162.

Glenn, P. J. 1995. "Laughing at and Laughing with: Negotiations of Participant Alignments through Conversational Laughter." Pp. 43–56 in P. ten Have and G. Psathas (Eds.), *Situated Order: Studies in the Social Organization of Talk and Embodied Activities*. Washington, DC: University Press of America.

Goudy, W. J., and H. R. Potter. 1975. "Interview Rapport: Demise of a Concept." *Public Opinion Quarterly* 39: 529–543.

Greif, I. (Ed.). 1988. *Computer Supported Cooperative Work: A Book of Readings*. San Mateo: Morgan Kaufman.

Haakana, M. 1999. "Laughing Matters: A Conversation Analytical Study of Laughter in the Doctor-Patient Interaction." Unpublished manuscript, Department of Finnish Language, University of Helsinki.

Heath, C. 1988. "Embarrassment and Interactional Organization." Pp. 136–160 in P. Drew and T. Wootton (Eds.), *Erving Goffman: An Interdisciplinary Appreciation*. Cambridge: Polity with Basil Blackwell.

Heritage, J. 1998. "Oh-Prefaced Responses to Inquiry." *Language in Society* 27: 291–334.

Hill, D. H. 1991. "Interviewer, Respondent, and Regional Office Effects on Response Variance: A Statistical Decomposition." Pp. 463–86 in P. P Biemer, R. M. Groves, L. E. Lyberg, N. A. Mathiowetz, and S. Sudman (Eds.), *Measurement Errors in Surveys*. New York: Wiley.

Houtkoop-Steenstra, H. 1996. "Probing Behaviour of Interviewers in the Standardised Semi-Open Research Interview." *Quality and Quantity* 30: 205–230.

Houtkoop-Steenstra, H. 1997. "Portrait of the Person-Oriented Survey Interviewer." Pp. 59–76 in A. Marcarino (Ed.), Analisi Della Conversazione e Prospettive di Ricerca in Ethnomethodologia. Urbino, Italy: Edizioni Quattro Ventia.

Hyman, H. H. 1954. *Interviewing in Social Research*. Chicago: University of Chicago Press.

Jefferson, G. 1978. "What's in a 'Nyem'?" *Sociology* 12: 135–139.

Jefferson, G. 1979. "A Technique for Inviting Laughter and Its Subsequent Acceptance Declination." Pp. 79–96 in G. Psathas (Ed.), *Everyday Language: Studies in Ethnomethodology*. New York: Irvington.

Jefferson, G. 1984. "On the Organization of Laughter in Talk about Troubles." Pp. 346–369 in J. M. Atkinson and J. Heritage (Eds.), *Structures of Social Action: Studies in Conversation Analysis*. Cambridge: Cambridge University Press.

Jefferson, G. 1985. "An Exercise in the Transcription and Analysis of Laughter." Pp. 25–34 in T. A. van Dijk (Ed.), *Handbook of Discourse Analysis*, Vol. 3: *Discourse and Dialogue*. London: Academic.

Jefferson, G., H. Sacks, and E. Schegloff. 1987. "Notes on Laughter in the Pursuit of Intimacy." Pp. 152–205 in G. Button and J. R. E. Lee (Eds.), *Talk and Social Organization*. Clevedon, England: Multilingual Matters.

Keller, E. F. 1985. *Reflections on Gender and Science*. New Haven: Yale University Press.

Latour, B. 1987. *Science in Action*. Cambridge: Harvard University Press.

Lave, J., and E. Wenger. 1991. *Situated Learning: Legitimate Peripheral Participation*. Cambridge: Cambridge University Press.

Lynch, M. 1993. *Scientific Practice and Ordinary Action: Ethnomethodology and Social Studies of Science*. Cambridge: Cambridge University Press.

Maynard, D. W. 1996. "On 'Realization' in Everyday Life: The Forecasting of Bad News as a Social Relation." *American Sociological Review* 61: 109–131.

Maynard, D. W., and N. C. Schaeffer. 2000. "Toward a Sociology of Social Scientific Knowledge: Survey Research and Ethnomethodology's Asymmetric Alternates." *Social Studies of Science* 30: 323–370.

McAdams, D., J. Jackson, and C. Kirshnit. 1984. "Looking, Laughing, and Smiling in Dyads as a Function of Intimacy Motivation and Reciprocity." *Journal of Personality* 52: 261–273.

Mishler, E. G. 1986. *Research Interviewing: Context and Narrative*. Cambridge: Harvard University Press.

Norrick, N. 1993. *Conversational Joking: Humor in Everyday Talk*. Bloomington, IN: Indiana University Press.

O'Donnell T., N. Adams, and K. Adams. 1983. "Heheh in Conversation: Some Coordinating Accomplishments of Laughter." *Western Journal of Speech Communication* 47: 175–191.

O'Muircheartaigh, C. A. 1997. "Measurement Error in Surveys: A Historical Perspective." Pp. 1–28 in L. Lyberg, P. Biemer, M. Collins, E. de Leeuw, C. Dippo, N. Schwarz, and D. Trewin (Eds.), *Survey Measurement and Process Quality*. New York: Wiley-Interscience.

Peräkylä, A. 2000. "Bales and Sacks: Two Traditions of Interaction Research." Unpublished manuscript, Department of Sociology and Social Psychology, University of Tampërë, Tampërë, Finland.

Platt, J. 1996. *A History of Sociological Research Methods in America 1920–1960*. Cambridge: Cambridge University Press.

Porter, T. M. 1995. *Trust in Numbers*. Princeton, NJ: Princeton University Press.

Poyatos, F. 1993. "The Many Voices of Laughter: A New Audible-Visual Paralinguistic Approach." *Semiotica* 93: 61–81.

Psathas, G. 1995. *Conversation Analysis: The Study of Talk-in-Interaction*. Thousand Oaks, CA: Sage.

Schaeffer, N. C. 1991. "Conversation with a Purpose — Or Conversation? Interaction in the Standardized Survey." Pp. 367–391 in P. P. Biemer, R. M. Groves, L. E. Lyberg, N. A. Mathiowetz, and S. Sudman (Eds.), *Measurement Errors in Surveys*. New York: Wiley.

Schaeffer, N. C. 2000. "Asking Questions about Threatening Topics: A Selective Overview." Pp. 105–122 in A. A. Stone, J. S. Turkkan, C. A. Bachrach, J. B. Jobe, H. S. Kurtzman, and V. S. Cain (Eds.), *The Science of Self-Report: Implications for Research and Practice*. Mahwah, NJ: Lawrence Erlbaum.

Schegloff, E. A., G. Jefferson, and H. Sacks. 1977. "The Preference for Self-Correction in the Organization of Repair in Conversation." *Language* 53: 361–382.

Schober, M. F., and F. G. Conrad. 1997. "Does Conversational Interviewing Reduce Survey Measurement Error?" *Public Opinion Quarterly* 61: 576–602.

Schuman, H. 1982. "Artifacts Are in the Mind of the Beholder." *American Sociologist* 17: 21–28.

Schuman, H., and G. Kalton. 1985. "Survey Methods." Pp. 635–97 in G. Lindzey and E. Aronson (Eds.), *The Handbook of Social Psychology*. New York: Random House.

Schuman, H., and J. Ludwig. 1983. "The Norm of Even-Handedness in Surveys as in Life." *American Sociological Review* 48: 112–120.

Shor, R. E. 1978. "The Production and Judgement of Smile Magnitude." *Journal of General Psychology* 98: 79–96.

Smith, T. W. 1982. "House Effects and the Reproducibility of Survey Measurements: A Comparison of the 1980 GSS and the 1980 American National Election Study." *Public Opinion Quarterly* 46: 54–68.

Suchman, L., and B. Jordan. 1990. "Interactional Troubles in Face-to-Face Survey Interviews." *Journal of the American Statistical Association* 85: 232–241.

Tartter, V. C. 1980. "Happy Talk: Perceptual and Acoustic Effects of Smiling on Speech." *Perception and Psychophysics* 27: 24–27.

Traweek, S. 1988. *Beamtimes and Lifetimes*. Cambridge, MA: Harvard University Press.

Turner, C. F. 1984. "Why Do Surveys Disagree? Some Preliminary Hypotheses and Some Disagreeable Examples." Pp. 159–214 in C. Turner and E. Martin (Eds.), *Surveying Subjective Phenomena*. New York: Russell Sage.

van der Zouwen, J., W. Dijkstra, and J. H. Smit. 1991. "Studying Respondent-Interviewer Interaction: The Relationship between Interviewing Style, Interviewer Behavior, and Response Behavior." Pp. 419–37 in P. P. Biemer, R. M. Groves, L. E. Lyberg, N. A. Mathiowitz, and S. Sudman (Eds.), *Measurement Errors in Surveys*. New York: Wiley.

Weiss, C. 1968. "Validity of Welfare Mothers' Interview Responses." *Public Opinion Quarterly* 32: 622–633.

Weiss, C. 1970. "Interaction in the Research Interview: The Effects of Rapport on Response." Pp. 17–20 in *Proceedings of the Social Statistics Section of the American Statistical Association* (August). Washington, DC: American Statistical Association.

West, C. 1984. *Routine Complications: Troubles with Talk between Doctors and Patients*. Bloomington: Indiana University Press.

Wilner, P. 1985. "The Main Drift of Sociology between 1936 and 1984." *History of Sociology* 5: 1–20.

Whalen, J., and D. H. Zimmerman. 1998. "Observations on the Display and Management of Emotion in Naturally Occurring Activities: The Case of 'Hysteria' in Calls to 9-1-1." *Social Psychology Quarterly* 61: 141–159.

Whalen, J., D. H. Zimmerman, and M. R. Whalen. 1988. "When Words Fail: A Single Case Analysis." *Social Problems* 35: 335–362.

Zimmerman, D. 1992. "The Interactional Organization of Calls for Emergency Assistance." Pp. 418–469 in P. Drew and J. Heritage (Eds.), *Talk at Work: Social Interaction in Institutional Settings*. Cambridge: Cambridge University Press.

CHAPTER 16

How Uniform Is Standardization? Variation Within and Across Survey Research Centers Regarding Protocols for Interviewing

Jocelyn S. Viterna and Douglas W. Maynard

INTRODUCTION

The quality of survey data relies heavily upon standardization in the survey interview. According to Fowler and Mangione (1990:14), "in all sciences, meaningful measurement occurs by applying the same procedure across a set of situations so that differences in the readings that result can be compared and interpreted as indicating real differences in what is being measured." To achieve the "same procedure" in survey research, interviewers are trained to expose each respondent to the survey instrument in a precise and replicable — that is, standardized — manner. This allows researchers to assume that any variation in the resulting answers is directly attributable to differences in what is being measured among respondents.

Recently, scholars have questioned whether or not the ideals of standardization are attainable given the human interaction of a survey interview (Beatty, 1995). Maynard and Schaeffer (2000; this volume, Chapter 1) argue that full standardization or *mechanical objectivity* (Porter, 1995) is never fully attainable because interviewers must constantly employ tacit, commonsense knowledge (Garfinkel, 1967; Polanyi, 1958) to negotiate and complete the interview. If this is true, then the question is not whether centers for survey research depart from standardization, because by necessity they will have to. The question is how much such departure occurs and to what extent it interferes with the quality of survey data.

Standardization and Tacit Knowledge: Interaction and Practice in the Survey Interview
Edited by Douglas W. Maynard, Hanneke Houtkoop-Steenstra, Nora Cate Schaeffer, and Johannes van der Zouwen
ISBN 0-471-35829-0.

In this chapter, we ask a question related to concerns about the attainability of standardization: "How uniform are concepts of standardization across survey centers?" We also explore how consistent survey centers are in their implementation of standardization. To anticipate our main finding, survey centers are inconsistent in their requirements for standardization. We account for this variation by reference to practitioner's use of tacit and commonsense knowledge. We extend this analysis by using a sociology of science framework and pointing to contingencies of survey interviewing and the influence of local cultures for devising solutions to these contingencies. Some survey centers appear to relax standardization guidelines purposefully and carefully, while still pursuing the goal of generating good-quality data. Other centers with the same goal are stricter in their adherence to standardization. By documenting and analyzing the disparities of standardization requirements, our aim is to add clarity to the debate about the importance and feasibility of standardization within the survey interview.

The variation among survey organizations in how their interviewers perform is a topic of importance in the field of survey research (and indeed to all social scientists who rely on survey centers to collect their data), yet few previous studies have offered insight into what variation occurs and how that variation affects data quality. Our analysis is an exploratory and preliminary attempt to address this gap. In this study, we document what interviewers are *trained* and *supervised* to do, not what they *actually* do on the job. Because our results are based almost entirely on training programs and procedures, we must be careful in extrapolating our findings to how training directives are enacted within the actual survey interview. However, based on a study by Fowler and Mangione (1990), we believe that interviewer training and supervision are crucial factors in determining whether and how standardization guidelines are implemented in actual interviews. We argue that the variation in standardization guidelines presented to an interviewer during training means different practices during the actual collection of data.

HISTORICAL EMERGENCE OF STANDARDIZATION

When interviewers were first used for academic research in the 1930s, standardization was not the norm. Researchers "viewed interviewers as an escape from the restrictions of closed, self-administered questionnaires and were anxious to use this potential freedom" (Beatty, 1995:148). When polling centers introduced standardization, academics were critical because it hindered the development of *rapport* with the respondent (Bradburn, Sudman, and Associates, 1979; Lazarsfeld, 1944; Richardson, Dohrenwend, and Klein, 1965). Previously, well-trained interviewers had been expected to standardize the *meaning* of the question and were given freedom to modify the question *wording* in the way they deemed best for each individual situation (Converse, 1987:40).

Although standardization was not initially the norm, it quickly began to achieve dominance as a research technique largely due to concerns about quality of data (O'Muircheartaigh, 1997:13–14). From its inception in 1941, the National Opinion Research Center (NORC) required interviewers to read questions verbatim (see Williams, 1942). By the 1950s two other pioneering survey research centers, the Bureau of Applied Social Research at Columbia University, and the University of Michigan Survey Research Center, had followed suit (Converse, 1987). Studies of interviewing by Herbert Hyman (1954) and his colleagues convinced many practitioners that standardized control would improve the reliability of survey data. In 1975, an influential study by Goudy and Potter (1975) added a theoretical basis to the practical appeal. Not only did they find no correlation between response rates and interviewer rapport, but they also found that an emphasis on rapport at the expense of standardization may actually bias the results of a study.

Today, standardization is prized in nearly all survey operations in the United States and in many countries around the world. Given the pervasiveness and ubiquity of survey research, which spans academic disciplines, national boundaries, government and business organizations, public and private sectors, and so on (Maynard and Schaeffer, 2000), it is important to discover whether official regard for standardization (*standardization in theory*) compares to what interviewers are trained and allowed to do (*standardization in practice*). Our study is confined to one nation (the United States) and one kind of survey center (university affiliated), but has wider implications.

DATA AND METHODS

We examine 12 academically based survey research centers. At five of these centers, the first author gathered data through participant observation in interviewer training sessions.[1] These sessions varied in length from 2.5 hours to 2 full days and included instructions on basic interviewing technique, CATI (computer-assisted telephone interviewing) commands and use, and general center regulations. Detailed notes were taken on what information was presented in the training, how it was reinforced through activities or exams, what questions were asked by participants, how they were answered, which interviewing procedures were corrected by supervisors during "mock" interviews, and which were allowed to pass without comment.

At four of the five centers visited, staff members who oversaw interviewer hiring, training, and supervision were interviewed. These interviews dealt with the history, budget, hiring procedures, ongoing training programs, interviewer monitoring and evaluation requirements, and the mission statement of the survey center. In addition, any discrepancies between what was taught during interviewer training and what was taught by accepted texts on standardization procedures were discussed. For example, staff members were asked how interviewers were taught to balance building rapport with maintaining professionalism or (if their training or manuals indicated this) why they advocated faster instead of slower rates of speaking.

Finally, the training manuals of these five centers, plus an additional seven training manuals from survey centers throughout the central and eastern United States,[2] were reviewed and coded to add breadth to the analysis. Inspection of the manuals from the sites visited, and participant observation at those sites, revealed a strong correspondence between the amount of coverage a topic received in the training manual and the emphasis that topic was given in the actual training program.[3] Therefore, we regard training manuals from nonvisited sites as a good indicator of training emphases at those sites and the additional manuals as reliable data for analyzing variation in interviewer training practices.

Data collection, transcription, and coding were completed by the first author in consultation with the second author. Coding involved organizing the data from each center (ethnographic notes, interview transcriptions, and manuals) around predesignated standardization guidelines, then determining how closely each particular center adhered to the standardization guidelines set forth in the literature. The analysis of these materials involved a bottom-up set of investigative procedures akin to analytic induction or grounded theory (Charmaz, 1983; Glaser and Strauss, 1967). We discuss the reasons for our coding decisions at particular points in the chapter.

STANDARDIZATION CONTINUUM

Variation was found among survey centers in nearly every facet of interviewer protocol. To appreciate patterning in such variation, we developed a continuum and portray two ideal types on it. The ideal type of survey center on the right of the continuum is characterized by strict standardization and a lack of interviewer autonomy. Here, interviewers are given very little information about the purpose of a study and are required to adhere to a detailed and complex set of rules designed to cover any situation that might arise during an interview. Interviewers speak only the words specifically written on the questionnaire or specifically approved in the training manual. They must follow regulations even in situations where personal judgment suggests that such adherence is confusing or irritating the respondent, or may even lead to inaccurate data.

In the ideal type of survey center on the left of the continuum, interviewers are expected to follow proper standardization techniques but are given permission, sometimes implicitly and sometimes explicitly, to relax the rules if they feel these techniques may lead to a wrong answer or to the respondent's refusal to complete the interview. Here, general training is secondary to the extensive and comprehensive project-specific instructions, when supervisors discuss the purpose and goals of each individual question and the study as a whole. The goal of obtaining accurate data takes priority over proper technique, and the interviewer is considered competent to determine when it is appropriate, and perhaps even necessary, to bend the rules. Our continuum was not derived from but is roughly related to the two poles that O'Muircheartaigh (1999:52–53) describes as having pervaded orientations to the survey interview: It is through interviewer autonomy that respondents are "empowered" (as one pole), while through "detailed specifications of (uniform) permissible interactions" standardization is achieved (as the other pole).

To locate a center on the continuum between the two ideal types, each one is classified according to five components of standardized interviewing procedure. Fowler and Mangione's *Standardized Survey Interviewing* guide (1990) identifies these components.[4] For each characteristic, we determine if a center's policies are more consistent with maintaining strict standardization or with emphasizing interviewer autonomy. If the policies define and advocate adherence to protocol, we assign a +1 to the center, thus moving that center one step closer to the standardized ideal type on the right. Likewise, if the policies are consistent with interviewer decision making, we assign a score of −1 to the center, and it moves one step further left, closer to the autonomous ideal type. In situations where no information is available on a particular interviewing component, we also assign a score of −1 to the center because it implicitly leaves that particular interviewing behavior to the interviewer's discretion.

In our later discussion, we add the scores for the 12 centers on each of the 5 components. This gives a condensed indication of where the center is located on the standardization continuum. We add the scores across centers for each component, which creates a summary score for each component. When the sum is near zero, this may indicate lack of consensus regarding how autonomous or standardized interviewers should be on that component. When the sum is strongly in the plus range, it indicates that the centers have consensus in the direction of standardization. When the sum is strongly in the minus range, centers are convergent in the direction of autonomy. As we will see, the tendency among our centers is toward autonomy.

Contrary to past discussions in the standardization debate (Beatty, 1995:150–151), we contrast *standardization* with *interviewer autonomy* rather than with *rapport*. Although strict standardization can (but does not necessarily) inhibit rapport, and greater autonomy can (but not necessarily) enhance rapport, developing rapport is not the only reason for relaxing standardization guidelines. For example, relaxed standardization guidelines often allow the interviewer to aid the respondent in understanding survey questions. For purposes here, it is sufficient to contrast standardization with interviewer autonomy.

Although some investigators may not agree that standardization is naturally opposed to interviewer autonomy, we argue that this is a helpful distinction, especially given the exploratory nature of our study. Granted, even the most standardized of interviewers must practice some autonomous decision making. For example, an interviewer must decide when a respondent's answer needs to be probed for clarification, and what probes will generate a codable answer without biasing data. The distinction, however, is that within a highly standardized center, even the autonomous choices allowed to the interviewer must be taken from a tightly constrained set of alternatives geared according to question type. In a less standardized center, the instructions for probing will be much less specific, and interviewers have more latitude in how they probe. Thus, we argue that for this preliminary look at standardization, interviewer autonomy provides a useful contrast to interviewer standardization, since standardization necessarily implies more limitations on the autonomous decision making of an interviewer.

ANALYSIS OF STANDARDIZATION COMPONENTS

To analyze variation in standardization, we examine five aspects of standardized survey interviewing procedures identified by Fowler and Mangione (1990) as relevant to the substantive portion of the survey interview:[5] pace, feedback, professional objectivity, probing, and monitoring procedures. For each component, we first discuss the existing literature for that particular component, then analyze the variation within the collection, and conclude with suggestions for how such criteria can be used to locate the center on the standardization continuum. The initial classification of each component is based on information obtained from center manuals. Supporting information from observations and interviews is given after the analysis of manuals and is clearly indicated in the text as separate from the data from the manuals.

Pace

Pace refers to the speed at which the interviewer reads the questions to the respondent in the substantive part of the interview. Although no direct link between pace and data quality has been documented, 9 of the 12 centers participating in this study include instructions for pace in their training manuals. Two theories of pace are consistently found in the data. On one hand, 5 centers follow the suggestion of Fowler and Mangione (1990:71) that "having interviewers speak slowly is one important and practical way in which researchers can increase the standardization across interviews." This suggestion is followed in manuals by such statements as, "Read at a slow, two-word-per-second pace to make sure all respondents concentrate on the questions and hear them clearly" (Center O)[6] or "Reading each question slowly and thoughtfully suggests to the respondent that he/she take the same care in answering the questions correctly" (Center C). Center supervisors indicate that insisting on a slow pace also reduces the possibility that the interviewer may change the wording of the question or make coding mistakes, thereby further increasing the level of standardization.

On the other hand, four centers direct their interviewers to speak either in normal, conversational tones or to use faster-than-normal rates of speech during the substantive part of the interview. Examples:

> Interviewers are judged as being more competent and professional, more friendly to the respondent, and less likely to be refused data when they have the following voice qualities: greater variation in pitch, relatively loud voice, *relatively fast rates of speaking*, and clearer pronunciation (our emphasis). (Center J)

> Speak slowly enough to ensure every word is heard but quickly enough to ensure the interview moves ahead promptly. (Center R)

The common thread among these centers is greater autonomy for the interviewer. The interviewer is not given a static rate (i.e., read at a two-word-per-second pace) and may be allowed to infer what respondents need or desire and then to act accordingly. Furthermore, the goals of slow pacing—to ensure fewer

Table 16.1 Reading Pace

Recommended Pace for Reading Questionnaire	Centers												
	1	2	3	4	5	6	7	8	9	10	11	12	Total
Questions should be read at a conversational pace OR at a faster-than-normal pace OR the pace of the interview is left to the interviewer's discretion.	X			X*				X*	X	X	X	X*	−7
Questions should be read at a slower-than-normal pace.		X	X		X	X	X						5
Standardization score	−1	+1	+1	−1	+1	+1	+1	−1	−1	−1	−1	−1	−2

mistakes and more consistent delivery across interviewers and thus increase standardization—may be lost when interviewers are allowed to personally define a "conversational" or "relatively fast" rate of speech.

The recommended pace for reading questions provides the first measure of where survey centers fit on the standardization continuum (see Table 16.1). For centers that specify a slow pace, and thus a more standardized means of collecting data, a +1 has been assigned, moving the center toward the right end of the continuum. For centers allowing a "normal' or "faster" pace, so that interviewers may decide what is best for the interview at hand, a −1 has been assigned, moving the center to the left. Centers that did not mention pace in their manuals or training sessions were also assigned a −1, but an asterisk is used to indicate that the interviewer's autonomy in regulating pace is implicitly (by omission of statement about pacing) and not explicitly allowed.[7] In the lower right-hand corner of the table, the negatives and positives are summed as an indication of overall consistency between centers on this topic.[8] The summary score for pace is −2, indicating low consistency among survey centers. Roughly speaking, a little over half the centers in this study allow the interviewer to determine the appropriate pace for the interview.

Feedback

In everyday conversations, if one asks a question it is common to acknowledge the answer by responding reciprocally, sometimes approving or disapproving of an answer, or registering it with an "oh." These *third-turn* comments in an interviewing sequence (i.e., QUESTION–ANSWER–ACKNOWLEDGMENT) are considered by ethnomethodologists to be the crucial mechanism by which participants recognize institutionalized activity frameworks (Heritage, 1984:290). In conversation, the content of the third turn after a question–answer sequence can indicate to the answerer a variety of things, such as whether the answer is heard, understood, accepted, correct, good, and so on. In institutional or organizational settings, therefore, the third turn is often regulated (see discussion in Chapter 1, this volume). For instance, in news interviews, after obtaining an answer from a subject, the host refrains from third-turn commentary, thereby indicating "neutrality" and acting as a conduit for the answer to reach an "overhearing" audience (Clayman, 1988; Heritage and Greatbatch, 1991). In educational testing interviews, administrators are permitted to use "okay," "thank you," and the like but are not to use evaluative terms indicating the correctness of an answer (Marlaire and Maynard, 1990; Maynard and Marlaire, 1992). By contrast, teachers asking *known information* questions in the classroom produce third turns that indicate whether a student's answer is right or wrong (Mehan, 1979).

The survey interview is most like the testing situation; survey designers accept the importance of third-turn utterances in acknowledging a respondent's answer but recognize that fine gradations in these responses potentially influence the respondent and create measurement error. A few studies demonstrate a significant relationship between the use of value-laden feedback and an apparent bias in the respondent's answers (Fowler and Mangione, 1990; Marquis, Cannell, and Laurent, 1972). The suggestion is that when respondents give answers appearing to be thoughtful, complete, and within the constraints of the given question, they are rewarded with a "value neutral" positive comment about their interviewing performance, not about the content of the answer (Cannell, Oksenberg, and Converse, 1977; Marquis, Cannell, and Laurent, 1972).

Ten of 12 centers in this study made specific reference to feedback use in their training materials. The proposed use of feedback varied markedly along three different measures: frequency, purpose, and content. First, feedback varied in its permitted frequency; only two centers specified how often feedback should be given. At one of these, interviewers were instructed to give feedback after every acceptable answer. At the other, interviewers were instructed to give feedback after 30–50 percent of the respondent's acceptable responses. The remaining 10 centers did not indicate when or how often feedback should be used.

In Table 16.2, centers that specifically indicate how often feedback should be used are given a score of +1, moving them further to the right on the standardization continuum because all respondents in these centers should receive roughly the same amount of feedback. Survey centers that do not regulate the frequency of feedback are assigned a value of −1, moving them further to the *autonomous* side of the continuum since the interviewer chooses how much feedback is desirable. Survey centers that do not mention feedback at all are again differentiated in the table with an asterisk and assigned a −1 since the omission of feedback directions implicitly allows the interviewer to choose the frequency of feedback. The score of −8 indicates that there is very high consistency in the regard for frequency of feedback usage across survey centers. Most centers allow the interviewer to determine how often feedback will be given.

A second mode of variation in feedback usage is found in how or whether a center defines its purpose. Of the 10 centers that recommended the use of neutral feedback, 4 indicated that feedback should be given after acceptable answers and withheld after unacceptable answers. These are given a score of +1

Table 16.2 Frequency of Feedback Usage

	Centers												
Frequency of Usage	1	2	3	4	5	6	7	8	9	10	11	12	Total
Training manual does not indicate how frequently feedback should be used.	X		X	X	X		X*	X*	X	X	X	X	−10
Training manual specifically indicates how often value-neutral feedback should be used.		X				X							2
Standardization score	−1	+1	−1	−1	−1	+1	−1	−1	−1	−1	−1	−1	−8

Table 16.3 Purpose of Feedback

	Centers												
Purpose of Feedback	1	2	3	4	5	6	7	8	9	10	11	12	Total
Training manual advocates use of feedback as a means of "building rapport" with the respondent OR makes no specific mention of feedback.			X	X	X		X*	X*	X		X	X	−8
Training manual advocates use of feedback as a way of "training" the respondent.	X	X				X				X			4
Standardization score	+1	+1	−1	−1	−1	+1	−1	−1	−1	+1	−1	−1	−4

in Table 16.3, indicating an orientation to standardization on the part of these centers. The remaining 6 centers stated only that the purpose of feedback is to build rapport with the respondent by demonstrating that the interviewer is listening, interested, and nonjudgmental. Thus we infer that feedback would be permitted for unacceptable as well as acceptable answers. These centers receive a score of −1 in Table 16.3 for their orientation to autonomy. The two centers not mentioning feedback in their manuals are also assumed to allow feedback for both acceptable and unacceptable answers and are scored −1. This results in a summary score of −4 and indicates that most centers (2/3) allow interviewers to use feedback as they see fit. Nevertheless, one-third of the centers advocate stricter use of feedback to shape respondent performance.

A third mode of variation in feedback use is found in the suggested feedback phrases themselves. Interestingly, many of the same phrases appeared in the majority of the survey manuals in this study with word-for-word replication from one center to the next.[9] Table 16.4 lists all feedback phrases from all manuals, along with the number of manuals that advocate use of that phrase.

Given the number of shared feedback phrases, it is surprising to find that the *suggested* feedback at some centers is the *forbidden* feedback at others. At three of the centers visited, interviewers were instructed that saying "yes" or "right" was forbidden because it indicated agreement. At one of these three centers, saying "okay" was also forbidden for the same reason. Yet as can be seen in Table 16.4, four other centers recommend in their manuals using "yes" and four recommend using "okay" when giving feedback.[10] In short, some centers advocate the use of feedback that, in the eyes of other centers, may compromise the data.

Recommendations about feedback phrases are difficult to use as a measure to locate a center on our continuum. Two conclusions, however, are warranted. First, there needs to be further research to see if and how feedback affects data

Table 16.4 Frequency of Suggested Feedback Phrases

Acceptable Feedback Options from Survey Center Interviewer Training Manuals	Number of Manuals Advocating Use of This Feedback Phrase
I see.	10
That's useful(helpful) information. That's useful (helpful) to know. It's useful to get your ideas on this.	7
It is important to find out how you (people) feel about this. It is important to get your ideas on this. It is important to get your opinion on that. It is important to find out what people think about this.	7
Thanks/Thank you.	6
Uh-huh/Uh-hmmm/Mm-hmmm	6
Okay/All Right	4
Yes	4
That's interesting	3
That helps our research a lot. That is helpful for our research. That is important for our research.	2
Right	1
Thanks, that's the sort of EXACT answer we need.	1
That's important to know	1

quality. Although survey centers and experts on survey design believe that types of feedback do affect respondents' answering (per discussion above), there is very little systematic investigation of this matter and almost none since the 1970s.[11] Second, survey center manuals do recognize the possibility of such an effect and offer sometimes conflicting advice about how interviewers should perform their feedback. This indicates a *leaning* toward standardization within centers and inconsistency across centers. The main form of agreement across centers is in the recommendation to use "I see."

In addition to finding variation in the feedback phrases listed in the manuals, we observed differences in the attention accorded to feedback in training programs. Three of the five centers visited gave extensive coverage to the issue of feedback both in the training manuals and in the training sessions, while two did not. Those centers that addressed feedback provided a list of suggested phrases for the interviewers to post in their carrels. At two of these three centers interviews were observed in progress and interviewers appeared to follow their training very closely in both centers. No feedback was heard that was not on the recommended list.

In the other two centers visited, the topic of neutral feedback received only one paragraph in the training manual. Further, neither of these centers made any mention of feedback in their training sessions, nor did they provide lists of "appropriate" feedback for interviewers to use while conducting an interview. It was in these two centers that strongly evaluative feedback was heard. In one instance, experienced interviewers said, "great," "very good," and "good," in mock interviews with their supervisors. Supervisors did not reprimand these interviewers, nor did they discuss inappropriate feedback throughout the rest of the training program or otherwise indicate that such feedback was a violation of good protocol. Indeed at one of these centers an experienced interviewer used phrases including "interesting," "good point," "yeah, that's true," and "no, these aren't the greatest questions" as third-turn utterances after the respondent's answers.

Two implications derive from our findings about feedback. First, lack of regulation at some centers may mean that interviewers are not exposing respondents to uniform interview situations. Second, even when feedback is tightly regulated, centers disagree about what "neutral" feedback is and how and when it should be used.

Professional Objectivity

Ten survey centers discuss in their manuals the need for *professional objectivity*, being *interpersonally neutral*, or establishing a *balanced rapport* with the respondent. Such discussions are consistent with Fowler and Mangione's (1990:48–49) recommendation. For an interview to be standardized, the interviewer must not "volunteer personal information to the respondent about life situations, views, or values" and "should be careful that the feedback provided to respondents does not imply any evaluation or judgment about the content of the respondent's answers."

Nevertheless, eight survey center manuals instruct interviewers to convey warmth and pleasantness in their tone of voice, two centers instruct their interviewers to be friendly, and one even suggests using humor to build better rapport with the respondent. To some extent, these suggestions to be both impersonal and friendly form a dilemma and can create strain for the interviewer.[12] This dilemma is acknowledged in five interviewer manuals. Two examples:

> Interviewers perform a delicate balancing act between being congenial with a respondent and maintaining a professional distance. A respondent will be more cooperative when the interview is a pleasant experience. However, maintaining a distance is essential to avoid biasing the interview. If a respondent thinks of the interviewer as a newly found "friend," there is the possibility that the answers may not be as truthful as they would be if a professional rapport were maintained. (Center C)

> In every interview, interviewers must play several roles: that of an individual who builds a friendly and trusting relationship with each respondent; and that of a technician who gathers information to select a respondent and who applies standard techniques to the interview. (Center N)

However, no manual specifies or explicates how interviewers are to deal with this dilemma. Rather, learning how to balance rapport and neutrality appears to be something interviewers learn on the job through listening to other interviewers around them and through feedback from their supervisors (Goodwin, 1998; Heil, 1999; Lavin and Maynard, this volume, Chapter 15).[13]

When survey staff were asked how they train interviewers to deal with the contradiction between remaining neutral and establishing rapport, they consistently asserted that interviewers learn what an acceptable level of rapport is through on-the-job experience. For example:

> It's not unusual for one of us to go over and... get with an interviewer and say, "well, that was ok, what you were doing was helping the flow of the interview and that sort of thing, but when they engage in talking about their dog, you know you really should have worked to disengage from that conversation and move on." We take it very much on a case to case basis. We don't have hard and fast rules about it, we don't tell people that if they laugh they're fired; likewise, we certainly discourage... them getting very friendly with people. (Center E)

In the four survey centers where the first author observed interviewers at work, she found wide variation in the definition of *professionalism*. In two centers, no emotional response was allowed from interviewers whatsoever. Supervisors corrected an interviewer for laughing at one center; and at the other, supervisors who heard interviewer laughter when monitoring would identify it as "nonstandard feedback" on evaluation forms. At a third center, however, laughter and casual banter were not sanctioned and were, accordingly, audible parts of interviews. Although interviewers did not share opinions or personal information, they did relate to the respondents in a friendly, even personal manner. At a fourth center, although the interviewers were more formal, some casual banter and discreet chuckles were noticeable. Intrigued, Viterna asked one of the interviewers if, for example, they were allowed to laugh at a respondent's joke. He replied, "Oh, yeah. You'd have to, no matter how lame the joke was." While he went on to point out that rapport-building mechanisms should never influence an

Table 16.5 Levels of Professional Objectivity

Professional Objectivity Guidelines	Centers												
	1	2	3	4	5	6	7	8	9	10	11	12	Total
Survey center instructs interviewers to be flexible, friendly, warm, and conversational when asking questions.	X*		X					X*	X	X	X	X	−7
Survey center instructs interviewers to be neutral, serious, and professional at all times.		X		X	X	X	X						5
Standardization score	−1	+1	−1	+1	+1	+1	+1	−1	−1	−1	−1	−1	−2

interviewer's remarks or alter the wording of the questions, he and other interviewers had learned that building rapport was more important than being formal and objective at all times.

To operationalize professional objectivity as an element of standardization, the wording in the professionalism section in each manual was analyzed. If a manual made reference to being serious, professional, speaking in a "matter of fact way," or, in general, keeping respondents "focused" or on "track," then a score of +1 was assigned. If the manual did not mention the issue of seriousness, professional distance, or focus, and instead emphasized warmth, friendliness, "communicating sympathy" or "building a friendly, trusting relationship," then we assigned a score of −1. Centers that did not discuss either issue were assigned a −1 as well, since in the absence of specific instructions for neutrality, we assumed that interviewers are permitted to be as engaging as possible with the respondent. See Table 16.5.

Probing

Our discussion has centered on how interviewers ask questions and respond to answers in a standardized fashion. Now we turn our attention to situations where the respondent breaches standardization by not answering a question in a way assumed by standardization. For example, a respondent may answer a question by saying "I don't know" or may give a response that does not fit into a designated response category. Given the likelihood of unpredictable situations, how are interviewers to obtain a "good" answer while still remaining standardized? Survey researchers have developed a well-known but complex procedure called *nondirective probing* to deal with such situations. In theory, nondirective probing allows interviewers to standardize the procedure for achieving a good answer by limiting and specifying the type of exchange that may take place, even if the actual exchange itself varies from situation to situation.

Probing Generally, Probing Closed Questions, Probing Open Questions

For purposes of clarity, we classify all probes into three categories: General probes, probes for closed questions, or probes for open questions. First, general probes are ones that can be used any time respondents express uncertainty and withhold an answer. For instance, if a respondent answers a question with "I don't know," the interviewer may try to elicit a specific answer by allowing an expectant pause, by saying "Generally,..." and repeating the question, or by assuring the respondent that "there are no right or wrong answers, we are only interested in your opinion." Furthermore, if a respondent asks for clarification of the meaning of the question, the interviewer is not allowed to give personal definitions, but

Table 16.6 Frequency of Suggested "General Probes"

Recommended General Probes from Survey Center Interviewer Training Manuals	Number of Manuals Advocating Use of This Probe
Repeat the question/repeat the answer categories.	10
Whatever it means to you./Whatever ________ means to you. Whatever you think of as ________. Anything you would call ________. Whatever interpretation seems best to you ________. It's important that the question be answered as best you can in terms of the way it's stated.	7
An expectant pause.	6
Generally, what would you say? Generally/Overall/For people in general/On average/On the whole/Usually/ Taking everything into consideration/In the country as a whole/ In most cases/ Most of the time/Generally speaking... (then repeat the question or the response categories).	6
In your opinion... (repeat question or response categories). Of course no one knows for sure (but we are interested in people's opinions). Of course there are no right or wrong answers. There are no right or wrong answers, just your ideas on it. We're just interested in what you think. We are just trying to get people's ideas on this.	5
What do you think? What do you expect?	2
I see. I understand, but I need to know... (repeat question).	2
We all hope, but...	1
Yes, but...	1

is expected to prompt the respondent into giving a codable response by saying, "Whatever interpretation seems best to you."

As with feedback, the same probes were often recommended by different survey centers. Table 16.6 lists all phrases found in the 12 interviewer training manuals that could be classified as general probes, along with the number of manuals that advocate use of each particular probe.

The second category of probes includes those that correspond to closed questions, which are designed to have a limited set of responses from which respondents must choose their answer. The most common probes for closed questions work to help the respondent fit her answers into one of the predesignated response categories. Take, for example, the hypothetical question, "How often do you typically brush your teeth: less than once a day, once a day, twice a day, three times a day, or more than three times a day?" If the respondent answers, "two or three times a day," the interviewer would then probe with "Would it be closer to two or three?" before coding the response. Table 16.7 lists all probes for close-ended questions as consolidated from the 12 interviewing manuals in this study. A high level of consistency of probes across centers is evident.

The third and final category of probes involves open questions, which ask the respondent to answer a question without predesignated answer categories. To obtain as much information as possible from the respondent, interviewers are often instructed to follow an answer to an open question with a probe such as "tell me more" or "why do you feel that way?" In addition to asking for more information, probes for open questions are often designed to obtain sufficient information for accurate coding. "Could you be more specific?" is a common probe for encouraging respondents to clarify their answers. Probes for open questions are summarized in Table 16.8.

Table 16.7 Frequency of Suggested Closed-Question Probes

Recommended Closed-Question Probes from Survey Center Interviewer Training Manuals	Number of Manuals Advocating Use of This Probe
Which would be closer (to the way you feel)? Which is closer? Which comes closest? What category comes closest to your feeling? Which figure do you think comes closest? Would it be closer to _______ or _______? I need to fit your answer into one of our categories. Please choose the one that's closest to the way you feel.	9
What is your best estimate/guess? What would be your best estimate?	2
Could you give me your answer in terms of (repeat response categories). We need a specific (number, unit, etc).	2

Table 16.8 Frequency of Suggested Open-Question Probes

Recommended Open-Question Probes from Survey Center Interviewer Training Manuals	Number of Manuals Advocating Use of This Probe
Are there any other reasons you feel that way? Do you have any other reasons for feeling as you do? (Are there) Any other reasons? (Is there) Anything else? What else (can you think of)? What others? Thinking about ________ a bit more, what else is there about it that comes to mind?	9
Please tell me more. Would/could you tell me (a little) more (about that)? Would/could you tell me more about your thinking on that? Could you give me more information about your thought on ________? I'd like to know more about your thinking about that.	7
How do you mean (that)? What do you mean (by that)? What/How do you mean exactly?	7
Could you be (a little) more specific (about that)? Please be more specific.	7
An expectant pause.	6
Repeat what the respondent has said.	4
What do you have in mind? Will/would you tell me what you have in mind.	4
Why do you say you feel like that? Why do you feel that way? We'd like to know how you feel.	4
I see. I understand, but I need to know... (repeat question).	2
What do you think causes that? Why do you think that is so?	2
I don't understand what you mean.	1
Ok/yes/uh-huh/that's interesting... (followed by expectant pause).	1

Overall, there are some widely accepted probes. That is, at least three-quarters of the centers agree that general probing should be done by repeating the question or answer categories, that probing for closed questions should be done with a question that asks for an answer that is "closest" to a predetermined category, and that probing for open questions should be done by asking for "other reasons" or "what else" in regard to an answer.

Beyond this agreement on probes and the need to implement nondirective probing—the importance of which is mentioned in 11 of the 12 centers in this study—training about probing *practices* differs greatly from center to center. Three measures of variation in probing practices demonstrate this lack of uniformity: the degree of specificity in regard to probing, the permissible quantity of probing, and the interviewer's autonomy in deciding whether or not to repeat all response categories.

Specificity of Probing

Deciding whether or not to probe the answer given by the respondent is the first issue interviewers face in the probing process. It is very difficult for survey center staff to design guidelines specific enough to cover all situations where probing might be necessary. Thus, we found very little information in the training manuals explaining what constitutes a complete response and what requires a probe. Instead, learning when to probe an answer appears to be an on-the-job task. Take, for example, a question such as, "Would you consider your professors helpful, neither helpful nor unhelpful, or not at all helpful," and imagine the respondent replied, "Oh, they are just super!" In one variation (at two of our centers), interviewers would be instructed simply to code "helpful" and proceed with the questionnaire. It would be taken as obvious that the respondent would choose this option, even if he did not use the exact wording of the response categories, and repeating the response options would only slow down the interview and irritate the respondent. In a second variation (one survey center), the interviewer would be instructed to say, "so that would be helpful?" and if the respondent answered yes, the interviewer would then code "helpful" and proceed. With this probe, the interviewer acknowledges the answer given by the respondent, and still gives a quick reminder that the respondent should answer with one of the predesignated responses.[14] In a third variation (one center), however, the interviewer would be required to repeat the categories: "So would you say helpful, neither helpful or unhelpful, or not at all helpful." In other words, interviewers at this center could not code "helpful" until re-reading all response categories and eliciting the respondent's unaided choice of one. Accordingly, even when an interviewing procedure such as probing is standardized with an instruction to probe inadequate answers, the local interpretation of such an instruction may implicate strategies that are not uniform from center to center.

The specificity of probing instructions differs in the manuals as well. Although we have organized probing phrases according to question type for the purposes of this chapter, not all survey centers provide the same or similar categorizations for their interviewers. In some survey centers, a short list of three or four probes is simply accompanied by directions to use the probes if the respondent provides an answer that is vague, unclear, or incomplete. The manual does not relate these probes to question type, nor does it limit interviewers to using only probes in the interviewing manual. Given the lack of direction in deciding which probe to use, these centers are scored a -1 on the continuum; interviewers are given much autonomy in determining how to probe a given situation.

Table 16.9 Probing Specificity

	Centers												
Probing Specificity	1	2	3	4	5	6	7	8	9	10	11	12	Total
Training manual does not categorize probes by use OR does not discuss probing at all.	X			X			X	X*	X	X	X		−7
Training manual explicitly dictates what are acceptable probes and which probes should be used for which situation.		X	X		X	X						X	5
Standardization score	−1	+1	+1	−1	+1	+1	−1	−1	−1	−1	−1	+1	−2

This is in contrast to survey centers that provide interviewers with a list of specific probing options organized by the type of question for which each probe is designed. In these centers, when interviewers believe a probe is necessary, they must select a probe from a list that will meet the needs of the particular situation and then read that probing phrase verbatim.[15] These centers are assigned a score of +1 because they establish specific guidelines regarding which exact phrase should be used in each type of situation. The results of this coding can be found in Table 16.9. As shown by the summary score of −2, there is strong variation across centers with respect to the specificity of probing guidelines. While five centers give very specific probing instructions, seven centers do not, and permit the interviewer to decide how best to handle inadequate answers.

How Many Times to Probe

A second measure of probing standardization concerns how many times interviewers are instructed to probe a respondent's answer before giving up and moving on with the interview. Four of the centers in this study give strict guidelines. Three state that the interviewer should only probe once so "it won't seem like you're badgering the R," and the fourth instructs its interviewers to probe twice. These centers all receive a score of +1 because the number of probes is prespecified.

On the other hand, eight centers allow interviewers significant autonomy in determining how many times they probe a respondent's answer. Two of these centers suggest probing only once but leave it to the discretion of the interviewer to determine if a second probe is a viable option without damaging rapport. In one of these, a trainer stated, "You must probe once and may probe twice, but if the respondent is obviously not going to answer, just code a refusal or a don't know and move on... it's not worth getting hung up on" (Center Y). Two other centers instruct interviewers to probe until they have received a complete answer or until the respondent has become frustrated. The remaining four centers in this category do not mention how often interviewers should probe a respondent's answer, and

Table 16.10 How Many Times to Probe a Respondent's Answer

	Centers												
Probing Frequency	1	2	3	4	5	6	7	8	9	10	11	12	Total
Training manual allows interviewer to decide how many probes would be beneficial given the specifics of the situation.		X	X*	X	X	X	X*	X*	X*				−8
Training manual explicitly specifies how many times an interviewer may probe a respondent's answer.	X									X	X	X	4
Standardization score	+1	−1	−1	−1	−1	−1	−1	−1	−1	+1	+1	+1	−4

are designated with an asterisk to indicate that the interviewer's autonomy comes from omission of rules rather than explicit permission to be autonomous. The results of this classification are found in Table 16.10. The summary score of −4 indicates some agreement among survey centers in allowing interviewers limited autonomy in deciding how often to probe an incomplete answer.

Repeating Response Categories

A final means of evaluating the standardization of probing concerns closed questions or those with predetermined response options. If a respondent does not choose an answer that matches the answer categories, then the interviewer must probe to find out which alternative is closest to the way the respondent feels. For example, if a question asked the respondent to rate her work environment as very good, good, fair, or poor, and the response was, "not very good," the interviewer would need to probe to find out which category is closest to the respondent's view. In such circumstances, Fowler and Mangione (1990:40) state that it is important to re-read all possible responses to the closed question because respondents "respond to the number of categories and the position of a category on a scale, as well as to the words, when classifying themselves. A truncated version of a set of responses is not the same stimulus, and it will affect the answers."[16] This line of thought is reflected in four survey center manuals where interviewers are instructed always to repeat the entire set of response categories, even if they feel certain responses have already been eliminated. These centers receive a score of +1 on the standardization continuum.

Four other centers instruct interviewers to re-read only those response options that have not already been clearly eliminated by the respondent. To demonstrate, in the above example about work environment, when the respondent answers "not very good," the interviewer would probe, "so would you say fair or poor?" This results in less overall standardization because interviewers have discretion to determine whether or not a response is codable and which responses have or

Table 16.11 Repeating Response Categories

Instructions for Repeating Response Categories	Centers												
	1	2	3	4	5	6	7	8	9	10	11	12	Total
The interviewer is allowed to determine which response categories need to be repeated to determine the correct answer when probing a close-ended question.	X		X		X*	X		X*	X	X		X*	−8
Training manual states that the entire set of response categories must be re-read when probing a close-ended question.		X		X			X				X		4
Standardization score	−1	+1	−1	+1	−1	−1	+1	−1	−1	−1	+1	−1	−4

have not been eliminated. These survey centers receive a score of −1 on the standardization continuum.

Table 16.11 shows that four centers did not take a position on how many response categories must be repeated when probing close-ended questions, therefore also leaving interviewers to decide the number of response categories to be repeated. Two-thirds of the centers in this study allow their interviewers to decide how many response categories must be repeated to code an appropriate response to a closed question. While survey centers lean toward autonomy in repetition, there still is variation, with some centers regulating the matter. Overall, centers agree that probing should be nondirective and prescribe common phrases to achieve this. However, they differ in the degree of specificity in using probes, how many times to probe, and whether and which response categories should be repeated.

Monitoring Procedures

The centralization of interviewing, made possible by CATI and other telephonic surveying, allows supervisors to monitor interviewers frequently and give immediate feedback regarding their technical performance. This is often cited as one of the greatest benefits of telephone survey interviewing over on-location, face-to-face interviews. In a room near to or part of the space where interviewers work, supervisors use headphones to listen to the interviewer and respondent, and may also watch, on their own computers, the interviewer's computer screen as she pages through the instrument.

According to Fowler and Mangione (1990:133), frequent and detailed supervision cannot take the place of good interviewer training. If an interviewer is inadequately trained before beginning her job, then there is little chance for any real improvement through monitoring and supervision. Still, the monitoring of interviewers is essential for maintaining high interview quality. The more experienced interviewers become, the more they rely on their own judgment about how to interview. Otherwise put, interviewing maturity leads to lower standardization levels (Bradburn, Sudman, and Associates, 1979; Fowler and Mangione, 1990).

Monitoring interviewing technique is only one way in which interviewers are evaluated in survey centers. Other ways include assessing efficiency (e.g., number of interviews per hour), cooperation (e.g., number of completions versus number of people to whom the interviewer spoke), and individual reliability (e.g., attitude, dependability, timeliness, and attendance). Ten of the 12 centers in this study have reported that interviewers are consistently evaluated on all these measures. Despite this claim, however, there is much variation. The frequency of online monitoring and whether or not the evaluation forms focus on standardization provide two key dimensions on which to measure uniformity (or lack thereof) in how evaluation is carried out among the centers.

For monitoring to be effective in encouraging standardized interviewing techniques, it should be frequent and without exact scheduling (so that interviewers know only *that* they will be monitored but not *when*). For example, at one survey center, monitoring is a part of every interview shift, and interviewers receive immediate feedback at least once or more on a shift about their interviewing technique and the way they handle specific problems. Knowing their words will be monitored at least once and possibly more on any shift, interviewers appeared disciplined in their interviewing. At another center, things are different. Throughout the three evenings the first author spent at this center, the monitoring room was never occupied and the lights were off, clearly signaling to interviewers that they did not need to be concerned about being observed and checked. Accordingly, they appeared less disciplined in their behavior during interviews, providing nonneutral feedback, spontaneous definitions, and casual banter in interaction with their respondents.

Frequency of Monitoring

Using information in the manuals and from a follow-up questionnaire, survey centers were classified according to the frequency of online monitoring. All centers that checked each interviewer at least once a shift received a +1 score. Centers that monitored less than once a shift per interviewer received a score of −1. Centers that did not report monitoring procedures in their manuals are indicated with an ND (no data) and are not assigned a value on the continuum, since we cannot infer interviewer autonomy without more knowledge of the monitoring practices in these centers. Table 16.12 demonstrates that most centers (7 out of 10 who reported) do not require once-per-shift monitoring of telephone interviews.

Of the 12 centers in this study, 9 reported that interviewers with more experience were monitored less frequently than newer interviewers. (No information was available for the other 3 centers). These centers all indicated that the minimum monitoring guidelines (i.e., once a shift, once a month, etc.) applied to all interviewers, including those with much experience, but new interviewers were given extra monitoring sessions and extra feedback whenever time permitted. Sometimes this practice was official and printed in the center's manual; other times it

Table 16.12 Frequency of On-line Monitoring

Frequency of On-line Monitoring	Centers												
	1	2	3	4	5	6	7	8	9	10	11	12	Total
Survey center does not require interviewers to be monitored at least once a shift.				X		X	X	X	X	X	X		−7
Survey center requires that each interviewer is monitored at least once a shift.	X	X			X								3
Survey center did not report monitoring frequency.			ND									ND	2(*)
Standardization score	+1	+1	ND	−1	+1	−1	−1	−1	−1	−1	−1	ND	−4

was informally stated by survey center supervisors. Either way, the practice is at odds with Fowler and Mangione's (1990) argument that more monitoring cannot make up for poorly trained interviewers, that more experienced interviewers tend to relax standardization guidelines more often than less experienced interviewers, and that supervision of experienced interviewers should not be relaxed.

Survey staff at the centers would probably agree with Fowler and Mangione (1990) that supervision does not take the place of solid interviewer training. Instead, extra monitoring sessions for newer interviewers may help them develop the *judgment* necessary to deal with situations that fall outside of what interview protocols cover. Furthermore, survey centers often regard more experienced interviewers as capable of such judgment:

> We're encouraging people, especially for experienced folks, to give it your best judgment, give it a try, and just take a note, and at the end of the interview you can make an outline, a thumbnail sketch or whatever of what happened and usually that's fine. Cause usually a lot of times when we get questions people will know the answers; they're just looking for a little reinforcement. (Center Y)

Content of Monitoring

Another approach for comparing the standardization of monitoring practices involves analyzing the content of the monitoring forms used by supervisors to evaluate interviews. Seven of the 12 centers in this study provided copies of their monitoring forms, which vary greatly in size and emphasis.[17] For the purposes of this study, we considered monitoring to encourage standardization when all the core factors highlighted by Fowler and Mangione (1990)—pace, feedback, probing, professional objectivity, and proper question reading[18]—were mentioned. Of these five factors, proper question reading and probing are included on all 7 survey centers' monitoring forms. Five centers include a component for professional behavior, 4 include a component for proper pace, and only 2 include

Table 16.13 Inclusion of Core Evaluative Factors on Monitoring Forms

Inclusion of Core Evaluative Factors	Centers												
	1	2	3	4	5	6	7	8	9	10	11	12	Total
Survey center is missing one or more of the five core evaluative factors on the monitoring form.				X	X	X	X			X	X		−6
Survey center includes all five core evaluative factors on the monitoring form.		X											1
Monitoring form is not available for survey center.	ND		ND					ND	ND			ND	5(*)
Standardization score	ND	+1	ND	−1	−1	−1	−1	ND	ND	−1	−1	ND	−5

a component for proper feedback. Survey centers that include all 5 of the factors on their monitoring forms are assigned a score of +1 on the standardization continuum. Those centers that omit one or more of the 5 core issues are assigned a score of −1.[19] The 5 centers that did not provide monitoring forms for analysis are again indicated by an ND signifying that no data was available for analysis. As shown by the consistency score of −5 in Table 16.13, the pattern is not to include all core standardization issues on the monitoring form.

Informally, supervisors recognize that fully standardized performance (as framed by their monitoring forms) is not possible:

> Usually I stress that it [the monitoring form] is based on a model interview that's perfect. The form is, at least. You know, there is an expected amount of error. I mean, it's not just the interviewer who's necessarily to blame for not doing a perfect session, it's also the respondent having difficulty with questions and having to steer them. A perfect [standardized interview] can be extremely difficult if you have to do some persuasion... they're refusing, things like that. (Center Y)

Accordingly, survey center staff recognize that at times complete standardization of interviewer behavior is impossible and that interviewers must use their judgment to determine how best to handle the interview.

In sum, monitoring interviewer technique is important for maintaining standardization within a survey enterprise, but survey centers vary greatly in how often they monitor their interviewers and whether or not they hold their interviewers responsible for key standardization practices such as appropriate feedback or interview pace. Furthermore, interviews with survey staff indicate that in some centers, supervisors view monitoring as an opportunity to help interviewers develop judgment, not just to reprimand them for improper technique. This shows a move toward developing interviewer autonomy on the job, even when such autonomy may not be evident by reviewing a center's formal guidelines for standardization.

DISCUSSION

Summing positive points for each procedure that favored standardization and negative points for each procedure that favored interviewer autonomy, it is possible to observe patterns intrinsic to specific survey centers and across them (see Table 16.14). We highlight four important tendencies exhibited in Table 16.14.

First, the additive picture of the standardization continuum for individual centers in Table 16.14 shows the extent to which they favor interviewer autonomy over standardization. Summing down columns, a score in the minus range indicates that on at least half the measures presented in this chapter, the center leans more toward autonomy than standardization. Ten out of 12 centers have this tendency. Only 2 centers (centers 2 and 6) are in the other direction and are strictly standardized on at least half of our measures.

Second, while the autonomy permitted interviewers is prevalent, it is also relatively constrained. As Figure 16.1 shows, of the 10 centers that fall to the left of the continuum's center, 5 are in the −1 through −3 range and are close to that center. Another set of 3 centers are at -5 on the continuum. This indicates that 8 centers are relatively standardized, rather than relatively autonomous, on at least some measures. Similarly, one center (6) is only at +1 on the continuum, which means that it is close to being balanced in terms of autonomy and standardization. In short, 9 out of 12 centers vary in their own orientations to standardization and autonomy. Only 3 centers are consistently one way or the other — 2 centers (8 and 9) strongly favor autonomy and one center (2) implements strict standardization.

Table 16.14 Total Standardization Scores by Center

Standardization Components	Centers												
	1	2	3	4	5	6	7	8	9	10	11	12	Total
1.1 Pace	−	+	+	−*	+	+	+	−*	−	−	−	−*	−2
2.1 Feedback: Frequency	−	+	−	−	−	+	−*	−*	−	−	−	−	−8
2.2 Feedback: Purpose	+	+	−	−	−	+	−*	−*	−	+	−	−	−4
3.1 Professional objectivity	−*	+	−	+	+	+	+	−*	−	−	−	−	−2
4.1 Probes: Specificity	−	+	+	−	+	+	−	−*	−	−	−	+	−2
4.2 Probes: Repetition	+	−	−*	−	−	−	−*	−*	−*	+	+	+	−4
4.3 Probes: Response categories	−	+	−	+	−*	−	+	−*	−*	−	+	−*	−4
5.1 Monitoring: Frequency	+	+	ND	−	+	−	−	−	−	−	−	ND	−4
5.2 Monitoring: Core factors	ND	+	ND	−	−	−	−	ND	ND	−	−	ND	−5
Total	−2	+7	−3	−5	−1	+1	−3	−8	−8	−5	−5	−3	

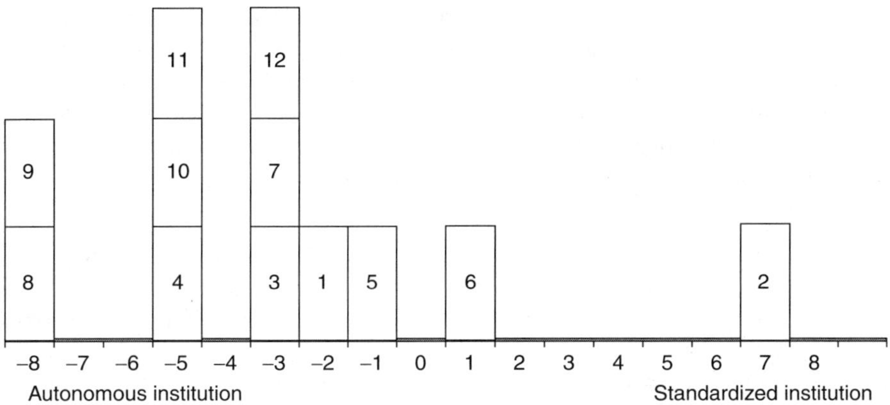

Figure 16.1 The standardization continuum. (Numbers in boxes identify institution.)

Third, this analysis demonstrates the variation across centers in how autonomy is granted. Recall that centers that neglected to mention a particular standardization measure were assigned a negative score for permitting interviewer autonomy, but marked with an asterisk to indicate that the autonomy was permitted implicitly (by lack of instruction) rather than explicitly (by official documentation). Table 16.15 provides a summary of the number of omitted guidelines related to those considered here—pacing, feedback, professional objectivity, probing, and monitoring—by center.

Interestingly, the two centers furthest to the left of the continuum—centers 8 and 9—are very different in their guideline specificity. It is logical to conclude that autonomy at center 8 may be attributed to underspecification or even under-management, while autonomy at center 9 more likely reflects a thoughtful position on the part of the survey center staff to allow their interviewers more individual discretion.

Fourth, Table 16.14 allows a look at standardization across standardization components and centers. By summing scores horizontally for each component, it is apparent that each one is negative. This indicates that not one of the practices (pace, feedback, etc.) is strictly standardized in even half the survey centers evaluated. That is, the majority of our centers favor interviewer autonomy on each component of standardization. Frequency of feedback is the component that is most frequently left to the interviewer's autonomous discretion, but feedback

Table 16.15 Omitted Guidelines across Centers

	Centers											
	1	2	3	4	5	6	7	8	9	10	11	12
Number of standardization guidelines NOT mentioned in the center's manual	1	0	1	1	1	0	3	7	2	0	0	2

purpose, probing repetition, and monitoring frequency scores are also strongly negative. This indicates that, regardless of the measure used, survey centers on the whole simply do not uniformly adhere to rigorous forms of standardization. Nevertheless, several summations move closer to the center of the continuum. These include pace, professional objectivity, and probing specificity. Accordingly, Table 16.14 also indicates the variation across centers on these measures.

CONCLUSION

Our exploratory investigation suggests that survey centers are internally variable in their approach to resolving the standardization and autonomy dilemma, and that the overall tendency across centers is toward autonomy rather than standardization. This raises two key questions: Why does there appear to be a tendency toward interviewer autonomy?[20] What does this tendency say about the quality of data obtained from different survey centers?

First, why are standardization principles administered the way they are within the survey enterprise? With a rich literature on how to standardize survey interviewing, why are less rigid forms of these techniques used in practice? From talking to center personnel, one reason may be that simplification of procedure eases the training and monitoring of interviewers. A more relaxed approach to survey interviewing may also add to an interviewer's on-the-job comfort, and thus lessen the pervasive problem of job turnover that plagues many survey centers. Survey centers with a high number of omitted guidelines have these concerns.

However, often the standardization components in this analysis are not simply ignored but rather carefully implemented to permit some interviewer autonomy. For example, interviewers are instructed to speak in normal conversational tones instead of a two word-per-second pace. And interviewers are told to use feedback to *establish rapport* rather than to *train* the respondent. Interviewers are allowed to probe as often as they deem necessary. And an interviewer's judgment is often given weight over technique when being monitored by supervisors. In short, survey center staff may be deliberate in how they institute principles of standardization in the direction of more interviewer autonomy. Centers 4, 10, and 11 are all located firmly on the autonomous side of the continuum, yet each of these center's manuals specifically addressed each component analyzed in this chapter.

Consistent with a theme in this volume, accordingly, an understanding of *tacit knowledge* may help explain why survey administrators permit increased autonomy for interviewers. It appears that some survey centers, rather than simply ignoring standardization or letting it slide, recognize the need for interviewers to exercise their commonsense as they interact with respondents. In other words, those survey centers that recognize broader areas of autonomy may recognize the contingencies of the interview process and believe that quality of data is improved when interviewers are taught to handle these contingencies with some degree of flexibility (Groves, 1989:404).[21]

Our speculations about reasons for relatively more autonomy should not imply an endorsement of any policy. Instead, we want to make sociological sense of patterns that appear in our data. A matter related to the recognition of tacit knowledge is what sociologists have found in regard to reproducibility and replicability in natural science. That is, studies of science and technology, including those of Collins (1985:57–58) and Shapin and Schaffer (1985:281), refer to the *capriciousness* of knowledge flow regarding methodology and have documented how difficult it is for practices and procedures to "travel" from one lab to another. We could hardly expect social science to be any less capricious in the flow of methodological knowledge than natural science. According to Converse's (1987) history of survey research in the United States, rules for standardization have proliferated to countless academic and polling organizations from a relatively few major centers, such as the National Opinion Research Center in Chicago, the Survey Research Center in Ann Arbor, and the Bureau for Applied Social Research in New York. While these or other survey centers may write extensive formulas for standardized interviewing, in the end it takes people to implement the formula, and they have to decide what counts as doing the "same" thing the formula prescribes in a place distanced from where the formula originates (Lynch, 1993). Such decision-making is overwhelmingly a situated activity that happens according to established routines in local cultures and in relation to circumstances of a potentially infinite variety that help condition how any survey center works to collect its data. Indeed, in the survey literature, this matter is discussed in relation to the influence on survey data of possibly distinctive practices involved in training and supervision at different survey organizations, that is, *house effects* (Hill, 1991; Smith, 1978, 1982; Turner, 1982).

Finally, to address our second question regarding quality of data, if researchers argue that greater autonomy may lead to improved data, and our findings indicate that survey administrators are endorsing autonomy, our study also suggests that the self-same survey, with identically scripted CATI questions, could be administered very differently depending on the survey center where it is placed. Interviewers could read questions at dissimilar paces, use disparate forms of feedback after respondents' answers, be more neutral and professional or more warm and friendly in demeanor, engage in distinctive probing practices, and be monitored more or less intensively. These are interactional matters and require investigations that treat them as such. Presently, despite Schaeffer's (1991; this volume, Chapter 4) call for "closer study of how conversational practices are enacted within the prescribed structure of the interview to create data," there is little research that addresses whether these specific interactional differences matter to the quality of data the survey collects.[22] Given the importance of reliability in measurement, research is needed to determine whether or not the same instrument at centers with disparate orientations to interviewer standardization and autonomy generate findings that are in substantial agreement. In the past, it has been difficult to discriminate what may be house effects from other influences, including of question order and of historical events occurring between different administrations of the same survey (Smith, 1978; Turner, 1982). Perhaps the time has come for more systematic efforts to determine how uniform measurement is from survey center to survey center and how consistent the results are from the use of identically written instruments in distinct organizations.

NOTES

1. The five survey centers in this group were selected because of convenience and proximity to Viterna's research base in Bloomington, Indiana. Given these constraints, the resulting mix among centers was quite diverse. The five centers had been in operation from as few as 2 years to as many as 57 years. Most of the surveys conducted by these centers were university affiliated, but many centers also conducted studies for local, state, and national government organizations, not-for-profit organizations, and even several marketing organizations. Resources varied drastically from center to center, from survey organizations with several hundred interviewers in multiple interviewing locations to a single cluster of rooms in a small building and 20 or 30 interviewers on the payroll.

2. We wrote to a dozen survey centers and used the manuals from the seven that responded.

3. For example, of the five centers the first author visited, two made little mention of feedback in their manuals and did not mention appropriate feedback at all throughout their training sessions. The remaining three centers all gave extensive coverage to feedback issues in both their manuals and during actual interviewer training. Likewise, four of the five centers specifically discussed when and how to probe in their manuals and provided lists of acceptable probes. These four centers also discussed probing extensively in their training programs. The fifth center only briefly mentioned probing in the manuals and provided no list of acceptable probes. During training at this center, trainees were told only that they were never to help the respondent understand a question by rephrasing or defining unclear parts. They never received information, verbally or in writing, discussing different situations that may require probing or what probes would be effective for various situations.

4. Fowler and Mangione (1990) is the most recent comprehensive source documenting standardized survey interviewing techniques. [Previous publications about interviewing include Bradburn, Sudman, and Associates (1979) and Cannell, Miller, and Oksenberg (1981).] Furthermore, Fowler and Mangione (1990) is cited in other recent sources on standardized survey interviewing (see, e.g., Lavrakas, 1993) giving it weight as an authoritative source for interviewing technique.

5. The substantive part of the interview contains the instrument's questions for gathering data. Specialized interviewing procedures are required in the substantive portion of the interview to ensure standardization, in contrast to the front end, or introduction, of the interview, where interviewers are characteristically allowed much more autonomy, and where the autonomy may improve interviewers' effectiveness at persuasion. See Houtkoop and van den Bergh (this volume, Chapter 9).

6. We identify survey centers with letters in our quotations and with numbers in our tables. The purpose of different identifiers is to ensure that our centers and their personnel remain anonymous.

7. We recognize that there is a fundamental difference between interviewer autonomy that is explicitly granted in a training program and interviewer autonomy that is implicitly given through the absence of a published procedure. One reflects a thoughtful position on behalf of survey center staff to permit autonomy, while the other may simply be a reflection of more relaxed management policy. Whether such distinctions matter substantively for data collection is not something we are able to address, and we simply collapse the explicit and implicit forms of orientation to pacing as both indicating autonomy.

8. A high absolute number indicates high consistency, with a high positive number indicating consistency in the direction of standardization and a high negative number indicating consistency in the direction of autonomy. A positive or negative total near zero indicates low consistency across centers.

9. Although personnel at several centers mentioned that their lists of "acceptable" feedback phrases may have been derived from practices used in the Center for Survey Research at the University of Michigan, no one was certain about when or where the phrases were initially created.

10. This is a total of seven centers, as three centers recommend "yes," three others recommend "okay," and a seventh center recommends both "yes" and "okay."

11. However, see the discussion of feedback in this volume, Chapter 1.

12. The contradiction between being "serious" (for purposes of standardization) and being "pleasant" (for purposes of rapport) often arises when respondents engage in displays of affect. For example, respondents may make a small joke or introduce laughter that operates to invite reciprocal laughter (Lavin and Maynard, this volume, Chapter 15). Or, a respondent may reveal a sad detail about her life circumstances, as when being asked her marital status, and she replies, "I'm recently widowed." If an interviewer has been instructed to show no emotion or reaction to a respondent's answer, no matter how affect-laden it may be, the strain can be intense.

13. See experimental research on interviewer style in Dijkstra (1987) and Dijkstra and van der Zouwen (1988).

14. In Chapter 1, this volume, see the discussion of predetermined clarifications, verification probes, tuning, and zeroing in as devices for nondirective probing in contexts where respondents have provided approximations that suggest, but do not exactly fit, answer categories.

15. For example, interviewers in Center O have a list of approved probes hanging in their cubicles. The list is organized into four major categories (closed probes, open probes, don't know probes, and refuse probes). The closed probes are further subdivided into probes for nonresponse, probes for two responses, probes for numbers, and probes for qualified answers. The open probes section is subdivided into probes for "anything else?," probes for limiting a response to one answer, and probes asking for more specificity.

16. As mentioned in Chapter 1 (this volume), research does confirm that the meaning of any individual category depends on the entire set of categories, suggesting the importance of re-reading all response options when probing (e.g., Schaeffer and Charng, 1991; Smit, Dijkstra, and van der Zouwen, 1997).

17. Interviewers are evaluated with as many as 22 different components of interviewing technique or as few as 5. Two of the forms in this collection are organized by the number of the question in the actual interview, allowing supervisors to make comments and evaluate procedure on a question-by-question basis. The other five monitoring forms are organized by topic; supervisors give an overall impression of pace or probing without necessarily evaluating each individual question asked by the interviewer to the respondent. A comparison of monitoring forms also finds great variety in which components of interviewing are included for evaluation. Some monitoring forms ask about the overall flow of the interview, if a pleasant tone was used, if time was wasted between calls, or if the interviewer correctly answered the respondent's questions. Others stick strictly to the basics of question asking: reading verbatim and coding responses correctly.

18. Proper question reading includes reading questions verbatim, neither adding nor omitting words, and asking every question on the survey in its proper order. All centers in this study require all questions to be read verbatim in the order they appear.

19. Of the seven centers for which we have data, only one includes all five factors identified as key in this chapter. Three include four of the five factors, two include three of the five factors, and one only includes two of the five factors. Although we considered making the cutoff point for coding a center as standardized at four out of five, we hesitated to do so because of the lack of consensus on which factors were excluded. For example, two of the three centers with four out of five factors were missing any sort of evaluative framework for proper use of feedback, and the third was missing an analysis of professionalism. Since Fowler and Mangione (1990) give equal weight to each aspect, we decided that only centers that included all key issues would get positive points. If we were to recode the data so that centers with four out of five would receive positive instead of negative points, centers 5, 6, and 11 would all move two spaces further to the right on the continuum in Figure 16.1 (discussed below).

20. We refer to a "tendency" toward autonomy although there actually may be an "increase" in autonomy. Beatty (1995:151) observes that during the 1970s there were "attacks on rapport." Most prominent were definitional problems inherent in the term *rapport* and lack of evidence that rapport (however defined) improved interviewing productiveness. The result was that "research on controlling interviewer variation through standardization... had new opportunities to flourish" (Beatty, 1995:151). Furthermore, interviewing manuals, such as the University of Michigan Survey Research Center's *General Interviewing Techniques*, published in 1983, contained extended attention to standardizing the behavior of interviewers compared to earlier manuals. On the basis of such evidence, Beatty (1995:153, our emphasis) refers to "the *increase* in survey standardization." We speculate that the 1980s may have

been standardization's heyday, and that interviewing practice has subsequently moved away from the stricter versions of standardization, and this has increased interviewer autonomy. It is because there is no concrete evidence of this increase that we discuss a tendency toward autonomy.

21. Recent studies lend credence to the idea that increased autonomy might improve data quality in some respects. Suchman and Jordan (1990) argue that strict standardization could obtain reliability at the expense of validity, and Schober and Conrad (1997) conclude that allowing interviewers freedom to clarify may improve response accuracy. But Dykema, Lepkowski, and Blixt (1997) find that not reading the question as worded was associated with accuracy in only 1 out of 10 questions they examined. Even then, the association was in the opposite direction from what was predicted (i.e., the association was negative).

22. However, in regard to more general matters such as the "style" of the interviewer, see the discussion in this volume, Chapter 1, and, in particular the studies of Dijkstra (1987), Dijkstra and van der Zouwen (1988), and Schober and Conrad (1997).

REFERENCES

Beatty, P. 1995. "Understanding the Standardized/Non-Standardized Interviewing Controversy." *Journal of Official Statistics* 11: 147–160.

Bradburn, N. M., S. Sudman, and Associates. 1979. *Improving Interview Method and Questionnaire Design: Response Effects to Threatening Questions in Survey Research.* San Francisco: Jossey-Bass.

Cannell, C. F., L. Oksenberg, and J. M. Converse. 1977. *Experiments in Interviewing Techniques: Field Experiments in Health Reporting: 1971–1977.* Hyattsville, MD: NCHSR.

Charmaz, K. 1983. "The Grounded Theory Method: An Explication and Interpretation." Pp. 109–126 in R. M. Emerson (Ed.), *Contemporary Field Research: A Collection of Readings.* Prospect Heights, IL: Waveland.

Clayman, S. 1988. "Displaying Neutrality in Television News Interviews." *Social Problems* 35: 474–492.

Collins, H. M. 1985. *Changing Order: Replication and Induction in Scientific Practice.* Chicago: University of Chicago Press.

Converse, J. M. 1987. *Survey Research in the United States: Roots and Emergence 1890–1960.* Berkeley, CA: University of California Press.

Dijkstra, W. 1987. "Interviewing Style and Respondent Behavior: An Experimental Study of the Survey Interview." *Sociological Methods and Research* 16: 309–334.

Dijkstra, W., and J. van der Zouwen. 1988. "Types of Inadequate Interviewer Behavior in Survey Interviews." Pp. 24–35 in W. E. Saris and I. N. Gallhofer (Eds.), *Data Collection and Scaling.* New York: St. Martins.

Dykema, J., J. M. Lepkowksi, and S. Blixt. 1997. "The Effect of Interviewer and Respondent Behavior on Data Quality: Analysis of Interaction Coding in a Validation Study." Pp. 287–310 in L. Lyberg, P. Biemer, M. Collins, E. de Leeuw, C. Dippo,

N. Schwarz, and D. Trewin (Eds.), *Survey Measurement and Process Quality*. New York: Wiley-Interscience.

Fowler, F. J., and T. W. Mangione. 1990. *Standardized Survey Interviewing: Minimizing Interviewer-Related Error*. Newbury Park, CA: Sage.

Garfinkel, H. 1967. *Studies in Ethnomethodology*. Englewood Cliffs, NJ: Prentice-Hall.

Glaser, B. G., and A. L. Strauss. 1967. *The Discovery of Grounded Theory: Strategies for Qualitative Research*. Chicago: Aldine.

Goodwin, T. 1998. "Managing Participation in a Local Work Collectivity: Finding Space, Eavesdropping, and Standardization in the Survey Research Center." Master's thesis, Department of Sociology, Indiana University, Bloomington, IN.

Goudy, W. J., and H. R. Potter. 1975. "Interview Rapport: Demise of a Concept." *Public Opinion Quarterly* 39: 529–543.

Groves, R. M. 1989. *Survey Errors and Survey Costs*. New York: Wiley.

Heil, T. 1999. "The Interview Was Really Good Overall?: Advice-Giving and Advice-Reception in a Supervisory Feedback Session." Master's thesis, Department of Sociology, Indiana University, Bloomington, IN.

Heritage, J. 1984. *Garfinkel and Ethnomethodology*. New York: Polity.

Heritage, J., and D. Greatbatch. 1991. "On the Institutional Character of Institutional Talk: The Case of News Interviews." Pp. 93–137 in D. Boden and D. H. Zimmerman (Eds.), *Talk and Social Structure*. Berkeley: University of California Press.

Hill, D. H. 1991. "Interviewer, Respondent, and Regional Office Effects on Response Variance: A Statistical Decomposition." Pp. 463–486 in P. P. Biemer, R. M. Groves, L. E. Lyberg, N. A. Mathiowetz, and S. Sudman (Eds.), *Measurement Errors in Surveys*. New York: Wiley.

Hyman, H. 1954. *Interviewing in Social Research*. Chicago: University of Chicago Press.

Lavrakas, P. J. 1993. *Telephone Survey Methods: Sampling, Selection, and Supervision*. Newbury Park, CA: Sage.

Lazarsfeld, P. F. 1944. "The Controversy Over Detailed Interviews — An Offer for Negotiation." *Public Opinion Quarterly* 8: 38–60.

Lynch, M. 1993. *Scientific Practice and Ordinary Action: Ethnomethodology and Social Studies of Science*. Cambridge: Cambridge University Press.

Marlaire, C. L., and D. W. Maynard. 1990. "Standardized Testing as an Interactional Phenomenon." *Sociology of Education* 63: 83–101.

Marquis, K. H., C. F. Cannell, and A. Laurent. 1972. "Reporting Health Events in Household Interviews: Effects of Reinforcement, Question Length, and Reinterviews." In *Vital and Health Statistics*, Series 2 (45). Washington, DC: U.S. Government Printing Office.

Maynard, D. W., and C. L. Marlaire. 1992. "Good Reasons for Bad Testing Performance: The Interactional Substrate of Educational Exams." *Qualitative Sociology* 15: 177–202.

Maynard, D. W., and N. C. Schaeffer. 2000. "Toward a Sociology of Social Scientific Knowledge: Survey Research and Ethnomethodology's Asymmetric Alternates." *Social Studies of Science* 30: 264–312.

Mehan, H. 1979. *Learning Lessons*. Cambridge, MA: Harvard University Press.

O'Muircheartaigh, C. A. 1997. "Measurement Error in Surveys: A Historical Perspective." Pp. 1–28 in L. Lyberg, P. Biemer, M. Collins, E. de Leeuw, C. Dippo, N. Schwarz, and D. Trewin (Eds.), *Survey Measurement and Process Quality*. New York: Wiley-Interscience.

O'Muircheartaigh, C. A. 1999. "CASM: Successes, Failures, and Potential." Pp. 39–62 in M. G. Sirken, D. J. Herrmann, S. Schechter, N. Schwarz, J. M. Tanur, and R. Tourangeau (Eds.), *Cognition and Survey Research*. New York: Wiley.

Polanyi, M. 1958. *Personal Knowledge: Towards a Post-Critical Philosophy*. Chicago: University of Chicago Press.

Porter, T. M. 1995. *Trust in Numbers: The Pursuit of Objectivity in Science and Public Life*. Princeton, NJ: Princeton University Press.

Richardson, S. A., B. S. Dohrenwend, and D. Klein. 1965. *Interviewing: Its Forms and Functions*. New York: Basic Books.

Schaeffer, N. C. 1991. "Conversation with a Purpose — Or Conversation? Interaction in the Standardized Survey." Pp. 367–391 in P. P. Biemer, R. M. Groves, L. E. Lyberg, N. A. Mathiowetz, and S. Sudman (Eds.), *Measurement Errors in Surveys*. New York: Wiley.

Schaeffer, N. C., and H-W. Charng. 1991. "Two Experiments in Simplifying Response Categories: Intensity Ratings and Behavioral Frequencies." *Sociological Perspectives* 34: 165–182.

Schober, M. F., and F. G. Conrad. 1997. "Does Conversational Interviewing Reduce Survey Measurement Error?" *Public Opinion Quarterly* 61: 576–602.

Shapin, S., and S. Schaffer. 1985. *Leviathan and the Air Pump: Hobbes, Boyle, and the Experimental Life*. Princeton: Princeton University Press.

Smit, J. H., W. Dijkstra, and J. van der Zouwen. 1997. "Suggestive Interviewer Behaviour in Surveys: An Experimental Study." *Journal of Official Statistics* 13: 19–28.

Smith, T. W. 1978. "In Search of House Effects: A Comparison of Responses to Various Questions by Different Survey Organizations." *Public Opinion Quarterly* 42: 443–463.

Smith, T. W. 1982. "House Effects and the Reproducibility of Survey Measurements: A Comparison of the 1980 GSS and the 1980 American National Election Study." *Public Opinion Quarterly* 46: 54–68.

Suchman, L., and B. Jordan. 1990. "Interactional Troubles in Face-to-Face Survey Interviews." *Journal of the American Statistical Association* 85: 232–253.

Turner, C. F. 1982. "Why Do Surveys Disagree? Some Preliminary Hypotheses and Some Disagreeable Examples." Pp. 159–214 in C. Turner and E. Martin (Eds.), *Surveying Subjective Phenomena*. New York: Russell Sage.

Williams, D. 1942. "Basic Instructions for Interviewers." *Public Opinion Quarterly* 6: 634–641.

PART 4

Interaction and Survey Data Quality

CHAPTER 17

Transcribing, Coding, and Analyzing Verbal Interactions in Survey Interviews

Wil Dijkstra

INTRODUCTION

Why Study Interactions in Survey Interviews?

For obtaining reliable and valid information from survey interviews, it is necessary that the participants, both the interviewer and the respondent, execute their respective tasks adequately. That is, the interviewer reads the question as worded in the questionnaire, the respondent understands the question and gives a scorable answer. See Example 17.1 (I means interviewer, R respondent).

Example 17.1

Question from questionnaire: What is your age?
Question–answer sequence:

I: What's your age?
R: Thirty-seven.
I: Thirty-seven, okay.

Such a sequence of utterances by interviewer and respondent, all belonging to one particular question from the questionnaire (or scripted question), is called a question–answer sequence (or QA sequence for short). A straightforward QA

Standardization and Tacit Knowledge: Interaction and Practice in the Survey Interview
Edited by Douglas W. Maynard, Hanneke Houtkoop-Steenstra, Nora Cate Schaeffer, and Johannes van der Zouwen
ISBN 0-471-35829-0

sequence as in Example 17.1 is called a paradigmatic sequence (see also Schaeffer and Maynard, this volume, Chapter 2), as its format looks as intended by the researcher.

Many QA sequences, however, look quite different and can be called nonparadigmatic. The question from Example 17.1, "What is your age?" was posed in a Dutch computer-aided telephone interviewing (CATI) survey on watching television. Example 17.2 shows what can happen with such a seemingly simple question.

Example 17.2

Question from questionnaire: What's your age?
Question–answer sequence (translated from Dutch):

I: What's your age, if I may ask?
R: My age?
I: Yes.
R: I'm an elderly man.
I: I already thought so, great.
R: But I'm all there.
I: It clearly shows, OK.

[End of sequence; interviewer enters 80.]

In this QA sequence, the respondent does not give an adequate (i.e., scorable) answer, and the interviewer fails to probe. Quite surprisingly, the interviewer enters 80.

Example 17.3 comes from the same survey:

Example 17.3

Question: Where did you watch TV yesterday?
Alternatives: at home, at work, someone else's home, restaurant/pub.
Question–answer sequence (translated from Dutch):

I: And where did you watch television yesterday?
R: uhmm well the news.
I: Mmm... and where do you watch that?
R: And games and quizzes and....
I: OK. And where do you watch them?
R: Well ehm... where I watch them?

R: It it starts at seven o'clock, and always the Wheel of Fortune.

I: Mmm. And do you look at home, or somewhere else?

R: Always at home.

I: OK.

[End of sequence; interviewer types: 1 ("at home").]

Apparently the respondent misunderstood the meaning of the question ("What did you watch on TV yesterday" instead of "Where did you watch TV yesterday"). Note that in attempting to correct the respondent's misunderstanding, the meaning of the question is subtly changed from "Where did you watch TV yesterday" to "Where do you watch TV in general."

Qualitative and Quantitative Approaches to Study of Interviewer–Respondent Interaction

Examples 17.2 and 17.3 show the importance of studying the interaction between interviewer and respondent. The scores entered eventually by the interviewer (80 and 1, respectively) do not indicate that there were any problems with these questions.

A distinction can be made between two approaches to the study of such interactions, a qualitative one and a quantitative one. In the qualitative approach, tape-recorded interviews are carefully transcribed, whereupon the transcriptions serve as the data for the analyses. Example 17.4 shows a typical transcription.

Example 17.4

```
IV:  .hh [(0.4) a:(gh)::nd how(gh) 'bout- ((cough)) how =
         [##
IV:  = 'bout to locate illegal aliens: is the census used for that.
FR:  m::::: probably
IV:  .h[hh _oh kay:: an how much have you heard or =
       [##
IV:  =rea(gh)d about the issue: of federal agencies sharing
     information about individuals? .hhh a gr[eat d ]eal=
FR:                                          [none   ]
IV:  =some not very much or nothing at all?
FR:  none
IV:  nothing at all?
     (1.0) ((typing?))
IV:  would that be nothing at all?
FR:  yes
IV:  oh(gh) kay(gh)
```

Such transcriptions give fairly complete information about what is going on between interviewer and respondent in the process of answering a survey question and make it possible to carefully study the cognitive and interactional aspects of the task. Information about the paralinguistic aspects of the speech (e.g., pauses, overlap, stutter) may often be very important for inferring what is going on in the minds of both participants. However, transcribing the texts using a transcription notation system like the one used here is an expert task that takes lots of time. Because this task is so time consuming, the number of QA sequences that are studied in this way is typically fairly low. This makes it difficult to discover patterns and regularities in QA sequences that can be ascribed to systematic causes or lead to systematic effects.

A more quantitative approach is interaction coding. In its simplest form, the utterances are coded directly from tape. Interaction coding yields sequences of codes as in Example 17.5.

Example 17.5

Code with description in parentheses:

1.1.1. (Question read as worded.)
2.1.5. (Respondent requests clarification of question.)
2.2.11. (Interviewer offers clarification.)
2.1.1. (Respondent answers adequately for coding or recording.)

[Adapted from Sykes and Collins, 1992.]

From such sequences of codes one can easily summarize, for example, the number of questions that are read as worded, the number of questions not read as worded, and which questions evoke relatively many requests for clarification. Because coding is far less time consuming than transcribing, this method is better suited to study large numbers of QA sequences. However, such a method is not very well suited to study the details of what is going on. For example, from the code 2.1.5 in Example 17.5, it is not clear what difficulties the respondent experienced in answering the question. Moreover, the analysis is primarily directed toward the occurrence of errors in verbal acts, not toward the course of interaction between interviewer and respondent. That is, analyses concern frequencies rather than sequences.

Requirements for the Study of Question–Answer Sequences

Examples 17.2 and 17.3 make clear that an invalid result of a QA sequence is not always caused by one clearly identifiable verbal utterance. Rather it is the interaction that is responsible for the eventual answer. Hence it is important to study the *whole* QA sequence, as is done in the qualitative approach. One main objective is to classify QA sequences and systematically investigate causes and effects of particular types of QA sequences. To this end we need to quantitatively analyze large numbers of QA sequences. A system that can achieve the goals of both quantitative and qualitative approaches must meet three requirements:

1. an appropriate coding scheme that represents the course of interactions in various types of survey interviews,
2. a convenient procedure for applying this coding system to verbal interactions in a fast but reliable way, and
3. a convenient technique for analyzing the coded QA sequences.

In this chapter we present an integrated system that meets these three requirements. First we will discuss the coding scheme. Next we will show how QA sequences can be transcribed quickly and how to code large numbers of QA sequences using all paralinguistic information. Finally, we will show how to analyze the sequential properties of coded QA sequences.

THE CODING SCHEME

Most coding schemes developed thus far have only limited use as they are mainly used to answer specific research questions. Especially the older coding schemes were developed to count the occurrence of particular behaviors, not to study the interaction (for a more detailed account of existing coding schemes, see Dijkstra, 1999). For example, the scheme developed by Blair (1978) is mainly directed toward interviewer errors. The scheme used by Cannell, Lawson, and Hausser (1975) only contains categories for interviewer behavior. Brenner's (1982) scheme contains codes for both interviewer and respondent but is nevertheless primarily concerned with interviewer errors (*rule-breaching actions*). But even more recent schemes (e.g., Sykes and Collins, 1992) are especially directed toward interviewer and respondent errors.

Our coding scheme is directed not only toward interviewer and respondent errors but to all kinds of interviewer and respondent behavior in order to gain insight into the interaction process from which the eventual answer results. To this end, we use a multivariate coding system. This means that each verbal utterance of interviewer and respondent is coded on a number of variables. Our coding system counts six such variables. It can easily be extended with more variables, for example to answer a particular research question. Here we will present only the six basic variables.

The Actor

The first coding variable designates the speaker, or *actor*. This can be the interviewer (coded as I), the respondent (R), or another, third person (T). In some cases it may be useful to differentiate between third persons, for example the partner (P) or a child (C) if one is studying the impact of third persons on answers given by the respondent.

Information Exchange

The second variable concerns the type of information exchange, called *exchange*. We distinguish between question (Q), answer (A), perception (P), request (R), comment (C), and detour (D). A question (Q) is a query to another actor to provide information, related to a scripted question. With an answer (A) the actor provides the information that is asked for.

Perception (P) concerns verbal acts that signify that the other actor's utterance is received or understood, for example by repeating that utterance. A request (R) signifies that an utterance is not understood: An actor can make a request to repeat or to elucidate an utterance by the other actor.

An actor can also comment (C) on an utterance of another actor, for example, "That's a difficult question" or "That's useful information." Finally, verbal utterances can depart temporarily from the mainstream of information exchange within a QA sequence. Such detours (D) can be related to the interview (e.g., "How many questions do I still have to answer?") or they may not be related to the interview at all ("Would you like a cup of tea?").

Relevancy or Distance

The third variable concerns the relevancy of the verbal utterance in view of the scripted question. This variable is called *distance* and describes the "distance" between verbal utterance and scripted question. If the interviewer asks such a question, the distance is zero (0). Similarly, the respondent's answer to such a question, for example the choice of a response alternative, has a distance of zero too.

If the interviewer asks to motivate such a choice, the distance of that question is 1. The motivation by the respondent also has a distance code of 1. The respondent may subsequently digress from the topic, for instance by giving examples. This is coded as distance 2. Distance 3 is used for utterances that are completely irrelevant to the scripted question.

Sometimes utterances refer to either previous sequences ("You said earlier...") or to future sequences ("We have some questions about that later on"). This is coded B (backward) or F (forward). Utterances that refer to a whole sequence or to the interview itself are coded G (general) on the distance variable.

Illustration

In Example 17.6 the first three coding variables are illustrated.

Example 17.6 (translated from Dutch)

Actor	Exchange	Distance	Verbal Utterance	Description
I	Q	0	I: How many contacts do you have with people in this neighborhood?	I poses question from questionnaire.
R	P	0	R: Hm-mm.	R signifies that question is perceived.
I	Q	0	I: Is that not at all, few, in between, many, very many?	I presents alternatives.
R	R	0	R: Please can you repeat that?	R requests repetition.
I	Q	0	I: Not at all, few, in between, much, or very much?	I repeats alternatives.
R	A	0	R: Few.	R gives answer.
I	P	0	I: Few.	I signifies that answer is perceived.
I	Q	1	I: Can you tell me why you have only a few contacts?	I asks motivation.
R	A	1	R: Oh well, they don't have any consideration for other people.	R gives motivation.
R	A	2	R: For example, last weekend. . .(etc.)	R explains motivation.

The QA sequence starts with the interviewer (I) posing a question (Q) from the questionnaire (0 on distance). Next, the respondent (R) signifies perception (P) of the question, followed by the interviewer's presentation of the response alternatives. One may observe that presenting the response alternatives is also coded as question (Q) on the variable exchange.

Subsequently, the respondent (the first R of RR0) requests (the second R) the interviewer to repeat the alternatives. The repetition itself is again coded as Q for question because it still has the function of a question. The respondent (R) selects an alternative (A), which is repeated by the interviewer (I), thus signifying perception (P). Note that all utterances so far concern the scripted question; hence they all have distance code 0.

Next, the interviewer (I) deliberately asks to motivate the selected alternative (Q), which is coded 1 on distance; and so on. The coded sequence can be represented by

IQ0 RP0 IQ0 RR0 IQ0 RA0 IP0 IQ1 RA1 RA2

From such a coded sequence we can *reconstruct* the course of actions, that is,

Interviewer reads out question from questionnaire.

Respondent perceives question.

Interviewer poses question from questionnaire.
Respondent makes request.
Interviewer poses question from questionnaire.
Respondent answers question.
Interviewer perceives answer.
Interviewer asks to motivate the answer.
And so on.

One sees easily that this reconstruction is still far from perfect. For example, the first and third verbal acts are coded similarly but are clearly different: We need more coding variables to account for such differences.

Specification of Information Exchange

The fourth variable, *specific*, further specifies the information exchange categories. In the case of questions, we distinguish between closed questions (C), yes/no questions (Y), open questions (O), and questions that ask to select from among a set of response alternatives (A). If a question is preceded by an introductory text, this is coded I at the fourth position. If the interviewer explains the meaning of a question because the question is not clear to the respondent, this is coded M.

Similarly, if the utterance being coded is an answer, an answer of "yes" or "no" to a yes–no questions is also coded Y. The selection of a response alternative is coded A, and so on. In addition, a refusal to answer is coded with r and "don't know" with b (for blank).

One should note the difference between meaning and motivation. For example, if an interviewer asks "What kind of work do you do?" the respondent might say, "I'm studying question–answer sequences." The interviewer may continue with "What are question–answer sequences," which is a request (coded R on exchange) to explain the meaning of the answer (code m). Alternatively, the interviewer may ask, "Why do you study question–answer sequences?" which is a question (coded Q on exchange); the respondent is asked to motivate the answer.

An actor can show his or her perception (coded P on exchange) of an utterance by repeating or summarizing it (code E, for echo) or by saying "hm–mm" or "OK" (coded n). In addition, we use the code s to indicate a silence and f to indicate a filled pause (like "eh"). Silences and filled pauses are viewed as separate events in this case.

Requests can be one of two kinds, either requests for repetition (d for duplicate) or requests to explain the meaning of a question or an answer (m). Finally, comments and detours can be either personal (p) or task oriented (t).

Adequacy

The fifth coding variable is an evaluation of the *adequacy* of verbal utterances. We distinguish between four different categories.

Adequate (A) — for example, the interviewer reads out the question as written; the respondent selects one of the response alternatives; an utterance of another actor is correctly repeated or summarized.

Invalid (I) — for example, the interviewer changes the meaning of a question; the respondent apparently misunderstands the meaning of a question.

Suggestive (S) — questions can also be posed suggestively, for example, when the interviewer suggests a particular response alternative to the respondent that is not justified by previous answers of the respondent.

The category mismatch (M) is used in two different cases. First, if a respondent's answer does not fit one of the response alternatives, this is called a mismatch (e.g., "not many" instead of "few" when "not many" is not an alternative). Second, if an utterance is incorrectly repeated by the other actor, this is also called a mismatch and coded M, for example, if the interviewer incorrectly repeats an answer of the respondent.

Direction

The sixth variable signifies the *direction* of utterances and is especially useful for the analysis of the selection of response alternatives. In case of response alternatives, the direction is primarily determined by the number of the alternative on the list. For example, if the respondent says "strongly disagree" and this is number one on the list, then the code is 1. Or the code 2 is assigned to direction if the respondent answers with "few," and "few" is number two on the list of alternatives. Similarly, if the interviewer suggests the alternative number 1, this is coded 1 on direction. In case of a mismatch, that is, the respondent does not select an alternative, but says for example "I certainly do not agree," it is unclear if this should be alternative 1 or 2, and this utterance can be coded with an l for "low" number. Similarly, the codes h (high) and m (middle) are used.

Code C is used if the utterance refers to a show card ("You can pick an alternative from this card"). Code A is used if the interviewer reads all alternatives. Code S (subset) is used if only part of the set of alternatives is read. For example, if the interviewer asks, "Would you say dissatisfied or very dissatisfied?" instead of all five alternatives ranging from "very dissatisfied" to "very satisfied," this is coded S. If only one alternative is mentioned by the interviewer, this alternative receives the number of the alternative on the list. For example, if the interviewer asks, "Would you say very dissatisfied?" this utterance is coded 1 on the variable direction (if "very dissatisfied" is the first alternative in the list). The code z is used if there is no direction, for example, if the interviewer reads a scripted question ("How satisfied are you with your house?"). Similar codes are used for answers given by the respondent.

An Application of the Coding Scheme

The Appendix gives an overview of the coding scheme. Generally code x in this scheme means "not applicable." Example 17.7, from Wisconsin interview 1, illustrates all six coding variables.

The interviewer starts the sequence with an introductory text, preceding the question proper. Both the introduction and the question are read verbatim. Because the question is a closed question, it is coded C on the specification variable. Subsequently, all alternatives are read as worded.

Next the respondent asks the interviewer to explain the meaning of a phrase in the question. The reply of the interviewer, "Well the question doesn't specify but it just says other government agencies," is not an explanation of the meaning of the question; rather it is a repetition of a part of the question and is coded in the same way as the original question.

Example 17.7 (From Wisconsin Interview 1)

Actor	Exchange	Distance	Specific	Adequacy	Direction	Verbal Utterance	Description
I	Q	0	I	A	z	I: As you know every 10 years there is a census of the population of the United States.	Introduction read as worded.
I	Q	0	C	A	z	I: How confident are you that the Census Bureau protects the privacy of personal information about individuals and does not share it with other government agencies?	Question read as worded.
I	Q	0	A	A	a	I: Very confident, somewhat confident, not too confident, or not at all confident?	I reads all alternatives as worded.
R	R	0	m	x	x	R: Share it with what other governments?	Requests elucidation.
I	Q	0	C	A	z	I: Well the question doesn't specify but it just says other government agencies.	Repeats (part of) question.
R	A	0	A	M	l	R: Oh probably very confident.	Mismatch.
I	P	0	n	x		I: Okay.	Closes sequence.

The respondent answers with "probably very confident." This answer does not match one of the response alternatives ("probably very confident" seems to be less confident than "very confident"); hence this is coded "mismatch." The direction of the answer is a low number ("very confident" is number 1; "somewhat confident" is number 2) and is coded as l ("el"). Finally, the interviewer says "oh kay" to signify that the answer is perceived.

Advantages of the Coding Scheme

According to this coding system, a verbal utterance is represented by a *code string*, consisting of codes on all variables. For example, the code string IQ0AAa (the third utterance in Example 17.7) signifies that the interviewer reads all alternatives as worded on the list. The sequence of code strings

IQ0IAz IQ0CAz IQ0AAa RR0mxz IQ0CAz RA0AMl IP0nxx

represents the verbal utterances of the QA sequence of Example 17.7. It should be noted that it is possible to reconstruct the verbal utterances very well from these codes, given the question and response alternatives from the questionnaire. According to the coded sequence, one can infer that the interviewer reads the introduction, the question, and the alternatives as worded in the questionnaire, at least without any significant alterations; the respondent asks for elucidation but the interviewer only repeats (part of) the question. The answer given by the respondent does not match one of the response alternatives but is nevertheless accepted by the interviewer.

Other advantages of this coding system are as follows:

1. The coding task is divided into a number of subtasks, each with only a few codes. This makes it easy to formulate decision rules and to describe differences between codes.
2. Coders have to remember only a limited number of codes.
3. Even if a coder errs on one coding variable, the codes on the other variables may nevertheless be valid.
4. Although the number of different codes is small, the number of different code *combinations* is very large. Even infrequently occurring types of utterances can adequately be described without "inventing" new codes. Suppose that the *respondent* asks the interviewer how confident she is about the Census Bureau protecting the privacy of personal information, this is easily coded as RQ0CAx.
5. As we will see in the next section, the coding system is very suitable for computer-assisted coding.

SEQUENCE VIEWER PROGRAM

Introduction

Sequence Viewer is a program written in HyperCard. One main objective of the program is to link the sound of the audiotaped interviews directly to the transcribed text and to use both transcriptions and sound to decide on appropriate codes. This has two advantages. First, no transcription notation system (like that in Example 17.4) is necessary to account for paralinguistic speech characteristics that may be sometimes essential to decide on the meaning of utterances. This makes the transcription task much easier and faster. Second, regardless of how carefully speech is described, the original sound will always provide more information, making the coding of the utterances more valid. In addition, the Sequence Viewer program is provided with numerous options to ease the tasks of transcribing the interview and coding the utterances. In this section we will describe how the Sequence Viewer program looks, how we can incorporate tape-recorded interviews and transcribe and code these interviews, and some of the additional options of the Sequence Viewer program.

Sequence Viewer Card

A Sequence Viewer file consists of a number of cards, called sequence cards, each corresponding to a particular QA sequence. For example, a (simple) file may consist of all QA sequences of one interview. Such a Sequence Viewer file is linked to a separate sound file containing the sound of that interview. Usually a Sequence Viewer file consists of a large number of different interviews, each linked to the corresponding sound file. On such a card we can observe the information given below (see Figure 17.1).

The sound bar at the card bottom represents a complete interview (in this case, Wisonsin number 1). With this bar we can play the interview (backward or forward and slow, normal or fast). We can go directly to any part of the interview using the slider on this bar. The current position of the sound is 38 units from the start of the interview, as we can see on the counter to the left of this bar.

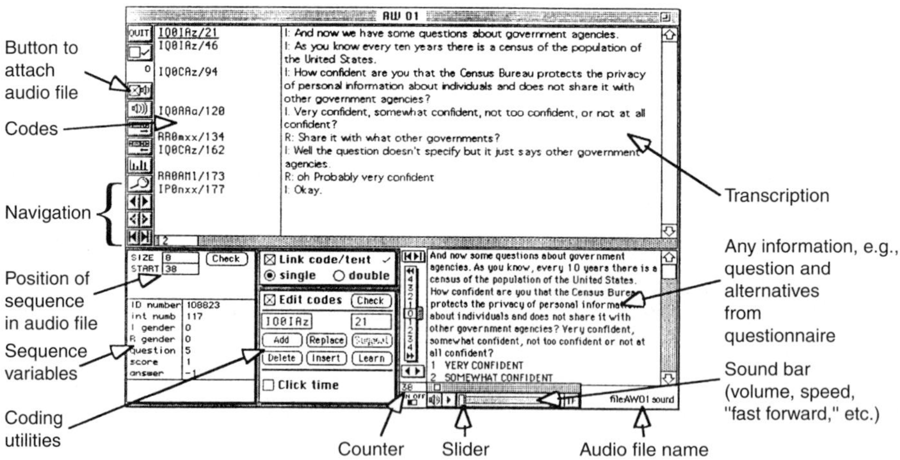

Figure 17.1 Sequence viewer card.

This is also the start of the QA sequence itself, as can be seen from the number 38 after "start" under the code field. Units are expressed as $\frac{1}{6}$ of a second, but the user can also define different units (e.g., $\frac{1}{60}$ of a second or 1 second).

The transcription appears in the field in the upper right corner. The codes are on the left. Each code string appears on the same line as the first line of the utterance in the transcription field. The utterances are coded according to the scheme described earlier. It should be stressed that a user can define a coding scheme completely different from the present one; for example, a coding scheme could consist of only one code rather than six.

Each code string is followed by a slash and a number. This number represents the time of the utterance in the QA sequence. For example, the number 21 after the first code string means that the first utterance ends 21 units after the start of this QA sequence (and hence $38 + 21 = 49$ units after the start of the interview). In this way the position of each utterance in the tape-recorded interview is known. This information is used to play the sound of a particular utterance; the user can just click the mouse on a particular utterance to hear it.

In the bottom left corner, the sequence variables are depicted, in this case the ID (respondent) number, the interviewer number, gender of interviewer and respondent, question number, score (i.e., the response as typed by the interviewer), and the answer as actually given by the respondent. Of course this response should be similar to the score. In the present case this is not true. The reason is that the respondent's answer ("*probably* very confident") is a "mismatch."

In the lower right corner, a field is available where we may put any information that seems relevant. In this case it is used to show the question and response alternatives from the CATI questionnaire. This may be useful if a coder needs to compare the question as posed by the interviewer to the scripted question.

We may immediately go to any QA sequence, for example by moving the sequence number indicator (with the number 2, meaning that the present sequence card is the second one) along this bar. In this way we have immediate access to each part of the interview. A Sequence Viewer file can consist of many hundreds or thousands of sequence cards. For example, the Wisconsin interviews each contain over 100 questions. A file containing all 23 interviews amounts to about 2500 sequence cards that provide immediate access to both the transcriptions and the sound of each utterance in these interviews.

A number of additional buttons with different functions also exist. We will discuss some of these buttons in the next sections.

Creating Sequence Viewer Cards

The sequence card described in the previous section is already complete. Of course, if one starts with a new card, it will be empty, e.g., no transcriptions and no sound bar. To fill an empty Sequence Viewer card with all this information entails the following steps.

1. Tape-recorded interviews are stored as sound files on disk (or video-recorded interviews as movie files); each interview is a separate file. One second of sound takes about 10 kilobytes of disk; 1 second of videotape takes about 100 kilobytes of disk. Hence, devices to store large amounts of data are necessary. If the number and size of these sound files are large, the files are written to CD-ROM; a CD-ROM can hold about 24 hours of monosound, sampled at 11 kHz, using appropriate compression techniques.

2. For each question of the questionnaire, a card is created. That is, the Sequence Viewer file now consists of as many sequence cards as there are questions in the questionnaire. Depending on the coding scheme used, we define, for example, the number of coding variables, their names (e.g., actor, exchange, if we use the coding scheme described earlier), and the allowable codes (e.g., I and R for coding the variable *actor*). Similarly we define number and names of the sequence variables (e.g., respondent number, interviewer number, question number). In addition, we put the question and the response alternatives as they appear in the questionnaire in the information field (in the lower right corner of the card; see Figure 17.1). All this information is similar for all interviews; hence we use this Sequence Viewer file as a "template" for each interview. We make a copy of this template for each interview.[1]

3. Each audio file (the tape-recorded interview) can be attached to a Sequence Viewer file. This is very easy: The Sequence Viewer program asks to select an audio file if the user clicks the audio attach button left of the code field. This causes the sound bar to appear. The name of the file appears to the right of the sound bar.

4. The first QA sequence is transcribed by the coder using the audio file. Each utterance is preferably preceded by I: or R:, designating the speaker. A great advantage over transcribing from tape recordings is that the program allows immediate access to any part of the audio QA sequence.

5. The second QA sequence starts at a different position on the sound bar. This position is reflected by the counter (see Figure 17.1) and can be entered by the coder in the field after "start" (in this way the sound will always start at the correct position if one opens a particular card in a next session). Now this QA sequence can be transcribed too. And so on for all QA sequences.

6. Next the utterances should be coded. Sequence Viewer has lots of options to ease the work of coding. We will discuss some of them in more detail in a separate section.

Coding with Sequence Viewer

Each utterance in the transcription field is assigned a code string in the code field; a particular code string belongs to a particular utterance; that is, code strings and

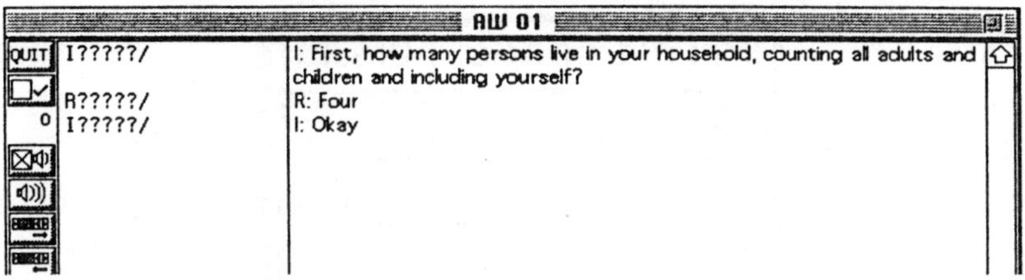

Figure 17.2 Utterances linked to "Empty" code strings.

utterances in the two fields are *linked.* Coding starts with linking so-called empty code strings to each utterance in the QA sequence. Linking is done by checking the 'Link code/text' checkbox (at the top of the Coding utilities field). The result for the first question of Wisconsin interview 1 looks like Figure 17.2.

Empty code strings consist of question marks; each question mark should be replaced by an appropriate code. Actually, the empty code strings are not completely empty. The first coding variable, which signifies the speaker, is already coded. The computer simply takes this information from the transcriptions: Each utterance in the transcription starts with I or R followed by a colon.

Next the coder has to play the sound of the whole QA sequence. This serves two objectives:

To check whether the transcriber has made any mistakes; if so, the transcriptions should be corrected.

To check whether the code for any utterance depends on how it is pronounced. For example, simple words like "yes" or "no" can be pronounced in very different ways, and this may affect how the utterance should be coded.

Now the actual coding can be done. The coder clicks on the first empty code string (belonging to the first utterance: "First how many persons live in your household, counting all adults and children, and including yourself?"). The empty code string now appears in the Edit field (see Figure 17.3). The first question mark is automatically selected and the coder can now type the appropriate code, in this case a Q for question. As a reminder for the coder, the name of the coding variable (*exchange*) and the allowable codes, appear in a so-called message box, under the sequence card. If the coder types a code that is not allowed (e.g., a B for exchange), this is not accepted by the program. This of course reduces typing errors.

The first utterance is coded Q for question, 0 (zero) because it is a scripted question (and not, e.g., a probe invented by the interviewer herself), C because it is a closed question, A because it is read as worded, and z because no particular answer is suggested. After typing the last code (z) all six codes are automatically inserted at the appropriate place in the code field and the program moves to the next empty code string; and so on.

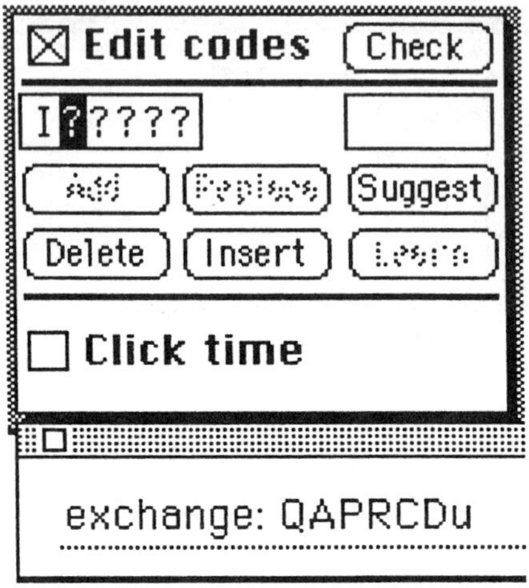

Figure 17.3 Coding an utterance.

Suggesting Codes

In a large survey, the same questions are posed to a large number of respondents. Each time a question is read verbatim by the interviewer, it is coded similarly. For example, each time an utterance in the transcription field looks like "First, how many persons live in your household, counting all adults and children, and including yourself?" it is likely to be coded as IQ0CAz.

We can define such associations between utterances and code strings in the Sequence Viewer program. To use such associations, the coder can click the Suggest button (see Figure 17.4). The program now picks from all defined utterances the one that resembles most the utterance to be coded and puts the code string associated with the defined utterance (e.g., IQ0CAz) at once in the Edit field. Moreover, the message box displays the percentage agreement as a measure of this resemblance.[2] Now the coder can click the Replace button to put the suggested code string in the appropriate place in the code field. Note that the process is still controlled by the coder, who may or may not accept the suggestion.

By using such suggestions, coding can be done both quickly and reliably. For example, the codes belonging to the introductions and questions in the example from Figure 17.1 can all be defined beforehand. When these texts are read as worded, the program's suggestions are always the same (with an agreement of 100 percent), and all the coder has to do is to click the Suggest and Replace buttons in succession. However, if the coder asks a suggestion for the first utterance in Figure 17.1, the message box will state "89% agreement" (see Figure 17.5). Now the coder knows that a difference exists between the utterance and the

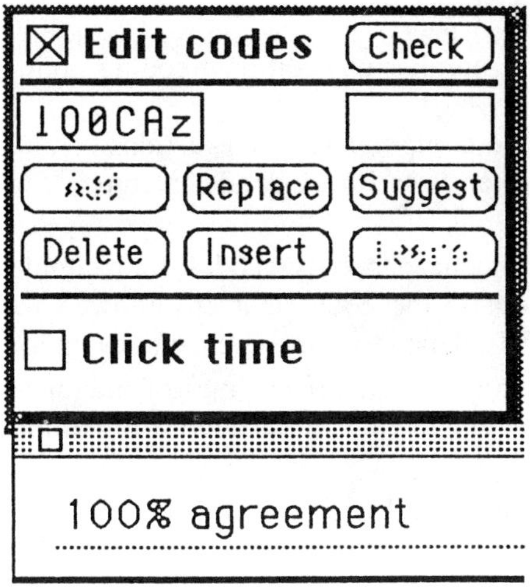

Figure 17.4 Suggesting codes.

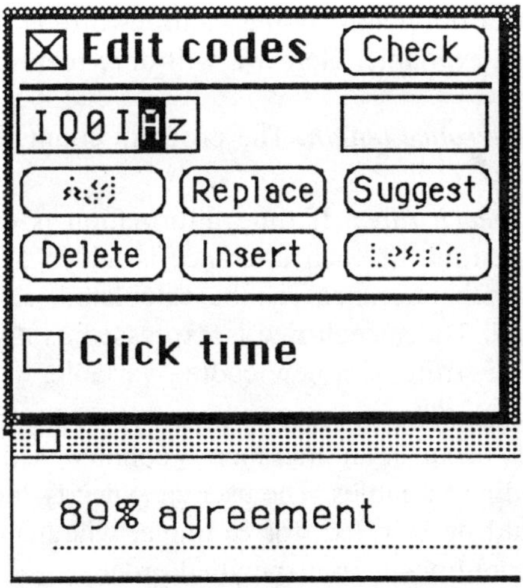

Figure 17.5 Imperfect agreement.

introduction as worded in the questionnaire. The interviewer appears to have added "we have" to the introduction, and the coder needs to decide whether to accept the code string suggested by the program or to change it. In this particular case the coder could change the code A to I if the coder judges that the addition changes the meaning of the question.

Additional Coding Options

Sequence Viewer has a number of additional coding options:

Coders can teach the computer new suggestions. This is especially useful if an interviewer often uses idiosyncratic utterances, for example to close a QA sequence.

One can define dependencies between codes; for example in the coding scheme discussed previously, the codes that are allowed for the coding variable specification partly depend on the exchange category.

Semiautomatic insertion of the offset times of utterances. All the user has to do is play the sound and click at the code string of an utterance the moment this ends.

Sophisticated options to find and replace codes. For example, it is very easy to replace the code x on direction by a z, but only if exchange is coded as R.

Other Options

In addition to the transcription and coding facilities offered by Sequence Viewer, many other options are available. Here we will only mention six:

1. *Calculation of utterance length.* The program counts the number of words in each utterance.
2. *Calculation of speech rates.* If time information is entered for utterances (i.e., if the offset time of each utterance is added to its code string), the speech rate of each utterance can be calculated, e.g., as the number of words per second. The speech rate is expressed as a whole number and is added to the code string as a new coding variable.
3. *Simple text analysis.* The number of times a word or a string of characters or a set of words occur in an utterance is counted. This number is assigned to one of the coding variables. The user can specify fairly complex requirements that should be fulfilled, for example, whether or not two words in the utterance should occur in a specified order.
4. *Frequency analysis.* Frequency tables of particular coding variables (e.g., *actor* or *exchange*) or sequence variables (e.g., *interviewer number* or *respondent gender*) are calculated.
5. *Import facilities.* For example, the sequence cards of another Sequence Viewer file are added to the present file.
6. *Export facilities.* Transcriptions with or without the associated codes are written to a text file. The codes are written as a separate file that can be read by the Sequence program for further analysis. Transcriptions and codes are

exported as ASCII files and so can be read by any platform (Macintosh, DOS, Unix).

THE SEQUENCE PROGRAM

The Sequence program (Dijkstra, 1994) is written to analyze sequential data. Sequence files consist of a number of sequences. Each sequence in turn consists of a number of coded utterances (a series of code strings as they appear in the code field of a sequence card in Sequence Viewer). In addition, each sequence is preceded by sequence variables to represent, for example, interviewer and question number and gender of interviewer and respondent. The sequence program allows us to perform a large number of analyses. In addition to the usual kinds of analyses, such as lag-sequential analysis, the program is especially suitable for studying characteristics of sequences as a whole, for example, by relating the occurrence of particular patterns in sequences or the degree of agreement with a "paradigmatic" sequence (sequences without deviations or errors) to context variables, or by performing a cluster analysis based on the agreement between all pairs of sequences in a file. In order to further study such characteristics of sequences, the user can add new sequence variables to the file, for example, to signify that a sequence belongs to a particular cluster. Questions like "Do interviewers differ with respect to the pattern of their QA sequences?" are easily answered by the program, for example, by calculating a cross-table of interviewer number by cluster number. Or one can calculate the average agreement between the sequences and the norm sequence for each interviewer. And so on.

Overview of Analyses in Sequence

The unit of analysis for the analyses performed by sequence may be the code string (or utterance) or the sequence and depends on the kind of analysis. For example, in a frequencies analysis (counting the number of times a particular code string occurs), the unit of analysis is the code string. If a cluster analysis is performed (based on the agreement between all pairs of sequences), the unit of analysis is the sequence.

The main options for analyses are as follows:

1. *Frequencies* counts the number of times utterances with a particular code occur. One can ask for a table to obtain the frequency distribution of the coded utterances for each value of a particular sequence variable (such as interviewer number or respondent gender to look if the frequency distribution of utterances is different for different interviewers or for male and female respondents).

2. *Next event and previous event* counts the number of times utterances with different codes follow (or precede) a specified coded utterance, with a lag between 1 and 9 and includes expected frequencies based on the overall occurrence of particularly coded utterances. For example, a next-event analysis may give the frequency distribution of utterances immediately following a mismatched answer of the respondent. The expected frequencies show if this distribution differs from the distribution expected by chance.

3. *Matrix* calculates a transition matrix with the frequency, or the probability that each utterance with a specified code is followed with a particular lag by another coded utterance. It also calculates z values, expressing whether the frequency of transitions differ from the frequency expected by chance.

4. *Tree* shows sequences as a tree structure, starting with an utterance with a specified code. It shows the number of times this utterance is followed by other utterances and the number of times each of these subsequent utterances are followed by what other utterances; and so on (see next section for an illustration).

5. *Path* shows sequences in the form of a number of different paths given a particular coded utterance to start with and a particular coded utterance to end with. It is very similar to Tree, except that each end point of a branch consists of an utterance with a specified code.

6. *Patterns* finds complex patterns of coded utterances in a sequence, for example, "find all sequences containing an inadequate response, *not* followed somewhere in the sequence by a correct presentation of the response alternatives by the interviewer, but instead by 'summarizing' the inadequate response by a particular alternative from the list." One may mark sequences containing the pattern, for example, to investigate whether the occurrence of particular errors is related to the interviewer (number).

7. *Count events or transitions*, for example, may count the number of times the respondent requests that the interviewer explains a question in a QA sequence or the number of times such a request is followed by an adequate explanation by the interviewer. The count is assigned to a sequence variable and can thus be used in further analyses. For example, the count can be used to investigate whether particular questions evoke relatively many requests for repetitions and the way the interviewer handles these requests.

8. *Count sequences* counts the number of sequences that are coded exactly the same.

9. *Compare sequences* calculates the degree of agreement between two sequences as a number between 0 (completely different) and 1 (exactly equal).

10. *Agreement matrix* calculates a matrix of agreement scores between all sequences.

11. *Cluster sequences* performs a cluster analysis on sequences.

12. *Reliability* performs a reliability analysis (using kappa as a measure) on utterances that are coded twice, for example, by two different coders.

Example Analysis with Sequence

A particularly useful analysis that quickly provides insight into the data is the "tree" analysis. To illustrate this analysis, 1227 QA sequences were coded using data from a survey among young adults who were posed large numbers of questions on education, work, and relationships with partners. The coded sequences concerned nine questions asked by 17 interviewers.

An interesting research question might be what happens after the respondent gives an answer that can be considered a mismatch to the scripted question. Such a research question can easily be answered by, for example, the tree procedure. The setup screen for this analysis looks like Figure 17.6. The Root event prompt requests the code string of the utterance we want to start with; that is, the code for mismatch answers RA0AM• is entered. At this time we are not interested in the direction (the sixth coding variable) of the utterances. So we only look at the first five coding variables. The fact that the sixth variable is neglected can be seen by the bullet (•). Lest we cannot see the tree for the branches, we attend only to the first four coded utterances after a mismatched answer (tree depth = 5). Finally we neglect transition frequencies between utterances that are smaller than 3. This means that transitions from a particular code string to another one are not shown if this transition occurs only once or twice.

The output of the tree analysis is shown in Table 17.1 (for reference purposes the lines are numbered; these numbers are not included in the actual output). This table can be read as follows. Line 1 shows that in 261 sequences the respondent gave a mismatched answer (code RA0AM•). These 261 mismatches are followed by signifying perception by the interviewer in 10 cases (line 2; IP0nx•), by correctly repeating the (mismatched) answer in 27 cases (line 3; IP0EA•), and so on. These numbers (10, 27, . . .) do not add up to 261 because transitions that occur fewer than three times are not included.

In 103 of the 261 sequences in which a mismatched answer occurs, the interviewer follows this answer with an adequate presentation of the response categories (line 8; IQ0AA•). In 82 out of these 103 sequences, the respondent

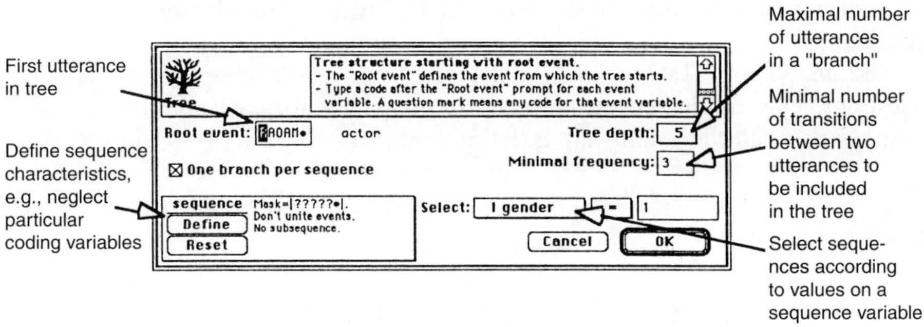

Figure 17.6 Setup screen for tree analysis.

Table 17.1 Output from Tree Analysis

1	→ 261 RA0AM•				
2		→ 10 IP0nx•			
3		→ 27 IP0EA•			
4			→ 5 RA0AA•		
5			→ 6 RA0AM•		
6		→ 18 IP0EM•			
7			→ 8 RA0AA•		
8		→ 103 IQ0AA•			
9			→ 82 RA0AA•	→ 13 IP0nx•	
10				→ 26 IP0EA•	
11				→ 7 IQ0AA•	
12					→ 7 RA0AA•
13			→ 11 RA0AM•		
14				→ 4 IQ0AS•	
15		→ 5 IQ0AM•			
16			→ 4 RA0AM•		
17		→ 26 IQ0AS•			
18			→ 19 RA0AA•		
19				→ 3 IP0nx•	
20				→ 3 IP0EA•	
21			→ 4 RA0AM•		
22		→ 5 IQ0EA•			
23		→ 3 IQ0EM•			
24		→ 23 RA1OA•			
25			→ 3 IP1nx•		
26			→ 3 IQ0AA•		
27			→ 3 IQ0AS•		

then selects an adequate answer (line 9; RA0AA•), but in 11 cases the respondent persists in giving a mismatched answer (line 13; RA0AM•).

It can also be seen that a mismatched answer of the respondent is followed by a suggestion from the interviewer in 26 sequences (line 17; IQ0AS•), and an adequate answer by the respondent is then provided in 19 sequences (line 18; RA0AA•); that is, the respondent then selects from the set of response alternatives. Needless to say these answers may be highly influenced by the interviewer's suggestion.

The analysis in Table 17.1 is collapsed over all 17 interviewers and nine questions. One might repeat the analysis separately for each question to study whether particular problems are question specific.

CONCLUSIONS

In the introduction three requirements to study QA sequences in survey interviews were mentioned: (1) how can the verbal utterances of interviewer and respondent be represented by a coding scheme; (2) how can QA sequences be efficiently transcribed and reliably coded; (3) how can the coded sequences be analyzed. In the preceding sections we described the procedures we developed to fulfill these requirements. By now, we have used these procedures to transcribe, code, and analyze thousands of QA sequences. An example of such an application can be found in van der Zouwen and Dijkstra (1996), where we were able to distinguish between four different types of sequences and to relate these types to their causes and to their effects on the obtained answers.

Apart from its capabilities to store and retrieve QA sequences from interview data, Sequence Viewer has many options to study interaction but is especially powerful in combination with the Sequence program. We will conclude with a number of interesting research questions that can easily be answered by using this combination.[3]

1. In Sequence Viewer one can assign to each code string the number of words of each utterance. Having done this, it is easy to use Sequence to answer questions such as "Is the length of a question related to the occurrence of requests to explain the meaning of a question?" or "Are longer questions more often interrupted than short questions? Does this in turn lead to more mismatched answers?"

2. Earlier we mentioned that pauses can be coded separately. For example, suppose there is a pause after a scripted question read by the interviewer. As such a pause can be assigned to the respondent, it can be coded RP0sxx/12, where the 12 means a pause of 12 units. With the Sequence program we can now investigate whether such reaction time latencies are related to the occurrence of mismatched answers or to requests to explain the meaning of the question.

3. We can assign the number of times particular words or phrases occur in an utterance to a coding variable. This in turn enables us to study whether the occurrence of words such as "probably" and "perhaps" is related to mismatches or requests for explanation. Moreover, this option also enables us to make finer distinctions between codes. For example, utterances like "Okay" or "Hm-mm" are usually coded as IP0nxx on the perception coding variable. However, one might view "Okay" as a *stop sign* but "Hm-mm" as a *continuer*. Questions of whether respondents actually stop after "Okay" and continue after "Hm-mm" can be answered in this way.

4. Also pauses by the interviewer can serve as continuers. One may compare the relative effectiveness of different continuers as a pause, "Hm-mm," or repeating an answer, both with respect to whether additional answers are obtained or not and to the time it takes before such an additional answer is given.

5. Of course, it is easy to investigate whether such relations are modified by interviewer and respondent gender, age, socio-economic status, and so on.

In this way Sequence Viewer and Sequence can serve as a bridge between the qualitative and quantitative approaches to the study of interactions in the survey interview. More information about both programs can be found at http://svn.scw.vu.nl. These programs are written for the Macintosh but can be used on a Windows machine by using a Mac emulator like SoftMac 2000 from Emulators, Inc. (http://www.emulators.com/).

NOTES

1. A useful device is to put the question itself also in the transcription field, especially if there are large numbers of interviews to be transcribed. Each time the interviewer reads the question as worded, nothing has to be changed; as far as only small differences exist, it is much faster to adjust the text than to type the whole text. From the example in Figure 17.1 it will be clear that this can make quite a difference: The first four utterances were already put into the transcription field.

2. To calculate the agreement between two utterances, a fairly complex algorithm is used. This algorithm is fully described in Dijkstra and Taris (1995). This measure depends on both the number of words that are common to both utterances and the similarity regarding the order of the common words. An agreement of 100 percent means that both utterances are exactly similar; an agreement of 0 percent means that both utterances have no single word in common.

3. Thanks to Hanneke Houtkoop for a number of these suggestions.

REFERENCES

Blair, E. A. 1978. "Nonprogrammed Speech Behaviors in a Household Survey." Unpublished Ph.D. dissertation, Department of Business Administration, University of Illinois.

Brenner, M. 1982. "Response-Effects of 'Role-Restricted' Characteristics of the Interviewer." Pp. 131–165 in W. Dijkstra and J. van der Zouwen (Eds.), *Response Behaviour in the Survey-Interview*. London: Academic.

Cannell, C. F., S. A. Lawson and D. L. Hausser. 1975. *A Technique for Evaluating Interviewer Performance*. Ann Arbor: University of Michigan.

Dijkstra, W. 1994. "SEQUENCE, A Program for Analyzing Sequential Data." *Bulletin de Méthodologie Sociologique* 43: 134–142.

Dijkstra, W. 1999. "A New Method for Studying Verbal Interactions in Survey Interviews." *Journal of Official Statistics* 15: 67–85.

Dijkstra, W., and T. Taris. 1995. "Measuring the Agreement between Sequences." *Sociological Methods and Research* 24: 214–231.

Sykes, W., and M. Collins. 1992. "Anatomy of the Survey Interview." *Journal of Official Statistics* 8: 277–291.

van der Zouwen, J., and W. Dijkstra. 1996. "Trivial and Non-Trivial Question-Answer Sequences; Types, Determinants and Effects on Data Quality." Pp. 81–86 in *Proceedings of the International Conference on Survey Measurement and Process Quality, Bristol, April 1995*. Alexandria, VA: ASA.

APPENDIX

The Coding Scheme

Actor	Exchange	Distance	Specification	Adequacy	Direction
			presents:		a: all
			I: introduction		i: interruption
Interviewer			Y: yes/no Q	A: adequate	l: low
Respondent	Q: question	0, 1, 2, 3	C: closed Q	M: mismatch	m: middle
Third person			S: statement	I: invalid	h: high
			O: open question	S: suggestive	s: subset
			A: alternatives		0, 1, 2,. . .
			M: meaning of Q		z: no direction
			Y: yes/no answer	A: adequate	l: low
			A: alternative	M: mismatch	m: middle
Interviewer			O: open answer	I: invalid	h: high
Respondent	A: answer	0, 1, 2, 3	M: meaning of A		0, 1, 2,. . .
Third person			b: don't know	x	x
			r: refusal		
					a: all
				A: adequate	l: low
Interviewer		0, 1, 2, 3	E: echo	M: mismatch	m: middle
Respondent	P: perception	G, B, F	(repeats other)		h: high
Third person					0, 1, 2,. . .
			n: notes		
			f: filled pause	x	x
			s: silence		
Interviewer			d: duplicate		
Respondent	R: request	0, 1, 2, 3	(requests repetition)	x	x
Third person		G, B, F	m: meaning		
			o: other		
Interviewer		0, 1, 2, 3	p: personal		
Respondent	C: comment	G, B, F	t: task	x	x
Third person					
Interviewer			p: personal		
Respondent	D: detour	G	t: task	x	x
Third person					
Interviewer					
Respondent	u: unintelligible	u	u	u	u
Third person					

CHAPTER 18

Testing Questionnaires Using Interaction Coding

Johannes van der Zouwen and Wil Dijkstra

INTRODUCTION

To ensure the comparability of the measurements of unobservable variables in surveys, such as opinions and attitudes, all respondents must react to the same "input" (Beatty, 1995; van der Zouwen, this volume, Chapter 2). This means that the questions have to be presented to and understood by all respondents in the way intended by the designer of the instrument.

The easiest way to reach equality of inputs is standardization of all actions of the participants. However, methodological research has made clear that the goal of standardization is not always reached in practice (Brenner, 1982). This observation has suggested the need for further investigation of the areas where and the conditions under which the basic requirement of survey research, that is, comparability of responses, is endangered.

One of the procedures to study what actually goes on in the process of data collection[1] is analyzing the interaction between interviewer and respondent. This analysis gives the survey methodologist information about the problems the participants in the interview have in executing their task adequately (Schaeffer and Maynard, 1996). Moreover, this analysis shows how participants, especially the interviewer, try to overcome these problems, for example, by motivating the respondents, by performing various checks, and by doing repair activities (Houtkoop-Steenstra, 1995, 1996; Sykes and Collins, 1992; van der Zouwen and Dijkstra, 1995, 1996). One can also observe in detail how some interviewers make serious mistakes (Dijkstra and van der Zouwen, 1988), mistakes that are hardly perceptible by inspecting and analyzing the completed instrument, debriefing the

Standardization and Tacit Knowledge: Interaction and Practice in the Survey Interview
Edited by Douglas W. Maynard, Hanneke Houtkoop-Steenstra, Nora Cate Schaeffer, and Johannes van der Zouwen
ISBN 0-471-35829-0

interviewers, or reinterviewing the respondents. In sum, interaction analysis can provide necessary and otherwise unobtainable information for the design and evaluation of surveys.

In this chapter we focus on interaction analysis as a means to evaluate questionnaires and to identify questions that cause trouble for the respondent, the interviewer, or both. We restrict ourselves to a particular form of interaction analysis, that is, analysis of question–answer sequences. As mentioned in Chapter 2 of this volume, we define a question–answer sequence as the string of all speech acts starting with the reading out of a question from the questionnaire by the interviewer and ending just before the next question from the questionnaire will be read out. The present chapter is also meant to illustrate the procedure described by Wil Dijkstra in Chapter 17 of this volume.

We will address the following two research questions:

1. Which characteristics of questions from a questionnaire used in standardized survey interviews decrease or increase the probability that problematic, or even inadequate, question–answer sequences will occur?
2. Does an analysis of the interactions in interviews identify the same or other questions of the questionnaire as being problematic or even inadequate, as compared with the questions identified by a jury of survey experts or resulting from an analysis of the (cognitive) difficulty of answering these questions?

Our research is related to the work done by Presser and Blair (1994). These authors performed an experimental comparison of four pretest methods of survey questionnaires: conventional pretests, behavioral coding of interviewer–respondent interactions, cognitive interviews, and expert panels. In the present study we only compare three methods of pretesting questionnaires, with strong emphasis on the analysis of the interaction during the interview, using a coding scheme that is more detailed than the behavioral coding used by Presser and Blair.

The first research question will be answered by analyzing data from a face-to-face survey; for the second research question we will use data from a telephone survey.

QUESTIONS AND SEQUENCES IN A FACE-TO-FACE SURVEY

Types of Sequences

To give an empirical answer to the first research question, we analyzed question–answer sequences originating from parts of face-to-face (PAPI) interviews with young adults conducted in the Netherlands in 1987. By the time of the interview the respondents were 18, 22, or 26 years old. To avoid large age differences between respondent and interviewers, we employed interviewers of about 25 years old. These interviewers received training during three full days before starting their fieldwork. During the fieldwork they were closely supervised by the field managers (cf. van der Zouwen and Dijkstra, 1995). For the analysis of question–answer sequences we chose 17 interviewers at random from the group of 72 interviewers. From the 25 interviews each of these interviewers conducted, we selected randomly about 11 interviews. From these 185 interviews we selected parts related to nine closed questions about leaving the parental home, becoming unemployed, satisfaction with the training one has received, and ideas about having children. Because not every respondent had to answer each of these nine questions, the number of sequences analyzed is "only" 1227.

The audiotapes of the selected parts of the interviews were transcribed into verbatim protocols and divided into series of verbal acts, which were coded by means of the computer-assisted coding procedure described by Wil Dijkstra in Chapter 17 of this volume. Next, the computer program SEQUENCE (Dijkstra, 1994) was employed to subsume each of the 1227 sequences under one of the categories of a typology.

The typology of sequences we used for this analysis consists of three main types:

1. *paradigmatic* sequences, which do not (substantially) differ from the model of standardized interviewing;
2. *problematic* sequences, in which such deviations do occur but are adequately "repaired" by one of the actors (usually the interviewer); and
3. *inadequate* sequences, in which this repair does not take place or is not accomplished.

Questions and Occurrence of Problems and Inadequacies

The distribution of the paradigmatic, problematic, and inadequate sequences varies substantially over the nine questions selected. The percentage of problematic sequences varies from 2 percent to 34 percent, and the percentage of inadequate sequences ranges from 3 percent to 46 percent. In Table 18.1 the nine questions are arranged in order of increasing percentage of inadequate sequences.

Three questions (Q146, Q46, and Q41) cause little trouble and lead only in a few cases to inadequate sequences. These three questions ask respondents to express their (dis)satisfaction with some aspect of life (training, job, and social relations) by choosing one alternative from show card 11. Most sequences involving these three questions follow the paradigmatic pattern: The question is read out correctly by the interviewer and the respondent mentions one of the alternatives printed on the show card. Sometimes the respondent adds an explanation to this particular choice. Only question Q41 causes some trouble if respondents are uncertain about which training is meant by the question: the present one, the last one, or all training received in the past. In these cases the interviewer is asked for help, and this sometimes leads to mistakes.

Questions Q97, Q152, and Q99 lead to a higher percentage of both problematic and inadequate sequences. For the remaining three questions, Q96, Q22, and Q21, the percentage of inadequate sequences is disturbingly high.

Table 18.1 Nine Questions from Face-to-Face Survey with Their Response Alternatives and Percentage of Paradigmatic, Problematic, and Inadequate Sequences Following These Questions

Question	Paradigmatic, % (1)	Problematic, % (2)	Inadequate, % (3)	*N* (4)
Q146: *How do you feel about not having a steady relationship? How satisfied or dissatisfied are you about that?* **Int: Use card 11:** (1) very dissatisfied; (2) rather dissatisfied; (3) neither satisfied, nor dissatisfied; (4) rather satisfied; (5) very satisfied	95	2	3	66
Q46: *How satisfied are you all in all with your present job?* **Int: Use card 11** (see Q146)	92	5	3	100
Q41: *How satisfied are (were) you all in all with the training you receive(d)?* **Int: Use card 11** (see Q146)	85	8	7	181
Q99: *Will you become more satisfied or less satisfied with life in general if you become (remain) unemployed?* **Int: Use card 9:** (1) much less satisfied; (2) somewhat less satisfied; (3) about the same; (4) somewhat more satisfied; (5) much more satisfied	76	14	10	171
Q97: *How do you feel about becoming (remaining) unemployed?* (1) very unpleasant; (2) unpleasant; (3) neither pleasant nor unpleasant; (4) pleasant; (5) very pleasant	60	24	16	171
Q152: *Do you think of having (another child or more) children in the future?* (1) definitely not; (2) probably not; (3) don't know as yet; (4) probably yes; (5) certainly yes	46	34	20	172
Q96: *Have you ever thought about how it would be if you became (remained) involuntarily unemployed?* (1) hardly ever; (2) seldom; (3) now and then; (4) often; (5) very often	53	18	29	170

Table 18.1 (*Continued*)

Question	Paradigmatic, % (1)	Problematic, % (2)	Inadequate, % (3)	*N* (4)
Q22: *How difficult will it be for you to find a reasonably priced place to live when you leave the parental home?* (1) will cause no problem at all; (2) will require only little effort; (3) will require quite a lot of effort; (4) will require very much effort; (5) will be nearly impossible	41	19	40	98
Q21: *How much trouble do you take to find living accommodations?* (1) no trouble at all; (2) a little trouble; (3) much trouble	35	19	46	98

To further investigate the observed strong relationship between the question and the types of the resulting question–answer sequences, we analyzed the problematic sequences and the inadequate sequences in more detail.

Problematic and Inadequate Sequences in More Detail

Most problematic sequences require some "tuning," or "correction," from the part of the interviewer; others contain irrelevant or redundant information:

1. *Tuning*. Before the response alternatives are presented, the respondent gives a "premature" answer that often does not match one of the response alternatives mentioned in the instrument (see also Chapter 2 of this volume). For example, in response to question Q97 ("How do you feel about becoming unemployed?"), many respondents spontaneously react with "pretty bad."

Interviewer and respondent have to do some extra work to find out which of the "scripted" alternatives (e.g., "very unpleasant" or "unpleasant") fits best with the original response. Some tuning occurs in 183 of the 1227 sequences, that is, in 15 percent (see columns 1 and 2 of Table 18.2).

2. *Correction*. A mistake made by the interviewer (e.g., a misreading, misinterpretation, or miscoding) is adequately corrected by either interviewer or respondent. This subtype is only present in 5 of all 1227 sequences.

3. *Irrelevancy*. The choice of a response alternative is preceded or followed by some kind of (superfluous) explanation by respondent or interviewer (2 percent of all sequences). Characteristic for this subtype is that the sequence is longer than necessary. For example, we recorded a sequence following the reading out of Q152 ("Do you think of having children in the future?"), consisting of 31 verbal utterances by interviewer and respondent, who were discussing the assumption behind the question, namely that it is up to the respondent whether to have another

Table 18.2 Distribution of Types of Question–Answer Sequences for All Nine Questions Together and for Questions Q21, Q22, Q96, and Q152 Separately

Type of Sequence	All Nine Questions (1)	(2)	Q21 (3)	Q22 (4)	Q152 (6)	Q96 (5)
Paradigmatic sequences	764	64%	35%	41%	46%	53%
Problematic sequences						
Tuning	183	15%	15%	15%	31%*	17%
Correction	5	0%	0%	0%	1%	1%
Irrelevancy	26	2%	4%	4%	3%	1%
Inadequate sequences						
Changing question meaning	38	3%	8%*	16%*	2%	5%
Changing response categories	69	6%	6%	6%	8%	15%*
Risky or wrong interpretation of answer	25	2%	13%*	1%	3%	3%
Suggesting an answer	42	3%	5%	9%*	4%	3%
Insufficient probing	30	2%	9%*	4%	1%	2%
Other	25	2%	7%*	3%	2%	1%
Total number of sequences	1227		98	98	172	170

child or not. This sequence did not contain errors, but it certainly produced information that was irrelevant from the viewpoint of the survey researcher.

Thus it appears that in most problematic sequences interviewers and respondents have to adapt the answer given by the respondent to the response categories specified by the researcher. This "tuning" activity mostly takes place when respondents have given an answer before they have heard or seen the response categories that belong to the question. "Correction" occurs only in a very few cases. Irrelevant sequences in which superfluous information is given are spread out "thinly" over all nine questions, just like corrections.

Within the class of inadequate sequences we can distinguish several types of inadequacies (see also Chapter 2 of this volume): (1) changing the meaning of the question; (2) changing the response categories; (3) risky or wrong interpretations of the responses given; (4) suggesting an answer; and (5) insufficient probing.

From column 1 of Table 18.2 it appears that the most frequently occurring subtype of inadequate sequences is the one in which the scripted response alternatives are incorrectly presented, that is, partly misread or omitted (69 sequences). Other common interviewer errors are suggesting an answer to the respondent (42 sequences), changing the meaning of the question (38 sequences), and insufficient probing of an adequate answer (30 sequences).

The right-hand part of Table 18.2 shows the distribution of deviations from the paradigmatic sequence for the four questions with relatively many deviations. An asterisk following an entry indicates that the observed proportion is more than twice the proportion for all nine questions together. For example, "changing the response categories" occurs in 69 of all 1227 sequences, that is, in 6 percent. But in the 170 sequences following question Q96 it occurs in more than 15 percent of these sequences.

Question Q152 requires more tuning than the other questions. Question Q96 has an incomplete set of response categories (the often appropriate response alternative "never" is missing), which often leads the interviewer to change the set of response categories. Questions Q21 and Q22 cause trouble at several points, resulting in a high percentage of inadequate sequences.

QUESTION DIFFICULTY AND INADEQUATE SEQUENCES

Components of Question Difficulty

We have seen that the distribution of types of question–answer sequences depends on the question that begins the sequence. To further explore this relationship, we analyzed the content of the question, the response categories, and the format of the question. The goal is to identify particular characteristics of a question that make that question more difficult to answer and may result in an increase of the number of inadequate sequences following that question. For the present analysis[2] we distinguished four components of question difficulty:

1. *Questions about Hypothetical Situations (Conditional Questions).* Some questions relate to 'hypothetical' situations, or contain a condition: "when you leave the parental home" (Q22), "if you became involuntarily unemployed" (Q96, Q97, Q99), or "in the future" (Q152). These questions are assumed to be more difficult to answer than those that concern a situation the respondent already has experience with (see also Converse and Presser, 1986; Schaeffer and Bradburn, 1989).

2. *Ambiguity of the Question.* It turns out that the wording of questions Q21 and Q96 is somewhat ambiguous. Question Q21 ("How much trouble do you take to find living accommodations?") is either interpreted as taking trouble "at present" or, in accordance with the preceding question (Q20, "For what reasons would you leave the parental home?"), as taking trouble at some moment in the future. Question Q96 ("Have you ever thought about how it would be if you became involuntarily unemployed"?) is interpreted with either a stress on the first part of the sentence ("Have you ever thought of..?") or on the last part (".. how it would be if you became involuntarily unemployed?"). Thus some respondents refer in their response to the frequency of *thinking* about this issue, while others refer to the *feelings* regarding becoming unemployed. In the latter case the interviewer has to do a lot of work to get an answer to the question intended by the designer of the instrument.

3. *Inadequate Range and Fit of the Response Alternatives Presented.* The response alternatives mentioned in the questionnaire have to fit well with the wording of the question, and they have to be spaced evenly over the subjective continuum specified by the question (Dobson and Mothersill, 1979; Spector, 1976). Otherwise respondents have difficulty in selecting an appropriate alternative.

In question Q21 the alternatives are quite unevenly distributed over the response scale: alternative 2 ("a little trouble") is much closer to alternative 1 ("no trouble at all") than to alternative 3 ("much trouble"). In question Q22 the distance between alternatives 2 and 3 is much larger than between 3 and 4. In question Q96 the alternative "never," which is mentioned spontaneously by many respondents, does not even appear on the list, while the distance between alternatives 3 and 4 is much larger than between 2 and 3. Furthermore, question Q96 is a yes–no question that has to be answered as if it were a multiple-choice question. The problem with the set of alternatives presented together with question Q99 is that the distances between the alternatives 3 and 2 and between 3 and 4 are less than the distances between 1 and 2 and between 4 and 5, respectively.

4. *No Use Made of Show Card.* Elsewhere (van der Zouwen and Dijkstra, 1996) we have mentioned the beneficial effect of the use of show cards, because show cards relieve the burden of both respondent and interviewer and may disambiguate ambiguities. For example, question Q146 actually consists of two different questions: Q146a: "How do you feel about not having a steady relationship?" followed by Q146b: "How satisfied or dissatisfied are you about that?" The introductory question Q146a provides an unintended interruption point, and the target question Q146b presents a choice between two categories, instead of the five categories actually available. If show card 11 is presented simultaneously with the reading out of question Q146, both problems are solved in one go. Show cards are used in questions Q99, Q41, Q46, and Q146; the other five questions were read-out questions. From Table 18.1 it appears that the four questions in which show cards are used have a higher percentage of paradigmatic sequences than the five questions in which no use was made of show cards.

Question Difficulty and Proportion of Inadequate Sequences

In Table 18.3 we have put together information about the proportion of inadequate sequences a particular question has "produced" and about the absence (0) or presence (1) of one or more of the four components of question difficulty just mentioned. Every component of question difficulty appears to be related to the percentage of inadequate sequences. A multiple regression analysis, relating for each of the nine questions the four components of difficulty to the proportion of inadequate sequences for that question, shows that components 3 ("range") and 4 ("show card") together explain 87 percent of the variance in the dependent variable (multiple $R = 0.93$). This means that the percentage of inadequate sequences following a particular question may be predicted quite well from information about the adequacy of the alternatives used and whether a show card is used or not.

Question Difficulty and Interviewer Variability

Another component of the interview that may threaten the ideal of standardization is the interviewer. From an earlier analysis (van der Zouwen and Dijkstra, 1996) it appeared that there are large differences with respect to the percentage of

Table 18.3 Relation between Four Components of Question Difficulty and Percentage of Inadequate Sequences

Question	Conditional Question (1)	Ambiguous Question (2)	Inadequate Range of Alternatives (3)	No Show Card Used (4)	Inadequate Sequences,% (5)
Q146: *How do you feel about not having a steady relationship? How satisfied or dissatisfied are you about that?*	0	0	0	0	3
Q46: *How satisfied are you all in all with your present job?*	0	0	0	0	3
Q41: *How satisfied are you all in all with the training you receive(d)?*	0	0	0	0	7
Q99: *Will you become more satisfied or less satisfied with life in general if you become unemployed?*	1	0	1	0	10
Q97: *How do you feel about becoming unemployed?*	1	0	0	1	16
Q152: *Do you think of having (more) children in the future?*	1	0	0	1	20
Q96: *Have you ever thought about how it would be if you became involuntarily unemployed?*	1	1	1	1	29
Q22: *How difficult will it be for you to find a reasonably priced place to live when you leave the parental home?*	1	0	1	1	40
Q21: *How much trouble do you take to find living accommodations?*	0	1	1	1	46
Correlation coefficient between difficulty component and percentage inadequate sequences	.27	.65	.71	.81	

paradigmatic sequences the interviewers produced: this percentage varies from 33 to 84 percent.

To disentangle the separate effects of the question and of the interviewer on the probability that the resulting sequence will be (in)adequate, we performed a "three-dimensional" analysis. It appears that for the least difficult questions (Q146, Q46, and Q41), there were only small differences between the 17 interviewers with regard to the proportion of inadequate sequences. The differences increased, however, when the questions became more and more difficult. Stated differently, standardization of interviewer behavior decreases as the difficulty of the task for the respondent increases.

IDENTIFYING PROBLEMATIC QUESTIONS IN CATI SURVEY

Introduction

Our second research question reads "Does an analysis of the interactions in interviews identify the same, or other, questions of the questionnaire as being problematic or even inadequate, as compared with the questions identified by a jury of survey experts, or resulting from an analysis of the (cognitive) difficulty of answering these questions?" (see also Presser and Blair, 1994).

To answer this question we use data from a telephone survey after "advertising use" conducted in the Netherlands at the end of 1995 and in the beginning of 1996 (see Smit, 1999). The 36 interviewers in this survey were all female first-year students at the University of Amsterdam following their introductory course in mass communication; these interviewers had no former interviewing experience; they received instruction on how to work with the CATI questionnaire. The 36 interviewers interviewed 1105 respondents; thus the average caseload was about 30 interviews. From these 1105 interviews we randomly selected 54 interviews. From these latter interviews we selected only those sequences following questions posed to all respondents; these questions were about television watching practices and about opinions and attitudes regarding television commercials. For each of these 37 questions we (1) assessed the cognitive difficulty of answering that particular question, (2) asked members of a jury of survey experts to evaluate the "quality" of the formulation of that question, and (3) analyzed the question–answer sequences belonging to that question by counting the number of deviations from the paradigmatic sequence.

Assessing the Cognitive Difficulty of Answering the Question

The difficulty of answering a survey question[3] consists of two components: the difficulty of that question and the (lack of) clarity of the task involved.

The difficulty of a question will (*ceteris paribus*) be greater if the following conditions hold:

- More cognitive operations are necessary to compute the response.
- It is necessary to retrieve information from long-term memory.

- Information about the question and the response alternatives has to be stored in short-term memory to adequately answer the question.
- The question relates to a "hypothetical," or future, situation.
- For open questions more precision of the response is required.
- For closed questions the response alternatives offered do not fit well with the question.
- For closed questions the response alternatives are unevenly distributed over the range of possible values of the variable that the question is meant to measure.
- No use is made of responding aid devices (like show cards, visual aids, ladder/barometer).
- The topic of the question involves problems of self-presentation (topic threat; differences with respect to social desirability of response alternatives).

The clarity of the task involved will decrease if

- The question is ambiguous,
- The question uses of unfamiliar words, and
- The question structure does not indicate when respondent's turn begins.

We have rated the 37 questions from the CATI questionnaire with respect to each of these 12 aspects of task difficulty. For every aspect it was decided whether the question concerned is significantly more difficult than a reference question, that is, a question that can easily be answered by any respondent, without any further explanation or stimulation by the interviewer (e.g., "How old are you?"). This means that the task difficulty score theoretically ranges between zero and 12 points. It appears that this CATI questionnaire does not include extremely difficult questions: None of these questions have a task difficulty score of more than six points[4]. Fourteen questions have a score of zero to two points, which means that they are relatively easy to answer. Eighteen questions have scores of three, four, or five points; they are not so easy to answer. Only five questions have a score of six points, which identifies them as rather difficult.

Four of these rather difficult questions regard the last time the respondent watched TV: q9 ("What percent of the time did you watch attentively?"), q10 ("You told me before for how long you watched TV. About how many blocks of commercials did you see during this period?"), q11 ("About how many commercials did you see during this period?"), and q12 ("What percent of these commercials have drawn your attention?"). Answering these questions requires at least retrieval of rather detailed information from long-term memory. Because the designer of this questionnaire was obviously aware of the difficulty of answering these questions, interviewers were instructed to help respondents to make the required estimates.

The fifth question (q17) is about the respondent's reaction when TV advertisements appear on the screen. It is part of a series of questions (q13–q19), a series introduced by the following introduction: "The next questions are about television advertising in general. I mention a number of reactions to commercials. Will you please indicate whether you react in this way every time, often, sometimes, or never, when advertisement appears at your TV screen?" Question q17 then reads: "You do something different for a while, but you leave the advertisement on: 4. every time; 3. often; 2. sometimes; 1. never." The interviewer has to explain, if necessary, that "doing something else" is meant as "doing something else near the TV-set, and *not* in another room."

Jury Ratings of Problematic Character of Survey Questions

Questions that induce problems in answering may worsen data quality. So it would be nice if a jury of experts could detect these questions in advance, or, stated more generally, if a jury could give a reliable and valid assessment of the "quality" of a question from a survey questionnaire. To test the feasibility of this idea, we asked 10 experienced survey researchers to evaluate the quality of the 37 questions from the CATI questionnaire. They were asked to rate each question on a scale going from 1 ("excellent question") to 7 ("worthless question"); for a comparable procedure see Presser and Blair (1994).

It turned out that the degree of consensus between the jury members was rather low: The standard deviation of the scores for some questions was more than two points on the seven-point scale. This indicates that members of the jury came to quite different assessments, probably because they partly used different criteria or weighted them differently.

Six out of the 37 questions obtained an average "problematic score" of 4.0 or more. Four of these, questions q9–q12, are the same as already identified in the preceding subsection as problematic, rather difficult, questions. The other two questions, q2 ("And on the days that you watch TV, how many hours or minutes on average do you watch TV?") and q22 ("I think that TV-commercials present meaningful information about the use of products by other people? 1. strongly disagree; 2. disagree; 3. agree nor disagree; 4. agree; 5. strongly agree?"), were rated earlier as "not so easy to answer" questions.

To describe the extent to which cognitive task scores and jury ratings lead to the same evaluation of the questions, we computed the rank order correlation between the scores on both procedures for all 37 questions. Spearman's rho is 0.35, indicating that there exists a positive but not very strong relationship between the two evaluations.

Identifying Deviations from Paradigmatic Sequence

In order to check these a priori assessments of the problematic character of the questions, we can look at the resulting sequences and see whether the interaction analysis identifies the same questions.

The coding of all verbal utterances included in the 1943 sequences analyzed makes it possible to identify all sorts of deviations from the paradigmatic sequence[5]. For the present interaction analysis we will restrict ourselves to four serious deviations from the paradigmatic sequence: incorrect question reading by the interviewer, not understanding the question by the respondent, directive questioning, and insufficient probing.

1. *Incorrect Question Reading.* This deviation from the paradigmatic sequence may seriously hamper data quality. Luckily this deviation hardly occurs in the CATI interview, except for two questions, q44 and q46. Both questions are scripted in such a way that the proper question sentence is followed by an explanation. For example, q46 reads "Are you working at the moment, that is, in a job, for a company, in government, or as self-employed?" Quite often this last part of the question is not read out, for example because the respondent already starts answering the question; see fragment 1:

Fragment 1 (q46)

```
I:     Hmm, are you working at the moment?
R:     No
I:     No
R:     I am answering the telephone. Hahaha
I:     Haha. Yes, Okay
R:     I work at home. Haha.
I:     Good, at home.
```

2. *Not Hearing and/or Understanding the Question.* The second deviation occurs when the respondent does not hear or understand the question and asks the interviewer for a repetition or an elucidation. Or the answer given indicates that the respondent has not understood the question properly. This happens in 6 percent of all sequences. The distribution of this type of deviation over the 37 questions is very uneven. One-third of all sequences starting with question q6 shows this problem. The question itself sounds very clear: "Where have you watched TV tonight?" But obviously many respondents do not understand "where" (in Dutch: *waar*) but "what" (in Dutch: *waarnaar*) and answer by mentioning a particular TV program; see fragment 2:

Fragment 2 (q6)

```
I:  Where have you watched TV tonight?
R:  The news and ugh..
I:  Goh..
R:  a few games of ''Lingo'' and ''I know better'' and that
    sort of things..
I:  Yes
```

3. *Directive Probing.* Another serious deviation from the paradigmatic sequence is a suggestive question or repetition of respondent's answer (see fragment 3). This error is also very unevenly distributed over the questions. It

turns out that one-third of the sequences that begin with question q21 contain this error:

Fragment 3 (q21)

```
I:      (reads statement) ''I find commercials funny to look at''
R:      No
I:      Haha, 'strongly disagree'
R:      Yes, absolutely, yes, you said that right.
```

4. *Insufficient Probing*. The interviewer is instructed to probe until an adequate response is given. However, it appears that interviewers, in about a quarter of the sequences, already type in a score in the computer, probably on the basis of some guesswork, but not on the basis of the information the respondent has provided as yet. Like the other deviations mentioned, this one is also unevenly distributed over the questions. See fragment 4, which presents question q2:

Fragment 4 (q2)

```
I:  And on the days that you[6] watch TV, how many hours or minutes do
    you watch on average?
R:  Well, I myself.. eh..work, I have a common forty hours-a-week job
R:  but my wife watches, let's say, about an hour or ten a day
I:  Ten? haha
R:  Yes
I:  Okay
    [Interviewer types in score 10 (=10 hours per day)]
```

Questions that are followed by sequences with one or more of these deviations may be identified as problematic.

It turns out that, although the questions themselves are on average not rated as difficult, deviations from the paradigmatic sequences are still quite common, in this survey especially "insufficient probing" by the interviewer. The distribution of the four deviations over the 37 questions is very uneven (in all four cases chi-square is highly significant; $p < .001$). It thus appears that particular questions are prone to particular deviations. Because not all questions were asked to all respondents, for every question we have divided the total number of observed deviations from the paradigmatic sequence by the number of sequences belonging to that question.

The seven questions that lead to over six deviations from the paradigmatic sequence per 10 sequences are the questions q14, q15, q18, q21, q26, q32, and q35.

Comparison of Three Procedures

In Table 18.4 we have summarized the outcome of the three procedures for identifying problematic questions. The table contains every question that is identified by at least one of the procedures as problematic. Such a question has a task assessment (TA) score of 6, a jury rating (JR) score of 5.0 or more, and/or an

Table 18.4 Fourteen Questions of CATI Survey Identified as Problematic by Either Cognitive Task Analysis (TA > 5), Jury Rating (JR > 5.0), and/or Interaction Analysis (IA > 0.60)

Question Identified as Problematic	TA (1)	JR (2)	IA (3)
q2: *And on the days that you watch TV, how many hours or minutes on average do you watch TV* ? [.. hours;.. minutes]	3	**5.1**	0.33
q9: *How many percent of the time did you watch attentively*? [. . . percent]	**6**	**5.6**	0.56
q10: *About how many blocks of commercials did you see during this period*? [. . . blocks]	**6**	**5.9**	0.48
q11: *How many commercials did you see during this period*? [. . . commercials]	**6**	**5.7**	0.40
q12: *How many percent of these commercials have drawn your attention*? [. . . percent]	**6**	**5.8**	0.42
q14: *I mention some possible reactions to advertisements on TV. Would you please indicate whether you do that every time, often, sometimes, or never, when commercials come on the screen? You switch to another channel*? [(4) every time; (3) often; (2) sometimes; (1) never]	4	3.9	**0.63**
q15 (idem): *You switch off the sound*?	4	3.5	**0.65**
q17 (idem): *You do something different for a while, but you leave the advertisement on*?	**6**	3.9	0.43
q18 (idem): *You leave the room*?	3	3.5	**0.63**
q21: *I find TV-commercials funny to look at.* [(1) strongly disagree; (2) disagree; (3) agree nor disagree; (4) agree; (5) strongly agree]	1	3.2	**0.69**
q22: *I think that TV-commercials present meaningful information about the use of products by other people.* (same alternatives as q21)	5	**5.1**	0.59
q26: *I find TV-commercials too glaring.* (same alternatives as q21)	1	4.5	**0.61**
Q32: *Are you in general negative or positive about radio advertising*? [(1) very negative; (2) negative; (3) neutral; (4) positive; (5) very positive]	2	4.1	**0.72**
Q35: *Are you in general negative or positive about advertising*? (same alternatives as Q32)	2	4.4	**0.65**
For the 1943 sequences following all 37 questions together	3.2	4.2	0.49

interaction analysis (IA) score of 0.60 or over. The scores that exceed these thresholds are printed in bold.

This brings us to the following three conclusions:

1. The fact that only 14 of the 37 questions appear in Table 18.4 means that 23 questions are rated as rather unproblematic by all three procedures used.

2. The problematic scores given by the jury (JR) lead to an identification of problematic questions that comes pretty close to the identification on the basis of the assessment of cognitive difficulty and clarity of the task (TA).
3. However, there are large differences between the two a priori assessments of the problematic character of a question (TA and JR) and the a posteriori assessment (IA). None of the seven questions identified as problematic by task assessment or jury rating produced relatively many deviations from the paradigmatic sequence. The word "relatively" must be stressed here: Most of the 37 questions lead to a considerable number of deviations: The average number of deviations per 100 sequences is 49. On the other hand, none of the seven questions that caused many deviations in the sequences were identified as problematic by either the TA or the JR procedure. These results call for a further analysis of the sequences belonging to the seven questions that caused many deviations.

Further Analysis of Seven Unexpected Problematic Questions

Table 18.5 presents the detailed information about the type of deviations that occur in each of the seven questions identified by interaction analysis (IA >

Table 18.5 Occurrence of Particular Deviation from Paradigmatic Sequence

Question	(a)	(b)	(c)	(d)	All
q14: *I mention some possible reactions to commercials on TV. Will you indicate whether you do that every time, often, sometimes, or never, when commercials come on the screen? You switch to another channel?* [(4) every time; (3) often; (2) sometimes; (1) never]	.02	.04	**.22**	.35	.63
q15 (idem): *You switch off the sound?*	.00	.02	.17	**.46**	.65
q18 (idem): *You leave the room?*	.00	.02	.20	**.41**	.63
q21: *I find TV-commercials funny to look at.* [(1) strongly disagree; (2) disagree; (3) agree nor disagree; (4) agree; (5) strongly agree]	.00	.00	**.33**	.35	.69
q26: *I find TV-commercials too glaring.* (same alternatives as q21)	.00	.06	.15	**.41**	.61
q32: *Are you in general negative or positive about radio advertising?* [(1) very negative; (2) negative; (3) neutral; (4) positive; (5) very positive]	.04	.06	**.22**	**.41**	.72
q35: *Are you in general negative or positive about advertising?* (same alternatives as q32)	.00	.05	.17	**.43**	.65
For all 1943 sequences following all 37 questions together	.03	.06	.14	.27	.49

Note: Columns are defined as (a) incorrect reading of the question; (b) not understanding the question; (c) directive questioning; (d) insufficient probing per sequence for each of the seven questions identified as problematic by the interaction analysis (IA Score > 0.60).

0.60). We mentioned above that incorrect reading of the question [column (a) in Table 18.5] does not occur frequently; it occurs most often with questions q44 and q46. We also mentioned that "not understanding the question" [column (b)] was particularly related to question q6, where "*Where* have you watched TV tonight?" was misunderstood as "*What* (program) have you watched at TV tonight?."

The deviations that really lead to a high IA score are (c) directive questioning and (d) insufficient probing. Especially question q21 ("I find commercials funny to look at") is prone to this first deviation. In fragment 3, presented above, we see how the interviewer "transforms" respondent's "no" into a "strongly disagree." The fact that the respondent agrees with the suggestion of the interviewer ("Yes, absolutely, yes, you said that right") is no guarantee that the response is not affected by the suggestion. The deviation from the paradigmatic sequence is not caused by inadequate question design, but by improper interviewer behavior. A further analysis of these suggestions made by several interviewers showed that all suggestions are in one direction: Commercials are not funny at all.

Insufficient probing (d) occurs in relation to questions q15 and q18. These questions look quite adequate at first sight: They were not identified as problematic by cognitive task analysis nor by the jury of experts. Only after hearing several interviews did we discover the problem with these questions: They belong to a series of seven questions asking about the frequency of various reactions to the appearance of TV commercials. Viewed apart, none of these seven questions is problematic. But together these questions cause problems for the respondent and the interviewer. For the answer to one question can make the reading out and the answering of the following question quite ridiculous: If the respondent has reported, in response to question q14 that he always switches to another channel, question q15, mentioning the option "switch off the sound," sounds rather strange. So if the respondent gives some type of reaction, the interviewer is very eager to "translate" that reaction into one of the four scripted response categories, even if the information provided by the respondent is not sufficient to do so. In these cases it is not the separate question that has caused the problem but the sequence of closely and even logically related questions.

Questions q32 and q35 lead to insufficient probing because in these telephone interviews the interviewer cannot use a show card. The question that has to be read out by the interviewer has only two response alternatives: negative or positive. But the interviewer has to score the response in the computer by entering a number representing one of the five response alternatives, ranging from 1 (very negative) to 5 (very positive). So every time a respondent has answered the question with negative, the interviewer has to further probe "Is that 'very negative' or 'ordinary negative'?" in order to choose between typing a 1 or a 2. This procedure leads to clumsy interactions, so interviewers seem to avoid this by interpreting the answer of the respondent in some way or the other. The story about question q26 is similar. Again, the formulation of the question and the response alternatives is not inadequate, but the fact that a show card is not available here leads to many inefficient interactions or to risky interpretations of the answers by the interviewer.

CONCLUSION

In this chapter it is argued that a detailed analysis of the interaction that takes place in survey interviews, looking for deviations from the paradigmatic question–answer sequence, may inform the survey researcher about difficulties interviewers and respondents have with the performance of their task.

It appears that the format and wording of a question from an instrument included in the questionnaire strongly affect the probability that a sequence starting with the question will be paradigmatic or not.

In a face-to-face survey the use of a show card to present the response alternatives will significantly increase the proportion of paradigmatic sequences. In telephone surveys the show card is usually not an option. Here the wording of the question and the related response alternatives become even more critical, and an assessment of the questions in terms of the problems they will probably induce become even more necessary. It was shown that an a priori assessment of the questions by judging the separate cognitive and interactional aspects of the task of interviewer and respondent comes close to a more global assessment by experts in the domain of survey research. Both procedures identified about the same questions as being problematic.

However, these problematic questions were *not* the questions that actually led to the largest proportion of deviations from the paradigmatic sequence. After closer inspection of these deviations, it is clear why these latter questions induce these unexpected problems. First, questions q14, q15, and q18 belong to a series of questions (q13–q19) that each may be formulated adequately; but the response ("I always do only A") to a preceding question may make posing the following questions of the series ("How often do you do B, C,..?") sound awkward, inducing the interviewer to reformulate the question, often in a suggestive manner, or to omit necessary further probing.

Second, questions such as q21, q26, q32, and q35, with alternatives that represent various degrees of acceptance of a statement (X) or degrees of appreciation of an attitude object (Y), which are asked in two steps ["Do you agree or disagree with X?"... "Is that strongly (dis)agree or ordinarily (dis)agree?" or "Are you in general negative or positive about Y?"... "Is that very negative or negative (respectively positive or very positive)?"], cause interactional problems if a respondent's reaction to the first question makes the second question sound rather redundant.

From a further analysis of the sequences it appears that different characteristics of the question or the questionnaire have different effects on the subsequent interaction:

The (cognitive) difficulty of the question may lead to longer reaction times (response latency; see Bassili and Fletcher, 1991; Bassili and Scott, 1996), to repeating of the question by the respondent, and to requests for elucidation of the question. However, as long as the respondent only mentions one of the scripted alternatives, after the question and the alternatives are read out by the interviewer, the question–answer sequence will be paradigmatic, not revealing these difficulties or errors made by respondents due to misunderstanding of the question or incorrect choice of the response category.

Inadequate response alternatives lead to "negotiations" between respondent and interviewer about the best choice of one of the scripted alternatives, to suggestive questions by the interviewer, or to insufficient probing.

Context effects: Responses to preceding questions may give the interviewer cues about the most likely response to the present question. These cues may lead to suggestive questioning or insufficient probing by the interviewer.

Our conclusion is that an interaction analysis as presented here, of interviews with a preliminary version of the questionnaire conducted with a small subsample, may point to problems in the questionnaire that will not be recognized by the a priori assessments (Foddy, 1996; Presser and Blair, 1994) or will not surface if only the response distributions are analyzed for the occurrence of response effects. On the other hand, interaction analysis is not a complete substitute for the other procedures used for the evaluation of questionnaires: Even a paradigmatic question–answer sequence may still transmit information based on misunderstanding of the question or errors in the selection of the response categories.

ACKNOWLEDGMENT

The authors thank Edith Smit (University of Amsterdam) for her kind permission to make use of data collected in her research project on opinions about TV commercials (Smit, 1999).

NOTES

1. For other procedures see Presser and Blair (1994) and Schwarz and Sudman (1996).

2. In the next section we will use a more elaborate inventory of components of question difficulty.

3. The present inventory of components of question difficulty is based on research on determinants of response errors in surveys (e.g., Schwarz and Sudman, 1996; van der Vaart, 1996). A more elaborate coding scheme for question difficulty has been used by Lessler and Forsyth, 1996.

4. It is interesting to note that all 37 questions have certain question characteristics in common, that is, have the same value on some of the 12 task difficulty variables. For example, none of the questions refers to a "hypothetical" or future situation (aspect 4), and in none of the questions is a responding aid device used (aspect 8). Moreover, generally the response alternatives fit quite well with the question, and they are evenly spread over the whole range of possible values

(aspects 6 and 7). Finally, the questions contain clear indications about proper turntaking (aspect 12).

The differences between the questions in the CATI survey regard other aspects than the differences between the nine questions in the PAPI survey mentioned earlier. In the PAPI survey the main differences regard aspects 4, 6, 7, 8, and 10. In the CATI survey the differences between the questions relate to aspects 1, 2, 3, 5, 10, and 11. This means that both data sets complement each other quite well.

5. The verbal utterances were coded with respect to the six variables mentioned in Chapter 17 of this volume. The intercoder reliability of the coding on all six variables together is rather high: kappa is 0.79.

6. Unlike English, Dutch makes a lexical distinction between the second-person singular personal pronoun *you* (the informal *jij* and the formal *U*) and the second-person plural personal pronoun (the informal *jullie* and the formal *U*). In all questions of this questionnaire "you" has the first-mentioned meaning, so respondent's information about his wife's behavior is inadequate and needs further probing.

REFERENCES

Bassili, J. N., and J. F. Fletcher. 1991. "Response-Time Measurement in Survey Research; A Method for CATI and a New Look at Nonattitudes." *Public Opinion Quarterly* 55: 331–346.

Bassili, J. N., and B. S. Scott. 1996. "Response Latency as a Signal to Question Problems in Survey Research." *Public Opinion Quarterly* 60: 390–399.

Beatty, P. 1995. "Understanding the Standardized/Non-Standardized Interviewing Controversy." *Journal of Official Statistics* 11: 147–160.

Brenner, M. 1982. "Response-Effects of 'Role-Restricted' Characteristics of the Interviewer." Pp. 131–165 in W. Dijkstra and J. van der Zouwen (Eds.), *Response Behaviour in the Survey-Interview*. London: Academic Press.

Converse, J. M., and S. Presser. 1986. *Survey Questions: Handcrafting the Standardized Questionnaire*. Beverly Hills, CA: Sage.

Dijkstra, W. 1994. "SEQUENCE, a Program for Analysing Sequential Data." *Bulletin de Méthodologie Sociologique* 43: 134–142.

Dijkstra, W., and J. van der Zouwen. 1988. "Types of Inadequate Interviewer Behaviour in Survey-Interviews." Pp. 24–35 in W. E. Saris and I. N. Gallhofer (Eds.), *Sociometric Research, Vol. 1: Data Collection and Scaling*. London: MacMillan.

Dobson, K. S., and K. J. Mothersill. 1979. "Equidistant Categorical Labels for Construction of Likert Type Scales." *Perceptual and Motor Skills* 49: 575–580.

Foddy, W. 1996. "The In-Depth Testing of Survey Questions: A Critical Appraisal of Methods." *Quality and Quantity* 30: 361–370.

Houtkoop-Steenstra, H. 1995. "Meeting Both Ends. Between Standardization and Recipient Design in Telephone Survey Interviews." Pp. 91–106 in P. ten Have and G. Psathas (Eds.), *Situated Order: Studies in the Social Organization of Talk and Embodied Activities*. Washington, DC: University Press of America.

Houtkoop-Steenstra, H. 1996. "Probing Behaviour of Interviewers in the Standardised Semi-Open Research Interview." *Quality and Quantity* 30: 205–230.

Lessler, J. T., and B. H. Forsyth. 1996. "A Coding System for Appraising Questionnaires." Pp. 259–292 in N. Schwarz and S. Sudman (Eds.), *Answering Questions: Methodology for Determining Cognitive and Communicative Processes in Survey Research*. San Francisco, CA: Jossey-Bass.

Presser, S., and J. Blair. 1994. "Survey Pretesting: Do Different Methods Produce Different Results?" Pp. 73–104 in P. V. Marsden (Ed.), *Sociological Methodology 1994*. Oxford: Blackwell.

Schaeffer, N. C., and N. M. Bradburn. 1989. "Respondent Behavior in Magnitude Estimation." *Journal of the American Statistical Association* 84: 402–413.

Schaeffer, N. C., and D. W. Maynard. 1996. "From Paradigm to Prototype and Back Again: Interactive Aspects of Cognitive Processing in Standardized Survey Interviews." Pp. 65–88 in N. Schwarz and S. Sudman (Eds.), *Answering Questions: Methodology for Determining Cognitive and Communicative Processes in Survey Research*. San Francisco, CA: Jossey-Bass.

Schwarz, N., and S. Sudman (Eds.). 1996. *Answering Questions: Methodology for Determining Cognitive and Communicative Processes in Survey Research*. San Francisco, CA: Jossey-Bass.

Smit, E. G. 1999. "Mass Media Advertising: Information or Wallpaper?" Ph.D. dissertation, University of Amsterdam, Amsterdam.

Spector, P. E. 1976. Choosing Response Categories for Summated Rating Scales. *Journal of Applied Psychology* 61: 374–375.

Sykes, W., and M. Collins. 1992. "Anatomy of the Survey Interview." *Journal of Official Statistics* 8: 277–291.

van der Vaart, W. (1996). "Inquiring into the Past; Data Quality of Responses to Retrospective Questions." Unpublished Ph.D. dissertation, Vrije Universiteit, Amsterdam.

van der Zouwen, J., and W. Dijkstra. 1995. "Trivial and Non-Trivial Question–Answer Sequences; Types, Determinants and Effects on Data Quality." Pp. 81–86 in *Proceedings of the International Conference on Survey Measurement and Process Quality in Bristol (UK)*. Alexandria, VA: American Statistical Association.

van der Zouwen, J., and W. Dijkstra. 1996. "The Impact of the Question on the Interactions in Survey-Interviews." Paper presented at the Fourth International ISA Conference on Social Science Methodology, July 1–5, University of Essex, Colchester United Kingdom.

CHAPTER 19

How Interviewers Make Coding Decisions

Tony Hak

INTRODUCTION: THE PHENOMENON

According to the paradigmatic version of the standardized survey interview, the interviewer presents a *stimulus* consisting of the question exactly as provided by the instrument with the complete list of response categories (if applicable). The respondent then replies by giving a *response* consisting of one of the response categories, which subsequently is recorded by the interviewer (see, e.g., Fowler and Mangione, 1990). This sequence can be represented by the three-part structure shown in Figure 19.1 in which *Q*? is the question as provided by the instrument, *a*, *b*, *c*,..., *n* is the complete list of response categories, and # designates the interviewer entering a number at the keyboard (usually a code for a specific response category). When the respondents' everyday categories differ from those in the list of response categories provided in the instrument, their task can be viewed as a *coding* task. Something that respondents know in their own terms must be mapped onto the categories that are provided by the instrument. Take the following (partially fictitious — see below) example in which the respondent is a platoon leader in the army:

Example 1

```
1  Iter   ↑Oh kay: and do you work for yourSELF (h)in a fa(h)mily
          business or for someone else:? {q359h}
2  Resp   I work for someone else
3  Iter   ((# 3 =someone else))
```

Standardization and Tacit Knowledge: Interaction and Practice in the Survey Interview
Edited by Douglas W. Maynard, Hanneke Houtkoop-Steenstra, Nora Cate Schaeffer, and Johannes van der Zouwen
ISBN 0-471-35829-0

```
1    Iter    Q? a, b, c,..., n
2    Resp    [one of a, b, c,..., n]
3    Iter    ((# one of a, b, c,..., n))
```

Figure 19.1

This respondent performs a coding task by choosing the appropriate category ("someone else") from the list provided in the instrument. In this paradigmatic version the interviewer's task is confined to providing the stimulus and the list of responses from which the respondent must make a choice. The respondent does the coding, not the interviewer. In this chapter a deviation from this paradigmatic version is discussed in which the interviewer—not the respondent—does the coding, that is, the mapping of the "information" offered by the respondent onto the response categories. Extract 1—an actual extract from which (the fictitious) Example 1 was derived—gives an example of this type of interviewer behavior.

Extract 1: AW02:706 {q359h}

```
Iter  ↑Oh kay: and do you work for yourself (h)in a fa(h)mily
      business or for someone else:? {q359h}
Resp  uh:: I work for the gov[ernment
Iter                          [((# 3 =someone else))
```

The respondent does not choose one of the three response categories provided ("self," "family business," or "someone else"), which then could be recorded by the interviewer, as in Example 1, but instead gives a "report" about his employer, "the government" (see Schaeffer and Maynard, this volume, Chapter 12). When questions are *closed* and response categories are provided, the paradigmatic version of standardized survey interviewing requires the interviewer to repeat the response categories provided in the instrument, as in Figure 19.2. (In this figure, part 2p is the respondent's "problematic" answer and part 1r is the paradigmatic "repeat" of the response categories.)

If the interviewer in extract 1 had behaved in a paradigmatic fashion, she would not have coded the respondent's answer ("for the government") as for "someone else." She would have repeated the response categories instead, as in the fictitious Example 2.

```
1     Iter    Q? a, b, c,..., n
2p    Resp    problematic answer [not one of a, b, c, ..., n]
1r    Iter    a, b, c,..., n ((repeat))
2     Resp    [one of a, b, c,..., n]
3     Iter    ((# one of a, b, c,..., n))
```

Figure 19.2

Example 2

```
1   Iter  ↑Oh kay: and do you work for yourself (h)in a
          fa(h)mily business or for someone else:? {q359h}
2p  Resp  uh:: I work for the government
1r  Iter  So would you say for yourself, in a family business,
          or for someone else?
```

Throughout this chapter, I will call interviewer behavior as represented in Example 2 *paradigmatic* and interviewer behavior that does not fit this paradigmatic pattern, such as in extract 1, *problematic* or *nonparadigmatic*. (I owe the term *paradigmatic* to Schaeffer and Maynard, 1996.)

The problem posed by questions such as Q359h ("Do you work for yourself, in a family business or for someone else?") is that the respondent's information about his situation (i.e., about his employer) needs to be coded into one of a limited set of categories defined by the investigator. Schober and Conrad (this volume, Chapter 3) discuss problems that arise when a respondent does not use the *legal* definition of the concepts used in the question, that is, when there is a miscalibration between the concepts as intended by the investigator and the ones used by the respondent. They also discuss how standardized interviewing might be made more flexible (*conversational*) to allow the detection and, if necessary, correction of such miscalibrations. Schober and Conrad assume that the respondent should perform the coding task, which is in accordance with the paradigmatic version of standardized interviewing. That is why, if necessary, the respondent should be helped by the interviewer to perform this task. Nonparadigmatic interviewer behavior, as presented in extract 1, is characterized by the fact that the interviewer performs the coding task that in the paradigmatic version belongs to the respondent.

This chapter will analyze in what circumstances this type of nonparadigmatic interviewer behavior, which I will call *immediate coding*, occurs and the tacit knowledge that is involved. To do so, I will compare immediate coding with a similar but different type of nonparadigmatic interviewer behavior, of which an example is provided by extract 2 (taken from the same interview as extract 1).

Extract 2: AW02:176 {q7d}

```
1   Iter  Sharing information between different agencies of
          Government when both want the same information .hhh saves
          time and money? but it ↑also means a loss in privacy for
          thee individual .hhh Do you think the benefits of ↑saving
          time and money outweigh the costs in privacy .hh or do you
          think the loss of privacy outweighs the benefits of saving
          time and money {q7d}
          (2.0) ((unidentifiable noises))
2p  Resp  I'd say it was the purpose of the census that (.) it's more
          beneficial duh:- (0.2) t' share the information
2rc Iter  .hhh eh- Okay so (0.4) you would think the benefits of
          saving time and ↑money outweigh the costs in privacy?
          (0.6)
2c  Resp  Yeah
3   Iter  ((# 1 =benefits of saving time/money outweigh costs
          in privacy))
```

```
1     Iter   Q? a, b, c,..., n
2p    Resp   problematic answer [not one of a, b, c,..., n]
3     Iter   ((# one of a, b,c,..., n))
```

Figure 19.3 Immediate coding

```
1      Iter   Q? a, b, c,..., n
2p     Resp   problematic answer [not one of a, b, c,..., n]
2rc    Iter   [one of a, b, c,..., n]? ((request for confirmation))
2c     Resp   ((confirmation))
3      Iter   ((# one of a, b,c,..., n))
```

Figure 19.4 Coding with confirmation

As in extract 1, the interviewer in extract 2 also performs a coding task by mapping the respondent's answer onto one of the response categories, but in this case she offers it as a candidate answer to the respondent before recording the code. (In the extract, part 2rc designates the interviewer's request for confirmation and part 2c designates the respondent's confirmation.) I will call this type of nonparadigmatic interviewer behavior *coding with confirmation*. These two types of nonparadigmatic interviewer behavior can be schematically represented as in Figures 19.3 and 19.4.

Coding with confirmation can be seen as a modified form of *immediate coding*. The difference between the two is that in the pattern represented in Figure 19.4 the interviewer's choice is offered to the respondent as a candidate response after or simultaneously with recording it (coding with confirmation). This offer (part 2rc) can be heard as a request for confirmation and the respondent can (optionally) do this (part 2c).

Coding with confirmation is a form of nonparadigmatic, suggestive *probing*. I use the term *coding with confirmation* in this chapter to emphasize the "coding" element rather than the "confirmation" or probing part. By using this terminology I want to draw attention to the processes of inference making, which are characteristic of both immediate coding and coding with confirmation rather than to the effects of probing on the respondent's response (as discussed in Smit, Dijkstra, and van der Zouwen, 1997).

INITIAL EXPLORATION: DISTRIBUTION OF THE PHENOMENON

This chapter analyzes a collection of 29 interviews conducted at the University of Wisconsin Survey Center in 1995, using a national amalgam survey with a random digit dialing design. During the period of data collection, each interviewer who worked on the survey was asked to record approximately two interviews; the interviews are not a probability sample, but they include both between-interviewer and within-interviewer variation. Interviews analyzed here were tape recorded at the University of Wisconsin Survey Center (UWSC) from the end of June through mid-July 1995. Each of 14 interviewers tape recorded two interviews, and two interviewers each recorded one interview. The recording failed for one interview, so there is a total of 29 interviews. Interviewers were instructed to request permission to record all interviews on their shifts until they had recorded two interviews. The interviews collected were part of the UWSC's Continuous National Survey (WISCON), a survey that uses a random digit dialing (RDD) design to interview a small replicate of a random sample of the continental United States every day. The survey instrument covered a wide range of topics, as illustrated in the interview transcribed in Appendix 2. Many of the interviewers employed on the CNS were, or had been recently, students at the University of Wisconsin, Madison. The initial data set used for the analysis in this chapter consisted of all question–answer sequences in six interviews, which had been transcribed in more detail (including the sound of strokes on the keyboard) than the remaining 23 interviews (i.e., AW01, AW02, AW03, AW04, AW21, and AW22). Initial analysis of this data was followed up by the "testing" of provisional hypotheses in the other 23, more "roughly" transcribed interviews.

Cases for the initial analysis were selected from the entire set of question–answer sequences in the interview proper, that is, after the respondent had consented to participating in the interview and after eligibility checks had been completed. The number of sequences per interview varies slightly because some questions were only relevant to some respondents, but on average there are 105 question–answer sequences per interview, making a total of about 630 question–answer sequences. About 9 percent of these sequences (58) were categorized as sequences of either immediate coding (as in Figure 19.3; 41 cases) or as coding with confirmation (as in Figure 19.4; 17 cases). Immediate coding and coding with confirmation are not distributed in the same way over the six interviews (Table 19.1) and over different types of questions (Table 19.2). Immediate coding is about 2.5 times more frequent than coding with confirmation and is more evenly distributed over the six interviews (see Table 19.1). This seems, at least in part, to be an interviewer effect. Instances of coding with confirmation cluster in the interviews AW01 and AW02, which both had the

Table 19.1 Distribution of Immediate Coding and Coding with Confirmation over Six Interviews

	Immediate Coding	Coding with Confirmation	Total
AW01 (Jane)	7	6	13
AW02 (Jane)	7	6	13
AW03 (Sue)	4	2	6
AW04 (Sue)	9	1	10
AW21 (Jill)	8	—	8
AW22 (Jill)	6	2	8
Total	41	17	58

Table 19.2 Distribution of Immediate Coding and Coding with Confirmation over Question Types

	Immediate Coding	Coding with Confirmation	Total
Yes/no questions	19	3	22
Two options (other types)	6	12	18
Other question types	16	2	18
Total	41	17	58

same interviewer (Jane). There is no apparent difference between the other two interviewers, Sue (AW03 and AW04) and Jill (AW21 and AW22), in this respect. AW03 and (to a lesser degree) AW22 also show a low number of instances of immediate coding.

In terms of distribution over question types, almost 70 percent of all instances of immediate coding and/or coding with confirmation occur in answers that follow questions with only two response options. In Table 19.2 this category of questions has been subdivided in two subgroups: yes/no questions and other questions with only two response options. Because of the high incidence of immediate coding and coding with confirmation in answers following yes/no questions and other questions with only two response options, I will first discuss these instances of interviewer coding, before discussing examples of the other instances.

YES/NO QUESTIONS

There are two types of yes/no questions in this data set. One type concerns *knowledge* or *beliefs* about facts, such as a series of five questions of the form "Do you think that the census is used for X?" (q5b-j). The other type concerns *opinions*, such as "Would you be willing to X?" (q9d). Both types of questions are read by the interviewer without explicit response categories (yes or no) that are instead projected by the form of the question. Yes/no questions show a rather high incidence of immediate coding (19 instances) and a relatively much lower incidence of coding with confirmation (3 instances). Therefore, a question to be answered in the analysis is not only why there is such a high incidence of nonparadigmatic interviewer coding but also why interviewers usually do not seek confirmation from the respondent for their choice from the response categories.

I first analyzed examples of paradigmatic question–answer sequences in the initial data set of six interviews and then developed hypotheses about reasons for (nonparadigmatic) interviewer coding and about the tacit knowledge that is involved. These hypotheses were then tested in the remaining 23 interviews. Let us first look at an almost perfectly paradigmatic (and as such exceptional) series of question–answer sequences: q5b-j in interview AW03 (extract 3).

Extract 3: AW03:82 {q005b-j}

```
 1  Iter People have different ideas about what the census is used
          for? .hhh I'm going to read you some of them (.) As I
          read each one .hhh ↑please tell me whether you think the
          census is used for that purpose .hhh to find a:reas in
          the country that need government help? .hh Do you think
          the census is used for that purpose? {q005b}
          (1.5)
 2  Resp n:: Gosh I d'n' kn(h)ow huh h[uh
 3  Iter                                [n:'kay:?((# 8 it =don't know))
    Resp huh [huh
    Iter     [.hhh ((cough))
    Resp Just uh (0.3) never given it much thought
 1  Iter .hhh ↑And is the census used to decide how many
          representatives each state will have in congress? {q005d}
 2  Resp Yes
 3  Iter ((# 1 =yes)) (0.8)
 1  Iter (tch) ↑And is it used  to see what changes have taken
          place in the United Sta:tes:? {q005f}
          (1.1)
 2p Resp hh huh huh huh .hhh (oh: gee) You can tell I haven't kept
          up huh [huh n:uh::?
    Iter         [hhh
 1r Iter You can also choose don't know if- if you don't know
          [any of thee answers to these]
 2  Resp [Oh   no   I-  I   really   d]on't kn[ow:
 3  Iter                                       [((# 8=don't know))
 1  Iter ↑Okay: .hhh ↑and do the pol↑ice and thee F B I use the
          census to keep track of trouble makers:? {q005h}
          (1.0)
 2p Resp Well I hope not
    Iter hhh
    Resp hhhuh
 1r Iter So(h)- so(h) would [that be: a yes or a [no (ma'am)
    Resp                    [huh                 [huh.hhh
 2  Resp huh huh.hhh We::ll: (m m) now I'm gonna say no: but
 3  Iter ((# 2 =no))
    Iter [.hhh
    Resp [I c- be wrong  eh:
 1  Iter (tch) ↑And how 'bout to locate illegal aliens? Is the
          census used for tha:t? {q005j}
          (0.9)
2p  Resp For what
1r  Iter .h To LOcate illegal aliens:?
          (1.6)
2   Resp .h Well I would- YES
          (0.6)
3   Iter [((# 1  =yes))
    Resp [I would hope so
```

Questions q5b-j are closed questions that project the response categories "yes" and "no" (which are not read by the interviewer). "Don't know" is also an allowed response, which is listed on the interviewer's screen. The respondent's answer to q5b is "I d'n' kn(h)ow," which is immediately recorded by the interviewer. The respondent's clarification "never given it much thought" is not hearably acknowledged by the interviewer but apparently does not lead to a change in the code that was recorded. Q5d is followed by the immediate response "yes" and the corresponding code is recorded accordingly. The initial response to q5f is a pause, followed by much hesitation ("hh huh huh huh .hhh") and the remark "you can tell I haven't kept hh huh huh huh up?" The interviewer mentions that the answer "don't know" is acceptable, which then is chosen by the respondent. Strictly speaking, this interviewer behavior is not paradigmatic—compare q5d in extract 4—but I do not consider this an instance of coding with confirmation because I restrict that category to interviewer coding of one of the core response categories—in this case "yes" and "no"—only. Next, in the answer to q5h, the interviewer reacts paradigmatically to the problematic response "well I hope not" by repeating the response categories. The respondent chooses "no," which is coded accordingly. Finally, the respondent requests a clarification of q5j, which is provided paradigmatically by a repeat of the requested part of the question. The answer "yes" is coded accordingly, after which the addition "I would hope so" is not taken into account. In conclusion, the interviewer (Sue) has recorded the codes for "yes," "no," and "don't know" when the respondent had mentioned those response categories. This has been done in an entirely paradigmatic way, with one exception (q5f) in which the interviewer has mentioned the availability of the response category "don't know" to the respondent after it had appeared that she had great difficulty in choosing either "yes" or "no." This sequence shows that Sue is able to handle this specific series of five questions in an entirely paradigmatic way. Let us now, by contrast, look at her other interview (AW04, extract 4) in this data set.

Extract 4: AW04:76 {q5b-j}

```
1   Iter  ((tch)).hhh People have different ideas about what the
          census is used for? .hhh and I'm going to read you some
          of them.  .hhh As I read each one (.) please tell me
          whether you think the census is used for that purpose.
          .hhh To find areas of the country that need government
          help..hh Do you think the census is used for that
          purpose? {q5b}
2p  Resp  I have no idea
3   Iter  ((# 8 =don't know))
1   Iter  .hh A:nd is the census used to decide how many
          representatives each state will have in congress? {q5d}
2p  Resp  .hhhh hhhhh I never thought about it.
1r  Iter  n'kay? So would that be: uh:m: (.) a yes no or don't
          [know ma'am]
2   Resp  [don't know]
    Iter  n'[kay? .hhh
2   Resp    [I don't know I guess
3   Iter  ((# 8 =don't know))
1   Iter  .h A:nd is:- is it used to see what changes have taken
          place in the United States? {q5f}
2   Resp  Probably [yes
```

```
3    Iter              [((# 1  =yes))
1    Iter    .hh And do the police and the F B I use the census to
             keep track of trouble makers? {q5h}
2p   Resp    I doubt it
3    Iter    ((# 2 = no))
1    Iter    .hh And how about to locate illegal aliens. Is the census
             used for tha:t? {q5j}
2p   Resp    N:ot that I know [o:f
3    Iter                     [((# 2 =no))
```

There are (only) two paradigmatic question–answer sequences in this set of five; namely the answer sequences to q5d and q5f. The interviewer's explicit mentioning of all response categories (including "don't know") in the first of these sequences (q5d)—"that be: uh:m: (.) a yes no or don't know ma'am"—is a more paradigmatic version than her behavior in a similar instance in the answer to q5f in AW03 above. The three nonparadigmatic instances are the immediate coding of the answer "I have no idea" to q5b as "don't know" and of the answers "I doubt it" to q5h and "not that I know of" to q5j as "no."

In these six interviews there are 16 other instances of immediate coding to yes/no questions. They all involve the immediate coding of a short phrase by the respondent as either "yes" or "no" or "don't know." Extracts 5 and 6 give two examples from other interviewers than the one discussed above.

Extract 5: AW01:100 {q5b}

```
1    Iter    Please tell me whether you think the census is used for
             that purpose: .hhh To find areas of the country? that need
             government help? (0.4) Do you think the census is used for
             that purpose:? {q5b}
             (1.3)
2p   Resp    I have no idea.
3    Iter    ((# 8 =don't know))
```

Extract 6: AW21:66 {q5f}

```
1    Iter    Is it used to see what changes have taken place in the
             United States? {q5f}
             (0.4)
2p   Resp    Uh↑hunh
3    Iter    ((# 1 =yes))
```

Table 19.3 presents a complete list of the phrases used by the respondents in these six interviews and of how these are coded by the interviewers. The list in Table 19.3 shows that it is difficult to make a distinction between more problematic and less problematic cases of immediate coding. On the one hand, "I have no idea" might be seen as a straightforward synonym of "don't know" and precisely for that reason it might be pragmatically unfeasible for the interviewer to react with the paradigmatic "would that be a yes no or don't know ma'am?," which is Sue's response in the answer to q5d in AW04 (extract 4). On the other

Table 19.3 Respondents' Phrases Immediately Coded by Interviewers

Respondents' Phrase	Interviewers' Immediate Code	Frequency
Um-hmm, mm hmm, uh^hunh	Yes	6
I'd say so	Yes	1
Right	Yes	1
That's what's supposed to be for	Yes	1
Probably	Yes	1
I suppose	Yes	1
I don't think so	No	2
I doubt it	No	1
Not that I know of	No	1
(I have) no idea	Don't know	4

hand, it is less evident (to me) why answers such as, for example, "I doubt it" and "not that I know of" can be seen as straightforward synonyms of "no."

My hypothesis about these 19 instances of immediate coding of a nonparadigmatic answer to a yes/no question is the following. In these cases, the interviewer finds it virtually impossible to repeat the set of response categories ("that be a yes no or don't know ma'am?") because, pragmatically speaking, the respondent has already provided one of them. In other words, in these cases the interviewer (tacitly) *knows* that the respondent's phrase *is* a "yes" or a "no" or a "don't know." For all practical purposes, the respondent has already done the coding that is required. The interviewer's task is, therefore, just to record the code chosen by the respondent.

This hypothesis can be "tested" in the remaining 23 interviews in the following way. If the hypothesis is correct—that is, if the respondents' phrases listed in Table 19.3 straightforwardly *mean* "yes" or "no" or "don't know"—the following should always be true for all interviews, without exception:

The phrases of Table 19.3, when given by respondents as an answer to a yes/no question, should never be followed by a paradigmatic interviewer's repeat of the response categories, *and*

The interviewer's code of a specific phrase should always be the same, that is, "yes" *or* "no" *or* "don't know," *and*

The interviewer should not seek a confirmation of this code.

The result of applying this test to the corpus of 29 interviews is that all three predictions proved to be correct. Not all phrases from Table 19.3 appear in answers following yes/no questions in the other interviews, but if they appear the interviewer immediately codes them in the same way as in the six interviews of the initial corpus.

We have discussed the 19 instances of immediate coding for yes/no questions in the initial corpus of six interviews. There are also 3 instances of coding with confirmation in this corpus. These cases concern phrases other than the ones listed in Table 19.3. One instance is the phrase "I have no qualms about that" in extract 7.

Extract 7: AW03:201 {q9f}

```
1     Iter Would you be willing to give your social security number
           to the census bureau? (°.hhh°) so they could match the
           information in your health care record? with thee
           information in their own files? {q9f}
2p    Resp Well (0.6) I have no qualms about that huh
2rc   Iter Okay so that would be a yes ma'am?
3c    Resp Ye:[s=
3     Iter     [((# 1 =yes))
```

In this extract, the interviewer treats "I have no qualms about that" in a different way than "I doubt it" and "not that I know of" in her other interview (AW04), which are immediately coded as "no." The interviewer infers that "I have no qualms about that" means "yes" but it is not a straightforward version of "yes" in the same way as "I doubt it" means "no." We can only speculate about the reasons for this. The interviewer might be unfamiliar with the term "qualms" or could want to make sure she heard the entire answer, for example, to distinguish between "I have qualms about that" and "I have no qualms about that."

The second instance of coding with confirmation in this corpus is the following one (extract 8).

Extract 8: AW03:636 {q270}

```
    Iter    .h And how about feel lonely {q0195}
            (2.8)
    Resp    m::
            (1.2)
    Iter    tch
            (0.5)
    Resp    Well I have a lot of that (0.6) I lost my husband
            (0.9)
    Iter    tch I'[m s]orry to hear that
    Resp          [( )]
            (0.2)
    Resp    Uh: hhhh (1.2) buht (0.4) Let's see m (.)
    Iter    .hh (0.4) hh
    Resp    My kids stay c-in (0.2) constant contact with me so I
            I'm not really that lonely (.) never
            (0.4)
    Iter    mkay So what number would you want to put on
            th[at
    Resp      [Oh: maybe two=
    Iter    =two?
    Resp    Yeah
            [data omitted]
1,3 Iter    .hh And have you ever been-oh yes you have been
            married {q0270} .hh an:d (0.4) you said your husband had
            just died? {q0275}
    Resp    No well he's been gone
```

In the last lines of extract 8, the interviewer codes the answer to q270 as "yes" and at the same time tells the respondent that she does so. The respondent accepts this. The rather simple explanation for this case of interviewer coding is provided by the first part of the extract in which the loss of a "husband" has been mentioned.

Arguably the most interesting case in this small sample of three instances is the one in extract 9 in which the interviewer does not accept the answer given ("no") and seeks confirmation from the respondent for choosing the response category "yes."

Extract 9: AW02:748 {q410}

```
1     Iter   A::nd now we'd like to ask you about health insurance
             .hh do you have any health insurance coverage? {q410}
             (0.9)
2     Resp   N::o I just have what the:- (.) the government provides
             (0.5)
      Iter   Okay so- (0.4) that i[s:-
2p    Resp                         [With the military uh I have
             somethin' I'm- (0.8) I'm not sure what the:-
             (0.6)
3     Iter   ((# 1 =yes))
      Resp   =It's just the standard (0.5) military
             (0.6)
      Iter   Okay [so-
      Resp        [benefits
             (1.0)
2rc   Iter   °y-° to a >certain extent< (0.3) you do have health
             insurance
2c    Resp   Yes
```

Although the respondent initially gives a (probably incorrect) negative answer ("no, I just have..."), the interviewer records an *affirmative* response on her screen (1 =yes) after the respondent has provided a "report" about his health insurance situation ("what the government provides with the military") and has expressed his uncertainty about how to answer this question ("I'm not sure what the..."). The respondent seems to be uncertain about whether, for the purposes of the survey, his situation counts as "having health insurance coverage." This might be due to the respondent noting an ambiguity: Is the question asking whether he "has" insurance or whether he has obtained it for himself, or bought it himself. Does he "have" it or was it "provided" for him? In any case, he seems to be unsure how to map his situation onto the response categories. He gives a report about his situation and leaves drawing the upshot to the interviewer (see Schaeffer and Maynard, this volume, Chapter 12) who "codes" it as an affirmative response. Whether this code is correct depends on what the investigator would consider "health insurance coverage," that is, it depends on coding instructions that are not provided with the question. I assume that the coding in extract 9 has been done validly (i.e., according to the investigator's concept of health insurance coverage) and reliably (i.e., another interviewer would have chosen the same category, assuming that interviewers know that the military provides health insurance).

The results of this analysis imply that, for all practical purposes, respondents' answers to yes/no questions as listed in Table 19.3 *are* versions of "yes," "no" and "don't know," and that the corresponding 19 instances of immediate coding do not count as forms of nonparadigmatic interviewer behavior. Obviously, this implies a considerable reduction of the number of cases of immediate coding in our sample. In contrast, the respondent's phrase "I have no qualms about that" appears to have no standard meaning which, therefore, must be confirmed by the interviewer. Among all 22 instances of interviewer coding after yes/no questions, only one instance of a genuine coding problem was found, that is, an instance in which a researcher's definition of a category (in this case his concept of "having health insurance coverage") is involved. Let us now look at the instances of interviewer coding in answers that follow other kinds of questions with only two response options.

OTHER TWO-OPTIONS QUESTIONS

In a large majority of the cases discussed in the previous section, the interviewer treats a phrase such as "I doubt it" as a version of "no." For all practical purposes such a phrase *is* a version of "no" and in that sense there is no coding involved. In the majority (12) of the 18 nonparadigmatic answers following other questions that require a choice between only two response categories, the interviewer explicitly seeks confirmation of her understanding of what the respondent has said (see Table 19.2). This gives the respondent an opportunity to confirm or deny such an interpretation. Typical examples are the answers in extracts 10, 11, and 12.

Extract 10: AW01:172 {q7d}

```
1     Iter   Sharing information between different agencies of
             Government when both want the same information (.) saves
             time and money? but it also mean some loss in privacy
             for the individual. .hhh Do you think the benefits? of
             saving time and money (0.4) outweigh the costs in privacy,
             .hh or: do you think the loss of privacy outweighs the
             benefits of saving time and money {q7d}
             (1.0)
2p    Resp   The first one
2rc   Iter   Okay benefits of saving time and money: outweigh: (0.5)
             costs in privacy
2c    Resp   °mm hmm°
3     Iter   Okay ((# 1 =benefits of saving time/money outweigh costs
             in privacy))
```

Extract 11: AW03:150 {q7d}

```
1     Iter   Sharing information between different agencies of
             government? .hhh when both want the same information?(0.2)
             saves time and money .hhh but it also means some loss in
             privacy for the individual .hhh Do you think the benefits
             of saving time and money: outweigh the costs in privacy?
```

```
               °.hhh° or do you think the loss of privacy outweighs
               benefits of saving time and money {q7d}
               (1.2)
2p    Resp     (hhh) (1.9) We::ll I think the privacy takes: (2.0) the
               loss of privacy's more important
2rc   Iter     n:'kay So you would say the loss of privacy outweighs
               benefits of saving time and money?
2c    Resp     Ye:[s
3     Iter        [((# 2 =loss in privacy outweighs benefits of saving
               time/money))
```

Extract 12: AW02:176 {q7d}

```
1    Iter   Sharing information between different agencies of government
            when both WANT the same information .hhh saves time and
            money? but it ↑also means a loss in privacy for thee
            individual .hhh Do you think the benefits of ↑saving time
            and money outweigh the costs in privacy .hh or do you think
            the loss of privacy outweighs the benefits of saving time
            and money {q7d}
            (2.0) ((unidentifiable noises))
2p   Resp   I'd say it was the purpose of the census that (.) it's more
            beneficial duh:- (0.2) t' share the information
2rc  Iter   .hhh eh- Okay so (0.4) you would think the benefits of
            saving time and ↑money outweigh the costs in privacy?
            (0.6)
2c   Resp   Yeah
3    Iter   ((# 1 =benefits of saving time/money outweigh costs
            in privacy))
```

In extract 10 the respondent points unambiguously to a category ("the first one") without formulating the answer in so many words. In extract 11 the respondent paraphrases the category provided in the instrument ("more important" for "outweigh"), and in extract 12 this appears to be the case as well ("more beneficial" for "outweigh"). One might hypothesize that the respondents in these cases are just avoiding the effort of repeating the whole nine-word response category — extract 10 ("the first one") points in that direction — but the answers that *are* given are equally elaborate. Furthermore, the same phenomenon occurs in answers following another set of questions in which the implied response categories are rather simple ("support" or "oppose"), as in extract 13.

Extract 13: AW02:518 {q120, q130}

```
1    Iter   Do you sup↑port or op↑po:se the following ↑measures to deal
            with aids .hh (0.4) Pro-hibit students with the aids ↑virus
            from attending public schools {q120}
            (1.0)
     Resp   I'd disagree with [that
2rc  Iter                     [.hhh so (.) do you oppose it?
2c   Resp   I:'d- I'd uh- (0.1) .hhh I would oppose b- (0.4) keeping
            people out of school cause of they had aids
3    Iter   ((# 2 =oppose)) Okay got it
1           .hhh A:nd how bout (.) develop? a government information
```

```
              program (0.4) to promote safe sex practices? such as the
              use of condoms {q125}
              (2.9)
2    Resp     I would oppose it cause I think they already got measures
              like that in place
3    Iter     ((# 2 =oppose)) Okay so you would oppose that
1             .hhh Permit insurance companies to test ↑applicants (.)
              FOR thee ai:ds virus: {q130}
              (1.3)
2p   Resp     .hhh I would uh:: be for that
2rc  Iter     .hhh So you would support it?
3             ((# 1 =support))
```

Respondents often provide paraphrases such as "disagree" (as in extract 13) for "oppose" and "I would be for that" for "support." In some cases the paraphrasing is ambiguous, as in extract 14.

Extract 14: AW01:442 {q120, q125}

```
1     Iter   Do you support or op↑po:se the following measures to deal
             with ai:ds .hhhh prohibit students with aids virus? from
             attending public school? {q120}
             (0.4)
2p    Resp   Absolutely no:t
             (0.3)
2rc   Iter   [Oppose?
3            [((# 2 =oppose))
             (0.5)
2c    Resp   Oppose
1     Iter   .hh Develop a government information
             prog[ram
2r    Resp       [oh wait uh- (0.3) oppose students from attending?
      Iter   .hhh well I'll re[peat] the question
      Resp                    [no  ]
2p           I absolutely think they should be able to go.
2rc   Iter   .hhh okay So you would su- (0.3) so you would oppose
              prohibiting students (.) with thee [aids-
      Resp                                       [right
      Iter   okay=
2c    Resp   =I would oppose prohibiting
             (0.4)
3     Iter   ((confirming oppose on the screen))
1            Got it. .hhh ah:: >how about< (.) Develop a government
             information ↑program to promote safe sex practices such as
             the use of condoms {q125}
             (0.5)
2p    Resp   You're talking to uh health teacher absolutely
2rc   Iter   uh Support?
             (0.7)
2c    Resp   °mm hmm°
3     Iter   ((# 1=support))
```

The examples given in this section are confined to only two question types, comprising six questions in the instrument. There are two questions of the type

"Does X outweigh Y or Y outweigh X?" (q7d and q7f). Six of the 12 answers in the corpus following these questions are nonparadigmatic (50 percent). There are four questions of the type "Do you support or oppose X?" (q120-135). Eight of 24 answers in the corpus following these questions are nonparadigmatic (33 percent). For comparison, instances of interviewer coding occur in only 9 percent of all question–answer sequences (58 out of 630).

In the majority of the cases of interviewer coding discussed here, the interviewer seeks confirmation of the category chosen. This means that in these cases the meaning of the answer given by the respondent is not obvious in the same way as the answer "I doubt it" to a yes/no question *is* a "no." We can only speculate about the reasons for this relative lack of transparency (to the interviewer) of what the respondent means. One possible explanation is the relative complexity of the two questions and, hence, of the answers caused, among other things, by double negatives (such as in "opposing prohibiting students from attending"). Apparently the interviewer does feel a need to ask for confirmation of inferences drawn from the respondent's answer. This confirmation follows in all cases, which means that the interviewer's inferences were correct all along.

INTERVIEWER AS CODER

In this section I discuss two of the remaining 18 instances of interviewer coding in the initial corpus. These 18 instances do not represent one single type or a small number of clearly identifiable types of interviewer coding. Each instance has its own characteristics. In this chapter I can discuss only two instances that contribute to the argument developed so far, which I will first summarize. (Note: Some of the other instances from my sample that I will not discuss here can be found in other chapters in this volume. See, e.g., extracts 1 and 5 in Schober and Conrad, this volume, Chapter 3).

Conclusions of the above analysis can be summarized as follows. In 19 of the 22 instances of interviewer coding that follow yes/no questions, the interviewer selected a response category without seeking confirmation by the respondent (immediate coding). This phenomenon has been interpreted as implying that interviewers, for all practical purposes, see the respondent's answer, though not literally one of the required response categories, as a version of such a response category ("yes," "no," or "don't know"). In 12 of the 18 instances of interviewer coding that follow other questions with only two response options, the interviewer asked for confirmation of an inferred response category (coding with confirmation). The latter phenomenon has been interpreted as implying that interviewers are not entirely certain about the correctness of their inference, although in practice these inferences are always correct. One explanation, complexity of the questions asked, is proposed for this uncertainty. Common to all cases discussed so far — with the exception of the instance in extract 9 — is that the interviewer's mapping of the respondent's answer onto a response category is commonsensical in the sense that it is not dependent on esoteric concepts defined by the investigator. Interviewers' interpretations (or "coding instances") rely on tacit everyday knowledge of the meaning of expressions such as "I doubt it" ("no") and "I would be for that" ("support"). A condition of such a successful mapping of an everyday expression (in the respondent's answer) onto another everyday expression (in the response categories) is that the response categories in the instrument provide an obvious equivalent of what the respondent appears to mean. I will now discuss an example in which such a response category is not available (see extract 15).

Extract 15: AW04:197 {q36}

```
1    Iter   eh Property or real estate taxes? {q32}
2p   Resp   .hhhh hhhh What was the middle choice?
1r   Iter   ((tch)) That was about right? ((pause)) or there's low
     Resp   ((cough))
     Iter   or excessively high
2p   Resp   Well there oughta be a just plain high uhm 'cause I don't
            think it's about right I don't think it's exCESSively high
            but (.) it damn sure isn't low. .hhh[hh
     Iter                                       [okay? uh:m=
     Resp   =pick one hhh[hh
1r   Iter                [hhhuh huh .hhhh We can put it in either
            category and I can put a note on the bottom if- if you wanna
            do that, so [.hhh
2p   Resp               [Yeah well just PUT IT excessively high and just
            put that I don't think it's really exCESSive it's just high.
3    Iter   n'kay:? ((pause; #)) ((tch)).hh ((pause)) A:nd let me write
            that down real fast so I can [input=
     Resp                                [okay
     Iter   uh- (.) heh- put that it in at thee e:nd.
            [.....]
1    Iter   .hhh And how about sales tax? {q36}
            (0.5)
2p   Resp   In Texas it's high
3    Iter   ((# 1 =excessively high))
```

The nonparadigmatic sequence is at the end of this extract (q36). On first sight, there seems to be no justification for coding "it's high" as "excessively high." A simple explanation for this apparent error is that the set of response categories does not provide for an option between "excessively high" and "about right." This problem has occurred also in the sequence following q32 (at the beginning of extract 15). There it has been solved by the respondent's proposal to "just PUT IT excessively high and just put that I don't think it's really exCESSive it's just high." Apparently the interviewer feels justified in applying the same solution here. There is one other case in another interview in which the interviewer immediately codes the answer "a little high" as "excessively high." In all these cases the response categories provided by the instrument do not provide for a category that is equivalent to the respondent's opinion, such as "(just) high." The problem arises because it is obvious to the interviewer (and, as in extract 15, to the respondent), in the everyday language they share, that the meaning of "just high" is neither (nearly) identical to "about right" nor to "excessively high."

The last instance I discuss here is similar to the one discussed at the beginning of this chapter (extract 1). It concerns the same question (q359h) in another interview (extract 16).

Extract 16: AW21:651 {q359h}

```
1    Iter    .hh Do you work for yourself in a family business or for
             someone else. {q359h}
2p   Resp    nahnah I work for (0.6) a large corporation=
3    Iter    =mhmm ((# 3 =someone else))
```

It is likely that, as in extract 1, the respondent would confirm the interviewer's choice of the category "someone else." Such a confirmation, however, does not have the same status as in the instances in answers that follow two-options questions such as discussed above. An interviewer can infer, and a respondent can confirm, that "more beneficial" and "outweigh" have equivalent meanings in the everyday language they share. There is not such an everyday equivalence between the word "corporation" (in the respondent's answer) and the response category "someone else." The mapping of the former onto the latter relies on knowledge of a social system of categories (and, probably, on knowledge of the investigator's intentions or definitions) rather than on a shared understanding of everyday pragmatic meanings of words. The respondent, if asked, could legitimately claim not to know how the investigator would classify his working for a large corporation. The mapping task of the interviewer in the instances discussed in the previous sections of this chapter consists of *inferring* (and *occasionally seeking to confirm) the meaning* of the respondent's answer. In contrast, the interviewer's task in sequences following q359h consists of coding an unambiguous meaning in terms of a category system.

CODER TRAINING FOR INTERVIEWERS?

The main conclusion of a case study of coder training (Hak and Bernts, 1996) was that a distinction must be made between two types of coding problems. There are problems in interpreting an answer (i.e., determining its "sense" or the respondent's intention) on the one hand and problems of interpreting the coding instructions (i.e., the rules for mapping meanings onto response categories) on the other hand. These two types are not only different analytically but are treated quite differently in coder training as well. The typical solution of the second type of problem (regarding the interpretation of the coding instructions) is theoretical training, that is, the specification of the instructions and the socialization of coders into its theoretical logic. The first type of problem, however, cannot be solved in this way. The reading of respondents' intentions in answers is an everyday competence that has no direct relation to professional views and instructions. In coder training, differences in interpretations of answers can only be superseded by an active dialogic search for a common ground between coders. This is the case because the respondent, who is the one person who would be able to decide unequivocally what the "meaning" of an answer is, is not present in the coding process.

The same distinction between two types of coding problems, and between two different ways of solving them, can be found in the interview data discussed in this chapter. There is, however, an interesting difference. In coder training (and, for that matter, in coding in general) the investigator is present as an informant and instructor whereas the respondent is absent, but in an interview the respondent is present as an informant whereas the investigator is absent. So, whereas in coder training the problem of what the "meaning" of an answer is cannot be solved (because of the respondent's absence), in the interview a solution to this problem can be found by seeking the respondent's confirmation of a specific response category. In coder training, the investigator can instruct the coders about the rules for mapping (unambiguous) answers onto response categories, but in the interview the investigator is not present. Therefore, the interviewer must write the answer in full (such as "just high, not excessively high") to give the researcher an opportunity to code the answer later on, after the interview (as in extract 15, q32). However, as the data discussed in this chapter show, it is not always the case that interviewers record such information. They might instead ask the respondent to agree with a specific way of mapping their answer onto the investigator's concept, such as in extract 9 ("so to an extent you have health insurance"), or they might immediately code the answer in a specific way (as in extract 1).

From the examples discussed in this chapter, it is clear that some questions are unnecessarily complex and that some sets of response categories are deficient. Other examples (such as the one on health insurance coverage) show that even straightforward questions can raise problems regarding the definition of concepts, either of concepts in the question itself or in the set of response categories. Interviewers can solve such problems in an ad hoc way by performing a coding task. This raises the question whether such nonparadigmatic behavior should be discouraged. Should interviewers be encouraged to record answers given by respondents in full—leaving the coding task to office coders or to the investigator? Or should they help respondents to perform the coding task by informing them about the investigator's definitions of concepts (as proposed by Schober and Conrad, this volume, Chapter 3). Or, in an alternative approach to the problem, should interviewer coding be encouraged and should interviewers be trained accordingly? An advantage of the latter approach is that this will sensitize interviewers to the kind of information needed for the purposes of the survey (e.g., about respondents' employer or insurance situation) and to stimulate them to extract such information. In contrast, in the *conversational* interview as proposed by Schober and Conrad, it would be left to respondents, who never will be properly "trained" as coders, to decide which information is pertinent to the question.

In either approach it is paramount to make a distinction between the two different types of problems that are involved in any coding. If a respondent presents a "problematic" answer (i.e., an answer that is not one of the categories provided in the instrument), the problem might be one of these two types. The first type is that the *meaning* or *sense* of the respondent's answer is not clear to the coder, this is, the interviewer. This is a problem of everyday understanding, which can be solved in an everyday manner, that is, by asking respondents to clarify their answers. The data discussed in this chapter show that interviewers do this frequently, by offering candidate meanings to respondents, and that respondents then might clarify or confirm what they have meant. The second type of problem occurs when the answer's meaning is evident but it is unclear how the answer must be mapped onto the response categories provided in the instrument. This is a problem of how to apply esoteric coding rules, based on the investigator's definitions of concepts. This problem cannot be solved by asking respondents to clarify what they mean but rather requires the investigator to clarify his or her concepts. This is what coder training consists of. Obviously, if interviewers would be allowed to code information given by respondents, they should receive proper coder training.

CONCLUSION: PROBLEMS IN MAPPING ANSWERS ONTO RESPONSE CATEGORIES

Several conclusions can be drawn from the data presented in this chapter. One is that there is a relatively high incidence (namely in about 9 percent of all question–answer sequences) of instances of nonparadigmatic interviewer coding. However, a large number of these instances appear, on analysis, actually *paradigmatic* in the sense that interviewers, for all practical purposes, see the respondent's answer, though not literally one of the required response categories, as a version of such a response category ("yes," "no," or "don't know"). The other conclusions of this chapter, as outlined below, pertain to genuine instances of interviewer coding. These are those instances of interviewer coding in which the interviewer is either not sure what the respondent means (and, therefore, asks the respondent for confirmation) or in which this (obvious) meaning has to be mapped onto the response categories as provided by the instrument.

A clear (but not new) conclusion of the above analysis is that inspection of question–answer sequences can be used as a diagnostic tool for assessing the *quality of questions*. The obvious example in the data discussed in this chapter is the question "Does X outweigh Y or Y outweigh X?" The extraordinarily high prevalence of nonparadigmatic answer sequences (50 percent) following this question is an indication that something is wrong with it. Interaction analysis does not only draw attention to problems but provides solutions as well. In this case inspection of answers in nonparadigmatic answer sequences shows that respondents may use a phrase such as "more important than" in their answer. This suggests that the question could be reworded as "What is more important: X or Y?" Similarly inspection of question–answer sequences can be useful to diagnose deficiencies in sets of response categories and to suggest solutions to such problems. The main example from the discussion in this chapter is the response set "excessively high," "about right," and "low." Inspection of answers given by respondents suggests that the category "(just) high" should have been included in the set of response categories. (Although it is not likely that an American respondent can be found who would consider a tax "excessively low," obviously that category could have been added as well.)

Another type of conclusion regards the allocation of coding tasks in the interview. It is generally recognized in the literature on the standardized interview (e.g., Sudman, Bradburn, and Schwarz, 1996) that one of the *cognitive* tasks of the respondent is to map answers onto response categories and that, by implication, selecting a response category is a form of coding. When it becomes apparent in the interview that a respondent has coding problems, the interviewer must solve them *or* must help the respondent to solve them. The latter option is the standard approach, which can take different forms, such as interviewers providing respondents with definitions of concepts in the conversational interview (Schober and Conrad, this volume, Chapter 3). The instances discussed above demonstrate that, at least occasionally, interviewers might opt for performing this coding task themselves. If interviewers are properly prepared to do this, for example, by providing them with some form of coder training as part of their interview training, the result might be both more valid and reliable than in current practice.

ACKNOWLEDGMENTS

This chapter analyzes a collection of 29 interviews conducted at the University of Wisconsin Survey Center in 1995. I thank the center for providing me with these interviews for analysis.

REFERENCES

Fowler, F., and T. Mangione. 1990. *Standardized Survey Interviewing*, Newbury Park, CA: Sage.

Hak, T., and T. Bernts. 1996. "Coder Training: Theoretical Training or Practical Socialization?" *Qualitative Sociology* 19: 235–257.

Schaeffer, N. C., and D. W. Maynard. 1996. "From Paradigm to Prototype and Back Again: Interactive Aspects of Cognitive Processing in Standardized Survey Interviews." Pp. 65–88 in N. Schwarz and S. Sudman (Eds.), *Answering Questions: Methodology for Determining Cognitive and Communicative Processes in Survey Research*. San Francisco: Jossey-Bass.

Smit, J., W. Dijkstra, and J. van der Zouwen. 1997. "Suggestive Interviewer Behaviour in Surveys: An Experimental Study." *Journal of Official Statistics* 13: 19–28.

Sudman, S., N. M. Bradburn, and N. Schwarz. 1996. *Thinking about Answers: The Application of Cognitive Processes to Survey Methodology*. San Francisco: Jossey-Bass.

CHAPTER 20

The Impact of Technology on Interaction in Computer-Assisted Interviews

Marek Fuchs

INTRODUCTION AND HYPOTHESIS

Computer-assisted interviewing is on the way to becoming a standard survey technique (Couper et al., 1998). More and more studies — both telephone surveys and those using face-to-face interviews — are conducted using computer-assisted interviewing techniques (CAI). Many of the large government surveys in the United States have completed the transition to CAI already. Even in Europe, we observe a shift toward computer-assisted interviewing (Fuchs, 1994, 1995; Laurie and Moon, 1997; Projektgruppe SOEP, 1998; Schneid, 1991).

In general, the movement toward CAI is evaluated positively. Researchers and field directors benefit from it (Nicholls and de Leeuw, 1996); and both interviewers (Couper and Burt, 1994) and respondents (Baker, 1992) show supportive or at least accepting attitudes. On the other hand, computer-assisted interviewing has introduced some additional problems into the interview situation. In the early years methodological research was mainly concerned with hardware and software problems (see Couper, Groves, and Kosary, 1989; Weeks, 1992, for overviews). But recent studies have dealt with interview duration (Martin, O'Muircheartaigh, and Curtice, 1993; Baker, Bradburn, and Johnson, 1995) and the usability of the computer-assisted instrument (Couper et al., 1997; Fuchs, 2000b; Fuchs, Couper, and Hansen, 2000; Hansen, Fuchs, and Couper, 1997, 1998). The present chapter contributes to this latter discussion.

Standardization and Tacit Knowledge: Interaction and Practice in the Survey Interview
Edited by Douglas W. Maynard, Hanneke Houtkoop-Steenstra, Nora Cate Schaeffer, and Johannes van der Zouwen
ISBN 0-471-35829-0

I argue that the technology used in a survey might have an impact on the survey process and the survey outcome. I assess so-called *technology effects* (Fuchs, 2000b; Fuchs, Couper, and Hansen, 2000) to determine their impact on interview duration, survey cost, nonresponse, and the resulting data.

What we usually call mode effects actually encompass various potential sources for observed differences: the means by which the instrument is delivered to the respondent; the co-presence, remote presence, or absence of an interviewer; the design of the interview protocol; the communication channels (and supplements to those channels) that are used to present individual items to the respondent. When we analyze *mode effects*, we examine how the sum of all of these affect the way participants understand the questions and the purpose of the interview in general [see Tourangeau and Smith (1996, 1998) for a general discussion of mode effects].

The question asked in the current chapter — what is the effect of whether or not an instrument uses computer-assisted technology? — does not fit neatly into this framework. In this chapter, I focus on the technology used to present the questionnaire to the respondent or to the interviewer in order to deliver it to the respondent. Consequently, I do not consider technology a dimension of mode effect because technology has its own impact on the interview situation and data quality that is independent from the sources and dimensions of mode effects mentioned above (Sanchez, 1992; Jenkins and Dillman, 1997). From this point of view, both paper and a computer are "technologies" that can be used to present a survey instrument (Figure 20.1). As a result, technology effects need their own theoretical framework and strategies of analysis that are different from those used to assess mode effects.

In a survey interview two actors (interviewer and respondent) interact to complete a task (conducting an interview). In fact, the interviewer administers a question–answer process [structured according to its own rules well conceptualized by the cognitive approach prominently represented by Sudman, Bradburn, and Schwarz (1996)] in an interaction following social and communicative norms that are different from but still similar to those applying to everyday conversations (Schaeffer, 1991, and Chapter 4, this volume). As a result, interviewers are in a double bind: They must follow standard interviewer procedures to guarantee standardization and at the same time they must both deploy and respond to interactional styles that are — to a certain extent — necessary to support the interviewer–respondent interaction (see also Chapter 1, this volume). Following the

	Technology	
Interview Mode	Paper Based	Computer Assisted
Personal interview	PAPI	CAPI
Telephone interview	PATI	CATI
Self-administered	Paper and pencil	CSAQ, audio-CASIC, video-CASIC

Figure 20.1 Mode effects and technology effects.

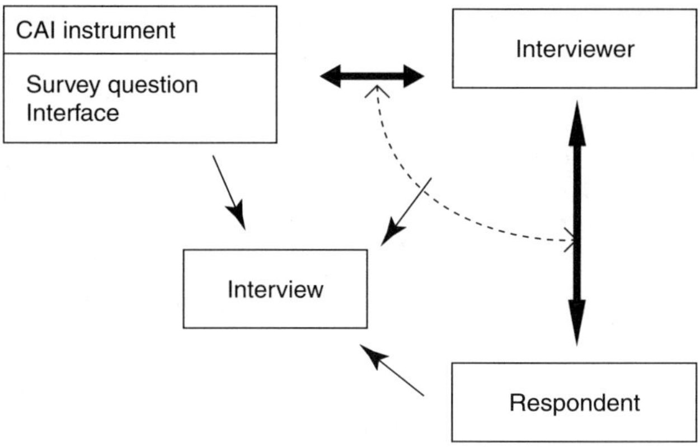

Figure 20.2 Participants and interactions in a CAI interview (Lepkowski et al., 1998).

latter set of rules may lead to behaviors that do to not match standard interviewer procedures completely (see Chapter 12, this volume).

Lepkowski et al. (1998) conceptualize the computer-assisted interview as a situation in which an additional actor — the computer — is involved (see Figure 20.2). When the interviewer uses a computer instead of a paper-based survey instrument, this third "actor" contributes to the outcome of the interaction and additional behavioral obligations occur. When administering the question, when searching information on help screens, when looking up additional information asked for by the respondent, when entering the data, and so forth, the interviewer relies on an interface — the way in which the instrument is implemented on the computer — that has its own rules, that demands and absorbs a great deal of attention, and thus has its own impact on the interview situation. Even though the paper questionnaire may also become a third party in an interview, it is less intrusive and more accessible during the course of an ongoing interviewer–respondent interaction. By contrast, the computer reaches a higher level of importance and a higher degree of presence compared to a paper instrument. As a result, a third set of computer-related communication rules is introduced into the interview situation. It is assumed that a tension — similar to the one between the rules that govern the cognitive question–answer process on the one hand and the communicative dimension of the interview on the other hand — also may exist between computer-related behaviors and respondent-related behaviors (this is indicated by the dotted line in Figure 20.2). Thus the way in which the CAI instrument is designed may affect the interviewer's behavior toward the respondent more than a paper instrument does.

When we assess technology effects, we focus on screen design and other aspects of the CAI interface. By contrast, effects related to the survey question itself and to the question–answer process (question wording, context effects, question order effects, and response order effects) are not included in this framework. Instead, when questions and response categories are identical — and thus invoke the same cognitive processes — but use different screen designs, we expect different interviewer behaviors and potentially an impact on data. This impact on the actors' behaviors and potentially on data quality is referred to by using the phrase *technology effect.*

Several aspects of the CAI interface can affect interviewer and respondent behaviors during all stages of the social process necessary to work through an item. From an analytical perspective, administering a question consists of at least five stages (Figure 20.3):

1. Interviewer: finding the next item, understanding the task associated with the question, for example, identifying question text and interviewer instructions
2. Interviewer: reading the question text
3. Respondent: understanding the question, generating an answer, providing the response
4. Interviewer: probing as instructed by the CAI system, searching for additional information requested by the respondent if needed, providing feedback
5. Interviewer: entering the data into the CAI instrument

Stage	Actor	Tasks	Impact of Technology	Possible Effects
1	Interviewer	Find next item, understand the task	Screen design, font, color; amount of information presented at a time; single-item vs. multiple-item screen	Incorrect administration of question; longer interview duration; periods of silence between items
2	Interviewer	Reading the item	Screen design, font and color of screen content; amount of information presented at a time; single-item vs. multiple-item screen	Stammer; monotonous reading of question; lack of "big picture"; asking items already answered; deviations from interview script
3	Respondent	Understanding question, generating and providing answer	(Direct: computer present in interview situation) Indirect: interviewer behavior modified due to effects described in stage 1 or 2	Annoyance due to redundant question asking; danger of break-offs; less effort to generate an answer
4	Interviewer	Searching for additional information on help screens, etc.	Presentation of definitions and other help contents on the screen	Difficulty finding a definition or explanation; delayed reaction to respondent inquiry; respondent develops non-standard understanding of item
5	Interviewer	Entering data into CAI system	Design of CAI interface; position and size of input field; logic of data capture	Difficulty while entering data

Figure 20.3 Sources of technology effects during the course of the interview.

During all five stages design characteristics of the CAI interface might influence the behaviors of the participating human actors:

1. Differentiating interviewer instructions and question text and understanding the task associated with a particular item can be easy or difficult from the interviewer's perspective depending on how the information is presented on the screen. Solving this task is more complicated for the interviewer when he or she needs to scroll to obtain all relevant information, when instructions and question text do not appear in the same place on all screens, when fonts are difficult to read, when too much or too little information is on the screen at the same time, and the like.

2. In addition, depending on how the interface presents information, the interviewer may change his or her behavior when reading the question text: The screen design may force the interviewer to spend more time and effort with the computer. As a result, he or she might spend less time and effort making sure that the respondent understands the question. The situation may develop in a way that the interviewer does not focus primarily on delivering the question appropriately, and so the reading of the question may be less informative to the respondent.

3. These interviewer behaviors might affect how the respondent generates and provides an answer to the actual question. Two effects need to be addressed separately: (a) respondents might react to nonstandard interviewer behaviors that are due to CAI problems and errors, for example, when a faulty screen design or a programming error makes it difficult for the interviewer to determine the task from the CAI instrument; (b) the respondent might react to a formally correct but highly segmented and monotonous administration of the questionnaire. That sort of administration of the question could occur because the interviewer does not see the "big picture," because the screen design does not allow the interviewer to avoid redundancies, and so forth.

4. When asked by the respondent to provide explanations about a question, the interviewer might find it difficult to obtain additional information on the screen or in the system's help menu.

5. Finally, the way in which the CAI interface is designed might facilitate or hinder correct data entry. The CAI interface also affects how easy it is for the interviewer to back up through the instrument in order to change data entered previously.

When discussing these issues, we need to differentiate between computer-related accidents due to faulty systems or programming errors on the one hand and interviewer behavior that is due to functional but poorly designed or misleading screens on the other hand. From our perspective, whether or not a screen design is considered good depends on its usability as a tool in the interview, considered as a social and communicative situation. Both types of problem deserve discussion and analysis. However, the remainder of the present chapter deals with the occurrences and the effects of interviewer behaviors caused by specific screen designs. So far, we cannot address all five stages mentioned above, and we are not able to provide empirical evidence for all effects. In addition, we do not know under which circumstances these effects on behavior might affect data quality. However, here we summarize some of the preliminary results of a line of research that deals with this kind of technology effect.

METHODS

In summarizing our usability research we draw on a laboratory experiment as well as on a field experiment. The laboratory experiment deals with the comparison of a set of identical questions under both paper assisted personal interviewing (PAPI) and computer assisted personal interviewing (CAPI) conditions. We assess the effects of the two technologies by comparing the way in which the items are administered in an interview situation. This comparison leads to a deeper understanding of how the interview situation is affected by design characteristics and technologies of the instruments. In the field experiment we compared four CAI screen designs applying the same questions and identical response categories. The results allow an assessment of the impact of different CAI design characteristics on interview duration, interviewer behavior, and respondent behavior.

Experiment 1

In the first experiment we compare the PAPI version with the CAPI version of the National Health Interview Survey (NHIS) instrument, an ongoing cross-sectional household study. CAPI was introduced in parallel with the ongoing paper-based survey during 1996, and by January 1997 the conversion of the NHIS to CAPI was complete. To evaluate the existing computer-assisted instrument already used in the field, the Survey Research Center (SRC) at the University of Michigan conducted a series of usability tests in the spring of 1997 (see Hansen, Fuchs, and Couper, 1997, 1998; Lepkowski et al., 1998). A total of 37 CAPI interviews were conducted by interviewers from the Bureau of the Census in SRC's usability laboratory. Also, 14 interviews using the PAPI instrument were collected. Nine interviewers participated in the test, with each interviewer doing 4–6 CAPI and 1 or 2 PAPI interviews. All the interviewers had experience on both the paper-and-pencil and CAPI instruments, having used the latter for at least 6 months in the field prior to the usability tests. Respondents were recruited from the area and paid for their participation. Respondents were randomly assigned to receive a CAPI or PAPI interview.

Several videotapes (capturing the interaction from different views) were recorded for each interview (see Hansen, Fuchs, and Couper, 1997). The results reported here are based on an additional coding of the previously analyzed videotapes. In this analysis we focus on items from the sociodemographic portion of the instrument (name, sex, DOB/age, relationship to reference person, race, etc.). This segment of the instrument is very similar across the two versions, permitting comparison. Each item was administered for every person in the household leading to a total of 2200 questions asked. In addition to the question–answer sequences, several relevant interviewer and respondent behaviors were coded for each item: searching for the next question on the screen or paper questionnaire (interviewer); reading the question text (interviewer); negotiating the meaning of question text and answer—asking for more information, providing an answer, probing, feedback, other task-related verbal and nonverbal contributions (interviewer and respondent); recording the response (interviewer); special behaviors, like digression; problems related to the questionnaire or the computer (interviewer and respondent); working with tools, for example, show cards, calendars, booklets (interviewer). This approach allows a detailed examination of the activities occurring in the duration of every single item and helps detect differences in the participants' behaviors even if the total duration of an item is equal in both technology conditions.

Experiment 2

The second experiment took place in Germany in March 1998. Immigrants of German origin from Poland, Rumania, and the former Soviet Union were surveyed. Starting February 28, 1998, and ending March 20, 1998, 15 interviewers completed 501 telephone interviews. All respondents received an advanced letter and were called by phone up to 15 times in order to complete an interview containing 95 questions.

Four versions of a small household roster with just three items per person (age, year of arrival in Germany, immigrant status) were included in the instrument (Figure 20.4). All instrument versions used the same question wording and identical response categories. However, the screen design differed considerably:

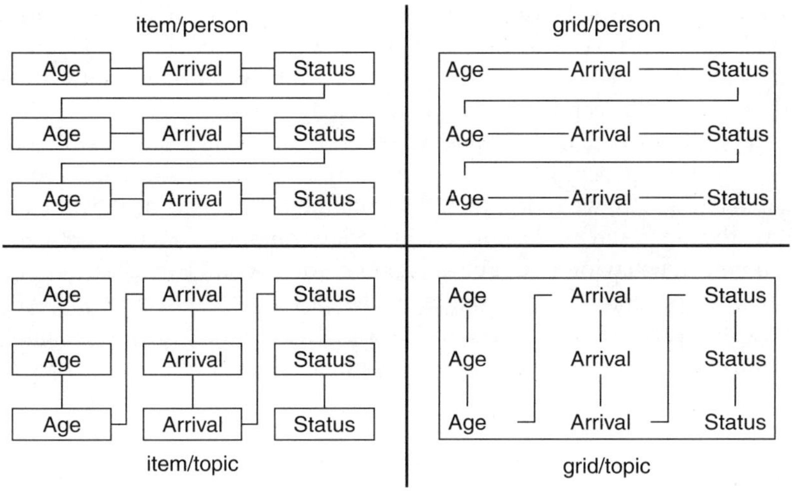

Figure 20.4 Four screen design versions tested in Experiment 2. Each label (age, year, status) displays an item, each box represents a screen, and the lines between items indicate the interview flow starting in the upper left corner.

(1) an item/person version, (2) a grid/person version, (3) an item/topic version, and (4) a grid/topic version. The versions differ in question order and in the way the questions are displayed on the screen. In the item-based versions one item is displayed on the screen at a time; in the grid versions all items are combined in a table and are visible on the screen at the same time. The topic-based versions ask one question for all persons in the household before administering the next item for all household members. In the person-based design all questions in this section of the instrument are asked for the first person in the household before the instrument displays the same questions for the second and all other persons in the household.

One of the four versions was randomly assigned to each interview—and thus to interviewers and respondents. We measured the total time needed for the household roster and the time spent on each single item in that section of all 501 interviews. In addition, 234 interviews were selected at random and the interviewer was videotaped working through the household roster section. The interviewer behavior and respondent behavior on the videotapes were coded using a coding scheme similar to the one used in experiment 1: A set of behaviors most likely to be affected by technology was extracted and coded for each question. This includes characteristics such as the degree to which the behaviors comply with standard interviewing procedures. The resulting data were combined with the time measurements to allow an assessment of the effects of different screen designs on interviewer behavior and respondent behavior as well as on interview duration.

RESULTS

First, we present selected results from the laboratory experiment comparing the PAPI and the CAPI version of the same questions (see Fuchs, Couper, and Hansen, 2000, for more detailed results). When analyzing these data, we focus on the effects of the technology (paper vs. CAI) on the interview situation. We provide preliminary data to determine whether or not using a screen with a certain design instead of a paper instrument with a certain design changes how an item is administered. Second, we report findings from the field experiment evaluating the effects of different screen designs on interview duration, interviewer behavior, and respondent behavior (see Fuchs, 2000b, for a complete description). The two experiments represent two stages in the process of introducing CAI: first, we assess the conversion from one technology to the other; then—assuming the conversion to CAI is completed—we evaluate the impact of different screen design solutions.

Experiment 1

A feature that might cause problems in a computer-assisted interview is the lack of flexibility. One of the advantages of a CAI instrument is the fact that an interviewer can rarely (if ever) skip a question. Although CAI instruments can make extensive use of skip patterns and filters, they apply a predefined question order. Usually, each item requires that the interviewer press the Enter key before the system will go on to the next screen. It is seen as an advantage that this rigid question order avoids any trouble the interviewer might have with the routing through the instrument, finding questions for specific respondents, following filters and skip patterns, and so on. He or she can abandon this administrative task and focus on the administration of the actual items. On the other hand, the way CAI systems operate makes the order of questions very strict and provides the interviewer with little flexibility in terms of item order.

Looking at the first substantive question of the instrument, we can demonstrate this effect. The question reads: "What are the names of all persons living or staying here? Start with the name of the person or one of these persons who owns or rents this home." The question wording is identical in both versions and the amount of information required from the respondent is the same. The respondent is expected to provide the first name, middle initial, and last name of all persons living in the household. Figure 20.5 shows an image of this screen from the CAPI version.

Comparing the results, we find it takes substantially more time ($p < 0.001$) to complete this question in the computer-assisted version (20.6 seconds) compared to the paper instrument (17.5 seconds). The values reported in Figure 20.6 represent the average duration of several interviewer and respondent behaviors for each instance in which this question is asked. The results show no significant difference in terms of time for the interviewer to find the next question and prepare to deliver it. Also, interviewers spend the same amount of time reading the question text under both conditions. Moreover, there is no significant difference between

```
Caseid: 005
Item: RPNAME@LNAME

-------------------------------------------------------------------------------
-RPNAME-
         What are the names of all persons living or staying here?
         Start with the name of the person, or one of the persons,
         who owns or rents this home.

         FR: BEFORE PROCEEDING, PLEASE MAKE SURE THE REFERENCE
         PERSON IS NOT AN ACTIVE ARMED FORCES MEMBER.

         PROBE FOR MIDDLE INITIAL IF NOT REPORTED.
         PRESS "ENTER" TO SKIP TO LAST NAME IF NO MIDDLE INITIAL.

         FIRST NAME:    John                                    (H)

         MIDDLE NAME:   A

         LAST NAME:     Smith
```

Figure 20.5 Experiment 1 example of text input on a multiple item screen. The screen dump is taken from the 1996 CAPI version of the National Health Interview Survey.

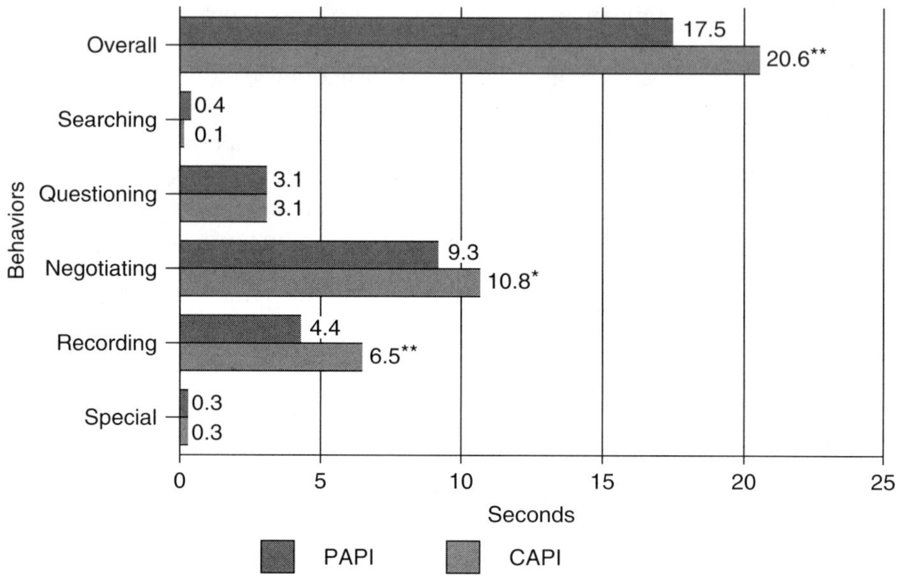

Figure 20.6 Experiment 1; $N = 160$; $^{*}p < 0.05$; $^{**}p < 0.01$. Duration of specific interviewer tasks on the item: "What are the names of all persons living or staying here?... ." See legend to Figure 20.5 for details regarding the question wording and the source of this item.

the two versions in terms of the time spent on additional actions or behaviors such as digression or dealing with problems related to the questionnaire or the computer and so on ("special" behaviors).

However, it is interesting to observe that the negotiations between interviewer and respondent (asking for more information, providing an answer, probing, providing feedback, other task-related verbal contributions and nonverbal turns) take significantly longer on CAPI than on paper. Also, recording the information takes longer under CAPI conditions (see Fuchs, Couper, and Hansen, 2000, for details). This is mainly due to the degree of flexibility provided by each version. The computer-assisted instrument is designed as a multiple-item screen that combines several input fields on one screen. The input is expected in a certain order: first name, middle initial, and last name. If the respondent provides the last name first, the interviewer has two options. (1) He or she can type dummy information for the first name followed by Enter and press Enter for the middle initial to move the cursor to the third input field. Then, he or she can record the last name and return to the first input field (by backing up twice) to type the first name and—after pressing Enter—the middle initial. (2) As an alternative, the interviewer can ask the respondent to provide the information in the order expected by the computer system. Whatever solution is chosen, it takes more time than in the paper instrument, in which the interviewer simply writes the information in the appropriate blank spaces regardless of the order in which it is provided.

The lack of flexibility of CAI demonstrated in this example applies to many screens with multiple input fields. This lack of flexibility occurs several times during the instrument. Looking at the sociodemographic module as a whole, we can see similar effects (Table 20.1). Both the paper-based and computer-assisted socioeconomic instruments collect about the same amount of core information.[1] On average, the CAPI interviews take 38 seconds longer ($p < 0.05$) than interviews conducted under the paper-based condition, which represents an increase of about 16.5 percent over paper. If we look at the time taken by the two persons acting as participants in the interview situation — interviewer and respondent — we compute almost the same proportional differences for each; both interviewers and respondents spend more time when using a computer-assisted instrument (given the small sample size, the difference for respondent time does not reach statistical significance).

If we assess the reasons for the longer duration of the computer-assisted instrument, we find some interesting differences. In the paper instrument, interviewers ask an average of 8.9 questions per household member for this series, compared to 11.1 in CAPI to collect the same information ($p < 0.001$). On the other hand, the time spent on each of these items is significantly shorter in the computer-assisted instrument. It takes an average of 8.9 seconds per question in the PAPI instrument compared to 7.6 seconds per item in the CAPI instrument ($p < 0.05$). This is explained in part by the fact that the two interactants take more conversational turns or actions to complete the task in the paper version (3.7) than in CAPI (3.3 turns or actions).

This leads to a more general discussion of two important characteristics of a CAI instrument that do not — or at least not to this extent — apply to PAPI instruments: In a CAPI environment the interviewer has an additional burden: the process of keying takes place in the interview situation. Usually, an interviewer reads a question, gets an answer, enters the data, presses Enter and then the next screen with the following question comes up. As a result of this procedure, the interviewer–respondent interaction is segmented. Compared to PAPI,

Table 20.1 Interview Duration of the Sociodemographic Section—Experiment 1[a]

	PAPI	CAPI	Difference
Overall duration	230 seconds	268 seconds	38 seconds*
Interviewer time	163 seconds	189 seconds	24 seconds*
Respondent time	67 seconds	80 seconds	13 seconds
Questions per household member	8.9	11.1	2.2†
Duration per question	8.9 seconds	7.6 seconds	−1.3 seconds*
Turns/actions per question	3.7	3.3	−0.4‡

[a] Mean duration adjusted for household size. While respondents were randomly assigned to the conditions, we did not control for differences in household size in this assignment. As a result, the average household size was 2.9 for the PAPI interviews and 3.3 for CAPI. Because the NHIS collects sociodemographic information on all eligible household members, we need to take these differences into account when examining time differences.

*$p < 0.05$; †$p < 0.001$; ‡$p < 0.01$.

interviewers cannot look ahead and anticipate the next upcoming question while recording the answers to the previous one, and they cannot start reading the next question before pressing Enter. So far we do not have quantitative evidence that this segmentation harms the data or the interview situation. But it has been argued that the interviewer loses the "big picture," and the relevance of questions may be unclear (House, 1985; Groves and Mathiowetz, 1984). Previous laboratory studies support this interpretation (Couper et al., 1997).

A small example — not taken from the NHIS but rather from the literature (see Oksenberg et al., 1992, for details) — demonstrates this segmentation effect: When asking all eligible household members about their race, it might happen (and in fact it happens quite often, see below) that the respondent provides not only the answer to the current question (e.g., "I'm African-American") but also to a related question: "We are all Black." Under certain conditions it is conversationally appropriate for a person to provide more information than she or he was actually asked for. With respect to Grice's (1975) assumptions regarding the cooperation of speakers and the principles of relevance and quantity, this behavior makes perfect sense: When asked for a single characteristic of all persons in the household (race in this case), respondents do not provide less information than the question asks for (see Sudman, Bradburn, and Schwarz, 1996 for a detailed discussion of Grice's 1975 approach).

When an interviewer uses a paper instrument, it is easy for her or him to make immediate use of any additional information provided by the respondent. In case the respondent answers, for instance, "We are all black," the interviewer can easily mark the appropriate check boxes for all household members at once or at least verify the information for all other household members instead of reading the questions as worded. This demonstrates the challenges the interviewer faces in attempting to work according to standard interviewing procedures while maintaining an adequate and responsive interaction with the respondent. When a respondent answers "We are all black" but the instrument presents the interviewer additional questions about the race of each household member, an interviewer has several choices: He or she can follow standard interviewer procedures and ask all subsequent items as worded — taking the risk that doing so will annoy, intimidate, or bore the respondent. Or the interviewer can skip the subsequent items or ask for verification. In our view some CAPI designs make it hard for the interviewer to make immediate use of this additional information — thus leading to more redundancy and to a higher proportion of interviewer behaviors indicating the interviewer's recognition of this redundancy. In contrast, in PAPI the interviewer is relatively less constrained by a specific design.

Results from our laboratory experiment 1 support our interpretation that interviewers react to the redundancy due to segmentation with certain modifications of the interview script: Even though under CAPI conditions interviewers deviate from the question wording to a significantly smaller proportion (4 vs. 17 percent), the proportion of items not administered is larger as is the number of questions that the interviewer verifies instead of reading as worded ("And your husband is African-American, too?"). In addition, interviewers can anticipate the answer

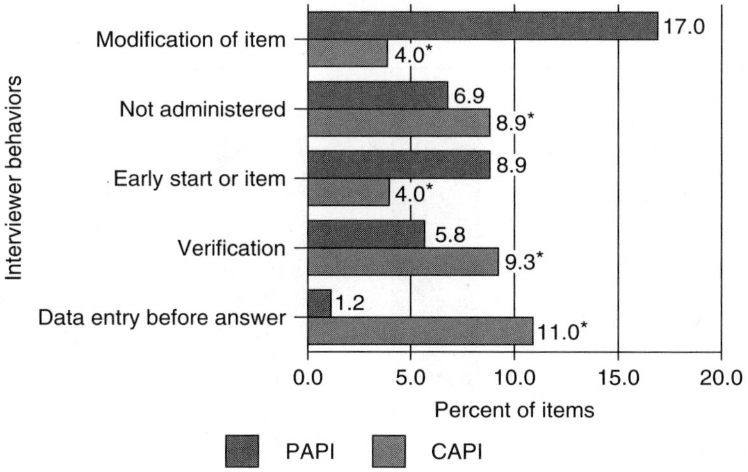

Figure 20.7 Experiment 1; displayed is the proportion of items affected by a certain behavior; $^*p < 0.05$.

to a redundant question to a higher degree and enter the response into the computer before the respondent actually provides the answer (11 vs 1.2 percent, see Figure 20.7).

In our view, under CAPI conditions interviewers develop a set of strategies to avoid segmentation and redundancy in the interview situation: They enter data without asking the respondent, they ask for verification of already known information, and they anticipate responses based on previous answers (they enter data before the respondent actually gives the answer). On the other hand, they are less likely to read the next upcoming question while entering the response to the previous one (they cannot look ahead and anticipate the next appropriate question). These results clearly indicate that design issues and technology can affect the interview situation even when instruments make use of identical questions.

Experiment 2

Our second experiment extends this approach: Comparing four screen designs that use the same technology (CATI) to ask an identical set of questions, we found a significant difference in interview duration as well as in interviewer and respondent behaviors. A detailed analysis of the data indicates that differences in administration time and behaviors are interconnected and caused by the screen design that structures the interview situation differently. Our first indicator for differences caused by the screen design is interview duration (Figure 20.8).

Interviewers needed 6.6 seconds per question in the item-based person version (which we use as the standard). In contrast, each question took 5.5 seconds in the grid-based topic version. This is a reduction of about 17 percent for the grid-based topic version. The two other versions are in between. It is important

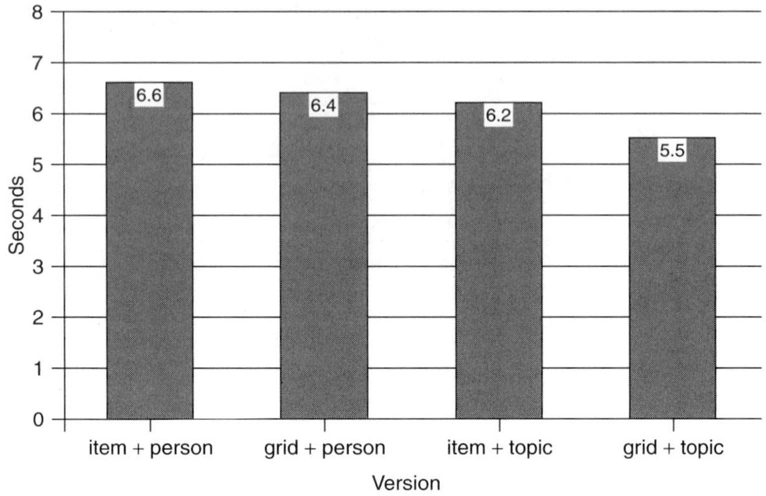

Figure 20.8 Experiment 2; $N = 501$; $p < 0.01$ interview duration per item by version; both factors — the screen design as well as the question order — contribute independently to the decrease in time (screen design: $p < 0.01$, one-third of total effect; question order: $p < 0.001$, two-thirds of total effect; no significant interaction).

to mention that both factors seem to contribute to the decrease in time used to complete the task. If we distinguish between the two factors, we end up with the following results: The two topic-based versions are significantly shorter than the two person-based versions and the two grid-based versions take significantly less time than the two item-based versions. The combined effect applies to the grid-based topic version and leads to the value of 5.5 seconds per item.

Analyzing the videotapes we can provide reasons for these differences at least in part: In the grid-based topic condition, both interviewers and respondents adapt differently to the interview situation compared to the item-based versions. When the interviewer asks a question for all persons in the household, the respondent recognizes the logic of the procedure very quickly. In quite a high proportion of all cases (about 30 percent) their reaction to this is "We all arrived in the same year" (meaning: "Don't ask me this question again and again"; see the example about race).

If the instrument follows a person-based design, the interviewer has to memorize this piece of information, and when the instrument comes to the next person he or she needs to remember: "Do not ask this question again, the respondent gave me the appropriate answer already!" In most cases, however, the interviewer just asks the question again. This is especially true when the instrument uses an item-based screen layout that gives no clues about the answers to the same question for the other household members. In a topic-based design, in contrast, the interviewer can easily adapt to such a situation: He or she just enters the same code for all persons in the household without asking the question repeatedly. Both the interviewer and the respondent get used to the questions, and so the question–answer process proceeds with fewer verbal contributions from the interviewer's side as well as from the respondent's. Both interviewer and respondent can anticipate the next question and the interview runs more smoothly. As a result the time used per item is substantially shorter, and the interviewer can provide respondent-oriented interviewer behavior.

The different versions tested in this experiment facilitate and support such behaviors to different degrees. From our results we can conclude that the grid-based topic version stimulates interviewers to deviate from the scripted interview to a higher degree than the other versions. As far as duration is concerned, this version allows the interviewer to make efficient use of information provided for all household members at once. Evidence from the video coding supports our interpretation of version-specific occurrences of time-saving interviewer and respondent behaviors.

By means of analyzing the videotapes (see Methods section), we observe quite a lot of interviewer behaviors that do not follow standardized interviewing procedures. About 78 percent of all items are read as worded, but interviewers do not administer 9.3 percent of all items to the respondent. In another 5 percent of instances, the interviewer does not read the question but instead provides an elliptic question form containing the relationship of the next person to the respondent (e.g., "... and your wife?").[2] In 5.5 percent of all cases the interviewer does not read the question but rather verifies the answer ("... and your wife is 32 years old?"). Some incomplete questions and wrong fills are also observed. In total we have about 22 percent of all items affected by at least one interviewer behavior that does not follow standardized interviewing procedures.[3] We will come back to the question of whether or not these behaviors help obtain valid measurements.

We draw the following conclusion from these particular findings: Most of these behaviors use some kind of conversationally appropriate short-cut, for example, the interviewer does not read the question text as worded, or he or she tries to make the conversation smoother and more suitable in terms of conversational rules. From our point of view this indicates that interviewers do not want to ask for information the respondent has provided already. They do not want to ignore the verbal contributions of the respondent, and they want to follow conversational rules. A side effect is that these behaviors are less time consuming than standard interviewer behaviors. The interviewer's priority therefore lies not with saving time but with customizing the question–answer process by taking into account behaviors by respondents that the computer-assisted instrument did not anticipate and cannot easily absorb.

Therefore we observe large differences between the four versions in the extent to which interviewers do not follow the scripted interview (Figure 20.9). Applying the grid-based topic version to an interview results in more than twice as many unscripted behaviors (27.4 percent of the items administered are affected) than the item-based person version (12.8 percent), which is the standard for most studies to date. Deviations from standardized interviewing procedures contribute to time savings ($p < 0.001$): Items affected by these unscripted interviewer behaviors take on average substantially less time (4.0 seconds) than

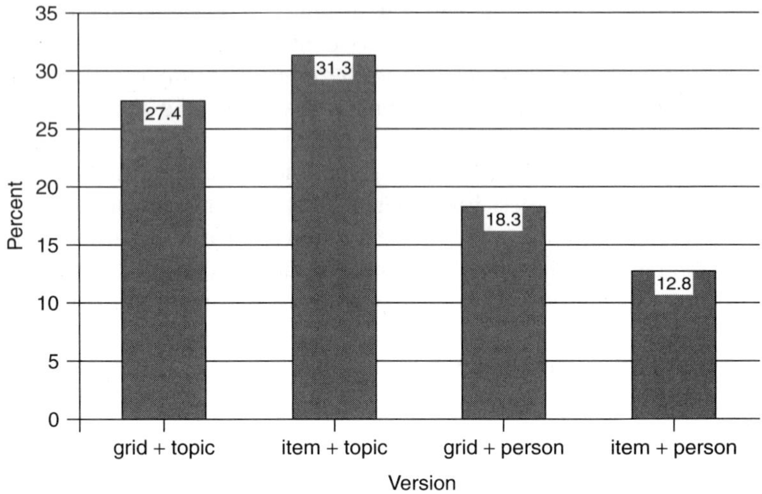

Figure 20.9 Experiment 2; $N = 234$; $p < 0.001$; proportion of items affected by interviewer behavior not following the scripted interview by version.

Table 20.2 Respondent Behavior by Version[a]

	grid + topic	item + topic	grid + person	item + person	Total
Respondent provides information for all persons in the household at once	38.2% (21)	45.3% (24)	29.5% (18)	10.8% (7)	29.8% (70)
No such behavior	61.8% (34)	54.7% (29)	70.5% (43)	89.2% (58)	70.1% (164)
Total	100.0% (55)	100.0% (53)	100.0% (61)	100.0% (65)	100.0% (234)

[a] $p < 0.001$ for the combined effect.

the items administered in a standardized way (6.8 seconds). This is true for not administering a question at all and also for verifications and—to a lesser extent—for other deviations from the interview script.

In addition, an analysis of the respondents' behavior shows that the topic design leads to a higher proportion of cases where the respondent provides at least once in the household roster section the information for all persons or a group of persons at once (e.g., "We all came in the same year;" "We all have the same legal status"). The proportion is even higher for the item-based topic version but does not differ significantly from the grid-based topic version. The results displayed in the table suggest that a topic-based question order increases the proportion of respondents that provide the answers concerning all persons in the household or a group of persons at once, whereas a person-based question order attracts fewer such shortcuts (Table 20.2).

It is surprising that results differ also for the two screen (grid vs. item) designs. The study was administered by telephone, the respondents not being aware of the screen design at all. The only possible explanation is based on the fact that the interviewers modify their behavior in concordance with the screen design, stimulating the respondent differently. So even though respondents cannot see the screen content, they perceive the structure of the instrument from the way the interviewer asks the questions. Accordingly, respondents, as well as interviewers, react to the screen design and the question order under the grid-based topic design in a way that facilitates the interviewer–respondent interaction and thus helps smooth the interview flow.

DISCUSSION

Our results show that CAI screen design potentially influences the participants' behavior. This might have an effect on the duration of the interview as well as on data quality. Based on our findings, we assume that certain aspects of the CAI interface (not related to the question or the response categories themselves) affect the interviewer behavior by providing him or her with ways to avoid segmentation and redundancy. As a reaction to the varying interviewer behaviors, the respondent might behave differently also. The screen designs differ in how they affect the context-based understanding of each question that both interviewers and respondents develop. In addition, some designs are easier to administer in the interview because they are easier to adapt to the flow of the interview.

Whether or not these differences of interviewer and respondent behavior have an effect on data quality is not yet clear. However, preliminary results reported elsewhere (Fuchs, 2000b) do not indicate any significant effect on data quality.[4] While these findings are based on a small-scale field experiment, we need additional data and thorough analysis to decide whether or not the screen design can affect nonresponse or substantive results. By contrast, we were able to demonstrate in this chapter that different screen designs lead to significant differences in the duration of the interviews and thus in survey cost. More important, however, we were able to show that changing the screen design might support a respondent-oriented interviewing style.[5]

To evaluate this approach we propose a theoretical framework that assesses CAI screen design from a new perspectives. A CAI instrument needs to meet the researcher's need for valid and reliable measures. It must also stimulate an appropriate cognitive process in the respondent's mind when she or he answers a question. Therefore, it is necessary to plan and evaluate the question–answer process incorporated in the CAI instrument. In addition, we need to understand the interview as a social situation. As a result, the CAI instrument needs to support the interviewer–respondent interaction in a way that allows smooth ongoing communication. Otherwise, a CAI instrument might influence the interviewer's behavior as well as the respondent's behavior negatively, which could increase survey cost and potentially harm data quality.

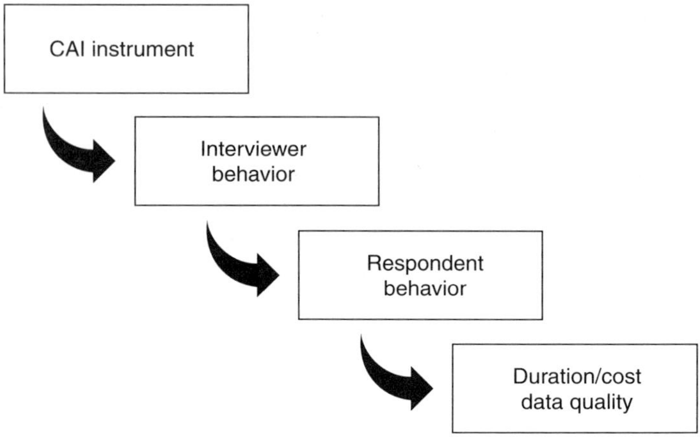

Figure 20.10 Assumed logic of technology effects.

The interviewer is not only supposed to follow standardized interviewer procedures but also expected to maintain a stable and unburdened communicative relationship with the respondent. To support this relationship, he or she might be required to behave in ways that are not considered standard interviewing behaviors. We showed in this chapter that different screen design solutions support or hinder the display of these behaviors by interviewers. So far we do not know which approach allows the best compromise between validity and reliability of the measurement process on the one hand and a smooth, short, and nonembarrassing interview flow on the other hand.[6] To find out how CAI screen design might harm data quality and how it might help save time, money, and interviewer effort, we need to conduct more usability studies. In addition, we need to learn more about the social and communicative requirements of an interview.

To assess the questions mentioned above we need more field experiments. Because we want to analyze the social dimension of the interview and its effects on interviewer behavior as well as on interview duration, laboratory experiments do not meet our needs completely. Of course, laboratory experiments allow a more controlled setting, reveal more detailed information about both participants, and — as a result — need a smaller number of cases. Still, without going into the field, we will never confront our prototypes and design solutions with real pressure to maintain and facilitate the interviewer–respondent interaction and the question–answer process at the same time. Usability testing should therefore be seen as a joint process of laboratory experiments and field tests.

NOTES

1. In addition to the common items, each instrument also contains items not implemented in the other version. For example, the paper instrument has check

boxes and other items for controlling the flow of the interview, whereas the CAPI instrument contains items for determining the selected respondent or for collecting information in greater detail than the paper instrument. For the purpose of the analysis reported in this chapter, we focus on comparable items only. The segments included in our data set thus collect the same amount and type of information in both instruments.

2. This elliptic question form explains the loop effect: Instead of repeating the full question text interviewers behave like ordinary speakers and use shortened elliptic forms that avoid redundancy. It is interesting to recognize that interviewers chose the same verbal expressions on their own that Moore and Moyer (1998a, 1998b) scripted in their experiments on question order.

3. Compared to other studies on interviewer behavior the values are considerable lower. For example, Oksenberg, Cannell, and Blixt (1996) report 41 to 37 percent of such interviewer behaviors when they apply behavior coding to the National Medical Expenditure Survey.

4. We have to recognize, however, that the items used in this experiment are basically sociodemographic questions. The answers to these questions are well known by the respondent, thus leading to a high data quality regardless of the structure of the communication. With attitude items we might observe notable differences.

5. The way in which respondents react to the specific question asked by the interviewers indicates that they appreciate the grid-based topic design. However, we do not have experimental data that supports this view.

6. While we have no evidence supporting the thesis that respondents forget the part of the utterance that is being left out, we assume that they do face difficulty in remembering all response categories.

REFERENCES

Baker, R. P. 1992. "New Technology in Survey Research: Computer-Assisted Personal Interviewing (CAPI)." *Social Science Computer Review* 10: 145–157.

Baker, R. P., N. M. Bradburn, and R. A. Johnson. 1995. "Computer-Assisted Personal Interviewing: An Experimental Evaluation of Data Quality and Costs." *Journal of Official Statistics* 11: 415–431.

Catlin, G., and S. Ingram. 1988. "The Effects of CATI on Costs and Data Quality: A Comparison of CATI and Paper Methods in Centralized Interviewing." Pp. 437–450 in R. M. Groves, P. P. Biemer, L. E. Lyberg, J. T. Massey, W. L. Nicholls II, and J. Waksberg (Eds.), *Telephone Survey Methodology*. New York: Wiley.

Couper, M. P., and G. Burt. 1998. "Interviewer Attitudes towards Computer-Assissted Personal Interviewing (CAPI)." *Social Science Computer Review* 12: 38–54.

Couper, M. P., R. P. Baker, J. Bethlehem, C. Z. F. Clark, J. Martin, W. L. Nicholls, and J. O'Reilly (Eds.). *Computer Assisted Survey Information Collection*. New York: Wiley.

Couper, M. P., M. Fuchs, S. E. Hansen, and P. Sparks. 1997. "CAPI Instrument Design for the Consumer Expenditure (CE) Quarterly Interview Survey." *Final Report*, University of Michigan. Ann Arbor, MI.

Couper M. P., R. Groves, and C. Kosary. 1989. *Methodological Issues in CAPI*. Pp. 349–354 in Proceedings of the American Statistical Association, Section on Survey Research Methods. Alexandria, VA: American Statistical Association.

Fuchs, M. 1994. Umfrageforschung mit Telefon und Computer. Einführung in die computergestützte telefonische Befragung. Weinheim: Psychologie Verlags Union.

Fuchs, M. 1995. "Die computergestützte telefonische Befragung. Einige Antworten auf Probleme der Umfrageforschung." *Zeitschrift für Soziologie* 23: 284–299.

Fuchs, M. 2000. "Interview situation in Computergestützten Befragungen — zur Wirkung von Technologie-Effekten." Pp. 71–92 in *Neue Erhebungsinstrumente und Methodeneffekte. Spektrum Bundesstatistik*, vol. 15. Stuttgart: Statistisches Bundesamt.

Fuchs, M. 2000b. "Screen Design and Question Order in a CAI instrument — Results from a Usability Field Experiment." *Survey Methodology* 26(2): 199–207.

Fuchs, M., M. Couper, and S. E. Hansen. 2000. "Technology Effects: Do CAPI Interviews Take Longer?" *Journal of Official Statistics* 16(3): 273–286.

Grice, H. P. 1975. "Logic and Conversation." Pp. 41–58 in P. Cole and J. L. Morgan (Eds.), *Syntax and Semantics*, Vol. 3: *Speech Acts*. New York: Academic.

Groves, R. M., and N. A. Mathiowetz. 1984."Computer Assisted Telephone Interviewing: Effects on Interviewers and Respondents." *Public Opinion Quarterly* 48: 356–369.

Hansen, S. E., M. P. Couper, and M. Fuchs. 1998. "Usability Evaluation of the NHIS Instrument." Paper presented at the Annual Meetings of the American Association for Public Opinion Research, May 1998, St. Louis, MO.

Hansen, S. E., M. Fuchs, and M. P. Couper. 1997. "CAI Instrument Usability Testing." Paper presented at the Annual Meeting of the American Association for Public Opinion Research, May 1997, Norfolk, VA.

House, C. C. 1985. "Questionnaire Design with Computer Assisted Telephone Interviewing." *Journal of Official Statistics* 1: 209–219.

Jenkins, C. R., and D. A. Dillman. 1997. "Towards a Theory of Self-Administered Questionnaire Design." Pp. 165–196 in L. Lyberg, P. Biemer, M. Collins, E. de Leeuw, C. Dippo, N. Schwarz, and D. Trewin (Eds.), *Survey Measurement and Process Quality*. New York: Wiley-Interscience.

Laurie, H., and H. Moon. 1997. "Converting to CAPI in a Longitudinal Panel Survey." Paper presented at the International Conference on Computer Assisted Survey Information Collection, December 1996, San Antonio, TX.

Lepkowski, J. M., M. P. Couper, S. E. Hansen, W. Landers, K. A. McGonagle, J. Schlegel, T. Wright, and F. Chevarley. 1998. "CAPI Instrument Evaluation: Behavior Coding, Trace Files, and Usability Methods." Paper presented at the Annual Meetings of the American Association for Public Opinion Research, May 1998, St. Louis, MO.

Martin, J., C. O'Muircheartaigh, and J. Curtice. 1993. "The Use of CAPI for Attitude Surveys: An Experimental Comparison with Traditional Methods." *Journal of Official Statistics* 9: 641–662.

Moore, J. C., and H. L. Moyer. 1998a. "ACS/CATI Person-Based/Topic-Based Field Experiment — Final Report." Center for Survey Methods Research, Bureau of the Census.

Moore, J. C., and H. L. Moyer. 1998b. "Questionnaire Design Effects on Interview Outcomes." Paper presented at the Annual Meeting of the American Association for Public Opinion Research, May 1998, St. Louis, MO.

Moyer, L. H. 1996. "Which Is Better: Grid Listing or Grouped Questions Design for Data Collection in Establishment Surveys?" Pp. 986–990 in *Proceedings of the American Statistical Association, Section on Survey Research Methods*. Alexandria, VA.

Müller, S., and P. Kesselmann. 1996. "Akzeptanz von computergestützten Erhebungsverfahren. Ein empirischer Vergleich mit der traditionellene Fragebogentechnik" *Zeitschrift für Forschung und Praxis* 18: 191–202.

Nicholls, W. L., and E. de Leeuw. 1996. "Factors in Acceptance of Computer-Assisted Interviewing Methods: A Conceptual and Historic Review." Paper presented at the Annual Meetings of the Section on Survey Research Methods of the American Statistical Association, May 1996, Norfolk, VA.

Oksenberg, L., T. Beebe, S. Blixt, and C. Cannell. 1992. "Research on the Design and Conduct of the National Medical Expenditure Survey Interviews." *Final Report, Survey Research Center*, Ann Arbor, MI.

Oksenberg, L., C. Cannell, and S. Blixt. 1996. "Analysis of Interviewer and Respondent Behavior in the Household Survey." Agency for Health Care Policy Research, No. 96-N016, U.S. Department of Health and Human Services.

Projektgruppe SOEP. 1998. "Funktion und Design einer Ergänzungsstichprobe für das Sozio-oekonomische Panel." Diskussionspapiere des DIW, 163, Berlin: Deutsches Instituit für Wirtschaftsforschung.

Sanchez, M. E. 1992. "Effects of Questionnaire Design on the Quality of Survey Data." *Public Opinion Quarterly* 56: 206–217.

Schaeffer, N. C. 1991. "Conversation with a Purpose—or Conversation? Interaction in the Standardized Interview." Pp. 367–392 in P. Biemer, R. Groves, L. Lyberg, N. Mathiowetz, and S. Sudman (Eds.), *Measurement Errors in Surveys*. New York, Wiley.

Schneid, M. 1991. Einsatz computergestützter Befragungssysteme in der Bundesrepublik Deutschland. ZUMA-Arbeitsberichte 91/20. Mannheim: Zentrum für Umfragen, Methoden und Analysen.

Sudman, S., N. Bradbrun, and N. Schwarz. 1996. *Thinking about Answers. The Application of Cognitive Processes to Survey Methodology*. San Francisco, CA: Jossey Bass.

Tourangeau, R., and T. W. Smith. 1996. "Asking Sensitive Questions: The Impact of Data Collection Mode, Question Format, and Question Context." *Public Opinion Quarterly* 60: 275–304.

Tourangeau, R., and T. W. Smith. 1998. "Collecting Sensitive Information with Different Modes of Data Collection." Pp. 431–453 in M. P. Couper, R. Baker, J. Bethlehem, C. Z. F. Clark, J. Martin, W. L. Nicholls II, and J. M. O'Reilly (Eds.), *Computer Assisted Survey Information Collection*. New York, Wiley.

Weeks, M. F. 1992. "Computer-Assisted Survey Information Collection: A Review of CASIC Methods and Their Implications for Survey Operations." *Journal of Official Statistics* 8: 445–465.

APPENDIX 1

Transcribing Conventions*

1. Silences

```
A:   I'm not use ta that.
          (1.4)
B:   Yeah me neither.
```

Numbers in parentheses indicate elapsed time in seconds.

2. Overlapping speech

```
A: Oh you do? R[eally  ]
B:             [Um hmmm]
```

Left-hand brackets mark a point of overlap, while right-hand brackets indicate where overlapping talk ends.

3. Latching

```
A:   I am absolutely sure.=
B:   =You are.
A:   This is one thing [that I=
B:                     [Yes?
A:   =really want to do.
```

Equal signs indicate where there is no gap or interval between adjacent utterances. Equal signs also link different parts of a speaker's utterance when that utterance carries over to another transcript line.

* Adapted from the work of Gail Jefferson. For reference see J. Maxwell Atkinson and John Heritage, pp. ix–xvi in *Structures of Social Action: Studies in Conversation Analysis*. Cambridge: Cambridge University Press, 1984.

Standardization and Tacit Knowledge: Interaction and Practice in the Survey Interview
Edited by Douglas W. Maynard, Hanneke Houtkoop-Steenstra, Nora Cate Schaeffer, and Johannes van der Zouwen
ISBN 0-471-35829-0

4. Creative spelling and contractions

```
A: Plea:se ↓don't cuz it'll PA:SS: °·hhhhhh° If not ah'b jis gunnuh
   move down ah'll haftih go get my clothes'n jis k-I'M uss
   gunnuh STAY PUT. (.) En gedda jo:b down here u(p) maybe I'n
   get o:n et (0.3) g-Ba:y. I don'know,
```

Creative spelling and contractions attempt to capture exactly how the words were pronounced. Abbreviations are spelled, for example, "tee vee" instead of "TV."

5. Sound stretching

```
B: I did oka::y.
```

Colon(s) indicate the prior sound is prolonged. More colons, more stretching.

6. Sound cutoff

```
A: This- this is true
```

Hyphens indicate an abrupt cutoff of sound.

7. Speech pacing

```
A: What is it?
B: >I ain't tellin< you
```

Part of an utterance delivered at a pace faster than surrounding talk is enclosed between greater than and less than signs. Talk at a slower pace is enclosed between less than and greater than signs.

8. Emphasis

```
A: I do not want it.
```

Underline indicates increased emphasis. Emphasis is done with some combination of a change in pitch, rise in volume, stretch of a sound, or stress on a vowel.

9. Intonation (or pitch contour)

```
A:  It was unbelievable. I ↑had a three point six? I ↓think.
B:  You did.
```

At the end of a phrase, a period indicates fall in tone, a comma indicates continuing intonation, a question mark indicates increased tone. Up arrows (↑) or down arrows (↓) indicate marked rising and falling shifts in intonation immediately prior to the rise or fall. An exclamation point can also be used for animated tone.

10. Higher volume

```
A: That's where I REALLY want to go.
```

Capital letters indicate increased volume.

11. Lower volume

```
A: °Yes.° That's true.
```

Material between degree signs is spoken more quietly than surrounding talk.

12. Breathing

```
A: You didn't have to worry about having the ·hh hhh curtains
   closed.
```

The *h* indicates audible breathing. The more *h*'s the longer the breath (a single *h* approximately equals 0.1 second). A raised or regular period placed before it indicates inbreath; no period indicates outbreath.

13. Breathiness and laugh tokens

```
A: Ok(h)ay. (0.3) °huh° huh hih
```

The *h* within a word or sound indicates explosive aspirations; for example, breathiness, laughter, and the like. "Huh," "heh," "hih," "hah" indicate laughter particles or tokens.

14. Creaky voice

```
A: ~Oh th~at's w't I'd l~i-:ke ↓t~uh h~ave ~is ~a↓ fresh
   ↓~one.
```

Creak is indicated by a tilda, ~, before the creaky part of the word. Creak sometimes occurs at the end of phrases, often occurs in very low voices, and sounds like the voice of a stereotypical "dying person."

15. Clicks

```
A1:  ·t·hhh We:ll I think it's sad thet they. . .
A2:  ·tch uh::m (0.5) condoning hhhhh ·hhhh. . .
```

Speakers often make audible clicks with the lips and/or tongue and roof of the mouth. Such clicks are often made inadvertently when the speaker opens his or her mouth to take an inbreath. In these a "·t" is used. When the click is longer, louder, and sounds more "intentional," a "·tch" is used.

16. Missing speech

```
A: Are they?
B: Yes because ...
```

Ellipses indicate where part of an utterance is left out of the transcript.

17. Candidate hearing

```
B: (Is that right?) (     )
```

Materials in single parentheses indicate that transcribers were not sure about spoken words. If no words are in parentheses, the talk was indecipherable.

18. Explanatory material

```
A: Well ((cough)) I don't know
```

Materials in double parentheses indicate audible phenomena other than actual verbalization.

19. Audible computer key clicks

```
FI:  uhsupp[ort?
           [#
FI:  . . . l'me go ahead anduh: (0.5) # (1.0)
```

Audible computer key clicks are indicated by a #. Square brackets indicate where a key click overlaps with talk.

APPENDIX 2

Transcript for Interview AW01 (Wisconsin Continuous National Survey)

Brief Version of Transcribing Conventions

`[]`	overlapping talk
`=`	latched utterances; no interval
`(1.0)`	timed silence in seconds
`(.)`	micropause (<0.2 sec)
`wor:d`	stretched sound
`word.`	fall in tone
`word,`	continuing intonation
`word?`	rising or questioning intonation
`word-`	abrupt cutoff
`↑word`	higher pitch
`↓word`	lower pitch
<u>`word`</u>	emphasis
`WORD`	louder talk
`°word°`	quieter talk
`.hhh`	inbreath
`hhh`	outbreath
`(gh)`	gutturalness
`>word<`	faster talk
`word(h)`	breathiness
`#`	interviewer's keystroke
`((note))`	transcriptionist's note
`(word)`	transcriptionist's uncertainty
`{q5d}`	Letters or numbers in squiggly brackets refer to survey instrument questions

Standardization and Tacit Knowledge: Interaction and Practice in the Survey Interview
Edited by Douglas W. Maynard, Hanneke Houtkoop-Steenstra, Nora Cate Schaeffer, and Johannes van der Zouwen
ISBN 0-471-35829-0

```
FR: he↑llo:
FI: .hh uh ↑hi: my name's: (jane smith) an' i'm calling
    from the university of wisconsin? as part of our
    national public opinion study? .hhh we're trying to
    reach people at their home telephone numbers? .hhh is
    this a residential number? {qconf}
    (0.4)
FR: yes
FI: .hh okay: a::nd to be sure i reached the number i
    dialed is this area code three oh three
    (0.5)
FI: (five five five) [seven (0.5) five one seven ] {qcnfr}
FR:                  [what are you calling about?]
FI: i'm calling from the university of wisconsin?
    .h [as
FR:    [fo:r what.=
FI: =part of a national opinion study?
    (1.0)
FR: .h uh concer:ni:ng what.
FI: .hh well it's:- it's a broa:d study it's what we call
    a mo:dular study so they're a bunch of different
    segments of questions each about (0.2) different
    things .hhh >some of the< things: in-volve things like
    politics: economic conditions and expectatio:ns: .hhhh
    jst (.) different things >like that< m- mostly:
    politics and government type [thing]s:
FR:                              [okay ]
FI: .hhh a:nd just to confirm your number it i:s (five five five)
    seven five five seven and area code three
    oh three {qcnfr}
    (0.3)
FR: Uh huh?
    [#
FI: [Oh kay: and because we want our results to represent
    the opinions of people all over the country? .hh we
    need to select one adult to speak with in your
    household [.hh a:nd to make this selection=
              [##
FI: =scientifically we first need to know how many adults
    eighteen or older live there .hhh [(ans)
FR:                                   [two
    (0.8)
FI: two:? (.) adults? [a:nd ] how many women? eighteen or=
FR:                   [mm hm]
FI: =older live there? {q0p}
FR: one
FI: okay one [woman (.) and how many men eighteen or=
             [#
FI: =older. {q0q}
    (0.5)
FR: one
FI: oh kay a:nd at randomly selected the woman of the
    ↑household as the person we would like to interview?
    for our important national study? .hhh a:nd may i
```

speak with the woman of the household? is that you?
{q0t}
FR: mm [hmm
[#
FI: .hh oh kay: we'd like to get your o↑pinions on some
important topics of current interest to the
uni[versity= {qexp2}
[#
FI: =of wisconsin researchers? a:nd before i begin i wanna
assure you that all of the information you give us .hh
is confidential and that none of it will be released
i[n any] way that would permit=
[((telephone tone))]
FI: =identification of you or your family? .hh and of
course your participation is voluntary? {qcnfd} .hhh
and we would also like to tape record this interview:
so that we can see how well the questions in the
interview work. .hhh uh do we have your permission? to
tape record the ↑interview. {qxxxx}
(0.5)
FR: sure
FI: oh kay: (.) a::nd >now we'll get started?< .hhh ↑first
(.) how many persons live in your househo:ld counting
all adults and children and including yourself. {q1}
FR: four
FI: .hhhh okay(gh): a::[::nd? now we have some questions=
[##
FI: =about government agencies. .hhh as you know:? every
ten years there is a census of the population of the
united states. .hhh how confident are you: (.) that
the census bureau protects the privacy of personal
information about individuals and does not share it
with other government agencies. .hhh very confident
(0.4) somewhat ↑confident (0.5) not ↑too confident?
(0.2) or not at all ↑confident. {q5}
(1.0)
FR: share it with what other governments
FI: (tch) .hh well the question doesn't specify: but (0.3)
it just says other government agen[cie]s
FR: [oh]
FR: probably very confident
(0.5)
FI: °oh kay° people have different ideas about what the
census is used for [and now] i'm going to read some=
FR: [uh huh]
FI: =of them? .hhh as i read each one (0.3) please tell me
whether you think the census is used for that purpose:
.hhh to find areas of the country? that need
government help? (0.4) do you think the census is used
for that purpose:? {q5b}
(1.3)
FR: i have no idea.
FI: .hhh okay? a:nd is the census used to decide how many
representatives each state will have in congress?

```
    {q5d}
    (1.8)
FR: no idea
FI: .hhh okay these are?- (0.3) they're just guesses, you
    know, we don't [(       )] expe(h)ct (h)you t(h)o know=
FR:                [uh huh::]
FI: =the: eh- solid hard fast truth it's just (0.3)
    whatever your perception may be: .hhh so: if you just
    wanna take a gue(gh)ss go ahea(gh)d. .hhh is it used
    to see what changes have taken place in the united
    states. {q5f}
FR: yes
    (0.3)  ((typing))
FI: oh kay(gh):. and do the police and f b i use the
    census to keep track of trouble makers? {q5h}
    (2.5)
FR: no idea
FI: .hh [(0.4) a:(gh)::nd how(gh) 'bout- ((cough)) how=
        [##
FI: ='bout to locate illegal aliens: is the census used
    for that. {q5j}
FR: m::::: probably
FI: .h[hh ↑oh kay:: an how much have you heard or=
      [##
FI: =rea(gh)d about the issue: of federal agencies sharing
    information about individuals? .hhh a gr[eat d]eal=
FR:                                         [none ]
FI: =some not very much or nothing at all? {q6}
FR: none
FI: nothing at all?
    (1.0) ((typing?))
FI: would that be nothing at all?
FR: yes
FI: oh(gh) kay(gh) an how concerned would you say you are
    a↑bout this issue? very concerned, somewhat concerned
    not very concerned or not at all concerned {q7}
FR: not at all conc[erned
                   [#
FI: ↑a:n' sharing information between different agencies
    of government when both want the same information
    (0.4) saves time [( )
FR:                  [but see now who's the different
    agencies of government.
FI: .hhhh we:ll (0.4) the- you know the government (0.2)
    is made up of different asiancies: one of which: is
    the census bureau
    (0.9)
FI: so (0.2) they're just talking abou:t
    (0.5)
FI: whatever: (0.4) the other government agencies are:
FR: within the united states
FI: >right<
FR: oh okay=
FI: =so w- y'know
```

FR: now i'm more clear
FI: ok(h)ay(h) huh huh .hhh AH: sharing information
between defferent agencies of government
[when both-]
FR: [i forgot] what was your question?
(0.3)
FI: Oh i'm- this is just the lead in
(0.3)
FR: oh o[kay
FI: [we're just starting. .hhh ah:: sharing
information between different agencies of government
when both want the same information (.) saves time and
money? but it also means some loss in privacy for the
individual. .hhh do you think the benefits? of saving
time and money (0.4) outweigh the costs in privacy,
.hh or: do you think the loss of privacy outweighs the
benefits of saving time and money {q7d}
(1.0)
FR: the first one
FI: okay benefits of saving time and money: outweigh:
(0.5)
FI: costs in privacy.
FR: °mm hmm°
FI: okay .hhh you've given me your general views on
government agencies sharing ↑information with
each other: and now i'd like to ask you some specific
questions about the census? .hhh the census bureau is
required by congress (.) to count the entire u s
↑population (.) once every ten years. .hhh census
tries to get information about each person in the
↑country (.) either by sending them a form [to=
[#
FI: =complete and mail in .hhh ↑or by sending someone to
their home to complete the form. .hhh each decade it's
becoming more difficult and co:stly for the census
bureau to locate a:nd collect information for each and
every person. {q9} .hhh changes to our nation's
↑health ↑care ↑system are now being discussed by the
president and congress .hhh a national health care
system could re↑cord some of the information now
col↑lected by the census, .hhh some people have
suggested (.) that this information should be given to
the census bureau (.) for use in the population census
.hh in order to reduce the cost of the census and make
the census simpler, .hhh how would ↑you feel: about a
national health system providing information about
you to the census bureau for use in the population
census. {q9e}
.h[hh
FR: [good idea
FI: (h)ok(h)ay(h) would you (.) favor it strongly (.)
favor it somewhat (.) oppose [it somewhat] or oppose=
FR: [stro:ngly]
FI: it stro:ngly

```
FR: strong[ly
          [#
FI: °okay.° favor it strongly.
FR: mm hmm
FI: .hhh would you be willing to give your social security
    number to the census bureau so they could match (.)
    thee information in your health care record with thee
    [infor]mation in their own fi(gh):les {q9f}
FR: [yeah ]
FR: yeah
FI: .hhh how willing would you be? to give your social
    secur[ity number? to the ce(h)nsu(h)s bureau so
FR:      [very
FI: that (0.3) if this would increase the accuracy of
    items in your files. .hhh so you'd be very willing,
    >somewhat willing? not very willing?< or not
    will[ing at all? {q9m}
FR:     [very
FI: very will[ing. .hhh okay how willing would you be: to=
             [#
FI: =give your social se↑curity number to a business you
    deal with? such as a de↑partment store: or a credit
    card company? .hhh if having this would increase the
    accuracy of items in your credit report (0.3) would
    you be very willing? [(.) somewha]t willing? not very=
FR:                      [    VERY   ]
FI: =willing or °not willing at a[ll° {q9p}
                                 [#
FR: very
FI: very willing. .hhh ↑next (0.2) do you think the
    present federal income tax system (0.3) is fair? to
    most people (0.2) or unfair to most people. {q28}
    (2.0)
FR: uh:::oh:::: hh
    (1.9)
FR: uh:::m wha(gh):- the first one was very fair?
FI: .hh >oh< just fair or unfair.
    (1.1)
FI: [to   mos]t people
FR: [(mm hmm)]
    (3.1)
FR: nothing in between huh? h[hh huh huh]
FI:                          [NO  MIDDLE] ROAD ON THI(h)S
    ONE
FR: .hhh uh:::m
    (1.2)
FR: unfair
FI: .hh[h okay: and the next ↑few questio:ns concern your=
       [#
FI: =feelings about the taxes you have to pay. .hh[h
FR:                                               [UH
FI: for ea(h)ch ty(h)pe of ta(h)x (.) please tell me
    whether you think the tax is .hhh excessively high:
```

```
    (.) about right (0.2) or low. .hh[h ]
FR:                                  [ok]ay
FI: first (.) federal income taxes?
    do you think federal income taxes are excessively
    high[: (0.5) about right]= {q30}
FR:     [excessively high   ]
FI: =>or low.< (0.4) [okay excessively high.
                     [#
FI: .hhh property or real estate t[axes {q32}
FR:                               [exCESSively h[igh
                                                [#
FI: okay? (.) social security taxes? {q34}
    (0.7)
FR: uh:::m (0.2) excessively high
    (0.3)
FI: a:nd sales taxes: {q36}
    (0.6)
FR: ec- uh:::
    (0.5)
FR: the second one
FI: about righ[t?
              [#
    (0.8) ((# #))
FR: oh no
FI: huh huh huh huh [huh
FR:                 [excessively high
FI: .hhh a::nd how about state income taxe[s. {q38}
FR:                                       [excessive[ly=
                                                    [#
FR: =high
FI: okay our next questions have to do with government?
    .hhh on a scale of one to ten (.) where one means very
    ↑poo:r and ten means excellent °.hhh° >how would you
    rate the< job the president is doing. {q49}
    (1.4)
FR: uh:::m:::
    (0.9)
FR: six
FI: .hhh and on the same one to ten scale? where one means
    very ↑poo:r and ten means excellent °.hh° >how would
    you rate< the job the congress is doing. {q49b}
    (1.0)
FR: uh:::
    (0.5)
FR: four
FI: .hhh on the same one to ten scale where one means very
    ↑poor and ten means excellent (.) >how would you rate
    the job< the supreme court is doing? {q49d}
    (0.8)
FR: uh:m five
FI: .hhh on the same one to ten scale where one means very
    poor and ten means excellent (.) >how would you rate
    the< job the military is doing? {q49f}
```

```
    (0.9)
FR: uh::m::
    (1.3)
FR: eight(h)
FI: eight ↑okay please tell me whether you agree: or
    disagree: with each of the following statements? .hhh
    the first statement is (0.4) people like me don't have
    any say about what the government does. {q50a}
    (1.8)
FI: and it's just agree or a disagree
    (1.2)
FR: uh::m uh-
    (0.5)
FR: agree
FI: .hhh government agencies (.) usually try to do: what
    is best for the ↑people {q50b}
    (1.4)
FR: agree
FI: .hh[h ↑a:nd i don't think public officials ↑care much=
       [##
FI: =about what people like me think. {q50c}
FR: agree
FI: °.hhh° [(0.4)] the government already knows more=
           [#### ]
FI: =about me than it needs to. {q50d}
    (1.7)
FR: well i don't know. (0.4) i don't know what they know
    about me hu::::h [(i   'on't)] really care what they=
FI:                  [okay       ]
FR: =know=
FI: =huh huh huh huh .hhh so (.) would you [(     )]ree?=
FR:                                        [uh::: ]
FI: =°disagree°
    (0.6)
FR: uh gree
    [#
FI: [okay(gh): .hh ↑and most people who go into public
     office (.) want to help others. {q50e}
    (2.0) ((unidentifiable sound))
FR: agree
    (0.8) ((#))
FI: n'kay. generally speak↑ing do you usually think of
    yourself as a re↑publican (0.4) demo[crat
FR:                                     [republican
FI: independent? or something else[:? re[publican?] {q56}
FR:                               [     [republi  ]can
                                  [#    [#
FI: .hhh >would you call< yourself a strong republican or
    a not very strong republican? {q56a}
    (1.3)
FR: strong re[publican
             [#
FI: are you a registered voter? {q58}
FR: yes
```

```
     [#
FI:  [.hhh ↑a:n' we're interested in how people are getting
     along financially these days. .hhh would you say that
     you and your family are better off or worse off
     financially than you were a year ago? or just about
     the same? {q72}
FR:  better
FI:  .hhh a[:nd now look]ing ahead do you think a year=
           [((typing))  ]
FI:  =from now: you will be better off [or wor]se off or=
FR:                                    [better]
FI:  =>just about the same as now< [better off. {q74}
                                   [# #
FI:  .hhh >now turning to business conditions< in the
     country as a whole (0.2) do you think that during the
     next twelve months we'll have ↑good times financially
     (0.3) bad times (0.3) or what {q76}
     (2.2)
FR:  or what?
FI:  mm hmm
     (1.4)
FI:  or what it's (.) whatever you wanna say eh huh huh
     (0.9)
FR:  uh::m
     (2.0)
FR:  not sure
FI:  oh kay .hhh ↑and would you say at the ↑present ↑time
     (.) ↑business conditions are better or worse than they
     were a year ago. {q78}
FR:  [better
FI:  [(or-)
FI:  or just about the same [(.) better .hhh >how about a<=
                            [ #  #
FI:  =year from now do you expect business conditions will
     be better or worse than they are at present or just
     about [the same? {q80}
FR:        [worse
FI:  .hhh ↑a:nd how about people out of work during the
     coming twelve months (.) do you think that there will
     [be mo]re unemploy(h)ment than now (.) [less]= {q82}
FR:  [worse]                                [more]
FI:  =unemployment >than now< or >about the same< [.hh[h
FR:                                               [   [more
                                                  [#  [#
FI:  more unemployment. .hhh during the next twelve months
     do you think that prices in ↑genera:l will go up (0.2)
     [go d]ow:n or stay where they are n[ow?  {q84}
FR:  [up  ]                             [
                                        [#
     (0.3) ((#))
FR:  up
FI:  do you expect that thee overall price increase (.)
     ↑during the next twelve months will be larger or
     ↑smaller than during the past twelve months or about
```

```
    the same? {q86}
FR: 'bout the same
    [##
FI: [.hhh ↑a:nd what about interest rates. do you think
    that during the ne[xt ↑tw]elve months in(h)tere(h)st=
FR:                   [u:p   ]
FI: =ra(h)tes will go up come down or >stay about the
    sa[me as] they are n[ow< (.) {q88} up .hhh there's a=
FR:   [up   ]           [
                        [#
FI: =lot of discussion these days about the disease aids
    (0.2) do you support or op↑po:se the following
    measures to deal with ai:ds .hhhh prohibit students
    with aids virus? from attending public school? {q120}
    (0.4)
FR: absolutely no:t
    (0.3)
FI: [oppose?
    [#
    (0.5)
FR: oppose
FI: .hh develop a government information prog[ram
FR:                                          [oh wait uh-
    (0.3) oppose students from attending?
FI: .hhh well i'll re[peat] the question
FR:                  [no  ]
FR: i absolutely think they should be able to go.
FI: .hhh okay so you would su- (0.3) so you would oppose
    prohibiting students (.) with thee [aids-
FR:                                    [right
FI: okay=
FR: =i would oppose prohibiting
    (0.4)  ((# #))
FI: got it. .hhh ah:: >how about< (.) develop a government
    information ↑program to promote safe sex practices
    such as the use of condoms {q125}
    (0.5)
FR: you're talking to uh health teacher absolutely
FI: uh support?
    (0.7)
FR: °mm hmm°
    [#
FI: [.hh a:nd permit insurance companies to test
    applicants: for the aids for the aids virus:? {q130}
    (2.7)
FR: (h)oh: bo:(h)y hhh you know there's a lot of
    controversy on tha:t
    (1.9)
FR: uh::m hh
    (6.4) ((unidentifiable noises))
FR: uh:::m i th- i think they should be able to yes
FI: so would that be support or oppo[se
FR:                                 [yeah support
```

```
FI: [(sup)]port .hhh have the government (.) pay all the=
    [ ##  ]
FI: =health care costs of aids patients? {q135}
    (1.6)
FR: uh:: hh support
    [#
FI: [.hhh ↑okay do you know anyone ↑personally: either
    living or who has died who came down with the disease
    aids? {q140}
FR: yes
    [#
FI: [.hhh and how many people have you known personally
    who have had aids. {q145}
    (0.6)
FR: that have had it or have died?
FI: .hhh either way(gh) (0.4) anyone livi:ng or who has
    died.
FR: uh::m:
    (4.0)
FR: two
FI: .hhh a:nd now i'm going to read a list of the ways you
    might have felt or behaved during the pa:st ↑week.
    .hhh please tell me >how many< days you've felt this
    way.
    (0.5)
FI: o:n >how many< days ↑during the past week (.) did you
    feel bothered by things that usually don't bother you.
    {q150}
    (1.6)
FR: none
FI: .hhh [on >how many< days during the past week (.) did=
         [#
FI: =you not feel like eating? your appetite was poor?
    {q155}
FR: no:ne
    [#
FI: [°.hhhh° feel that you could not shake off the blues
    even with help from your family or friends? {q160}
FR: no:ne
FI: °.hhh° have trouble keeping your mind on what you were
    doing. {q165}
FR: no:ne
    (0.6)
FI: feel depressed. {q170}
FR: no:ne
FI: feel that everything you did was an effort. {q175}
FR: no:ne
FI: °.hhh° feel fearful. {q180}
    (0.6)
FR: no:ne
FI: °.hhh° on >how many< days during the past week (.) was
    your sleep restless: {q185}
FR: no:ne
```

```
     [#
FI:  [.hhh o:n >how many< days ↑during the past week (.)
     did you talk less than usual {q190}
     (0.6)
FR:  no:ne
FI:  °.hhh° feel lonely {q195}
     (0.4)
FR:  no:[ne
        [#
FI:  feel sa:d= {q200}
FR:  =no[ne
        [#
FI:  feel you could not get going. {q205}
     (0.4)
FR:  no[ne
       [#
FI:  .hhh ↑okay ↑now ↑for ↑a ↑different ↑topic (.) are you
     currently married and living with your husband? {q265}
FR:  yes
FI:  in what year did you and your husband get married? {q267}
FR:  nineteen seventy eight
     [#        [#
FI:  [seventy [eight a:nd in what month (.) did you get
     married {q267a}
     (0.4)
FR:  may
FI:  .hhh [may:: and is this your first marriage or have=
          [#
FI:  =you been married before? {q268}
FR:  first
FI:  .hhh ↑okay what about your husband (.) is this your
     husband's first marriage (.) [or has he been married=
FR:                               [yeah
FI:  =bef[ore {q269}
         [#
     (0.4)
FI:  h[is fir]st?
FR:   [yeah  ]
     [#
FR:  [°mm hmm°
FI:  (.hh)
     (0.6)
FI:  ((cough)) do you have any sons or daughters or
     stepchildren living in your household? {q302}
FR:  two
FI:  .hh[h okay(gh) [:: a::::n:d? what is thee age of your=
        [#          [#
FI:  =younger child living with you {q315}
     (0.5)
FR:  eigh:t(h)
     (0.3)  ((#))
FI:  a:nd [how old is-]
FR:       [oh   no   i]'m sorry she just [turned ni[ne
                                         [#        [##
```

FI: ↑oh [well congratulations to] her .hhh and what is=
        [((typing))   ]
FI: =thee age of your older child {q320}
FR: eleven
FI: eeleven .hhh ↑an' now we'd like to ask about your
    spouse's employment status (.) did your spouse work at
    a paid job last week= {q358}
FR: =yes
FI: .hhh a:nd about how many ↑hours per week does he work
    in an average week? {q358c}
    (0.7)
FR: uh::m
    (0.6)
FR: .hh hhh
    (0.9)
FR: we're both teachers eh: h[hhh so it-] (0.3) it=
                             [okay so   ]
FR: =varies .hhh uh::m (.) uh(gh): normal work week i-
    (1.2)
FR: for teachers is like forty bu:t (0.2) we always end up
    (0.2) probably working °.hhhh° fifty sometimes sixty
    depending on what's going on at school
    (0.4)
FI: okay s:o what would you say [(   )
FR: [well we're out for the
    summer (.) [so ah
FI: [ok(h)a(h)ay .hhh ah:[:
FR: [uh:m probably: fifty
    (0.5)
FI: f:ifty[(gh): okay(gh) and >now we'd would like to ask=
          [#
FI: =about< y↑our ↑employment ↑status .hhh did you do any
    work (.) for pay (.) last week {q359}
    (1.1)
FR: °.hh° well
    (0.6)
FR: i'm still getting paid but school's out (.) so
    (0.7)
FI: okay s:o (.) would you say (0.4) i mean (0.3) [it's-
FR: [well
FI: it's your c[all
FR: [i got paid (.)
FR: [for work (.) but i was]n't at work
FI: [ok(h)ay(h) huh huh huh huh]
FI: .hhh o↑kay (.) a:nd how many hours did you work last
    week {q359a}
    (2.0) ((unidentifiable noise))
FR: we:ll do you want me to pretend i'm still in school?
FI: ↑no (.) jus:t- (0.3) fur-
    (0.5)
FI: i mean j'st (.) the honest truth
    (0.7)
FR: well (.) then (.) for my husband too zero
FI: .hhhh okay well we wer- on your husband's question we

were asking the average week .h[hh] and for this one=
FR: [oh]
FI: =is actually (.) for you for la:st week (0.3) so last
week was zero hours?
(1.0)
FR: zero
FI: okay: i'm just gunna (0.2) put a note in my computer
to explain this:
(2.5) ((typing))
FR: i mean did a lot uh work at ↑home
FI: .hhh [o↑kay
FR: [huh huh
(0.5)
FI: i- we all do °y'know° .hhh okay so you got
pai[:d but (.) di:d (.) n:ot (.) work]
[((typing))]
(0.6) ((typing))
FI: [because o:f summer::] v:a::cation
[((typing))]
(2.5) ((typing))
FI: .hh (0.2) ↑oh kay(gh)::::
(0.8) ((typing))
FI: a(gh)::::nd i:s:: this: thee uh(gh)::: ((3 whistles))
(0.4) uh >is this the number of hours that you usually
wor[k< {q359b} (0.3) no? (0.2) okay so: .hhh how many=
[#
=hours per week do you usually work {q359d}
(.)
FR: fifty
FI: (°fi[fty°) okay .hh and what was your job last week=
[##
FI: =that is what ↑kind of work were you: doing (0.3)
quote doing {q359e}
FR: nothing
[#
FI: [o:kay: and (.) what's your- (0.4) just give me the
name of your job
(0.7)
FR: teacher
FI: [teach↑er:: at what grade level?]
[((typing))]
(0.5) ((# #))
FR: high school
FI: [.hhh high s::: ↑choo:::l .hhh] any: area of=
[((typing))]
FI: =specialty:?
(0.7)
FR: physical education [and health]
[((typing))]
(0.4)
FI: [h y:: s °i c a ↑l° edu ca tion: .hhh a::nd health]
[((typing))]
(2.0) ((typing))
FI: .hh a:nd what were your most important activities or

duties on this jo(gh)::b hh {q359f}
(2.8)
FR: well you mean of teaching?
FI: .hhh ye- well i mean i know i just need you to tell me
(.) so i can put it in the computer
(0.5)
FI: what do you do
(0.7)
FI: in your jo(gh):b what do you do teaching
(1.1)
FR: (h)oh:: [well] tha- that could take a long time
FI: [hhh]
FI: okay [so-
FR: [you mean you wanna know all the things that i
teach?
FI: no: let's say top three
(1.0)
FR: [uh:::m:
FI: [jst-
(0.6)
FI: what are your: activitie[s
FR: [in health? i- eh:: my top
[three things i probably teach are: .hhh drugs(h)]
[((typing))]
(1.2) ((typing))
FR: [sex education]
[((typing))]
(1.8) ((typing))
FR: [uh::::m]
[((typing))]
(0.5) ((typing))
FR: c p r?
(1.0) ((typing))
FI: okay?
(1.4) ((typing))
FR: [a::nd in physical education:]
[((typing))]
(0.6) ((typing))
FR: [.hhh oh gosh i teach every sport]
[((typing))]
(0.5) ((typing)
FI: okay? [so teaching sports that's: good] enough
[((typing))]
FR: [okay
[# #
(4.0) ((typing)
FI: .hhh a::nd do you work for yourself in a family
business or for someone else:? {q359h}
FR: for [someone [else
[# [
FI: [.hhh have there been ↑any times during
the past twelve months when you did not have a job and
you were looking for work {q359q}
FR: no

```
FI: .hhh a:nd now we'd like to ask you a ↑couple
    ↑questions about exper↑iences you may have had as a
    victim of crime .hhh during the past twelve months (.)
    did anyone break in↑to or [some]how illegally (.) get=
FR:                           [yes ]
FI: =into your home and steal something {q400}
    (0.8) ((#?))
FR: uh:m not my [home (0.3) but my c[ar
                [#                  [#
FI: okay so it was no:t your ho:me .hhh a:nd during the
    past twelve months did anyone ↑take something directly
    from you by using force such as a stick up mugging or
    threat {q401}
FR: no
FI: .hhh a:nd now we'd like to ask you about health
    insurance .hhh do you have any health insurance
    coverage {q410}
FR: yes
FI: .hhh [and now i'll ask about your insurance coverage=
         [###
FI: =for various ↑categories of health care services? .hhh
    do you have insurance coverage for hospitalization?
    {q411}
FR: yes=
    =#
FI: prescription drugs {q412}
    (0.7)
FR: ye[s
      [#
FI: physician's visits {q413}
FR: yes=
    =#
FI: .hh long term care {q414}
    (0.8)
FR: yes
    [#
FI: [.hhh a:nd now i'll ask some questions about ↑future
    uncertain out comes .hhh in ↑each case: try to think
    about the whole range of possible outcomes and think
    about how likely they are to occur during the next
    twelve months:? .hhh in some of the questions i'll ask
    about the percent chance of something happening and
    the percent chance must be a ↑number from zero to one
    hundred .hhh numbers like two or five percent may be
    almost no ↑chance °.hhh° twenty percent or so may mean
    (0.2) not ↑much chance .hhh a forty five or ↑fifty
    five percent chance may be: a pretty even chance .hhh
    eighty percent or so: (0.2) may mean a very good
    chance and a ninety five or ninety eight percent
    chance may be almost certain .hhh the percent chance
    can ↑also be thought of as the number of chances out
    of one ↑hundred .hhh now i'll begin asking the
    ↑questions let's start with the weather where you live
    .hhh what do you think is the per↑cent chance that it
```

```
     will rain tomorrow {q420c}
     (1.2)
FR:  uh:::m:
     (1.8)
FR:  ten percent
FI:  .hhh and what do you think is the percent chance that
     someone will break into your home? and steal
     something? during the next twelve months {q430}
     (0.6)
FR:  (h)eh uh uh(gh):: in my home?
FI:  yes
FR:  uh:::
     (0.7)
FR:  ten percent
FI:  .hhh what do you think is the percent chance that
     someone will take something directly from you by using
     force such as a stick up mugging or threat? (.) during
     the next twelve months? {q431}
     (0.5)
FR:  zero percent
FI:  .hhh a:nd now think about your health insurance
     coverage twelve months from now .hhh what do you think
     is the percent chance that you will have health
     insurance coverage twelve months from now {q440}
     (0.4)
FR:  hunderd per[cent
                [###
FI:  if you ↑were to have health insurance coverage twelve
     months from now (.) what do you think is the per↑cent
     chance that the ↑coverage would be at least as
     com↑ple:te as your current health insurance coverage
     {q442}
FR:  hundred
     [###
FI:  [.hhh ↑a:n' now we'd like you to think about your
     employment prospects over the next twelve months .hhh
     what do think is the percent chance that you will lose
     your job (0.2) during the next twelve months {q451}
FR:  zero
FI:  .hhh a:nd if you were to lose your job during the next
     twelve months what is the percent chance that (.) the
     job you eventually find and accept would be at least
     as good as your current ↑jo:b (.) in terms of wages and
     benefits: {q452}
FR:  hundred percent
     [###
FI:  [.hhh hundred percent >and what do you think is the<
     percent chance that you will leave your job
     voluntarily during the next twelve months {q453}
FR:  zero
FI:  .hhh ↑okay now i'd like to ask some questions about
     your own personal ↑income prospects in the next twelve
     months .hhh what do you think is the lowest amount
     that your own total income from all sources could
```

possibly ↑be: in the next twelve months (.) before
taxes {q460c}
(2.6)
FR: just mine?
FI: yes (0.4) separate [from your husband's=
FR: [uh::
FR: =thirty eight thou[sand
[# #
FI: .hh
(0.9) ((typing)
FI: okay: a::[nd what do you think is the highest amount=
[##
FI: =that your own total ↑income from all sources could
possibly be [in] the next twelve months [bef]ore taxes=
FR: [oh] [oh]
FR: =thirty eight thousand {q460e}
(0.4)
FI: would be the highest?
(1.0) ((typing)
FR: mm ↑hmm
FI: and the lowest
FR: mm hmm.
FI: okay? so(h) th(h)at's (h)what it's gonna ↑be i guess
.hh[hh a::[:n'
[## [
FR: [well-
FR: it's gonna be the sa::me so
FI: okay got it .hhh still thinking about your own total
income before taxes in the next twelve months .hh what
do you think is the percent chance that your own total
income before taxes will be under thirty thousand
dollars {q461k}
(0.6)
FR: zero per[cent
[#
FI: .hhh what do you think is the percent chance that your
own total income before taxes will be under thirty
five thousand dollars {q461l}
(0.7)
FR: say that one again please?
FI: what do you think is the percent chance that your own
total income before taxes will be under thirty five
thousand dollars? {q461l}
(1.8)
FR: zero per[cent
[##
FI: .hh and what is the percent chance that your own total
income before taxes will be under forty thousand
dollars in the next twelve months {q461n}
(2.6)
FR: ↑say that one again?
FI: .hhh what [is the percent chance that your own total=
FR: [huh
FI: =income before taxes will be under forty thousand

(0.7)
FR: before taxes? {q461n}
FI: mm hmm
(2.0)
FR: uh::
(1.3)
FR: so what you're asking is forty- (0.4) a:m i gonna make
forty thousand dollar
FI: .hhh well they're asking- (0.6) i mean just what the
question says what is the percent chance that your
income before taxes is gunna be under forty thousand
(0.6)
FI: in the upcoming twelve months {q461n}
FR: zero percent
(0.6)
FI: eh- [(0.4) okay (.) s:o
[#
(0.6)
FI: there's- (0.4) do you think it will be under forty
thousand
(0.6)
FR: yes
FI: then you would want a high percent ['cause there's=
FR: [oh
FI: =a good chance [(it'll be-)
FR: [hundred percent
FI: oh(gh) kay(gh)[.hhh it's a confusing section i know=
[## #
FI: =a lot of people get real messed up (h)on th(h)at .hhh
okay what was your age at your last birth day? {q800}
(0.8)
FR: thirty nine
FI: thir:ty: ni:ne and the next questions are about your
edu↑cation .hhh do you have a high school diploma
{q804}
FR: mm hmm
FI: .hhh did you ever attend a college university
vocational or a ↑technical school(gh) {q806}
(1.1)
FR: no
FI: a:(gh):nd is your race white black asian a[merican=
FR: [whi:te
FI: =indian or something else? {q848}
FR: white
FI: are you of hispanic origin such as mexican >american
[latin american puerto rican or< cuban? {q850}
FR: [no
FI: and what is the zipcode at this residence? {q852}
FR: (xxxxx)
FI: [(x) [(x) [(x) [(x) [(x) [.hhh a::(gh):nd=
[# [# [# [# [# [#
FI: =°i guess that's correct° a::nd what is your religious
preference? if any? {q858}
FR: christian

FI: °.hhh° any denomination {q858a}
(0.9)
FR: ah:: lutheran
FI: (°.hhh°) a:nd how often do you tend religious
servic:es:? {q860}
FR: once a week
FI: once per week .hhh a(gh):::nd >what about< listening
or watching radio or t v religious programs .hhh about
how many times did you do this during the last month
{q862}
(0.6)
FR: zero
[#
FI: [.hhh a:nd are you a citizen of the united states
{q864}
FR: yes
[##
FI: [.hhh and now we need to ask some questions to be sure
we do our probability sampling properly .hhh are there
any other telephone numbers? which can be used to
reach this household? {q866}
(2.0)
FR: uh: in the summer no
FI: .hhh a::nd do you sometimes use call forwarding to
have calls made to other numbers come to this phone
{q870}
FR: no
FI: .hhh a:nd just roughly what was your spouse's total
income from all sources in the past twelve months?
before taxes .hhh be sure to include income from work
government benefi[ts
FR: [forty thou[sand
FI: [pensions and all other
sources {q892}
FR: [forty thousand]
[((typing))]
(0.5) ((typing)
FI: .hh a:nd just roughly what was your own total income?
from all sources in the past twelve months before
taxes be sure to include income from >work government
benefits [pensions and all other sources<] {q894}
FR: [around thirty eight i think]
FI: okay .hhh
(0.6) ((typing))
FI: ↑a::(gh):nd just let me check to make sure: i haven't
missed anything? {qchck} .hhh o↑KAY this concludes the
interview: again let me assure you that all of the
information will be kept confidential and will be used
only for routine statistical research purposes?
{qthnk}.hhh and now that you know: that the
interview:- what the interview was about we would like
per↑mission to use the tape recording for research
we're doing about communication and for educational
purposes? .hhh may we use the taped interview for our

```
     research? {qxxxx}
     (0.9)
FR:  ah: where a:t
FI:  .hhh at- at the university of wisconsin it won't leave
     a:nd (0.2) just s:o- (0.4) just so you know after
     (0.3) i finish the interview there will be: (.) no
     identifying information about you (.) no phone number:
     no name (.) no nothing
FR:  that's fine
FI:  .hhh o↑kay well thank you very much for your
     participation and [your ↑time
FR:                    [this-
FR:  now this isn't gunna go into a thing where .hhh you
```

Author Index

Subject Index

WILEY SERIES IN SURVEY METHODOLOGY

Established in Part by WALTER A. SHEWHART AND SAMUEL S. WILKS

Wiley Series in Survey Methodology covers topics of current research and practical interests in survey methodology and sampling. While the emphasis is on application, theoretical discussion is encouraged when it supports a broader understanding of the subject matter.

The authors are leading academics and researchers in survey methodology and sampling. The readership includes professionals in, and students of, the fields of applied statistics, biostatistics, public policy, and government and corporate enterprises.

BIEMER, GROVES, LYBERG, MATHIOWETZ, and SUDMAN · Measurement Errors in Surveys
COCHRAN · Sampling Techniques, *Third Edition*
COUPER, BAKER, BETHLEHEM, CLARK, MARTIN, NICHOLLS, and O'REILLY (editors) · Computer Assisted Survey Information Collection
COX, BINDER, CHINNAPPA, CHRISTIANSON, COLLEDGE, and KOTT (editors) · Business Survey Methods
*DEMING · Sample Design in Business Research
DILLMAN · Mail and Telephone Surveys: The Total Design Method, *Second Edition*
DILLMAN · Mail and Internet Surveys: The Tailored Design Method
GROVES and COUPER · Nonresponse in Household Interview Surveys
GROVES · Survey Errors and Survey Costs
GROVES, DILLMAN, ELTINGE, and LITTLE · Survey Nonresponse
GROVES, BIEMER, LYBERG, MASSEY, NICHOLLS, and WAKSBERG · Telephone Survey Methodology
*HANSEN, HURWITZ, and MADOW · Sample Survey Methods and Theory, Volume 1: Methods and Applications
*HANSEN, HURWITZ, and MADOW · Sample Survey Methods and Theory, Volume II: Theory
KISH · Statistical Design for Research
*KISH · Survey Sampling
KORN and GRAUBARD · Analysis of Health Surveys
LESSLER and KALSBEEK · Nonsampling Error in Surveys
LEVY and LEMESHOW · Sampling of Populations: Methods and Applications, *Third Edition*
LYBERG, BIEMER, COLLINS, de LEEUW, DIPPO, SCHWARZ, TREWIN (editors) · Survey Measurement and Process Quality
MAYNARD, HOUTKOOP-STEENSTRA, SCHAEFFER, VAN DER ZOUWEN · Standardization and Tacit Knowledge: Interaction and Practice in the Survey Interview
SIRKEN, HERRMANN, SCHECHTER, SCHWARZ, TANUR, and TOURANGEAU (editors) · Cognition and Survey Research
VALLIANT, DORFMAN, and ROYALL · Finite Population Sampling and Inference: A Prediction Approach

*Now available in a lower priced paperback edition in the Wiley Classics Library.